D1566909

INVENTIONS

Books by R. Buckminster Fuller
available from St. Martin's Press

Critical Path
Tetrascroll: A Cosmic Fairy Tale
Grunch of Giants

INVENTIONS

THE PATENTED WORKS OF

R. BUCKMINSTER FULLER

ST. MARTIN'S PRESS • NEW YORK

NOTE

In some cases the text of the original patents as printed by the U.S. Patent Office has been changed to correct typographical errors. In addition, the "claims" section appended to each patent has been dropped to avoid excessive repetition.

Library of Congress Cataloging in Publication Data

Fuller, R. Buckminster (Richard Buckminster), 1895–1983
Inventions: the patented works of R. Buckminster Fuller.

Includes index.
1. Patents—Catalogs. 2. Fuller, R. Buckminster (Richard Buckminster), 1895–1983 I. Title.
T223.C2F84 1983 620'.00272 83-9797
ISBN 0-312-43477-4
ISBN 0-312-43478-2 (lim. ed.)

First Edition

10 9 8 7 6 5 4 3 2 1

Design by Dennis Grastorf

CONTENTS

INTRODUCTION

GUINEA PIG B*

I AM NOW CLOSE TO 88 and I am confident that the only thing important about me is that I am an average healthy human. I am also a living case history of a thoroughly documented, half-century, search-and-research project designed to discover what, if anything, an unknown, moneyless individual, with a dependent wife and newborn child, might be able to do effectively on behalf of all humanity that could not be accomplished by great nations, great religious or private enterprise, no matter how rich or powerfully armed.

I started out fifty-six years ago, at the age of 32, to make that experiment. By good fortune I had acquired a comprehensive experience in commanding and handling ships, first as a sailor in Penobscot Bay, Maine, and later as a regular U.S. naval officer. The navy is inherently concerned with not only all the world's oceans, but also the world's dry land emanating exportable resources and import necessities and the resulting high seas commerce. The navy is concerned with all vital statistics. I saw that there was nothing to stop me from thinking about our total planet Earth and thinking realistically about how to operate it on an enduringly sustainable basis as the magnificent human-passengered spaceship that it is.

Planet Earth is a superbly conceived and realized 6,586,242,500,000,000,000,000-ton (over 6.5 sextillion tons) spaceship, cruise-speeding frictionlessly and soundlessly on an incredibly accurate celestial course. Spaceship Earth's spherical passenger deck is largely occupied by a 140-million-square-mile "swimming pool," whose three principal widenings are called oceans.

Upon the surface of the "swimming pool," humanity is playing high-profit gambling games with oil-loaded ships. The largest of all such ships in all history is a quarter-mile-long tanker of 580,000 tons. At top speed it can cross the 3,000-mile-wide Atlantic Ocean in six days. That 3,000-mile, six-day tanker distance is trav-

*"B" stands for Bucky: "Guinea Pig B" was the last manuscript Fuller finished before his death and it constituted Bucky's final rendering of accounts with God and with his fellow man for the course of of his life. He intended it to be read with his collected patents, his honorary doctorates, and the *Basic Biography* prepared by his office, which lists all his primary and secondary social functions, academic appointments, awards, publications, exhibitions, museum shows, major domes, and keynote addresses—i.e., as much of his life as could be objectively listed, quantified, and tallied. (The *Basic Biography* is available from the R. Buckminster Fuller Institute, 3501 Market Street, Philadelphia, PA 19104, for five dollars.) Although he died a few months before publication of this book, Fuller oversaw all stages of the work except the final reading of the galleys. —Ed.

eled every two and one-half minutes by the eleven-quadrillion-times-heavier Spaceship Earth, which has been moving at this fast rate for at least seven billion years with no signs of slowing or "running out of gas." As it travels around the Sun at 66,000 m.p.h., it also rotates at an equatorial velocity of 1,000 m.p.h.

The units of time and energy expenditures as "matter" or "work" necessary to structure, equip, and operate all the transcendent-to-human contrivings, biological and chemical organisms and equipment, and their "natural" operational events, including the time-energy units invested in creating and operating volcanoes, earthquakes, seaquakes, and tornadoes, as well as to accomplish this fully equipped and complexedly passengered planet Earth's 66,000-m.p.h. cosmic-highway-traveling speed, stated in the terms of time and energy expended per each ton-mile accomplished at that speed, produces a numerical figure of a staggering magnitude of energy expending.

This staggering energy-expenditure figure for operating planet Earth is in turn utterly belittled when compared to the sum of the same units of time-energy expenditures for structuring, equipping, integrally operating, and moving all of the asteroids, moons, and planets as well as the stars themselves of each of all the known approximately 100 billion other star systems of our Milky Way galaxy, as well as of all the asteroids, moon, and planets, and stars of all the approximately 100 billion star systems of each of all the other two billion galaxies thus far discovered by Earthians to be present and complexedly interacting and co-intershunting with Spaceship Earth in our astro-episode neighborhood of eternally regenerative scenario Universe.

All of these "really real" cosmic energy expenditures may be dramatically compared with, and their significance considered in respect to the fact that, the total of all energy used daily—95 percent wastefully—by all humans for all purposes aboard Spaceship Earth amounts to less than one-millionth of 1 percent of Spaceship Earth's daily income of expendable energy imported from the Universe around and within us.

A vast overabundance of this Earthian cosmic energy income is now technically impoundable and distributable to humanity by presently proven technology. We are not allowed to enjoy this primarily because tax-hungry government bureaucracies and money-drunk big business can't figure a way of putting meters between these cosmic energy sources and the Earthian passengers, so nothing is done about it.

The technical equipment—steel plows, shovels, wheelbarrows, boilers, copper tubing, etc.-—essential to individuals' successful harvesting of their own cosmic energy income cannot be economically produced in the backyard, kitchen, garage, or studio without the large-scale industrial tools' production elsewhere of industrial materials and tools-that-make-tools, involving vast initial capital investments. If big business and big government don't want to amass and make available adequate capital for up-to-date technological tooling, people will rarely be able to tap the cosmic energy income, except by berry-, nut-, mushroom-, or apple-picking and by fishing.

Volumetrically, S.S. Earth is a 256-billion-cubic-mile spherical vessel of 8,000-miles beam (diameter), having at all times 100 million of its approximately 200 million square miles of spherical surface always exposed to the Sun—with the other hemisphere always in nighttime shadow—bringing about enormous atmospheric, temperature, and pressure differentials and all their resultant high-low weather-produced winds that create all the waves thunderously crashing on all the trillions of miles of our around-the-world shorelines. The atmospheric temperature differentials in turn induce the electromagnetic potential differentials that transform the atmosphere into rain- and lightning-charged clouds from Sun-evaporated waters of the three-fourths-ocean-and-sea-covered Earth.

Gravity's ability to hold together the planet itself, its waters, its biosphere, and other protective mantles, as well as to pull the rain to Earth, thus combines with the photosynthetic capability of the Earth's vast vegetation to harvest solar radiation and store its energies in a manner readily and efficiently convertible into alcohols. The alcohols (four types) constitute the "Grand Central Reservoir" of cosmic-radiation- and gravity-generated energy in its most immediately-convertible-into-human-use state—for instance, as high-octane motor fuels, synthetic rubber, and all other products misnamed "petro-chemical products" by their exploiters, whose petroleum is in reality a time-, pressure-, and heat-produced by-product of alcohol.

It was reliably reported in February 1981 that many thousands of individuals in the U.S. have developed their private distilling equipment and are producing their own alcohol directly from the vegetation's photosynthesizing of the Sun's radiation into hydrocarbon molecules, which may be converted into the alcohol with which these moonshiners are now successfully operating their automobiles.

For exploratory purposes we will now tentatively adopt my "working assumption," later described in detail, that humans are present on Spaceship Earth only because they have an ultimately-to-become-operative, critical function to perform in Universe—a function of which humanity, in general, is as yet almost totally unaware. Assuming for exploratory purposes that this cosmic function of humanity is in due course demonstrably proved to be valid leads to the discovery that this ultimate functioning has involved the investment of important magnitudes of cosmic energy over a period of billions of years, to develop a celestial "incubator planet" having a regeneratively sustained, exactly controlled environment, successfully accommodating the complex development and maintenance of the hydraulically structured (60 percent water) and 98.6-degrees-Fahrenheit-tuned human organisms. We will now also assume that this biosphere's omni-controlled environment can successfully nurture, grow,

Radiation is disintegrating because it is held together only at this end of each of its energy-manifesting vectors. Disintegrating, the vectors can be angularly aimed, ergo focused.

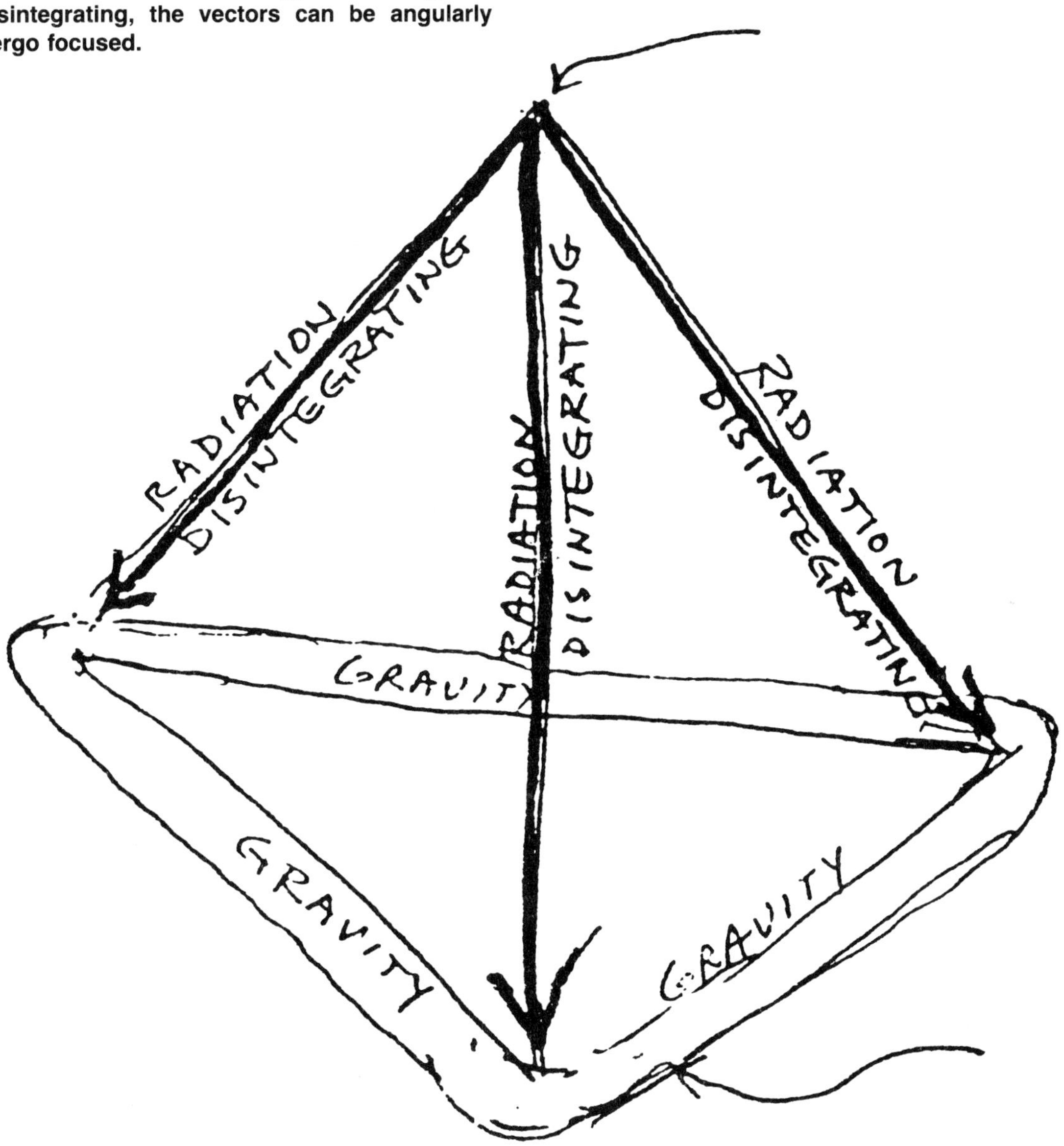

Gravity is inherently integrated as a closed system with no ends, ergo is an inherently closed system having twice the coherence integrity of equally energy-vectored radiation.

and develop such an abundance of ecological life support as to be able to accommodate the only-by-shockingly-wasteful-trial-and-error-education of humans in preparation for their ultimately-to-be-employed-and-maintained semi-divine functioning in Universe, as local Universe information harvesters and local Universe problem solvers in support of the integrity of eternally regenerative scenario Universe, provided, however, that, if after the billions of years of their development, those humans living right now can and do pass their final exam for graduation into this semi-divine functioning.

We have next to assume that this crucial test may well lead to humans erroneously pressing the buttons that can now release so much destruction as—within only a few days after tomorrow—to terminate further human life aboard this planet.

Since destruction of humans in Universe would by our working assumption seem to be cosmically undesirable and to be accomplishable only by the anti-intelligent use of the mega-mega-concentrates of energy that humans have learned only recently how to produce and explosively detonate, it is vitally worth our while to stretch our conceptual faculties to understand the physical potentials and possible mystical significance of our most comprehensive inventory of cosmological and cosmogonical information.

The omni-conserved, nonsimultaneously intertransforming energies of eternally regenerative Universe consist most simply of a plurality of omnimagnitude syntropic convergences here and entropic divergences there. The syntropic convergences integrate as matter. The entropic divergences disintegrate as radiation. The convergently associative function is gravitationally integrative and inherently nondivisible and nondifferentiable until convergently realized as matter.

The radiated disintegration of matter, on the other hand, is inherently differentiable and subdivisible, ergo assignable to a plurality of distinctly separated, vastly remote interminglings with illions of other systems' entropically separated-out atomic constituents, thereafter individually to intermingle tentatively and in progressive syntropy as one of myriads of entirely new star systems.

Radiation is inherently focusable and shadowable; gravity is inherently nonfocusable and shadowless.

All the stars are atomic-energy-generating "plants." The star Sun is a hydrogen-into-helium intertransformative regenerator of radiation. The sun operates internally at a heat of 26,000,000° Centigrade—a scale on which 0° represents the freezing point and 100° the boiling point of water.

The cosmic design problem—with the solution of which we are now concerned—was that of employing a safe way of providing the right amount of exclusively star-emanating energy for developing and maintaining the critical-function-in-Universe-serving humans on the only planet having the right environment for both protecting and ecologically supporting those delicately, intricately designed humans, together with their physically and metaphysically critical functionings.

The next star nearest to planet Earth—beyond the Sun—is 300,000 times farther away from the Earth than is the Sun. In solving the "humans on Earth" problem, it was therefore necessary to employ the cosmic facilities in such a manner as to transmit the appropriate amount of the specific radiation constituents in nonlethal concentration from the Sun, the star nearest to the human-incubating planet Earth—the planet having not only the most propitious environmental condition for humanity incubating, but also maintaining the *exact* vast complex of close-tolerance-of-physical-error limits within which humans could survive.

Humans cannot impound enough energy by sunbathing to keep them alive and operative. Planet Earth's safe importing of Sun radiation must first be impoundingly accomplished by vegetation in energy quantities adequate to supporting not only the vegetation and the humans but all the other myriads of species of life altogether constituting the regenerative ecologic system. To do this, the angular fan-out concentration of energy intensity must be accomplished by attaining adequate distance from the Sun, and thereby to arrive at protoplasmically tolerable increments of energy exactly sustained for conversion by botanical photosynthesis into hydrocarbon molecules, which thereafter can be assimilated metabolically by all other living biological organisms. This means that the Sun's surface energy radiation must be transformingly programmed to angularly deconcentrate during its eight-minute, 92-million-mile passage from Sun to Earth to arrive in nonlethal increments and at nonlethal temperatures.

The farther away a cosmic radiation source is, the less concentrated the radiation. Only about one-billionth of the radiation given off omnidirectionally by the Sun is so aimed as to impinge directly upon the surface of planet Earth.

With all the space of Universe to work with, the great designing wisdom of Universe seems to

have found it to be essential that the energy organized as the predominantly water-structured substance that is all biological life be maintained at a distance apparently never greater than 95 million miles and never less than 91 million miles away from the atomic-energy-generating plant of the Sun's initial magnitude of radiation concentration.

Sad to say, those of the present Earthian power-structure's scientists who assert that they can safely bury atomic radiation wastes within the ever-transforming structure of planet Earth are rationalizing information critical to human continuance aboard planet Earth. They and their only-selfishly-motivated masters are gambling the future of all humanity to win only the continuing increase of their personal economic power control for the few remaining years of only their own lives.

While the totality of atomic wastes now Earth-buried or ocean-sunk is as yet of a feasibly rocketable total bulk and weight for a plurality of blastings-off, we may still send such atomic wastes back into the Sun where they can once more become safely exportable.

Humans who think they are better designers than God are the most to be pitied of liars—the liars who are believingly convinced by their own only-self-conveniencing fabrications.

On rereading what I have just said about humans falsifying critical information, I am retrospectively shocked at my making those two negative citations.

Firstly, my positive information, which if comprehended can help toward realization of physical success for humanity, when followed by my negativism, reduces the credibility of my wisdom and thereby reduces the value of my positive information.

Secondly, we observe from experience that nature has its own checks and balances, accelerators and brakes, temporary side-tracking, overload circuit breakers, self-starters, transformers, birth and death rates, and complex overall evolutionary gestation rates.

Nature's biggest, most important problems take the longest to solve satisfactorily because they can be solved only by lessons humanly learned through trial-and-error mistake-making; and, most important, by those who make the mistakes and their self-recognition and public acknowledgment of their errors and their only-thereby-learned-from positive clues to effective solutions of evolutionary problems—in the present instance, the problem of how best to abruptly terminate further atomic energy development for human use as fissionally or fusionally generated aboard planet Earth. The problem is one of immediate and direct concern to each individual of our four billion humans, as well as to all the potential many yet to be born and to God. Its satisfactory solution can be arrived at only through major design-science initiative-takings that produce far superior technologies to render spontaneously obsolete the previous undesirable technology.

Vividly, I recall an occasion when I was about ten and my father heard me call my own brother a fool. My father said, "Bucky, I do not take everything I read as being reliable, not even when I read it in the Bible. But I find by experience that statements in the Bible are far more often reliable than are declarations printed elsewhere. There is a statement in the Bible that you should remember: he who calls his brother a fool shall be in danger of hell's fire."

In these critical times let us no longer make the mistake of identifying as fools those with whom we disagree.

The late twentieth century's confused, fearful human chitchat about an Earthian energy crisis discloses the abysmal state of ignorance within which we Earthians now struggle. All this results in realization of the almost absolute futility of the disintegrated ways in which humanity's present leaders cope with this one-and-only-available Spaceship Earth and its one-and-only-available Universe and with the problems of human survival, let alone attainment of physical success for all humanity within the critical time limit.

For all those who wished to observe them—I being one such person—all the foregoing concepts were already apparent in 1927, though in far less obvious degree. The sum of all these facts made it clear to me in 1927 that no matter how much I could, did, or might as yet accomplish, as a human problem-solver employing only an artifact-designed industrial-production revolution on Spaceship Earth—seeking by techno-economic obsolescence, rather than by political reform, to make physically obsolete all ignorantly incapacitating reflexing of humanity—I could not possibly make more of a mess-of-it-all than that being made by the behind-the-scenes absolutely selfish world power-structures' puppeting of the 150 "sovereign" prime ministers, their national legislatures, and their in-turn-puppeted generals and admirals, and the latter's omni-intercompetitive commanding of our one-and-only spaceship, with each five-star "admiral general" looking out only for his own sovereignly-escutcheoned stateroom and all the starboard-side admirals trying to find a way of sinking all the port-side admirals without the winning admirals getting their own feet wet, let alone being drowned.

I saw that there was nothing to stop me from studying—hopefully to discover, comprehend, and eventually employ design-wise—the integrated total family of generalized principles by which nature operates this magnificent, human-passengered, spherical spaceship as entirely enclosed within an external set of physically unique, spherically concentric environmental zones altogether producing the critically complex balance of intertransformative energy conditions essential to maintaining an omniregenerative planetary ecology—all accomplished in local Universe support of eternally omni-interregenerative Universe itself—by means of planet Earth's syntropic, biochemical capability to photosynthetically convert stellar radiation (primarily that of the Sun) into hydrocarbon-structured vegetation that in turn is converted as "food" into all manner of biological proliferatings, ultimately—after aeons of enormous heat and pressure treatment produced by deep-Earth burial—to be converted into fossil fuels.

This Earthian energy impounding and conserving altogether constitutes a cosmic accumulation of energy ultimately adequate—billions of years hence—to produce "critical mass" for self-starting its own "all-out" atomic energy generators and thus itself becoming a radiation-exporting star.

I saw that this planet Earth's organic-biochemical interstructuring is (1) tensionally produced only by triple-bonded, no-degrees-of-freedom, crystalline interarrayings of atomic events, (2) compressionally structured only by double-bonded, flexibly jointed, pressure-distributing and omni-stress-equalizing, hydraulic interarrangements of atomic events, and (3) shock-absorbingly structured, single-bondedly and pneumatically, by gaseous interarrangement of atomic events.

I saw that approximately one-half of all the mobile biological structuring consists of water, which freezes and boils within very close thermal-environment limits, the physical accommodation of which limiting requirements is uniquely maintained in Universe only within the biosphere of Spaceship Earth—that is, so far as human information goes.

And I repeat for emphasis that I saw in 1927 that there was nothing to stop me from trying to think about how and why humans are here as passengers aboard this spherical spaceship we call Earth. Return to this initial question has always produced for me the most relevant and incisive of insights. Therefore I hope all humanity will begin to ask itself this question in increasingly attentive earnestness.

I also saw that there was nothing to stop me from thinking about the total physical resources we have now discovered aboard our ship and about how to use the total cumulative know-how to make this ship work for everybody—paying absolutely no attention to the survival problems of any separate nations or any other individual groupings of humans, and assuming only one goal: the omni-physically successful, spontaneous self-integration of all humanity into what I called in 1927 "a one-town world."

I knew at the 1927 outset that this was to be a very long-distance kind of search, research, and development experiment, probably to take at least one-half a century to bring to fruition, with no capital backing. At any rate, I want you to understand now why I had no competition undertaking to solve all human physio-economic problems only by an environment-improving, artifact-inventing-and-developing revolution, which inadvertently produced its recognition by the media, which incidental news publishing is the only reason you know about me—especially since I have been only inadvertently producing news-provoking artifacts for fifty-six years. All that news has failed to induce any sincerely sustained realistic competition with my efforts and on my economic premise of non-money-making—but hopefully sense-making—and only-by-faith-in-God-sustained objectives.

Because it was a very large undertaking, I didn't know that I would be here to see it through for all those 56 years. I was born in 1895. The life insurance companies' actuarial life-expectancy for me was 42 years. I was already 32 when I started the project. I have been amazed that things have worked out to the extent that they have, that I am as yet vigorously active, and that I have been able to find so many relevant things that a little individual can do that a great nation and capital enterprise can't do.

Yet I am quite confident there is nothing that I have undertaken to do that others couldn't do equally well or better under the same economic circumstances. I was supported only by my faith in God and my vigorously pursued working assumption that it is God's intent to make humans an economic success so that they can and may in due course fulfill an essential—and only mind-renderable—functioning in Universe.

Assuming this to be God's intent, I saw that if I committed myself only to initiating, inventing, and full-scale prototyping of life-protecting

and -supporting artifacts that afforded ever more inclusive, efficient, and in every way more humanly pleasing performances while employing ever less pounds of materials, ergs of energy, and seconds of time per each accomplished function, a young public's enthusiasm for acquisition of those artifacts and youth's increasing satisfaction with the services thereby produced might induce their further development and multiplication by other significance-comprehending and initiative-taking young humans.

I saw that this ever-multiplying activity could lead ultimately to full-scale, world-around, only-for-industrial-mass-production-prototyped artifacts. I saw that this mass-production-and-distribution of livingry service could provide an adequate inventory of public-attention-winning-and-supporting artifacts, efficient and comprehensive enough to swiftly provide the physical success for all humanity.

This would terminate humanity's need to "earn a living," i.e., doing what others wanted done only for others' ultimately selfish reasons. This attending only to what needs to be done for all humanity in turn would allow humanity the time to effectively attend to the Universe-functioning task for the spontaneous performance of which God—the eternal, comprehensive, intellectual integrity usually referred to as "nature" or as "evolution"—had included humans in the grand design of eternally regenerative Universe.

It was clear to me that if my scientifically reasoned working assumptions were correct and if I did my part in successfully initiating, and following-through on realizing, the previously recited potential chain-reactive events, I would be supported by God in realistic, natural, but almost always utterly surprising-to-me ways. I therefore committed myself to such initiations, realizations, and followings-through.

Because I knew at the 1927 outset of the commitment that no one else thought my commitment to be practical or profitable, I also knew that no one would keep any record of its evolvement—should it be so fortunate as to evolve. Since I intended to do everything in a comprehensively scientific manner in committing myself to this very large-scale experiment (which, as already stated, sought to discover what a little, unknown, moneyless, creditless individual with dependents could do effectively on behalf of all humanity that—inherently—could never be done by any nations or capital enterprise), I saw clearly that I must keep my own comprehensive records—records being a prime requisite of scientific exploration. This I have done. It has been expensive and difficult both to accomplish and to maintain. It is comprehensive and detailed. I speak of the record as the "archives." They consist of:

A. The "Chronofile," which in 1981 consisted of 750 12″ × 10″ × 5″ volumes. These volumes contain all my correspondence, as well as sketches and doodles made during meetings with others, and also back-of-envelope and newspaper-edged notes, all maintained chronologically—in exact order of inbound and outbound happenings—all the way from my earliest childhood to the present keeping of such records as induce discovery of what to avoid in future initiatives

B. All the drawings and blueprints I have been able to save of all the design and full-scale artifact-inventing, -developing and -testing realizations

C. All the economically retainable models

D. All the moving picture and television footage covering my work

E. All the wire and tape recorded records of my public addresses

F. All the affordable news-bureau and clipping-service records of articles or books written by others about me or my work

G. All the posters announcing my lecturing appearances as designed and produced by others

H. A large conglomeration of items (for instance, over 100 T-shirts with pictures of my work or quotations of my public utterances) produced and distributed by students at many of the 550 universities and colleges that have invited me to speak; collection of awards, mementos, etc.

I. All the multi-stage copies of the manuscript and typescript versions of my twenty-three formally published books and many published magazine articles

J. Over 10,000 4″ × 5″ photo negatives and over 30,000 photographs, all code-listed, covering my life and work; also 20,000 35-mm projection slides

K. My own extensive library of relevant books and published articles

L. All my financial records, including annual income tax returns

M. All the indexes to the archival material

N. All the drafting tools, typewriters, computers, furniture, and file cabinets for an office staff of seven

O. A large collection of framed photos, paintings, diplomas, cartoons, etc.

P. Biographical data, published periodically (approximately every three years), summarizing all developments of my original commitments

Q. The "Inventory of World Resources, Human Trends and Needs"
R. The World Game records

The archives' collected public record now consists of over 100,000 newspaper and magazine articles, books, and radio and television broadcasts about me or my work, unsolicitedly conceived and produced by other human beings all around the world since 1917.

The prime public record of my more-than-half-century's fulfillment of my commitment has been realized in the working artifacts themselves—the 300,000 world-around geodesic domes, the five million Dymaxion World Maps, the many thousands of copies of each of my twenty-three published books—and, most important of all, within the minds and memories of the 30,000 students I have taught how to think about how to design socially needed, more efficiently produced artifacts.

I do not now employ, and never have employed, any professional public relations agents or agencies, lecture or publishing bureaus, salespeople, sales agencies, or promotional workers. As indicated earlier, I am convinced that nature has her own conceptioning, gestation, birth, development, maturization, and death rates, the magnitudes of which vary greatly in respect to the biochemistry and technological arts involved. The most important evolutionary events take the longest.

Since maintenance of the updating and safety of the archives is as yet my responsibility, they are not open to the public, though scholars from time to time are allowed to view them and be shown items of special interest to them. Because I avoid employing any professional agencies, the magnitude of my development is not kept track of and publicly reported by any of the professional agency associations. For instance, I am not included in the annual statements appearing in the news regarding the public speakers most in demand. Therefore every three years or so my office updates my "basic biography," as it is called, to be distributed to those who ask for information.

Because we are now entering upon the 1927-initiated half-century period of realization, it is now appropriate to make public exposure of the record in order to encourage youth to undertake its own mind-evolved initiatives.

I am therefore publishing herewith:

1. A compilation of my U.S. patents, with photographs of each of the realized artifacts covered by each patent.
2. The compendium of honorary doctorate citations. Since I did not graduate from any college or university and since I have not amassed riches and made generous financial gifts to the schools, I am confident that none of the doctorates were conferred upon me as a financial benefactor but only in recognition of my on-campus academic activities, as a visiting lecturer, a research project initiator and director, or as an appointed professor. Here my work could be intimately judged for its educational value, wherefore the awarding of honorary doctorate degrees to me constitutes an objective assessment of the magnitude and validity of my working knowledge and of its usefulness to the educational system and society. Like the public record established by my patents, these doctorates can serve as critical appraisal of the historical relevance, practicality, and relative effectiveness of my half-century's experimental commitment to discover what, if anything, an individual human eschewing politics and money-making can do effectively on behalf of all humanity.

I hope that the record so documented (and your hoped-for-by-me close examination of it) will serve as an encouragement to you as individuals to undertake tasks that you can see need to be attended to, which are not to the best of your knowledge being attended to by others, and for which there are no capital backers. You are going to have to test a cosmic intellectual integrity as being inherently manifest in the eternally generalized scientific and only mathematically expressible laws governing the complex design of Universe and of all the myriads of objective special-case realizations.

I am presenting all these thoughts and records because I think we are coming to a very extraordinary moment in the history of humanity, when only such a spontaneous, competent, and ultimately cooperative design-science initiative-taking, on the part of a large majority of human individuals, can ensure humanity's safe crossing of the cosmically critical threshold into its prime and possibly eternal functioning in the macro-micro cosmic scheme.

All humans have always been born naked, completely helpless for months, beautifully equipped but with no experience, therefore absolutely ignorant. This is a very important design fact. I do not look at such a human start as constituting a careless or chance oversight in the cosmic conceptioning of the intellectual integrity governing the Universe. The initial ignorance of humans was by deliberate cosmic (divine) design.

We know that before humans are born forth, naked from their mothers' wombs, their protoplasmic cells are chromosomally (DNA-

RNA) programmed to produce each human in incredibly successful regenerative detail.

If we study both the overall integrated system and the detail-design features of our own physical organism—for instance, our optical system and its intimacy with our brain's nervous system—we realize what miraculously complex yet eminently successful anticipatory-design-science phenomena are the humans and the cosmic totality of our complex supportive environment.

If I accidentally scratch-cut myself when I am 3, nature goes instantly to work and repairs the cut. I don't know, even now at 88, how to repair my own cell-structured tissue, and I certainly didn't know how at 3 years of age. We understand very little. Obviously, however, we are magnificently successful products of design in a Universe the complexity and intricacy of whose design integrity utterly transcends human comprehension, let alone popularly acceptable descriptions of "divine design."

Therefore the fact that we are designed to be born naked, helpless, and ignorant is, I feel, a very important matter. We must pay attention to that. We are also designed to be very hungry and to be continually rehungered and rethirsted and multiplyingly curious. Therefore we are quite clearly designed to be inexorably driven to learn only by trial and error how to get on in life. As a consequence of the design, we have had to make an incredible number of mistakes, that being the only way we can find out "what's what" and a little bit about "why," and an even more meager bit regarding "how" we can take advantage of what we have learned from our mistakes.

Suppose a hypothetical 3-billion-B.C. "you," being enormously hungry, ate some invitingly succulent red berries and was poisoned. The tribe concluded, "Berries are poisonous. People can't eat berries." And for the next thousand years, that tribe did not do so. Then along came somebody who showed them they could safely eat blueberries.

At any rate it took a long, long time for humanity to get to the point where it was inventing words with which to integrate and thereby share the lessons individually learned from error-inducing experiences. Humans could help one another only by confessing what each had found out only by trial and error. Most often "know-it-all" ego blocked the process.

After inventing spoken words, humanity took a long time more to invent writing, with which both the remote-from-local-community and the dead gave the remote-from-one-another beings information about their remote-in-time-and-space error-discovering and truth-uncovering experience. Thus does evolution continually compound the wisdom accruing to error-making and -discovering experiences, the self-admission of which alone can uncover that which is true.

Informed by the senses, the brains of all creatures unthinkingly reflex in pursuing only the sensorially obvious and attractive wilderness trails, waterways, and mountain passes along which may exist fallen-leaves-hidden pitfalls.

This fact made possible the effectiveness of the pitfall trap—falling into the pit at 90 degrees to the 180 degrees line of sight spontaneously followed in pursuit of the obvious. This straightaway reflexing is frequently employed by nature to give humans an opportunity to learn that which humanity needs to learn if humanity is ever to attain the capability to perform its cosmically assigned, spontaneous, intellectually responsible functioning. If ego is surmounted, mind may discover and comprehend the significance of the negative event and may thereby discover the principles leading not only to escape from the entrapment but discovery of what is truly worthwhile pursuing, but only as a consequence of the mind's exclusive capability to discover and employ not just one principle but the synergetically interoperative significance of all the human mind's thus-far-discovered principles.

Up to a very short time ago—that is to say, up to the twentieth century—humans were only innately wise but comprehensively information-ignorant. We have had to discover many errors to become reasonably intelligent.

For instance, when I was young those humans who were most remote from others had to travel a minimum of six months to reach one another; approximately none of them ever did so. That is why Kipling's "East is East, and West is West, and never the twain shall meet" seemed so obviously true to the vast millions who read or heard his words. That has all changed. Today, any one of us having enough credit to acquire a jet air-transport ticket can—using only scheduled flights—physically reach anybody else around the world within twenty-four hours, or can reach each other by telephone within minutes.

The furthestmost point away from any place on our approximately 25-thousand-mile-circumferenced Earth sphere is always halfway around the world, which is 12,500 great-circle miles away in any direction from the point at which we start. Flying the shortest distance from the exact North Pole to the South Pole, any direction you first head in will be "due south." If you keep on heading exactly south you will find yourself following a one-great-circle meridian of longitude until you get to the South Pole. Furthermore, to reach your half-way-around-the-Earth, furthest-away-from-you point—which is always 2,500 miles away—flying in a Concorde supersonic transport, cruising efficiently at Mach 2 (approximately 1,400 m.p.h.), and including stopover refueling times, you will reach that furthestmost half-way-around-the-world-from-where-you-started point in half a day.

Humanity, which yesterday was remotely deployed by evolution, is now being deliberately integrated to make us all very intimate with one another and probably ultimately to cross-breed us back into one physically similar human family.

When I was young we were extremely ignorant about other people. I was told that people even in the next town were very dangerous: they "drink whiskey and have knives. . . . you had better not go over there." I was 7 years old when the first automobile came into Boston. When the owner of one of those excitingly new 1902 automobiles drove me over into one of those "dangerous" next towns, I could see no one who appeared to be more "dangerously threatening" than any of my very nonthreatening hometown neighbors.

I was 8 when the Wright brothers first flew. I was 12 when we had the first public wireless telegraphy. I was 19 when we had our first "world" war.

I was born in an almost exclusively walking-from-here-to-there world, a Victorian world in which we knew nothing about strangers. The assumption was that all strangers were inherently very unreliable people. When I was young, 95 percent of all humans were illiterate. Today over 60 percent of the three times as many humans present on Earth are literate. All this has happened unpredictedly in only one lifetime. The majority of the older humans of today are as yet apprehensive of strangers and pretend nonrecognition while the majority of those 30 years and younger tend to welcome strangers, often with open arms.

Something very big has been going on in my particular generation's lifetime where all the fundamental conditions of humanity are changing at an ever-accelerating acceleration rate.

When I was young, not only were 95 percent of human beings illiterate but their speech patterns were also atrociously difficult to understand. I had two jobs before World War I. The men I worked with were very skilled but their awkwardly articulated, ill-furnished vocabularies were limited to about 100 words—50 percent of them blasphemous or obscene. Primarily they let you know how they felt about matters by the way in which they spit—delightedly, amusedly, approvingly, or disgustedly. They were wonderfully lovable and brave human beings but that swearing and spitting was the most articulate and effective expressive language they had. Their pronunciations varied not only from town to town and from one part of a town to another but also from house to house and from individual to individual.

Something extraordinary has happened. Only within our last-of-the-twentieth-century time, approximately everybody has acquired a beautiful vocabulary. This did not come from the schooling system but from the radio and TV, where the people who secured the performing jobs did so by virtue of their common pronunciation, the clarity of their speech, and the magnitude of their vocabularies. They no longer spoke with the myriad of esoteric pronunciations of yesterday. People were introduced to a single kind of language. This brought about primary common-speech patterns. The necessity for pilots of airplanes operating daily around the world from countries all around the world to have a common language has swiftly evolved into a common language. Olympic Games, athletics in general, and frequently televised world affairs have all been accelerating the coming of a to-be-evolved world language. That the language most commonly used in 1983 is English is unfortunate and untrue. What we call English was not the language of long-ago-vanished Angles and Jutes. It is the most crossbred of all the world-around languages of all the world-around people who on their ever westwardly and mildly northwestwardly colonizing way have historically invaded or populated England. "English" now includes words from all the world's languages and represents an agglomeration of the most frequently used and most easily pronounceable words.

Since throughout at least three million years people did not understand languages well, they were relatively ignorant. Their group survival required leaders. Thus through the comprehensively illiterate ages we came historically to require powerful leaders—sometimes as physical warriors, sometimes as spiritual warriors, sometimes as unique individual personalities. Statesmen steered great religious organizations or great government organizations that led the affairs of illiterate, uninformed humanity.

We have come now to a completely new mo-

ment in the history of humans at which approximately everybody is "in" on both speech and information.

Humanity has been coming out of a group-womb of permitted ignorance made possible only by the existence of an enormous cushion of natural physical resources with which to learn only by trial and error, thus to become somewhat educated about how and why humans happen to be here on this planet. Despite a doubling of world population during my lifetime, humanity has at the same time gone from 95 percent illiterate to 60 percent literate.

Until I was 28 we Earthians knew only about our own Milky Way galaxy. In 1923 Hubble discovered another galaxy. Since that time we have discovered two billion more of them. That explosively accelerated rate of expansion of our astronomical-information-acquisitioning typifies the rate of popular increase of both scientifically general and technically special information during my lifetime.

We are living in a new evolutionary moment in which the human is being individually educated to do the free-standing human's own thinking, and each is thereby separately becoming extraordinarily well informed. We have reached a threshold moment where the individual human beings are in what I consider to be a "final examination" as to whether they, individually, as a cosmic invention, are to graduate successfully into their mature cosmic functioning or, failing, are to be classified as "imperfects" and "discontinued items" on this planet and anywhere else in Universe.

We are at a human examination point at which it is critically necessary for each of us individually to have some self-discovered, logically reasonable, experience-engendered idea of how and why we are here on this little planet in this star system and galaxy, amongst the billions of approximately equally star-populated galaxies of Universe. I assume it is because human minds were designed with the capability to discover from time to time the only-mathematically-stateable principles governing the eternal interrelationships existing between various extraordinary phenomena—a capability possessed by no known phenomenon other than humans.

How and why were we given our beautiful minds with their exclusive access to the scientific principles governing the operational design of eternally regenerative Universe?

As an instance of the human mind's capability, we have the integration of the unplanned, only evolutionarily combined works of Copernicus, Kepler, and Galileo informedly inspiring Isaac Newton into assuming hypothetically that the interattractiveness of any two celestial bodies varies inversely as of the second power of the arithmetical distances intervening—if you double the arithmetical distance intervening, you will reduce the interattractiveness to one-fourth what it had been. Astronomers applied his hypothesis to their celestial observations and discovered that this mathematical formulation explained the ever-mobile interpositioning behaviors of all known and measuringly-observed celestial bodies. Thus, Newton's hypothesis became adopted as a "scientifically generalized law."

In dramatic contradistinction to the brain's functioning within only directly nerve-sensing limits, the human mind has the capability, once in a great while, to discover invisible, soundless, unsmellable, untasteable, untouchable interrelationships eternally existing between separate, special-case cosmic entities, which eternal interrelationships are not manifest by any of the interested entities when considered only separately and which interrelationships can be expressed only mathematically and constitute eternal cosmic laws—such as Newton's discovery that the interattractiveness existing invisibly between any two celestial bodies always varies inversely as the second power of the arithmetically expressed distances intervenes those bodies. Such Universal laws can be expressed only mathematically. Mathematics, we note, is *purely* intellectual. Altogether these laws manifest the eternal intellectual integrity of Universe that I speak of as "God."

Human intellect (mind) has gradually discovered a number of these extraordinary, generalized, non-sensorially-apprehensible eternal principles. We human beings have been given access to at least some of the design laws of the Universe.

We don't know of any other phenomenon that has such a faculty and such an access permit, wherefore we may assume that we humans must be cosmically present for some very, very important reason.

As far as I know, we humans haven't thought or talked very much about how and why we are here as either a desirable or a necessary function of Universe. We have talked a great deal about the great mystery of being here. But the majority of our public talking centers on the egotistical assumption that human politics and the wealthy are running the Universe, that the macrocosmic spectaculars are an accessory amusement of our all-important selfish preoccupations, and that Universe's microcosmic invisibles are exclusively for corporate stockholders' money-making exploitation—though always individually discovered or invented but only industrially developed, funded exclusively by the research departments of great corporations as initially funded by humanity's taxes-paid military defense expenditures after production rights are transferred to prime corporations.

We humans are overwhelmed because we are so tiny and the Earth is so big and the celestial systems so vast. It is very hard for us to think effectively and realistically about what we feel about the significance of all we have learned about Universe.

At any rate, we are now at a point where we have to begin to think realistically about how and why we are here with this extraordinary capability of the mind. Our remaining here on Earth isn't a matter of the cosmic validity of any Earthian economic systems, political systems, religious systems, or other mystical-organization systems. Our "final examination" is entirely a matter of each of all the individual human beings, all of whom have been given this extraordinary, truly divine capability of the mind, individually qualifying in their own right to continue in Universe as an extraordinary thinking faculty. If all of humanity as a cosmic invention is to successfully pass this final examination, it would seem to be logically probable that a large majority of all Earthian individuals make and do the passing for themselves and for the remaining numbers.

Are we individually going to be able to break out from our institutionally misconditioned reflexing? Are we going to be able to question intelligently all the things that we have been taught only to believe and not to expose to the experiential-evidence tests? Are we going to really dare to make our own behavioral strategy decisions as informed only by our own separate experimentally or experientially derived evidence? The integrity of individually thinking human beings—as mind—vs. brain-reflexing automatons is being tested.

My own working assumption of why we are here is that we are here as local-Universe information-gatherers and that we are given access to the divine design principles so that we can therefrom objectively invent instruments and tools—e.g., the microscope and the telescope—with which to extend all sensorial inquiring regarding the rest of the to-the-naked-eye-invisible, micro-macro Universe, because human beings, tiny though we are, are here for all the local-Universe information-harvesting and cosmic-principle-discovering, objective tool-inventing, and local-environment-controlling as local Universe problem-solvers in support of the integrity of eternally regenerative Universe.

To fulfill our ultimate cosmic functioning we needed the telescope and microscope, having only within the present century discovered that almost all the Universe is invisible. I was two years old when the electron was discovered. That began a new era. When I entered Harvard my physics text had a yellow-paper appendix that had just been glued into the back of the white-leaved physics book. The added section was called "Electricity." The vast ranges of the invisible reality of Universe constituted a very new world—for humanity, a very different kind of world.

When I was born, "reality" was everything you could see, smell, touch, and hear. That's all there was to it. But suddenly we extended our everyday doings and thinkings, not linearly but omnidirectionally into the vast outward, macro-ranges and inwardly penetrating to discover the infra-micro-tune-in-able ranges of the invisible within-ness world.

We began to discover all kinds of new chemical, biological, and electromagnetic behaviors of the invisible realm so that today 99.999 percent of the search and research for everything that is going to affect all our lives tomorrow is being conducted in the realm of reality nondirectly contactable by the human senses. It takes a really educated human to be able to cope with the vast and exquisite ranges of reality.

So here we are—as human beings—majorly educated and individually endowed with developable capabilities of getting on successfully within the invisible ultra-macro and infra-micro world.

Here we have found that each chemical element has its own electromagnetic spectrum wavelengths and frequencies that are absolutely unique to that particular element-isotope. These wavelengths and frequencies are nondirectly tune-in-able by the human senses but are all spectroscopically differentiable and photographically convertible into human readability.

Repeating to emphasize its significance, we note that exploring with the spectroscope, photo-telescope, and radio telescope, humans only recently have traveled informationally to discover about two billion galaxies with an average of 100 billion stars each, all existing within astronomy's present 11.5 billion-light-years' radial reach in all directions of the vastness around us.

Taking all the light from all those galaxies

we have been able to discover spectroscopically—and have inventoried here on board planet Earth—the relative interabundance of each of the unique categories of all the chemical elements present within that 11.5 billion-light-years realm around us. That little Earthian humans can accomplish that scale of scientific inventorying makes it possible to realize what the human mind can really do. One can begin to comprehend what God is planning to do with the humans' cosmic functioning.

In our immediate need to discover more about ourselves we also note that what is common to all human beings in all history is their ceaseless confrontation by problems, problems, problems. We humans are manifestly here for problem-solving and, if we are any good at problem-solving, we don't come to utopia, we come to more difficult problems to solve. That apparently is what we're here for, so I therefore conclude that we humans are here for local information-gathering and local problem-solving with our minds having access to the design principles of the Universe and—I repeat—thereby finally discover that we are most probably here for local information-gathering and local-Universe problem-solving in support of the integrity of eternally regenerative Universe.

If our very logically and experientially supported working assumption is right, that is a very extraordinarily important kind of function for which we humans were designed ultimately to fulfill. It is clearly within the premises of the divine.

We note also that when nature has a very important function to perform, such as regenerating Earthian birds to play their part in the overall ecological regenerating, nature doesn't put all her birds' eggs in one basket. Instead she provides myriads of fail-safe alternate means of satisfying each function.

Nature must have illions of alternate solutions in this cosmic locality to serve effectively the local information-gathering and local problem-solving in local support of the integrity of an eternally regenerative Universe—should we humans fail to graduate from our potential lessons-learning games of ever more exquisite micro-invisible discrimination and ever greater political, economic, educational, and religious mistake-making.

When in 1927 I started the experiment to discover what a little individual might be able to do effectively on behalf of all humanity, I said to myself. "You are going to have to do all your own thinking." I had been brought up in an era in which all the older people said to all the young people. "Never mind what you think. Listen to what we are trying to teach you to think!" However, as experience multiplied, I learned time and again that the way things often turned out evidenced that the way I had been thinking was often a more accurate, informative, and significant way of comprehending the significance of events than was the academic and conventional way I was being taught, so I said, "If I am going to discard all my taught-to-believe reflexings, I must do all my own thinking. I must go entirely on my own direct experiential evidence."

I said, "It has been an impressive part of my experience that most human beings have a powerful feeling that there is some greater and more exaltedly benign authority operating in the Universe than that of human beings—a phenomenon they call 'God.' There are many ways of thinking about 'God' and I saw that most children are brought up to 'believe' this and that concerning the subject. By the word *believe* I mean 'accepting explanations of physical and metaphysical phenomena without any supportive physical evidence, i.e., without reference to any inadvertently experienced information-harvesting or deliberate-experimental-evidence expansion of our knowledge.'"

I then said to myself, "If you are going to give up all the beliefs and are going to have to go only on your own experiential evidence, you are going to have to ask yourself if you have any experience that would cause you to have to say to yourself there is quite clearly a greater intellect operating in the Universe than that of human beings."

Luckily I had had a good scientific training, and the discovery of those great, eternal, scientific principles that could only be expressed mathematically—mathematics is purely intellectual—made me conclude that I was overwhelmed with the manifests of a greater intellect operating in Universe than that of humans, which acknowledgment became greatly fortified by the following experiential observations.

Scientific perusal of the personal diaries, notes, and letters written by three of history's most significant scientific discoverers—just before, at the time of, and immediately after the moment of their great discoveries—finds that the written records of all of them, unaware of the others' experiences, make it eminently clear that two successive intuitions always played the principal role in their successful accomplishing of their discoveries: the first intuition telling them that a "fish was nibbling at their baited hook" and the second intuition telling them how to "jerk" their fishing line to "hook" and successfully "land" that fish.

The records of these same three scientists also make clear that to start with there must be human curiosity-arousal followed by an intuition-excitation of the human that causes the individual to say, "There is something very significant going on here regarding which I have yet no specific clues, let alone comprehension, because special-case evidences of generalized principles are always myriadly and only complexedly manifest in nature." To qualify as a scientifically generalized principle the scientifically observed and measured special-case behavioral manifests of the generalized interrelationship principles must always be experimentally redemonstrable under a given set of explicitly and implicitly controlled conditions.

Scientifically generalized principles must be repeatedly demonstrable as being both exceptionless and only elegantly expressible as simple, three-term mathematical equations, such as $E=mc^2$.

Constantly intercovarying, ever experimentally redemonstrable interrelations, which have no exceptions, are inherently "eternal."

As a consequence of the foregoing, I then said, "Brain is always and only coordinating the information reported to us by our senses regarding both the macro outside world around us and the micro world within us, and recording, recalling, and only reflexively behaving in response to previous similar experiencing—if any—or if 'none' to newly imagined safe-way logic. No one has ever seen or in any way directly sensed anything outside the brain. The brain is our smell-, touch-, sound-, and image-ination tele-set whose reliability of objective image formulation has been for all childhood so faithful that we humans soon become convinced that we are sensing directly outside ourselves, whereas the fact is that no one has ever seen or heard or felt or smelt outside themselves. All sensing occurs inside the brain's 'television control zone.' The brain always and only deals with temporal, special-case, human-senses-reported experiences. Mind is always concerned only with multi-reconsidering a host of special-case experiences that have intuitively tantalizing implications of ultimately manifesting the operative presence of an as-yet-to-be-discovered eternal principle governing an invisible, unsmellable, soundless, untouchable eternal interrelationship and that complete interrelationships' possibly-to-be-discovered, constantly covarying interrelationships' rate of change. Such principles, whether discovered or not, are intuitively held to always both embrace and permeate all special-case experiences."

I then said to myself, "I am scientifically convinced that the thus-far-discovered and proven inventory of unfailingly redemonstrable generalized principles are a convincing manifest of an eternal intellect governing the myriad of nonsimultaneously and only overlappingly occurring episodes of finite but non-unitarily conceptual, multi-where and multi-when eternally regenerative scenario Universe.

"In so governing the great Universe's integrity, cosmic intellect always and only designs with the generalized principles that are inherently eternal.

"That all the eternal principles always and only appear to be comprehensively and concurrently operative may prove to be an eternal interrelationship condition, which latter hypothesis is fortified by the fact that none of the eternal principles has ever been found to contradict another. All of the eternal principles appear to be both constantly interaccommodative-intersupportive and multiplyingly interaugmentative.

"When you and I speak of *design,* we spontaneously think of it as an intellectual conceptualizing event in which intellect first sorts out a plurality of elements and then interarranges them in a preferred way.

"Ergo, the "eternal intellect"—the eternal intellectual integrity—apparently governs the integrity of the great design of the Universe and all of its special-case, temporal realizations of the complex interemployments of all the eternal principles."

So I said, "I am overwhelmed by the evidence of an eternally existent and operative, omnicompetent, greater intellect than that of human beings."

Consequently, I also said in 1927, "Here I am launching a half-century-magnitude program with nobody telling me to do so, or suggesting how to do it." I had absolutely no money and my darling wife (who has now been married to me for 66 years) was willing to go along with my thinking and commitments. I said to myself, "If I, in confining my activity to inventing, proving and improving, and physically producing artifacts suggested to me by physical challenges of the a priori environment, which inventions alter the environment consistently with evolution's trending, whereby I am doing that which is compatible with what universal evolution seems intent upon doing—which is to say, if I am doing what God wants done, i.e., employing my mind to help other humans' minds to render all humanity a physically self-regenerative and comprehensively intellectual integrity success so that humans can effectively give their priority of attention to the ongoing local Universe information-gathering and local problem-solving, primarily with design-science artifact solutions, which will altogether result in comprehensive environmental transformation leading to conditions so favorable to humans' physical well-being and metaphysical equanimity as to permit humans to become permanently engaged with only the by-mind-conceived challenges of local Universe—then I do not have to worry about not being commissioned to do so by any Earthians and I don't have to worry how we are going to acquire the money, tools, and services necessary to produce the successively evoluting special-case physical artifacts that will most effectively increase humanity's technological functioning advantage to an omnisuccess-producing degree."

This became an overwhelming realization, for it was to be with these artifacts alone that I was committing myself to comprehensively solve all humanity's physio-economic problems.

I then said, "I see the hydrogen atom doesn't have to earn a living before behaving like a hydrogen atom. In fact, as best I can see, only human beings operate on the basis of 'having to earn a living.' The concept is one introduced into social conventions only by the temporal power structure's dictums of the ages. If I am doing what God's evolutionary strategy needs to have accomplished, I need spend no further time worrying about such matters."

"I happen to have been born at the special moment in history in which for the first time there exists enough experience-won and experiment-verified information current in humanity's spontaneous conceptioning and reasoning for all humanity to carry on in a far more intelligent way than ever before.

"I am not being messianically motivated in undertaking this experiment, nor do I think I am someone very special and different from other humans. The design of all humans, like all else in Universe, transcends human comprehension of "how come" their mysterious, a priori, complexedly designed existence.

"I am doing what I am doing only because at this critical moment I happen to be a human being who, by virtue of a vast number of errors and recognitions of such, has discovered that he would always be a failure as judged by society's ages-long conditioned reflexings—and therefore a "disgrace" to those related to him (me)—in the misassuredly, eternally-to-exist "not-enough-for-all," comprehensive, economic struggle of humanity to attain only special, selfish, personal, family, corporate, or national advantage-gaining, wherefore I had decided to commit suicide. I also thereby qualified as a "throwaway" individual who had acquired enough knowledge regarding relevantly essential human evolution matters to be able to think in this particular kind of way. In committing suicide I seemingly would never again have to feel the pain and mortification of my failures and errors, but the only-by-experience-winnable inventory of knowledge that I had accrued would also be forever lost—an inventory of information that, if I did not commit suicide, might prove to be of critical advantage to others, possibly to all others, possibly to Universe." The realization that such a concept could have even a one-in-an-illion chance of being true was a powerful reconsideration provoker and ultimate grand-strategy reorienter.

The thought then came that my impulse to commit suicide was a consequence of my being expressly overconcerned with "me" and "my pains," and that doing so would mean that I would be making the supremely selfish mistake of possibly losing forever some evolutionary information link essential to the ultimate realization of the as-yet-to-be-known human function in Universe. I then realized that I could commit an exclusively "ego" suicide—a personal-ego "throwaway"—if I swore, to the best of my ability, never again to recognize and yield to the voice of wants only of "me" but instead commit my physical organism and nervous system to enduring whatever pain might lie ahead while possibly thereby coming to mentally comprehend how a "me"-less individual might redress the humiliations, expenses, and financial losses I had selfishly and carelessly imposed on all the in-any-way-involved others, while keeping actively alive in toto only the possibly-of-essential-use-for-others inventory of my experience. I saw that there was a true possibility that I could do just that if I remained alive and committed myself to a never-again-for-self-use employment of my omni-experience-gained inventory of knowledge. My thinking began to clear.

I repeated enlargingly to myself, "If I go ahead with my physical suicide, I will selfishly escape from my personal pain but will probably cause great pain to others. I will thereby also throw away the inventory of experience—

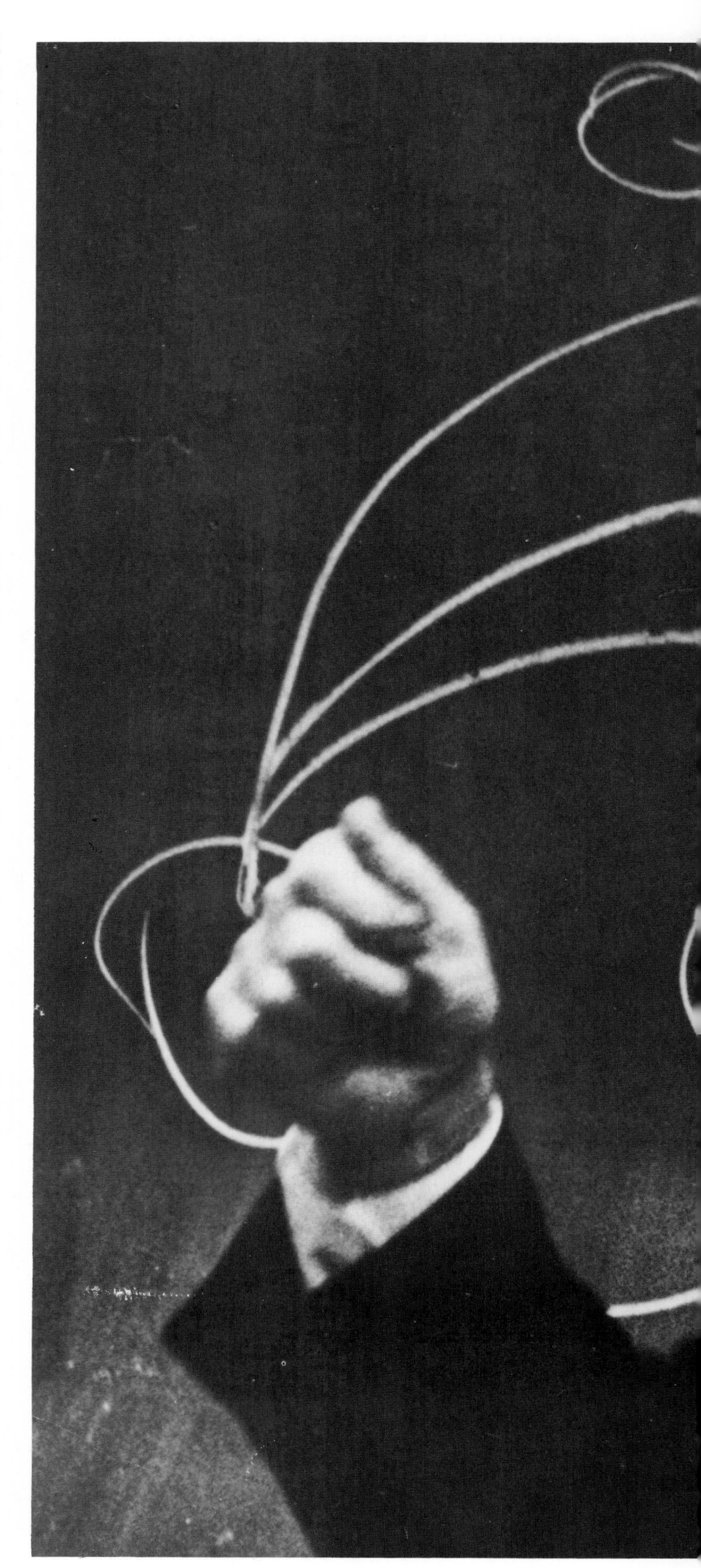

which does not belong to me—that may be of critical evolutionary value to others and even may be said to belong not to me but only to others.

"If I take oath never again to work for my own advantaging and to work only for all others for whom my experience-gained knowledge may be of benefit, I may be justified in not throwing myself away. This will, of course, mean that I will not be able to escape the pain and mortification of being an absolute failure in playing the game of life as it has been taught to me."

I then found myself saying, "I *am* going to commit myself completely to the wisdom of God and to realization only of the advantages for all humanity potentially existent in what life has already taught and may as yet teach me." I found myself saying, "I *am* going to commit myself completely to God and to realization of God's apparent intent to assign semi-divine functioning to an as-yet-to-qualify-for-such-functioning humanity. To qualify for such local-Universe's evolutionary adjustings, humans themselves must intelligently discover and spontaneously employ their designed-in potentials and themselves realize the sustainable success of their evolutionarily scheduled physio-economic potential."

From that time, 56 years ago, I have had absolute faith in God. My task was not to preach about God, but to serve God in silence about God. Because such commitment to faith is inherently a "flying blind" commitment, I have often weakened in my confidence in myself to comprehend what it might be that I was being taught or told to do. Because I am a human and designed like all humans to learn only by trial and error, I have had many times to do the wrong things in order thereby to learn what next needed to be done. Making mistakes can be and usually is a very dismaying experience—so dismaying as to make it seemingly easier to "go along with unthinking custom."

If I had not in 1927 committed "egocide," I would probably have yielded long ago to convention and therewith suicide of my "only-for-all-others" initiative.

Friends would say, "You are being treacherous to your wife and child, not going out to earn a living for them. Come over here and we will give you a very good job." When, persuaded by their obvious generosity and concern, I did yield, everything went wrong; and every time I went "off the deep end" again, working only for everybody without salary, everything went right again.

I was convinced from the 1927 outset of this new life that I would be of no benefit whatsoever to the more than two billion humans alive in 1927 if I set about asking people to listen to my ideas and endeavoring to persuade them to reform their thinking and ways of behaving. People listen to you only when in a dilemma they recognize that they don't know what to do and, thinking that you might know, ask you to advise them what to do. When they do ask you and you have only a seemingly "good idea" of what they might do, you are far less effective than when you can say "Jump aboard and I'll take you where you want to go"—or "Jump aboard that vehicle and it will take you to where you want to go." This involves an inanimate artifact to "jump aboard."

Quite clearly I had to address not only the specific but also the comprehensive problem of how to find ways of giving human beings more energy-effective environment-controlling artifacts that did ever more environment-controlling with ever less pounds of materials, ergs or energy, and minutes of time per each realized functioning, until we attained the physically realized techno-energetic ability to do so much with so little that we could realize ample good-life support for everyone, hoping that under those more favorable physical circumstances humans would dare to be less selfish and more genuinely thoughtful toward one another, instead of being lethally and subversively competitive for a share in the existing misassumed-to-be-fundamental inadequacy of life support, perpetually to be extant, on our planet. Every time I recommitted myself so to do, everything went well again.

So we—my wife and family—have for 56 years realized a series of miracles that occur just when I need something, but not until the absolutely last second. If what I think I need does not become available I realize that my objective may be invalid or that I am steering a wrong course. It is only through such non-happenings that I seem to be informed of how to correct both my grand strategy and its constituent initiations.

I can't make plans of how to invest that which I don't have or don't know that I am going to have. I cannot count on anything. During all these last 56 years I have been unable to budget. I simply have to have faith and just when I need the right-something for the right-reasoning, there it is—or there they are—the workshops, helping hands, materials, ideas, money, tools.

Throughout the last 56 years, I have been able to initiate and manage a great many physical-artifact developments, well over $20 million direct-expenditures' worth of artifact-prototype-producing-and-testing physical research and development. For the last twenty years my income has averaged a quarter of a million dollars a year and my office overhead, travel, invention research and development, and taxes have also averaged a quarter of a million dollars. I am always operating in proximity to bankruptcy but never going bankrupt. While I owe all the humans of all history an unrepayable debt of only-by-experience-winnable knowledge, I don't owe anybody any money and have never consciously and deliberately gained at the expense of others.

I have no accrued savings of earnings. Income taxes take away even the most meager cushioning of funds. I have no retirement fund. I am on nobody's retirement roll. My wife's and my own social security combine to $9,000 per year. The tax experts tell me that the base of the U.S. government taxing theory relates to capitalism and its initially-amassed-dollars-investments in physical production facilities and the progressive depreciation of those physical properties. I operate in a differently accounted world in which there is nothing that I can depreciate wherewith to accrue further-initiative-funding capability. The "know-how" capital capability with which I operate is always appreciating at an exponentially multiplying rate.

There is always an inventory of important follow-through tasks to be accomplished and a number of new, highly relevant critical initiatives to be taken.

Over the years there have also had to be a number of important errors to be made and important lessons to be learned. From the outset of my 1927 commitment and the first twenty subsequent years, there were individuals who were altruistically inspired to support my commitment. For instance, in 1928 a lawyer friend of mine gave me his services for nothing. Seeing the advantage for me of incorporating my activity, he took a few shares in my corporation to pay for his very prodigious services with a hope of incidental personal profit (see my book *4 D Timelock*). Fortunately, the individual investments of those who sought to help me were as monetarily small as they were altruistically large, for their direct profits were never realized and their investments seemingly lost, unless they felt greatly rewarded, as many did, to discover that they had helped to launch an enterprise that as years went on seemed to be ever more promising for all humanity.

The last such "ultimately lost," friendly financial backing occurred during my design and production of the Dymaxion Dwelling Machine's mass-production prototype for Beech Aircraft in Wichita, Kansas. It was there, in 1944, 1945, and 1946, that I produced in full working scale the Dymaxion House, which weighed the three tons that I had estimated and published it would weigh when I designed it in 1927–29. This Beech-Dymaxion realized weight of three tons proved the validity of both my structural and economic efficacious theory and its important technological advance over its conventional, equi-volumed and -equipped counterpart residence of 1927, which weighed 150 tons. It was the 1927 design-initiative discovery that I could apparently by physical design reduce fifty-fold the weight of materials necessary to produce a home, given certain operating standards, that gave me a "rocket blast-off" as an increase of confidence in my theory of solving humans' economic problems by producing ever more performance with ever less energy investments. The confirmation realized at Beech Aircraft seventeen years later was a second-stage rocket acceleration of my only-by-artifact problem-solving initiative.

Produced on the premises of Beech Aircraft, this 1944 mass-production dwelling machine prototype, along with its 100 percent spare parts, was ordered and paid for by the U.S. Air Force. The prototype, along with its 100 percent spare parts, cost $54,000 and after World War II was turned over by the government to a privately organized two-hundred-thousand-dollar corporation formed by about 300 subscribers—averaging a $666 gamble. This corporation hoped to organize the mass production of the physically realized and three-times-refined government-paid-for prototype. Beech Aircraft had its production engineering department plan the tooling and complete an estimate of producing the Dymaxion Dwelling Machine at a rate of 20,000 per year. Beech then made a firm offer to produce them at that rate for $1,800 each, delivered in Wichita minus the kitchen equipment and other electrical appliances to be provided by General Electric on a rental basis of $200 a year. Beech required, however, that a ten-million-dollar tooling cost be provided by outside finance. This was not raised because there existed no high-speed, one-day, "turnkey"—no marketing, distributing, and installing service industry. Prepayment checks for 37,000 unsolicited orders had to be returned. The hopefully-into-mass-production gamble of the private corporation occurred despite my two-fold warning that (1) experience had by then taught me that the gains accruing to my work apparently were distributable only to everybody and only as techno-economic advantage profits for all humanity and (2) that while I was producing an important prototype dwelling machine suitable for mass production at a low cost, there existed as yet no distribution and maintenance service industry (and that the latter would require a development period taking another third of a century).

No one has ever won their direct monetary gains by their investment-for-profit bets on my artifact-inventing and -developing concepts. I have always been saddened by this because the backers' motives were most often greatly affected by a personal affection for me. Fortunately, this gambling on the financial success of my work despite my warnings seems altogether to have ceased a third of a century ago.

With several hundred thousand geodesic domes, millions of my world maps and books, and structural and mechanical artifacts now distributed and installed around the world, I have realized new and ever greater degrees of technical advantaging of society and its individuals. I have continually increased the knowledge of the means of accomplishing across-the-board and world-around advances in technological abilities to produce more economically effective structures and machinery that do ever more work with ever less pounds of materials, ergs of energy, and seconds of time per function—all accomplished entirely within the visible structural inventions and invisible (alloying, electronic, etc.) realms of technology. Twelve years ago I had learned enough to be able to state publicly that humanity had now passed through an evolutionary inflection point whereafter, for the first time in history, it is irrefutably demonstrable that a ten-year design revolution that employs all the physical resources of humanity (now majorly invested in killingry technology) and transforms weaponry scrap into livingry technology, can, within ten design-science-revolution years, have all humanity enjoying a higher standard of living—interminably sustainable—than any humans have ever experienced, while concurrently phasing out all further everyday uses of fossil fuels or atomic energy. We can live handsomely using only our daily income of the Sun's and gravity's multi-way-intergenerated energies.

This omni-humanity eco-technical breakthrough opportunity involves the successful inauguration of the industrial mass production of the dwelling machines and all other geodesic and tensegrity, air-deliverable, move-in-today, environmental controls.

The technology for producing the dwelling machines, their air deliverability, their energy harvesting and conserving, and their prolonged autonomy of operation, has now reached the service industry launching stage. The dwelling machine service industry will not sell houses but will only rent them (as with rental cars or telephones). Much of the dead U.S. automobile-manufacturing industry can and probably will be retooled to produce the dwelling machines that will be needed to upgrade the deployed phases of living of four billion humans. The environment-controlling service industry will provide city-size domes for protecting and housing humanity's convergent activities and the dwelling machine will accommodate all humanity's divergent activities.

My part in this half-century-long technological development has all been fulfilled or occurred as an absolute miracle. From what I have learned I can say—at this most critical moment in all the history of humanity—that whether we are going to stay here or not depends entirely on each and all of our individually attained and maintained integrities of reasoning and acting and not on any politically or religiously accepted or power-imposed socioeconomic-credo-system or financial reforms.

At the outset of my 1927 commitment to exploring for that which only the individual could do effectively for all humanity, while depending entirely on the unpromised-to-him backing of his enterprise only by God, it became immediately evident that if indeed the undertaking became affirmatively supported by God, it would entail many extraordinary physical and metaphysical insights regarding both human and cosmic affairs. I asked myself, "(1) Can you trust yourself never to turn to your own exclusive advantage the insights entrusted to you only for the realization of benefits for all humanity and Universe itself? (2) Can you also be sure that you will never exploit your insight by publicly declaring yourself to be a special 'son of God' or a divinely ordained mystic leader? (3) Can you trust yourself to remember that you qualified for this functioning only because you were an out-and-out throwaway? (4) Can you trust yourself to reliably report these facts to others when they applaud you for the success of the experiment with which you were entrusted?"

Fortunately, I can, may, and do report to you that I have never broken that trust nor have I ever been tempted to do so.

I do not deem it to be a breaking of that trust when—entirely unsolicited by me or by my family or by any of those working with me on my staff—individuals or organizations outside my domain come spontaneously to me or my staff to employ me as a speaker, author, architect, or consultant. It is these unsolicited, uncontrived, spontaneous short-engagement employments of me in one role or another, plus—on rare occasions—an unsolicited outright gift to me of money, materials, tools, working space, commissions for designs, orders for specific products, etc., which altogether uncontrived employments and unsought gifts I have classified for you as the "miracles," always unforeseeable-in-advance, which have financed or implemented my technical initiatives.

I hope this book will prove to be an encouraging example of what the little, average human being can do if you have absolute faith in the eternal cosmic intelligence we call God.

INVENTIONS

1▲STOCKADE: BUILDING STRUCTURE (1927)

U.S. PATENT—1,633,702

APPLICATION—OCTOBER 8, 1926

SERIAL NO.—140,234

PATENTED—JUNE 28, 1927

MY FATHER DIED ON MY TWELFTH BIRTHDAY. He had not been able to communicate with me since I was ten. My mother, her friends and my father's, and my relatives all tried to make me conform to the most acceptable type of reliable employee of whoever might offer me a job after my education was finished. All of them said, "Never mind what you think, listen. We are trying to teach you." Because I knew how much they loved me, I did my best to pay no attention to my own thoughts.

When I became engaged to Anne Hewlett, her father, a leading architect of New York, was the first grown individual to tell me to pay attention to my own thoughts, which he said he found to be constructive, inventive, and order-seeking. When I came out of the navy after World War I, I was given an excellent job in the national meat-packing house of Armour and Company, in which I had been employed before the war. They made me assistant export manager. I was doing very well in that job when a man named Eddie MacDonald asked me to resign from Armour and take a job with him.

Eddie MacDonald had been my commander in the navy when he was in command of the Naval Aviation Training Unit at Norfolk, Virginia, early in World War I. He had earned the U.S. Medal of Honor in the pre–World War I Battle of Santa Cruz, Mexico, and had also been head of the Yale University flying team, to which he was loaned by the Navy Department in anticipation of the World War I flying requirements of the country. Eddie had saved the life of one of his students, Truby Davidson, son of one of the senior partners of J.P. Morgan. With World War I over, J.P. Morgan put Eddie in as head of various corporations that had expanded during the war and might have continuing promise in peacetime, but were as yet question marks. Eddie gave me a job very high in a corporation he had just been made president of, not telling me he might be discontinuing it, which he did, and I was suddenly out of a job.

At that time, my father-in-law, Monroe Hewlett, had invented what seemed to be a useful and ingenious method of producing reinforced concrete buildings and I undertook to develop a company to produce the material and introduce it to the building world.

I did get four factories going to produce the Stockade materials, and succeeded in building 240 buildings in the large residential–small commercial field.

UNITED STATES PATENT OFFICE

James Monroe Hewlett and Richard Buckminster Fuller, of Lawrence, New York, assignors to Stockade Building System, Inc., a corporation of New York

BUILDING STRUCTURE

In the patent of James Monroe Hewlett, No. 1,604,097, dated October 19, 1926, there is claimed a wall structure involving the use of fibrous, substantially non-absorbent blocks. The present invention relates to a building structure preferably utilizing such blocks, and while the present structure is illustrated in said Hewlett patent it is not claimed therein, as the same constitutes the joint invention of the present applicant's rather than the sole invention of Hewlett.

According to the present invention, a wall is formed which comprises two distinct elements, one being a weight-carrying structure and the other a wall filling which also serves as the spacing medium for floor beams and the like which are supported by the weight-carrying structure. The filling medium is composed of blocks which preferably are fibrous, with large interstices between the fibers. Such blocks are substantially non-absorbent, but they have very little strength to resist crushing under strains; under a load of any substantial weight they would be readily compressed and therefore when we refer to these blocks as "compressible," we mean blocks which would be squashed or compressed under building loads. These blocks are formed with appropriate openings so that the weight-carrying structure of concrete can be cast within them. Also the blocks are of such a nature that they can readily be cut with an ordinary saw to permit the insertion of floor beams and the like into the side of the wall so that such load elements may rest upon the load-carrying structure.

The present invention can readily be understood by reference to the said Hewlett patent and to the accompanying drawings in which the figure is substantially similar to Fig. 1 of said Hewlett patent.

The method of producing the blocks here illustrated is fully described in the Hewlett patent, but it may be stated briefly that these blocks consist of fibrous material, such as excelsior, coated with plaster and loosely compacted into a mold so that the fibres will stick together but will form a block with relatively large interstices between the fibres. In the accompanying drawing, it will be noted that three forms of block are illustrated. The blocks B may, for example, be 16 inches long, 8 inches wide and 4 inches thick. These blocks have two openings through them to receive vertical concrete columns 7 as illustrated at the right hand side of the drawing where the blocks are shown in sections. These columns may, for example, be about 4 inches in diameter. It is to be understood that the appropriate reinforcing rods are used in these columns, as for example described in said Hewlett patent.

The second type of blocks designated as B′ are much thicker than Blocks B and have the openings for the vertical columns 7 and likewise have registering channels to

FIGURE 1

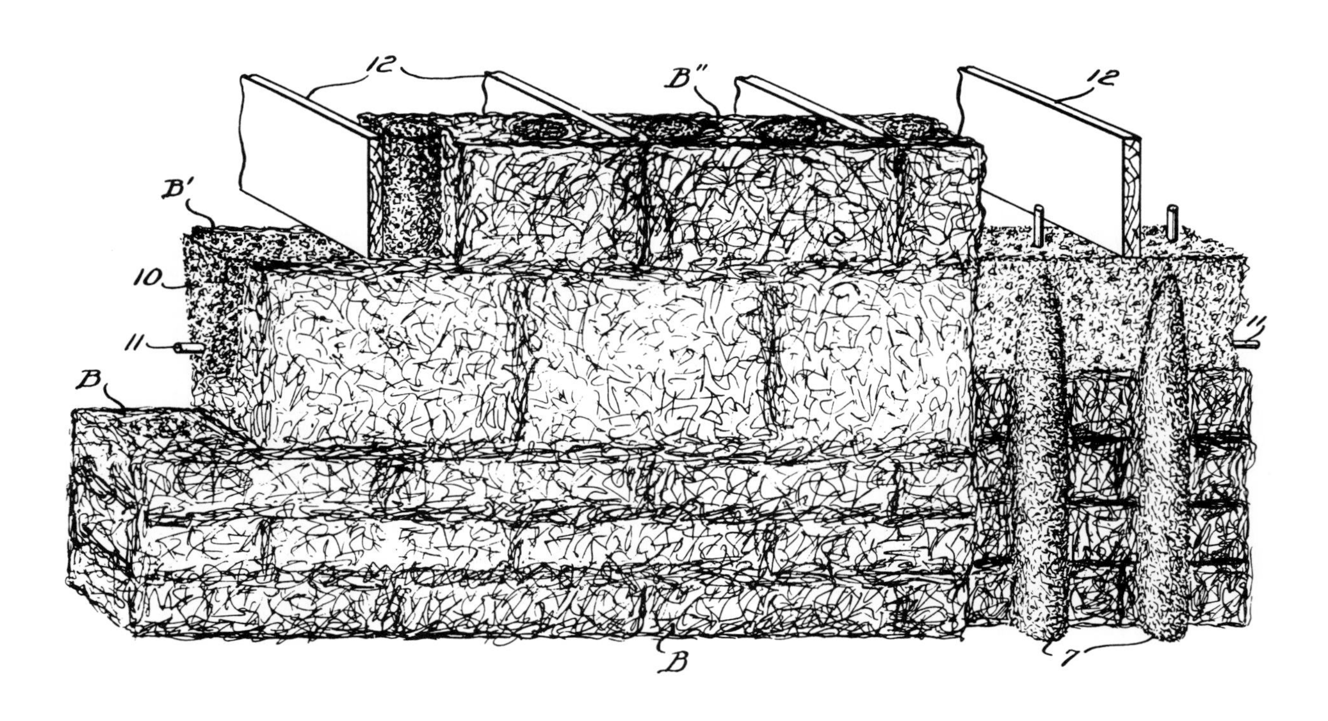

form a horizontal concrete beam or girth 10. The horizontal beams 10 are preferably provided with longitudinal tie-rods 11 and the reinforcements for the vertical columns 7 are preferably made to extend into the horizontal beams 10, thereby tying the two together. The beams 10 must receive all the load and distribute it to the columns so they must be strong, preferably having a depth considerably greater than the diameter of the columns 7. For example, they may be 8 inches deep.

In constructing a house or similar structure, the blocks B are made with the joints broken in every course. At each story, a course of blocks B′ is provided, so positioned that the upper face of these blocks will be immediately below the floor beams 12. After the vertical columns 7 and horizontal beams 10 are poured (preferably with tie-rods extending up to connect additional vertical columns 7, as is well understood in the art) a set of blocks B″ is placed on top of blocks B′. The blocks B″ are exactly like the blocks B except that they are thicker, preferably being made the same thickness as the depth of the floor beams 12. As the blocks B″ are put in place, notches are cut on the inside to receive the ends of the floor beams 12. By this arrangement the blocks B serve as spacing members for the floor beams, and make a very tight, air-proof packing around the ends of the floor beams. The wall is then continued up until the next story where the operation may be repeated. If desired, two or more courses of blocks B may be used in place of blocks B″ with their sides appropriately notched to receive the ends of floor beams 12. If the usual pitched roof is used, the top of the wall structure may end with the girth blocks B′ and the horizontal beams 10, the roof beams being carried by the beams 10 either directly or through the medium of a wooden nailing plate.

It will be understood that the girth blocks B′ may be used in other points, if desired; for example, if there is a large window opening, the girth blocks B′ may be placed directly over such opening, in which case the horizontal beam formed in these girth blocks will act as a lintel beam.

While this invention has been described as used with the fibrous blocks of the aforesaid Hewlett patent, it may be used with other types of block which are compressible and therefore not adapted to carry building loads.

2▲STOCKADE: PNEUMATIC FORMING PROCESS (1927)

U.S. PATENT—1,634,900

APPLICATION—DECEMBER 31, 1924

SERIAL NO.—758,991

PATENTED—JULY 5, 1927

THE MATERIAL NECESSARY TO PRODUCE the Stockade System had to be developed and the method of production discovered. This became one of my first tasks and I did invent much machinery in connection with the production. One was the pneumatic forming process. I had to develop a method of distributing the very expensive cementing material over the surface of wood excelsior so thinly there would not be wasteful clots, and of handling the wet, sticky excelsior in such a manner that it would be evenly felted together. The process is disclosed in the patent. While I did much inventing of technologies, the "biscuit-forming" process was the only one I felt was worth patenting.

UNITED STATES PATENT OFFICE

Richard B. Fuller, of Lawrence, New York, assignor to Stockade Building System, Inc., a corporation of New York

MOLD FOR BUILDING BLOCKS AND PROCESS OF MOLDING

The present invention relates primarily to a process and mold for making building blocks for use in wall structures such as that described in United States Patent No. 1,450,724, granted April 3, 1923, to James Monroe Hewlett. The particular type of block which is to be produced in these molds is now set forth in Hewlett application, Ser. No. 88,522, filed February 16, 1926. These blocks are made up of loosely compacted fibrous material impregnated with a binder. I have found that such blocks can be efficiently made and given a proper degree of compactness if the prepared fibre is carried by an air blast into a mold so constructed that the air current will give the fibres a light but relatively uniform pressure. For the manufacture of the blocks referred to a particularly suitable fibrous material is shredded wood fiber, known as excelsior, but other fibrous material, as straw may be used, the main consideration being that it is obtainable at a low cost. A suitable binder is a mixture of magnesium oxide and magnesium chloride, another is slaked lime, the qualities of which are improved by the addition of sugar.

In the accompanying drawings which form a part of this description.

Figure 1 is a general view showing the several instrumentalities involved in the means and method embodying this invention.

Fig. 2 is a perspective of the body of the mold.

Fig. 3 is a perspective of the bottom plate of the mold.

Fig. 4 is a perspective of the support for the cores and mold.

Excelsior is obtainable in compact bales. These are broken-up by hand and the material is fed to an ensilage cutter 10, which may be of standard design such as is used by farmers in cutting up fodder and delivering it into silos. The ensilage cutter comprises a feed belt, a revolving cutter for cutting the fibers into short lengths and a is forked over to ensure thorough distribution, or the binder in solution may be put in first and the excelsior projected into it by the air blast with such force that it is completely submerged and thoroughly moistened. In this

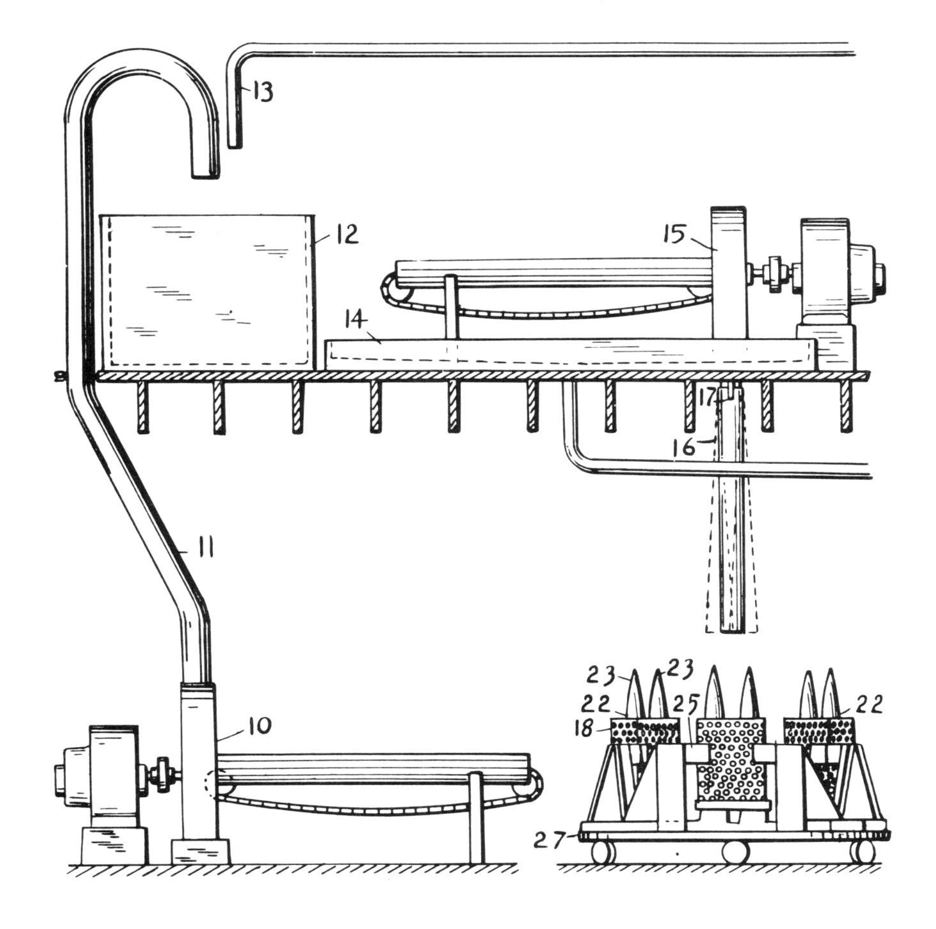

fan blower. The excelsior is here handled dry, and this ensilage cutter is preferably located where the excelsior is received, which is preferably in a building separate from the others of the plant to reduce the fire hazard to the others.

From this ensilage cutter, the broken-up excelsior is blown through a conduit 11 the mouth of which is over a bin 12. A sufficient amount of binder in solution is delivered onto the excelsior from a pipe 13, and the excelsior case it is forked out onto a tray 14 where the surplus solution is permitted to drain off. While sufficient of the binder is still adhering to the fiber it is fed through a second ensilage cutter 15 by which it is again broken up and delivered with a powerful blast of air downward through a vertical pipe 16. The pipe is pivoted at 17 so that the lower end can be swung slightly. This pipe is without bends, except for the very slight bend in swinging the lower end since the material is now in such a condition

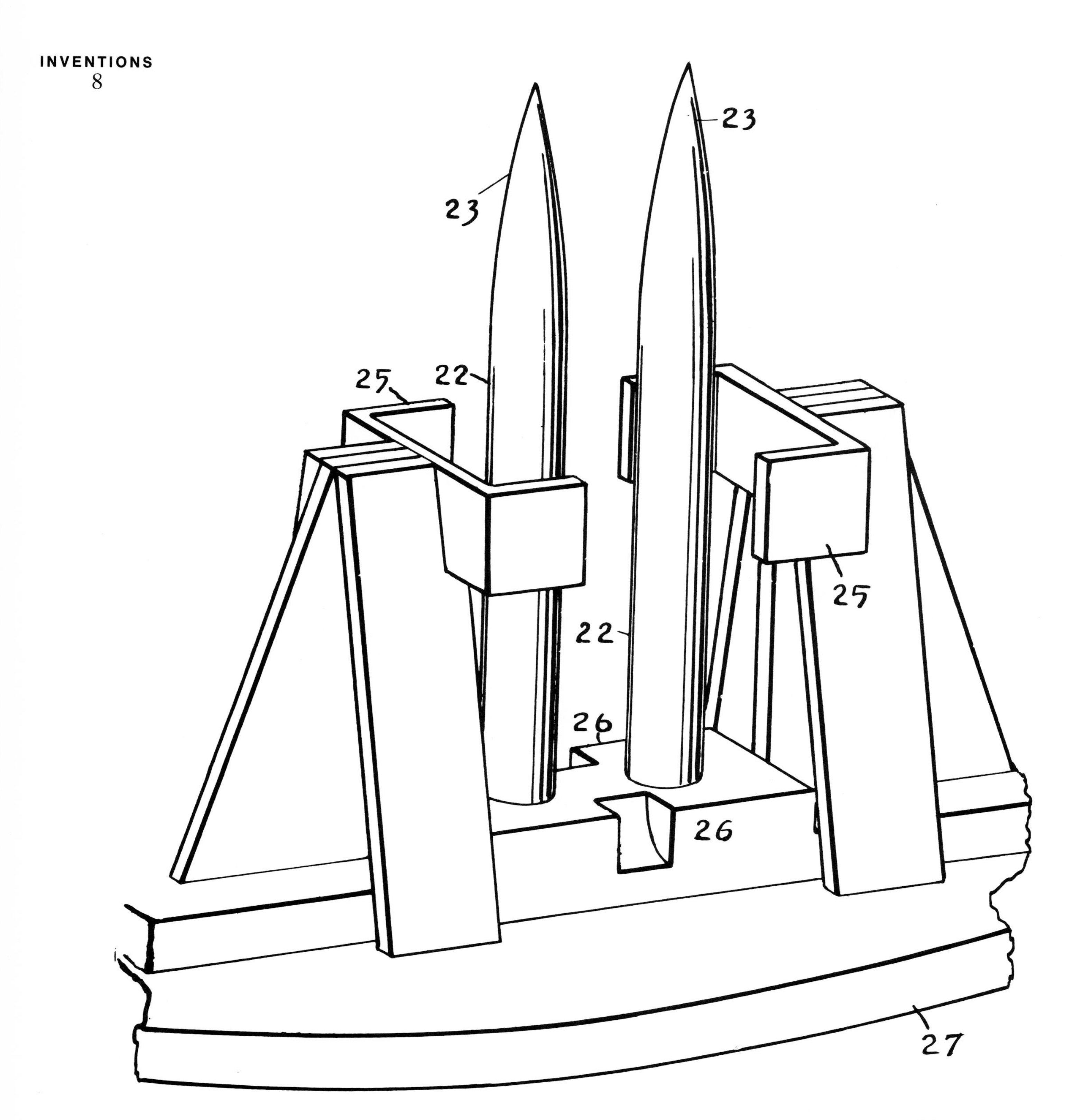

FIGURE 4

that it would tend to lodge at bends and clog the passageway. An open mold with a body portion 18 is located under the end of the pipe, and the pipe can be swung sufficiently to discharge into all parts of this mold. The body portion is formed of sheet iron of one-sixteenth of an inch gauge and is perforated with one-half inch holes which are sufficiently close together to remove about fifty percent of the sheet. The ends of the sheet are overlapped and held together by buttons 19, 19 which are carried by one of the ends and pass through eyes 20, 20 in the other end. The eyes are in the form of slots, each

FIGURE 2

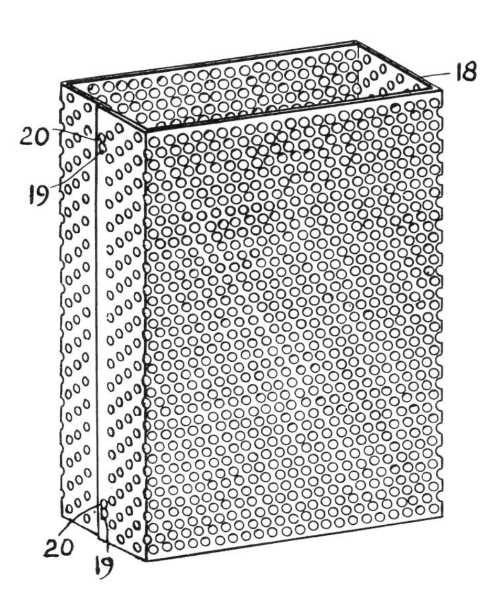

enlarged at one end so that by sliding the meeting edges endwise the buttons can be brought to the enlarged ends of the eyes and passed through to connect or disconnect the ends. A bottom plate 21 completes the exterior of the mold. This has flanges to engage the outside of the lower edges of the body portion and hold the body portion in shape. The bottom plate has two large circular holes through which pass upstanding cores 22, 22. These cores extend upward through the mold and terminate in streamline points 23, 23 above the mold. Suitable dimensions for the manufacture of building blocks are eight by sixteen inches in cross section and twenty-two inches in height. The molds are dipped in oil before use to prevent rust and the adhesion of the molded material.

A suitable support 24 rigidly carries the cores and has means in the form of guides 25, 25 to hold the upper end of the mold evenly spaced from the cores. The guides are not of sufficient extent to materially obstruct the perforations in the mold body. Proper positioning with respect to the cores at the lower end of the mold is ensured by the bottom plate. Notches 26, 26 in the base of the support are provided so that hooks from a crane can engage under the mold to lift it off when filled. The supports are mounted on a turn table 27 so that the molds can be quickly brought into place for filling.

When filling the mold, an attendant swings the delivery pipe so as to be sure that the material will be evenly distributed. Ordinarily, no tamping will be necessary; the material has already been broken up in such a way that there will be no solid masses and the air blast will pass through the upper portion of the material which is already in place and escape through the perforations in the sides of the mold, thereby compacting the fibres just enough so that they will stick together. The air current

FIGURE 3

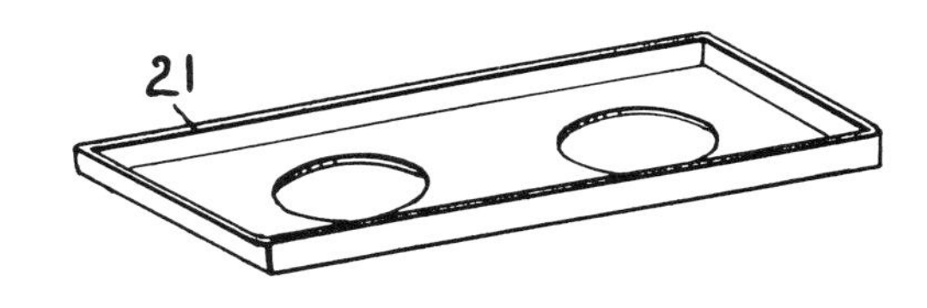

may also serve to dry the binder a little so that the fibres will remain in the position which they assume under the influence of the pressure of the flowing air stream.

In this connection it may be noted that owing to the stream-line shape of the cores 22, the air current will not be deflected from the sides of these cores, but the material directly adjacent the sides of the cores will be subjected to the action of the air stream and properly compacted. Of course if a mold is, in part, filled improperly, a small amount of tamping or manual arrangement of the material may be necessary. After the molds are filled, they are removed and the contents dried in any desired way as in a kiln. After the contents has dried, the molds are unhooked, opened and removed from the molded blocks, and each block from the mold is sawed apart to form five blocks, each four inches in height and eight by sixteen inches in horizontal dimensions.

3▲4D HOUSE (1928)

WHEN, AFTER GETTING UP 240 Stockade System houses in the Eastern half of the United States, I discovered that I could not make money by introducing so new a process into the building world, I became interested in how we might use aircraft technologies to produce dwelling machines for humanity. I was convinced by my experience in the building world that its activities were thousands of years behind the development of environment controlling for the sea and the sky. The 4D House was a consequence of my attempt to bring that advanced technology to bear on human environment controlling. I was producing tensionally cohered structures, and the 4D House was the result. My patent attorney felt that it would be much more convincing to the patent examiners if the same principles were employed but resulted in a rectilinear structure of conventional appearance. While the patent application embodied all the principles of the 4D House, it looked like a conventional house.

I knew very little about the world of patenting. In the case of my first two patents, the joint one with my father-in-law and mine for the manufacturing process, all the work was done by my patent attorney, who did not consult with me after the first disclosure. It is the formal procedure of attorneys dealing with the U.S. patent office to file applications for patents that first make a philosophical disclosure of the state of the art in which the invention is operative, then carefully describe the invention with accompanying drawings, then list a series of claims of what the inventor feels is the most economical statement of that which he feels is his unique invention. Then after the list of claims is filed, the patent office examiner sends back what is called the first rejection, rejecting a number of the claims but allowing one or two. The attorney and the inventor have the opportunity to make two more resubmissions of the stated claims, hoping to restate them acceptably to the examiners. The examiner indicates which of the claims might be restated in a final submission. In all, there are four such exchanges between the claiming inventor and the patent examiner. The patent attorney I had for the 4D House changed partnership and moved out of town. He did not tell me that the first rejection by the patent office was anything but a rejection. I did not know that subsequent resubmission of the patent was possible; I just assumed it was a final rejection and let it go. I knew I had a filing in the patent office of the invention and that it would serve as a first disclosure of my discovery from a scientific viewpoint.

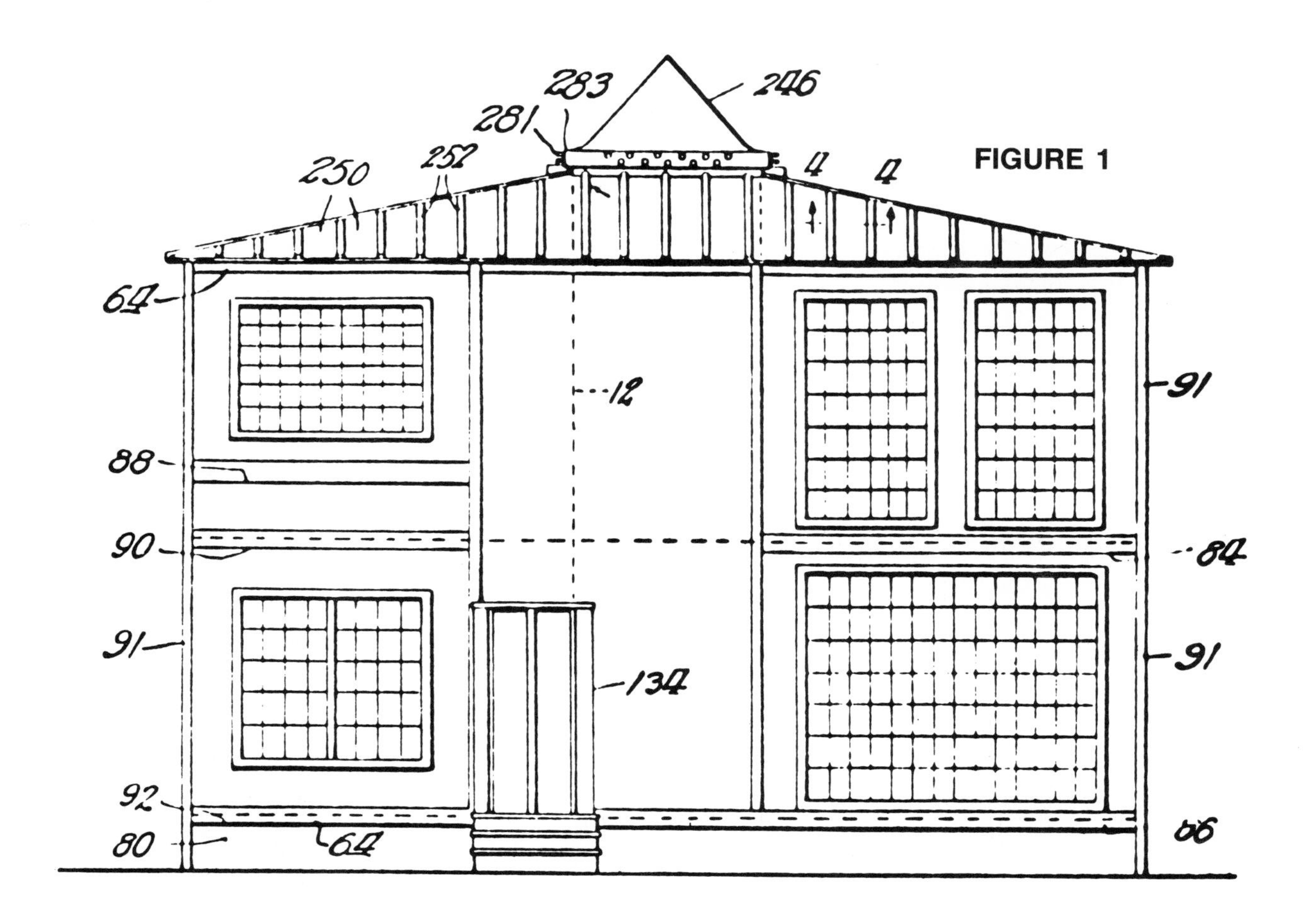

THE 4D HOUSE*

My invention relates to buildings and the erection thereof and includes among its objects and advantages the application of mass-production methods facilitated by changes in the building itself of such a nature as to make its completed parts capable of convenient transportation.

In the accompanying drawings:

Fig. 1 is a front elevation of a two-story house according to the invention.

Fig. 2 is a central sectional view to the same house.

Fig. 3 is a plan view of the framework supporting the second floor.

Fig. 4 is a detail section on line 4–4 of Fig. 1.

Fig. 5 is a floor plan of the first floor of the house.

Fig. 6 is a plan of the second floor of the house.

Fig. 7 is a detail section indicating one method of assembling a corner joint between the panels forming the outside wall.

Fig. 8 is a vertical section of the outer edge of the ceiling of second floor and roof.

Fig. 9 is a similar section of the outer edge of the floor of the second floor.

Fig. 10 is a similar section of the outer edge of the floor of the first floor.

Fig. 11 is a section on line 11–11 of Fig. 8, chiefly in plan view.

Fig. 12 is a section on line 12–12 of Fig. 10.

Fig. 13 is a vertical section of the partition forming the ceiling of the first story and the floor of the second story.

Fig. 14 is a vertical section of the central column in a plane at right angles to that of Fig. 2.

Fig. 15 is a section on line 15–15 of Fig. 13.

Fig. 16 is a side elevation.

Fig. 17 is a plan view of a unitary one-piece bathroom.

Fig. 18 is a section of a floor construction.

Fig. 19 is a section of a construction for an inside partition.

Fig. 20 is a section of a construction for an outside wall.

---

*Application abandoned by Buckminster Fuller, leaving prior art evidence in patent office files.

FIGURE 2

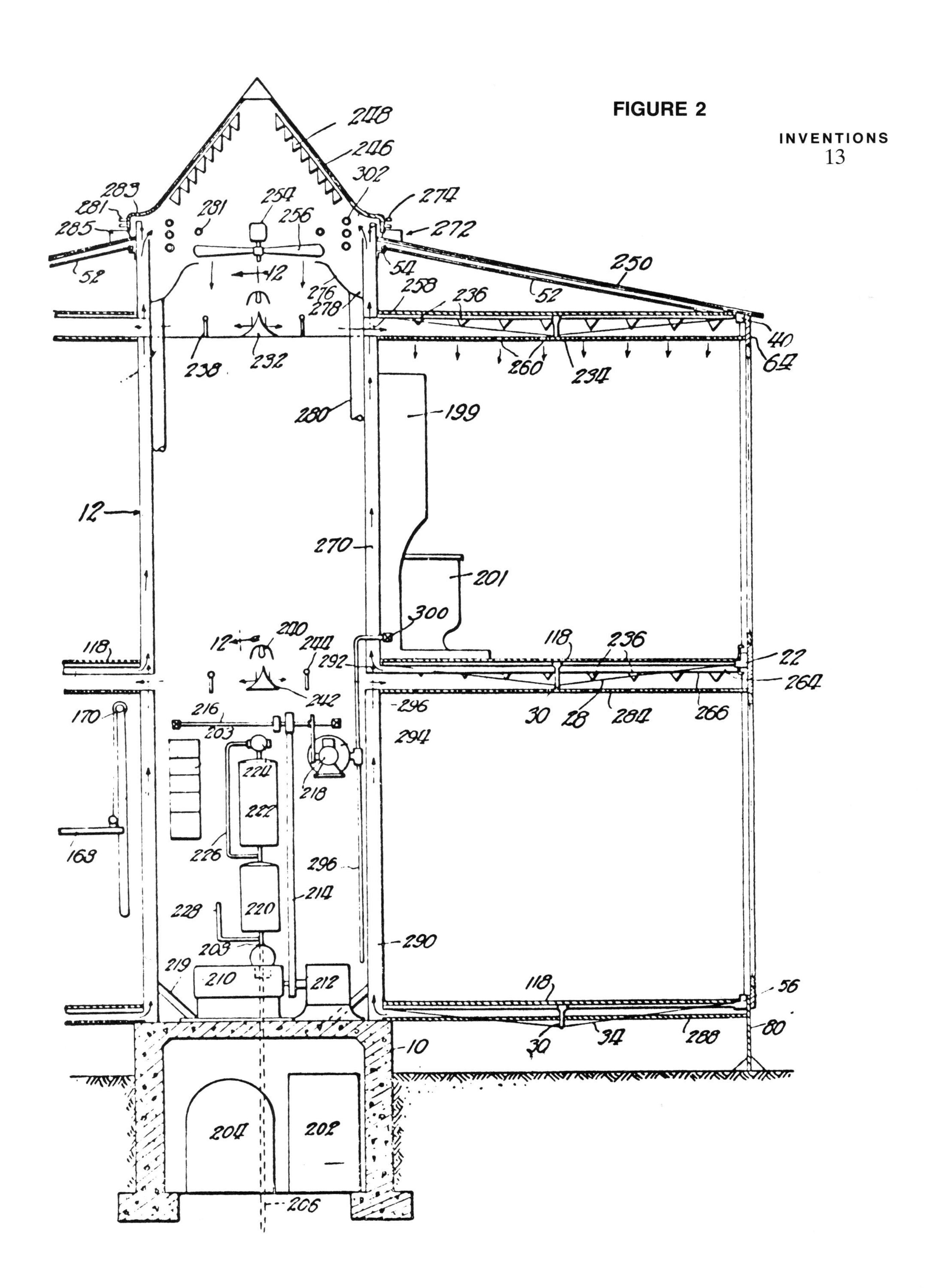

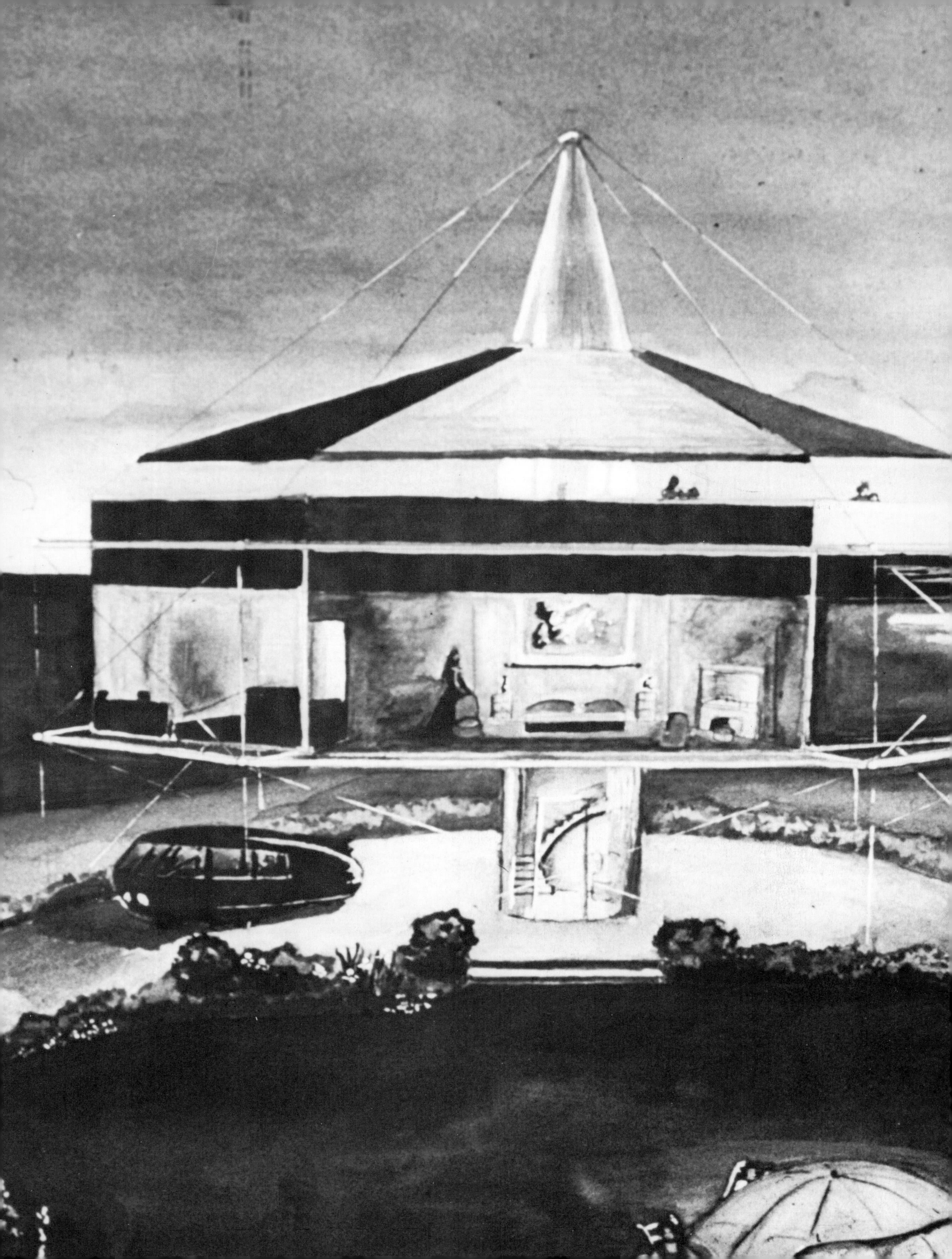

Fig. 21 is an enlarged section through the outside connection of the ventilating system.

Fig. 22 is a side elevation of an improved type of beam.

Fig. 23 is a section on line 23–23 of Fig. 22.

Fig. 24 is a section of a fitting for the end of a beam of Fig. 22.

Fig. 25 is a detail horizontal section through a vertical wall structure.

FOUNDATION

Referring now to Figs. 1 and 2 it will be seen that in the embodiment of a building according to the invention selected for illustration, the only foundation required is a central concrete housing or caisson 10 penetrating only a short distance below the surface of the ground and requiring a relatively negligible amount of excavation before pouring.

SUPPORT

This caisson is surmounted by a central mast indicated as a whole by the reference character 12. At the level of each floor or ceiling, this mast is the center of a plurality of radiating members in the form of load-carrying beams supported at both ends. Referring to Fig. 3, which shows the supporting framework for the second floor in plan view, the end plates 14 are each provided with five receiving sockets for the metal tubes 16 radiating to the ends, and the similar side plates 18 carry the sockets for four similar metal tubes 20 radiating to the sides of the house. The tubes 16 and 20 are braced and united at their outer ends by a peripheral rim 22 in the form of a tube having openings for receiving the tubes 16 and 20. These openings are elongated as indicated at 24 in Fig. 12 to accommodate tubes entering at various angles, and a short lip or lug 26 projects from the edge of the hole both above and below the tube 20 to provide a better bearing surface. Each tube is braced and stiffened by a tension rod 28 held in spaced relation at the center of the span by a strut 30. This tension member may be connected by suitable turnbuckles 32 to a U-shaped clevis 34 passing through a transverse bore in the end of tube 20.

The frame at the level of the ceiling for the second floor differs from that just described in that the tube 36 may have much thicker walls because it carries a heavier load. It also extends in to receive the tie-rod 38 in a vertical bore. (See Fig. 8.) The tie-rod 38 not only brings up the weight load from the first and second floors but anchors the rod 36 against endwise movement with respect to the peripheral rim 40. The rim 40 differs from the rim 22 in receiving the rods 36 on its horizontal center line, and in having a separate cast section 42 where it receives the rod 36 fastened in alignment with adjacent portions 44 by a snugly fitted inserted plug 46 at each end of the casting. The casting carries a short tubular portion 48 receiving the tube 36 and 50 on either side of this tubular portion, affording connections for relatively heavy tension rods 52 connected at 54 to the central column.

The frame at the level of the first floor may be identical with that at the level of the second floor except that the rim 56 is somewhat narrower in radial dimension to accommodate a different side wall construction. It is also vertically apertured where it receives each bar 20 to receive the lower end of a tie-rod 58 passing through the rim and also through a vertical bore in the end of the rod 20. The upper end of the tie-rod 58 is connected through to the lower end of the tie-rod 38 by a casting 60 having hooks 62 through which the ends of the tie-rods pass. It will be apparent that the weight load on all the beams at their inner travels directly down the column 12 to the caisson 10. A portion of the floor load from the first floor will be carried up through tie-rods 58, through the casting 60, through the tie-rods 38 to the end of the cantilever truss bound by beams 36 and tension rods 52. A portion of the floor load on the second floor will be carried by the casting 60 and up through the tie-rods 38 to the end of the truss formed by the tension members 52 and the beams 36. It will thus be apparent that the entire weight of the house and its contents may be supported by the central caisson 10.

short studs 72 along the upper edge of the rim 40. By tipping the panel outward from the position shown in Figs. 8, 9 and 10 the studs 72 may be readily slipped into the holes and the panel is then permitted to swing down to the position shown.

At the second-floor level the panel is held against outward movement by a simple tie-bolt 74 and wing nut 76, taking over an anchor 78 on the rim frame for the second floor.

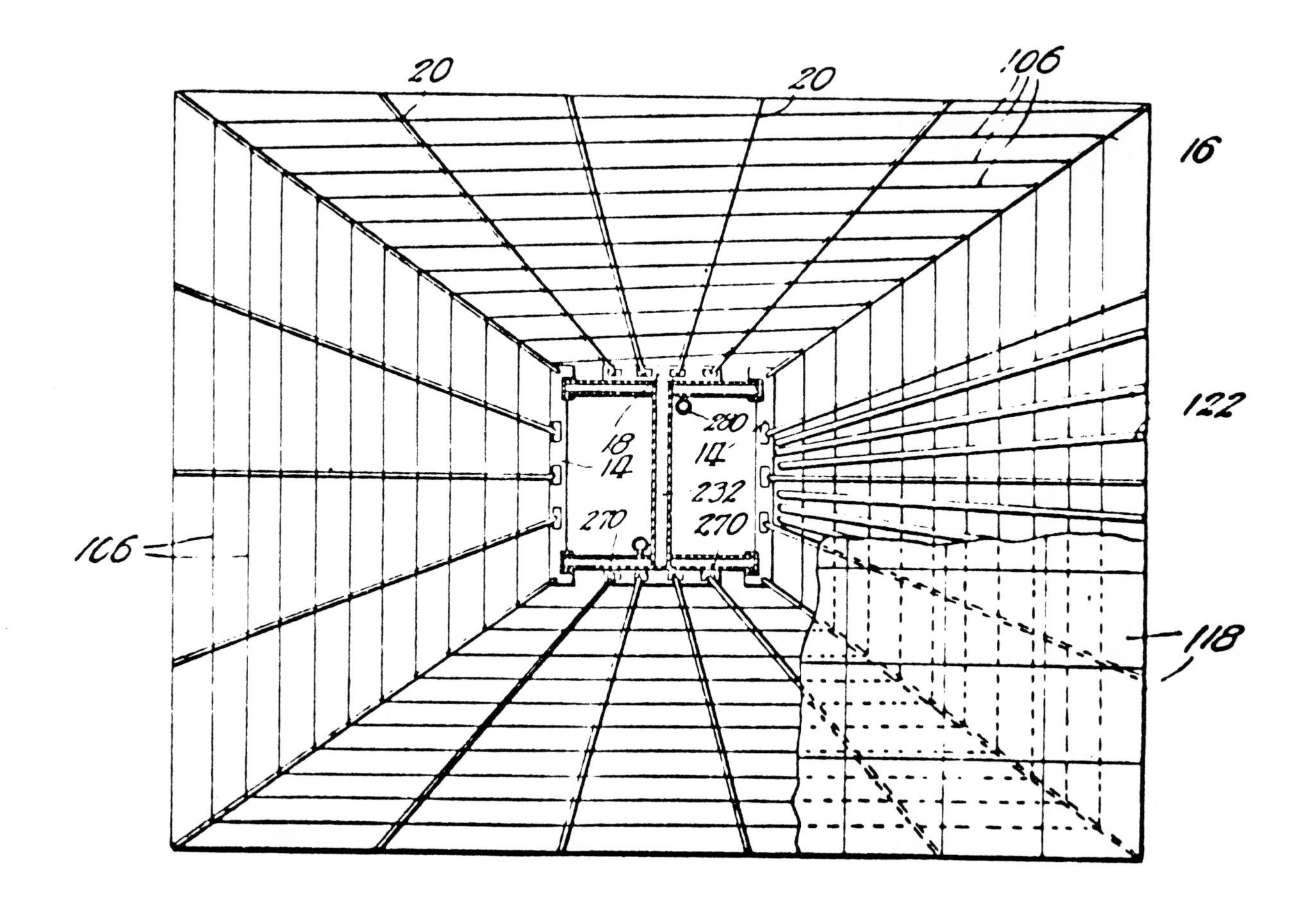

FIGURE 3

SIDE WALLS

The side walls illustrated in Figs. 1, 2, 8, 9 and 10 comprise flat thin panels each in the form of a sheet metal grill 64 with cross pieces about one foot apart and insulating and protecting panels 66 inserted in each opening. Where the panels are transparent this amounts to a window structure. It will be obvious that they may be made translucent or that they may be opaque and of a color identical with the metal frame so as to present a uniform appearance to the eye.

Figs. 8, 9 and 10 are sections in a vertical plane just to the right of the sliding double door in Fig. 1. Over this portion of the front wall of the house a single panel is mounted. The edge of the sheet-metal frame is crimped together at 68 and spot-welded to an edging 70 having a plurality of apertures to enable it to take over a set of

At the level of the first floor I first mount an additional panel 80 on the sill 56 in precisely the same way as the main panel is mounted on the sill 40. The panel 80 extends downward a few feet and its edge may enter the ground for appearance's sake and to keep animals out of

FIGURE 4

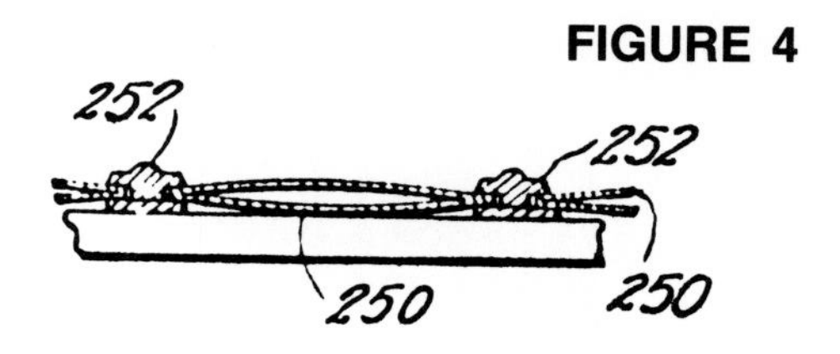

FIGURE 5

FIGURE 6

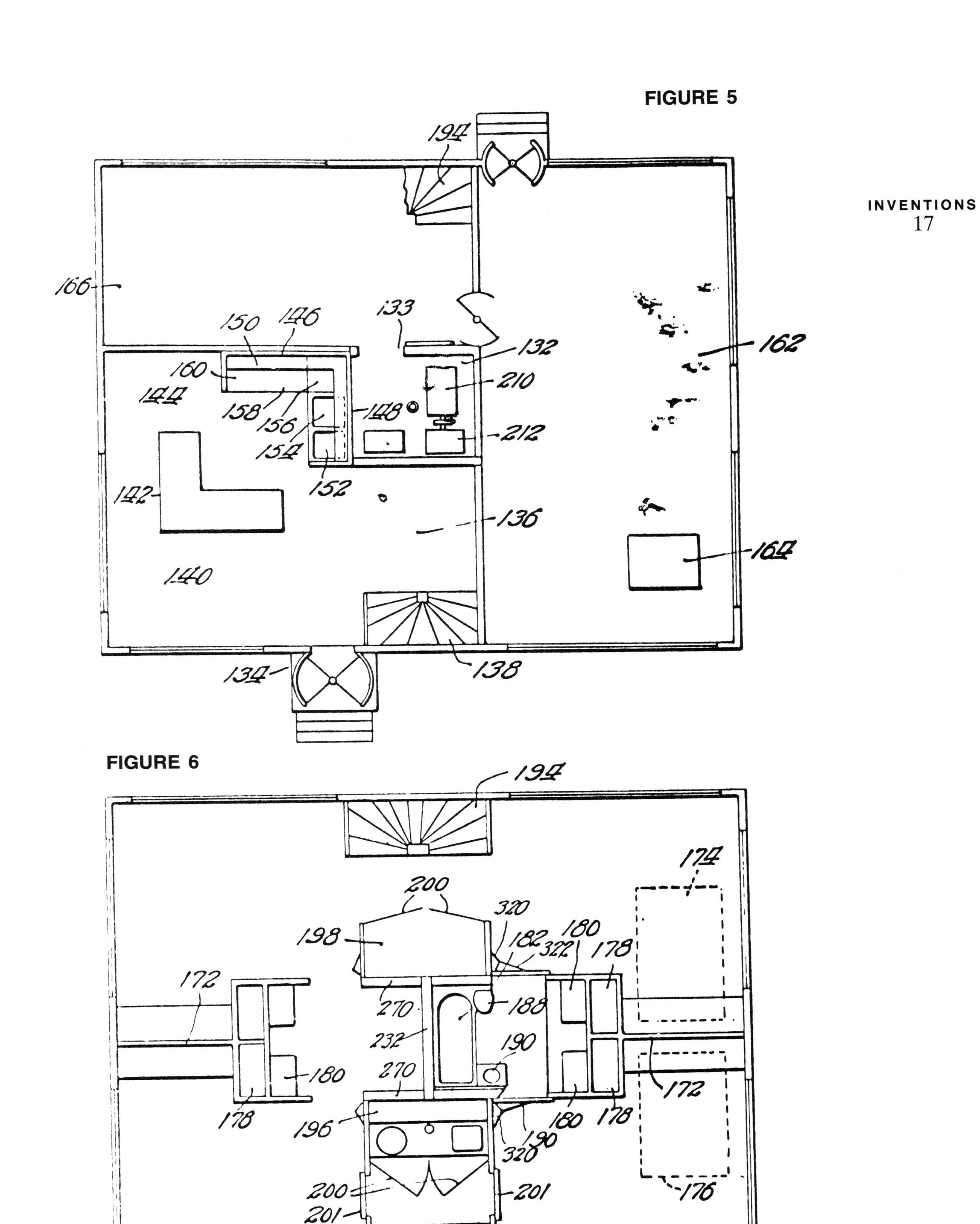

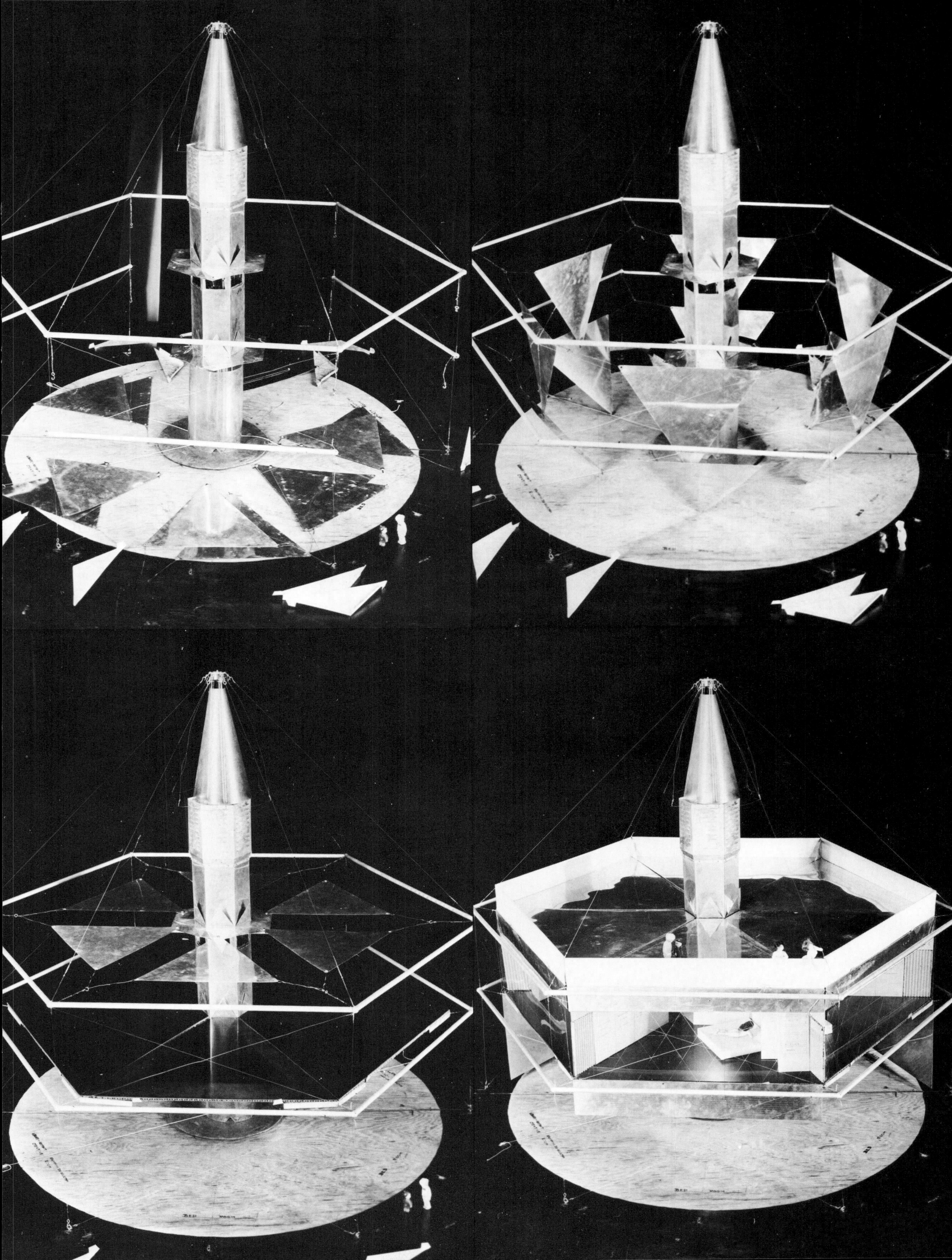

the space between the first floor and the ground. It carries no weight. The lower edge of the main panel is connected to the sill 56 by additional tie-bolts 74 precisely as at the second floor. Between each panel and the members against which it rests, I interpose a packing. I have

FIGURE 7

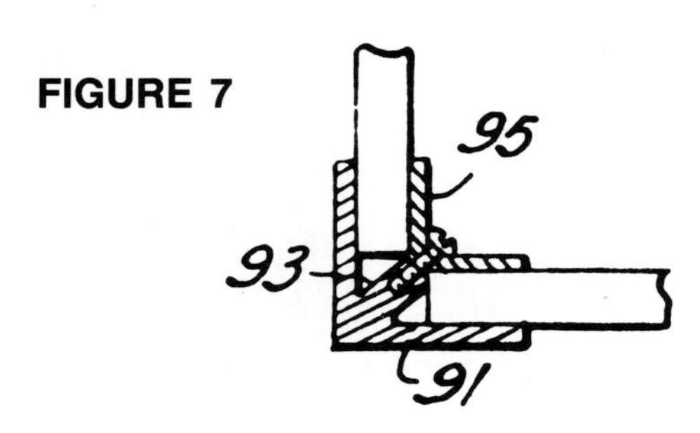

illustrated flat strips of pneumatic tubing 82 laid in place between the sill 40 and the upper edge of the main panel, between the sill 22 and the middle portion of the main panel, between the sill 56 and the upper edge of panel 80, and between the upper edge of panel 80 and the lower edge of the main panel. After the parts are in place, inflation of the packing will clamp everything firmly together.

At the right side of the house illustrated in Fig. 1, I show a second-story panel having its lower edge at 84 and a different first-story panel having its lower edge at 86. In this portion of the house the joint beside the sill 22 will be precisely analogous to the joint shown beside the sill 56 in Fig. 10.

At the left side of the house shown in Fig. 1, I have illustrated a separate window panel hanging from the sill 40 and having its lower edge at 88. The next unit is a short panel having its lower edge at 90 and below that is the panel for the dining room or grille having its lower edge at 92. Such a construction permits complete removal of the panel terminating at 88 in warm weather, leaving this upstairs room open to the air and enclosed by a railing made up of the panel with its lower edge at 90.

The adjacent edges of the panels may be simply battened together. In Fig. 17 I have illustrated a corner batten comprising an outer strip 91 with a central rib 93 to receive the fastening screws for the inner strip 95.

The side wall unit illustrated in Fig. 20 has panels 94 differing from the panels 66 in having their edges concave as at 96. Each pane or panel 94 is enclosed in a fabric edging drawn in around wire hoops 98 on each side and then out again, much as a piece of fabric for embroidery is drawn in a stretching frame. The free edges of this fabric edging 100 may be used to tie each pane to the adja-

FIGURE 8

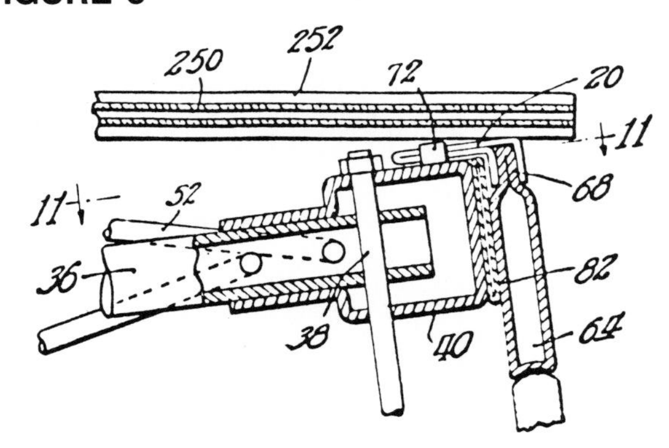

FIGURE 9

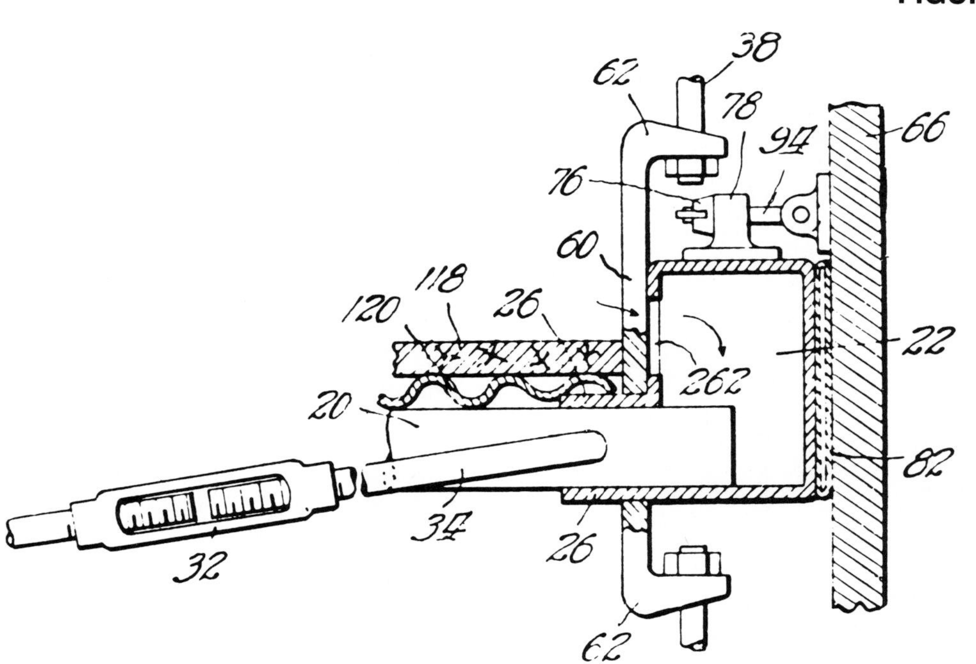

cent panes by stitching the edges of the edging strips together as at 102. Before this is done pneumatic tubing 104 is laid between the opposing edges of the panes and connected to a suitable inflation valve. After a plurality of such panes have been united to form a large panel, the introduction of air under pressure into the tubing 104 will brace and stiffen the structure into a semi-rigid panel which can be mounted and fastened in place in the same way as the panel illustrated in Figs. 8, 9 and 10.

FLOORING

To build up a floor suitable for use I first run a mesh of tension wires 106 (see Fig. 3) over the load-carrying beams 16 and 20. These wires may be made up of a single strand arranged in a spiral. Above these wires (see Fig. 18) I stretch a sheet of canvas or heavy tarpaulin 108. A plurality of pneumatic floor mats are provided with a module of three feet in plan view. Thus an ordinary mat will be three feet square in plan view, although units six feet square or three feet by six or nine feet may readily be employed. The pneumatic mats illustrated are about one inch thick and hold to a flat shape by tension cords 110 suitably embedded in the rubber lining 112. The whole rubber unit is encased in a fabric sheet 114 of heavy canvas.

To build up the floor, the entire surface of the tarpaulin 108 is covered with mats and suitable rugs or matting 116 are laid on top of the pneumatic mats. This provides a floor surface ready to walk upon. In the construction of Fig. 15, the matting 118 is made up in units three feet square and laid over a corrugated metal panel 120 of the same size to make up a unit panel of flooring. The wires 106 are bowed to a catenary curve by laying strips 122 between the wires and the panels of flooring, and the positioning of the panels on top of the wires and strips completes the floor.

INSIDE PARTITIONS

The inside partitions between the rooms may be of construction identical with the panels of Fig. 8 or Fig. 20. I have illustrated a pneumatic partition made up of a plurality of rubber tubes 124 encased in a fabric sheathing 126 to form a unit not unlike the body protector worn by a baseball catcher. A flat and ornamental tapestry surface may be given to such a partition by enclosing the entire unit in a fabric bag having walls 128 of tapestry of such other material as may be desired for a proper decorative effect within the room. Obviously, where the partition subdivides two rooms of different kinds, the surfacing 128 may be of different sorts on opposite sides of the bag. Thus one side might be a light blue tapestry suitable for a bedroom, and the other side might be a white waterproof oilcloth or linoleum suitable for a bathroom wall. The sides of the bag may be brought together at the top and stitched to form a welt 130 by means of which the wall may be suspended in place.

In any surface I may secure a snug assembly by the instrumentalities illustrated in Fig. 25. The tightening

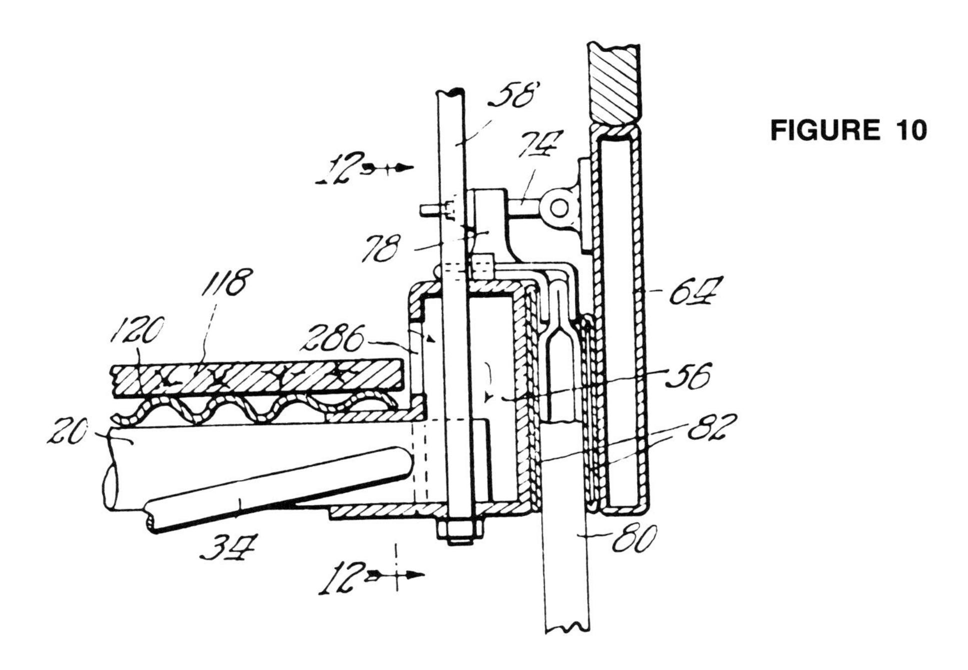

FIGURE 10

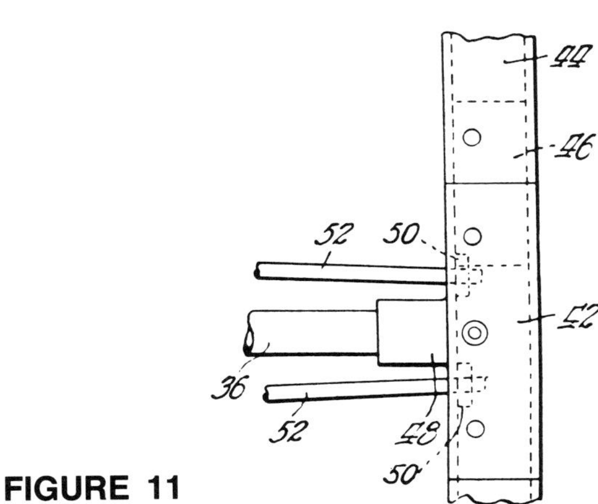

FIGURE 11

FIGURE 13

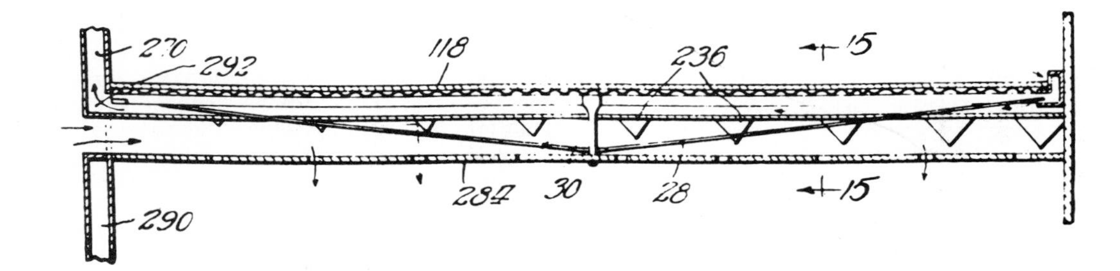

means comprises a pneumatic bladder 135 in a flexible protecting casing having the flat shape indicated in dotted lines before inflation. One or more of the vertical joints between adjacent panels 137 may have a tightener slipped in place. After the panels are set in place, the inflation of the tighteners will set up a snug gripping engagement at all the joints. By varying the inflation the dimensions of the joint may be varied. In this way, by using several tighteners standard units may be set into spaces just large enough to receive them, or expanded to fit properly in larger spaces.

FIGURE 12

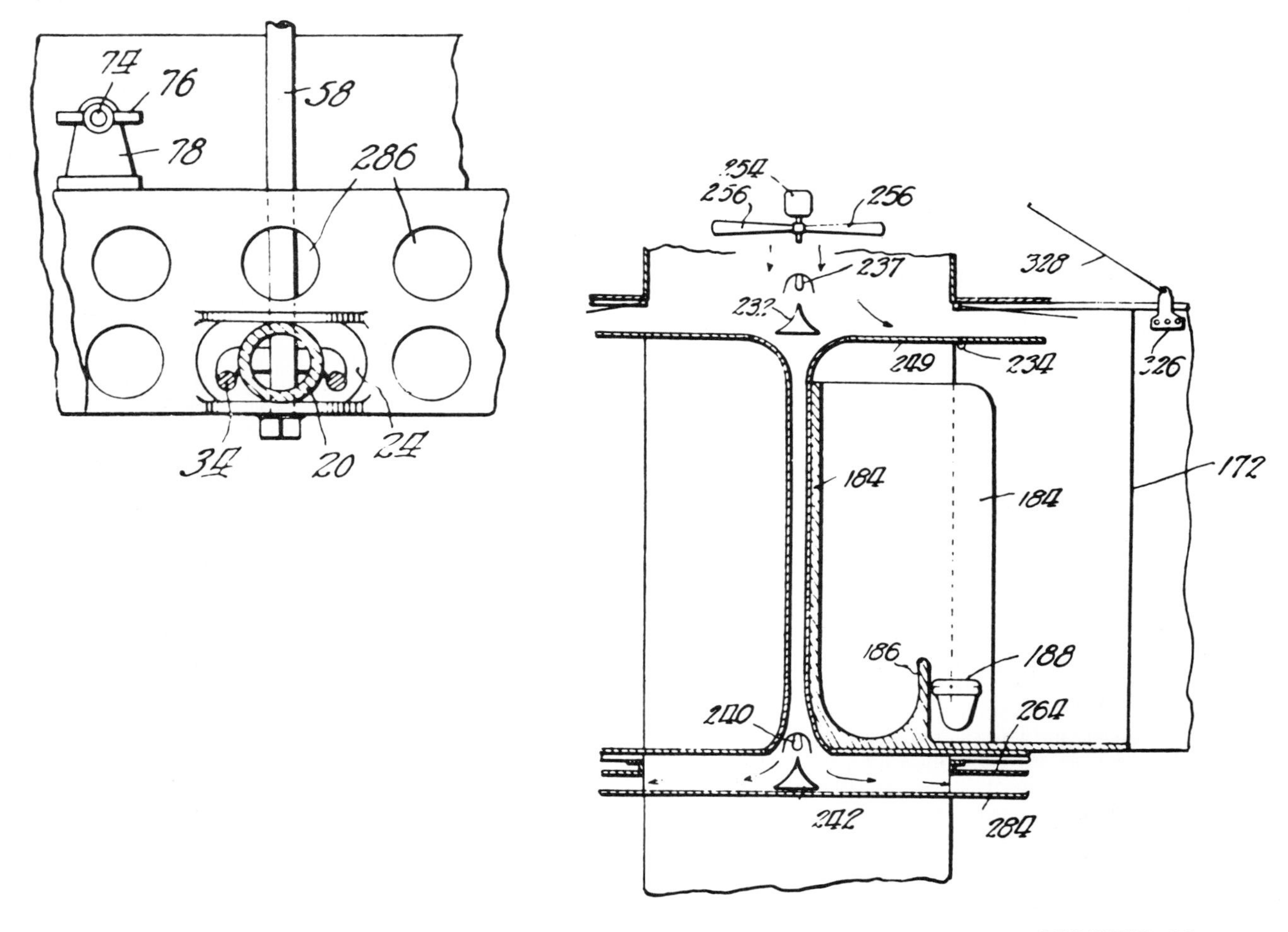

FIGURE 14

FIRST-FLOOR PLAN

Referring now to Fig. 5, I have indicated a central storage space for the power plant and service units at 132 on the first-floor level inside the column 12. Access to this space is from the garage through a door at 133. The revolving front 134 may open into a space at 136 serving as a hallway and means of access to the front stairway 138. This area may or may not be partitioned off from the grille located at 140. For the grille I have indicated an L-shaped table 142 and a unitary kitchen unit 144.

FIGURE 15

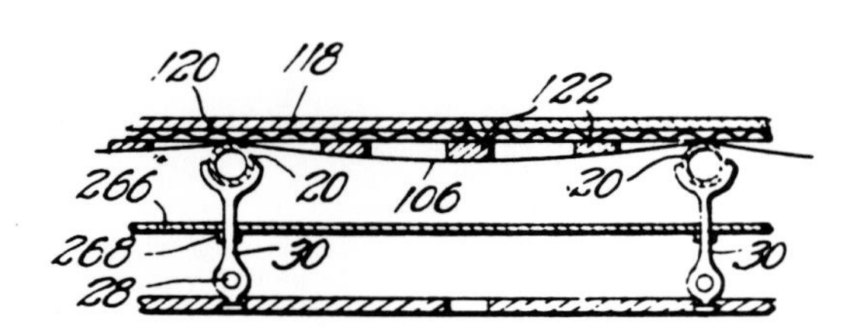

The kitchen unit may be assembled at a central point or factory, complete and ready for use except for connecting to local gas and water pipes, etc. It comprises a floor panel, two modules on each side, with vertical supporting surfaces at 146 and 148, which extend to the ceiling and carry the upper storage cabinet indicated at 150. At customary levels all the units necessary for complete kitchen or grille service are built into place. These may include an oven 152, stove unit 154, dishwashing machine 156, worktable 158 and sideboard 160.

FIGURE 16

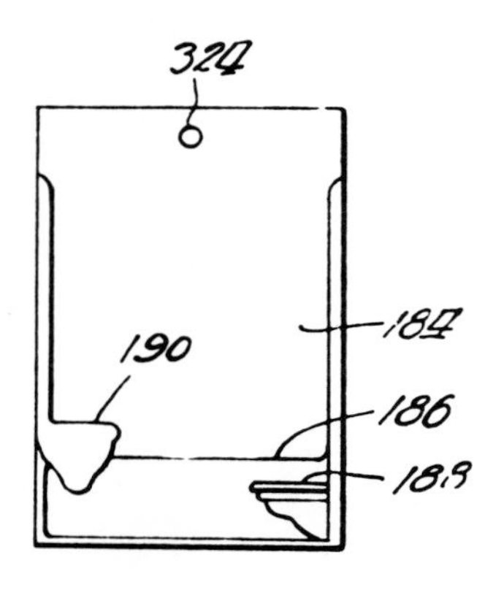

I have indicated a living room 162 extending all across one end of the house. In addition to the usual pieces of furniture, this room may be provided with a combined unit 164 including a radio receiving set, a motion picture machine, a Dictaphone, steel safe for valuables, a phonograph, a television receiving set, and any other desired conveniences. I have indicated a garage space 166 where a car may be stored. The vertical distance between floors is such that additional storage space for trunks and the like may be made available by having a false floor 168 (see Fig. 2) movable vertically by means of hoists 170 in the garage. This can be lowered to provide easy and convenient access to trunks and the like supported thereby, and then raised to a sufficient height to afford ample clearance for the storage of a car beneath the same.

SECOND-FLOOR PLAN

Referring now to Fig. 6, I have illustrated a second floor that may be made up to have four bedrooms and two baths. From an appropriately positioned load-carrying beam 36 I suspend a rigid partition 172. This partition forms part of a complete unit assembled as such at the

FIGURE 17

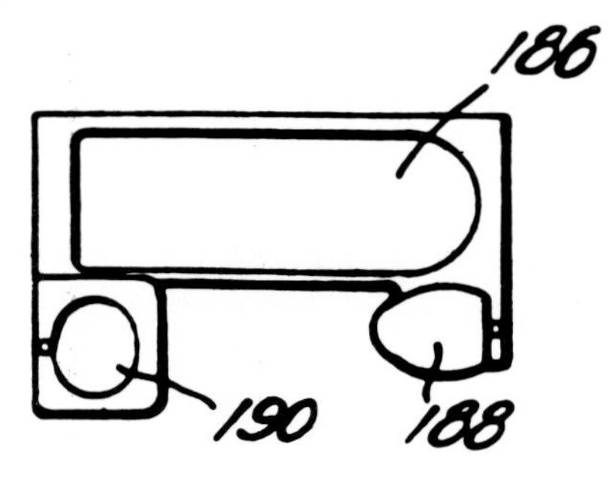

central point or factory. The unit includes two beds 174 and 176 foldable into a vertical position out of the way or out into the positions indicated in Fig. 6 for use. The partition also supports the wardrobes 178 and other suitable units 180, such as dressing tables facing the bathroom at 182. It will be noted that above the second-floor level the central mast changes from a box section to an H section providing the space at 182 inside the H into which may be set the bathroom unit indicated in Figs. 16 and 17. This is a one-piece unit including a side wall 184 extending to the ceiling level, a bathtub 186 beside the side wall, a toilet seat 188, and the lavatory 190. All this is preferably assembled with a special flooring unit extending out to the position indicated in dotted lines at 190 in Fig. 6 and sent out from the factory ready to be set in place, and when set in place ready for use except for connecting up piping. The space between the corner bedrooms at the front and rear of the house is occupied by the front stairway 138 and the rear stairway 194, both of which are semi-circular, and by utility units located at 196 and 198 with a hall between to afford means of connection between the bedrooms. Where a single family occupies the entire house, one of these units may be a laundry unit comprising a drier 199 (see Fig. 2) and complete washing and ironing equipment at 201, and the other might be a miniature carpenter shop or radio laboratory for the children. Where two families occupy one house both units may be laundry units. In connection with a laundry unit in such a location, I prefer to employ overhead rolling doors 200 in front of the utility unit and sliding doors 201 at each end of the passage across in front of the rolling door. These doors may be arranged so that when they are both thrown open the laundry unit and the space in front of it constitute a separate enclosure shut off from the bedrooms on either side.

POWER AND SERVICE

The central mast 12 is of a size that can readily be shipped and transported on a railroad car or a motor truck. It has built into it at the central point or factory practically all the equipment indicated in Fig. 2 except for some minor items such as the reflecting surfaces for the lighting system and the liquid for the storage batteries 203, which might interfere with handling during shipment. The preparation of the terrain involves only a very little leveling and the installments of the caisson 10 with a fuel oil tank 202, a septic tank 204, and a pipe connection at 206 running down to a well or to the city water supply. The mast has built into it the upper section 208 and the piping and tanks associated therewith. I have also indicated a diesel engine 210 direct-connected to an electric generator 212 and connected by a belt 214 to a countershaft 216 from which the air compressor 218 and any other desired power machinery equipment may be driven. This unit is fastened into the mast by braces 219, and shipped out with the mast.

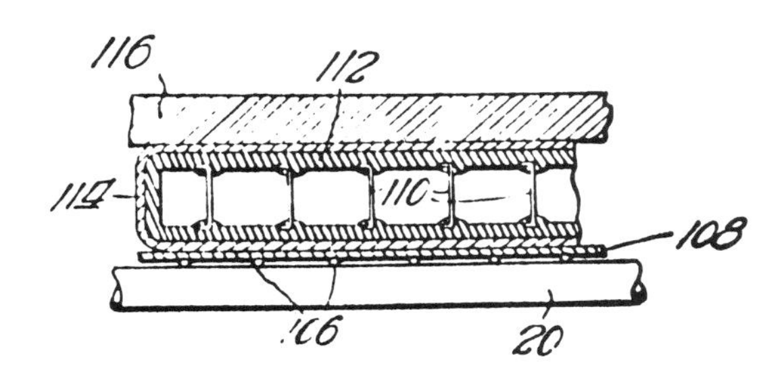

FIGURE 18

WATER

The single water inlet at 208 may be connected through at pump 219 and a water softener 220 and a heater 222 to a mixing valve 224 provided with a bypass 226 around the heater. The mixing valve may readily be connected to the bathroom unit to deliver water at any desired temperature for the bathtub and lavatory. A side tap at 228 may provide unsoftened water for drinking, air washing, and the toilets.

LIGHT

It will be noted at the outset that the roof and side walls disclosed may be constructed to admit many times the light now available in an ordinary dwelling. The artificial lighting system indicated includes a powerful central light 230 level with the ceiling of the second story just above a conical reflecting surface 232 shaped to reflect light horizontally just above the ceiling panels 234 of the second story. These ceiling panels are translucent or semi-transparent and above them I mount a series of reflectors 236, each reflector extending down a trifle lower than the one next it and nearer the source of light. In this way the flood of light emanating from the reflector 232 is distributed over the entire ceiling panel and sheds a diffusing light through the rooms below. Shutters 238 have been indicated for cutting off the light to any portion of the second floor, so that each room has its own selective light control. These shutters may be combined with colored glasses, making it possible to change the color and intensity of the illumination at will.

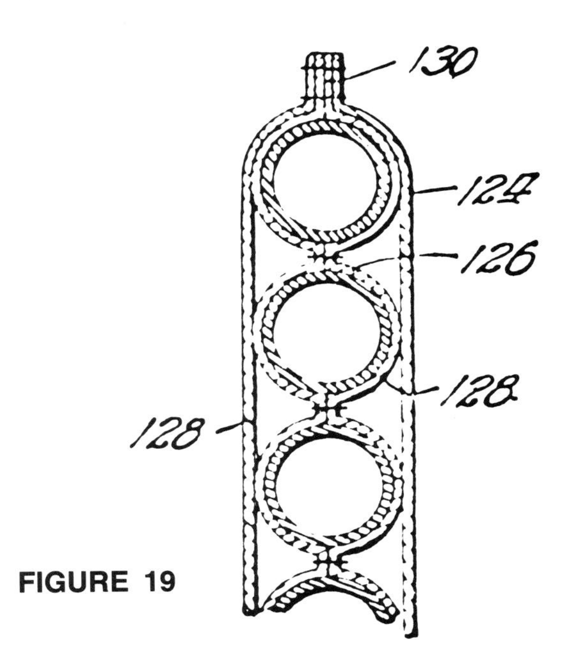

FIGURE 19

A similar central lighting unit including the light source 240, reflector 242 and control-shutters 244 is provided at the level of the floor for the second story and the light is distributed by reflectors 236 which may be identical in construction and function with those for the upper story.

The top of the central mast is surmounted by a cone 246 provided with a serrated lens structure 248 for focusing sunlight in the space immediately above the reflector 232 where a considerable amount of the same will be re-

FIGURE 20

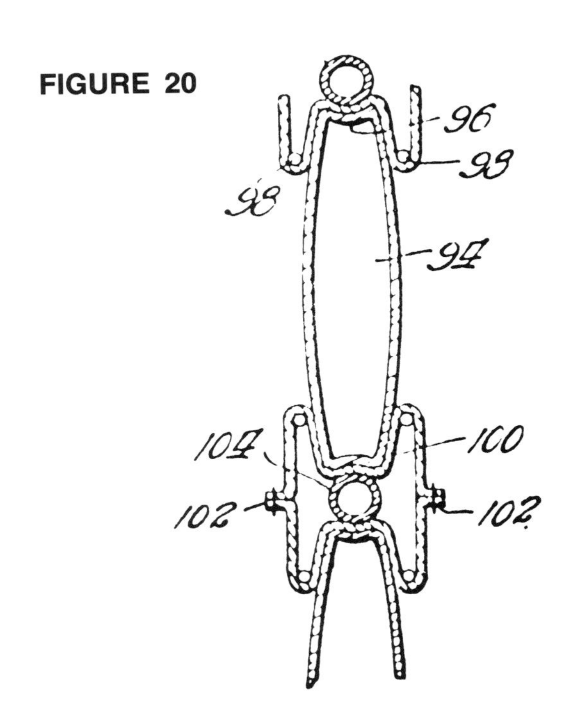

flected out against the reflectors 236. A shutter 249 in the ceiling of the bathroom may be removed to permit a flood of almost direct sunlight to shine directly in. The roof may be built of transparent vacuum panels 250 supported by battens 252 so that a flood of sunlight may pass

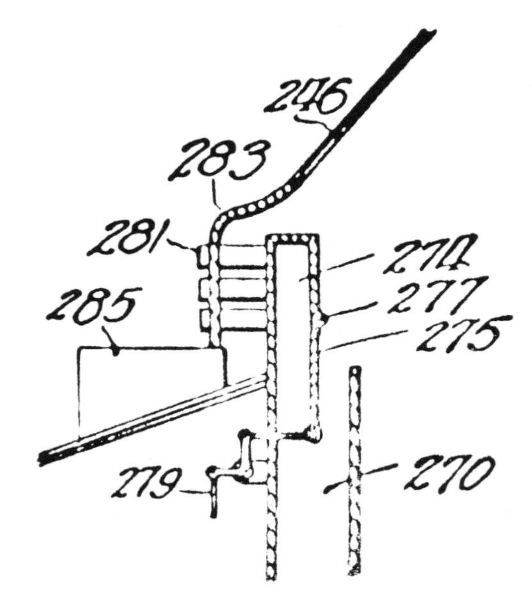

FIGURE 21

through and illuminate the translucent ceiling 234 of the entire upper story.

VENTILATION AND CLEANING

The house is provided with power means, indicated as an electric motor 254 driving a fan 256, for generating throughout the house.

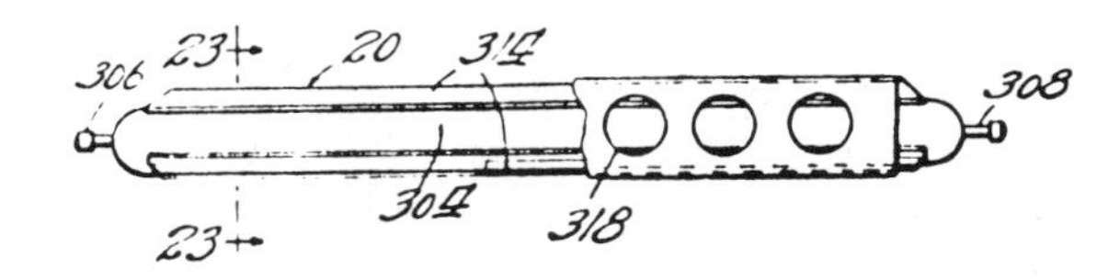

FIGURE 22

The circulation for the upper story is from the fan downward to the level of the lighting units 237 and 232 (which will be cooled by the circulation of air over them and incidentally afford an appreciable amount of heat for the house) and then out through louvers 258 into the space above the second-floor ceiling 234. From here it passes downwardly through a large number of small apertures indicated at 260 in the form of a descending bank or

slowly moving mass to the level of the floor. From the floor level it is withdrawn through the apertures 262 in the sill 22 (see Fig. 9) and passes back at 264 through a conduit defined by the flooring for the second floor and by a horizontal baffle or partition 266 separating the space between the ceiling for the first story and the floor for the second into two levels. The baffle or partition 266 may readily be held at the proper level by connections at 268 (see Fig. 15) with the struts 30. The air returning under the floor passes up at 270 through the double walls of the mast to the extreme top at 272, where it is deflected inwardly and returned to the suction side of the fan 256.

Additional air from outside may be mixed with the air circulating in the building at this point. Referring to Fig. 21, I have illustrated a box 274 mounted at the outer edge of the upper end of the passage 270. The lower edge of the box may be in the form of a flap 275 manually adjustable about the hinge 277 by control connections 279. Horizontal tubes 281 extend outward from the box 274 to discharge air into the outer atmosphere. The cone 246 is continued over the box 274 in the form of a shelter plate 283 through which the tubes 281 pass. The air ejected through the tubes 281 may be replaced by an influx through a suitable chemical air gas filter 285. This filter may operate under ordinary conditions to eliminate traces of sulfur dioxide, odors from stockyards, or other undesirable contamination in the air. It may also be constructed so as to be easily changed over to function as a protection against poison gas. The air entering through the filter passes upward around the tubes 281 and between the box 274 and the cone 246 to the exhaust side of the fan.

The rotation produced by the fan 256 will throw such solid particles as may be brought up by the returning airstream to the outside of the baffle or dust guard 276 to accumulate in a pile at 278, from which the accumulation may be removed by flushing at infrequent intervals through the outlets 280. In winter weather when the air is dry it may be moistened by the spray 281, which also washes down the accumulated dust.

The circulation for the lower story is identical in principle with that just described for the upper. The air from the pressure side of the fan 256 passes down the central portion of the H as indicated at 282 in Fig. 14 and then horizontally out between the baffle 264 and the ceiling 284 for the lower story. It leaves the rooms through the apertures 286 in the sill 56 and passes back under the flooring and above the lower baffle or cover 288 and upward at 290 to join the air returning from the second story at 292 (see Fig. 2). The lower cover 288 should have the same thermal insulation properties as the side wall panels.

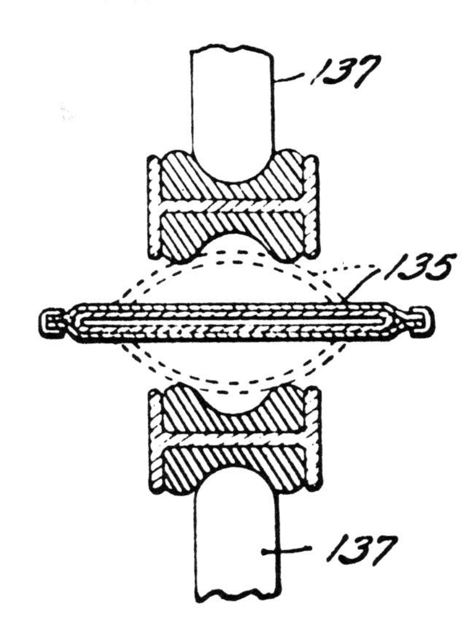

FIGURE 25

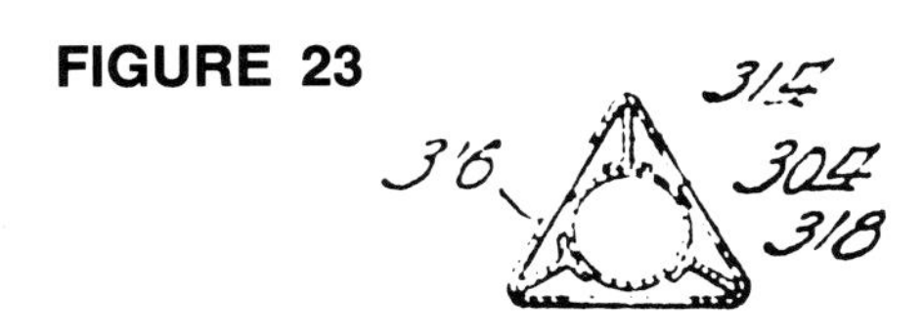

FIGURE 23

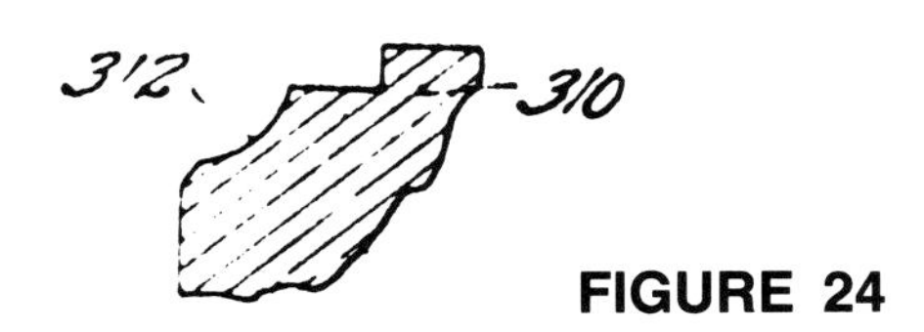

FIGURE 24

It will be apparent that the floors of both the first and second stories are subject to a continuous removal of dust and dirt by the air passing out through the apertures 262 and 286. For cleaning I prefer to employ a pressure system. Referring to Fig. 2, it will be noted that the air tank 294 has been connected by piping 296 to an outlet 298 on the first floor and a similar outlet 300 on the second. These terminals may be quick detachable coupling units such as those employed by the Alemite and other systems for dispensing fluids, and the housewife may clean the floor or furniture simply by coupling a length of tubing to the nozzle of the air pressure connection and blowing all the dust and dirt on the floor through the apertures 262 and 286 to be carried up and deposited in the dust accumulators 276.

In cold weather it will be found most advantageous to add the necessary heat to the air circulation indicated at the top on the suction side of the fan 256. The lights at 238 and 240 may supply all or nearly all the heat necessary even in ordinary winter weather. I have indicated an auxiliary heating means in the form of a coil of piping 302 adapted to be connected with the exhaust pipe of the diesel engine 210, or with any suitable steam generator or the like.

In a dwelling intended for quick and easy transportation and erection and subsequent moving from place to place with relative great facility, the weight of the structure itself becomes a significant item. Referring to Fig. 2, I have illustrated a beam for use as one of the beams 16 and 20 of Fig. 3 or as part of the central mast in case the mast is a built-up structure. The beam comprises a tube 304 of relatively large diameter so as to get the metal as far as possible from the neutral axis. To avoid too large a cross section of metal, it may be necessary to make this tube with a very thin wall. I have illustrated a valve 306 by means of which the tube may be inflated with air at relatively high pressures. The tube may carry a relief valve 308 at the other end adapted to open when, as in the case of fire, some temperature rise raises the pressure above the bursting strength of the metal. The necessary load-carrying connections at the ends of the tube may be provided by metal heads 310 having sockets 312 shaped to fit the end of the tube.

Where such a beam is to support flooring or carry the weight of a suspended partition, it may be provided with additional stiffening and attachment means in the form of radial fins 314 of which I have illustrated three in Fig. 23. The edges of these fins may be braced against buckling by thin sheet-metal shields 316 cut away as at 318 for lightness. Such units can be made materially lighter and easier to handle and ship than an ordinary beam or truss.

Another weight-saving expedient is indicated in Figs. 6 and 14. The floors of the utility units 196 and 198 have beveled shoulders at 320 to abut shoulders 322 on the bathroom units. All four units are hung on the mast 10, as by attachment means at 324 (see Fig. 16). The abutments at the bottom provide complete alignment means for the units independent of the mast. The partition 172 may abut the outer edges of the flooring of the bathroom units 182 and be supported primarily by a tension connection at 326 so that the weight of the partition 172 and associated parts will cause it to bear snugly against the bathroom unit. The concentrated load may be carried by an auxiliary tension member 328 running direct to the point of attachment 326.

Without further elaboration, the foregoing will so fully explain the gist of my invention that others may, by applying current knowledge, readily adapt the same for use under various conditions of service.

4▲DYMAXION CAR (1937)

U.S. PATENT—2,101,057

APPLICATION—OCTOBER 18, 1933

SERIAL NO.—694,068

IN GREAT BRITAIN—SEPTEMBER 8, 1933

PATENTED—DECEMBER 7, 1937

FIGURE 1

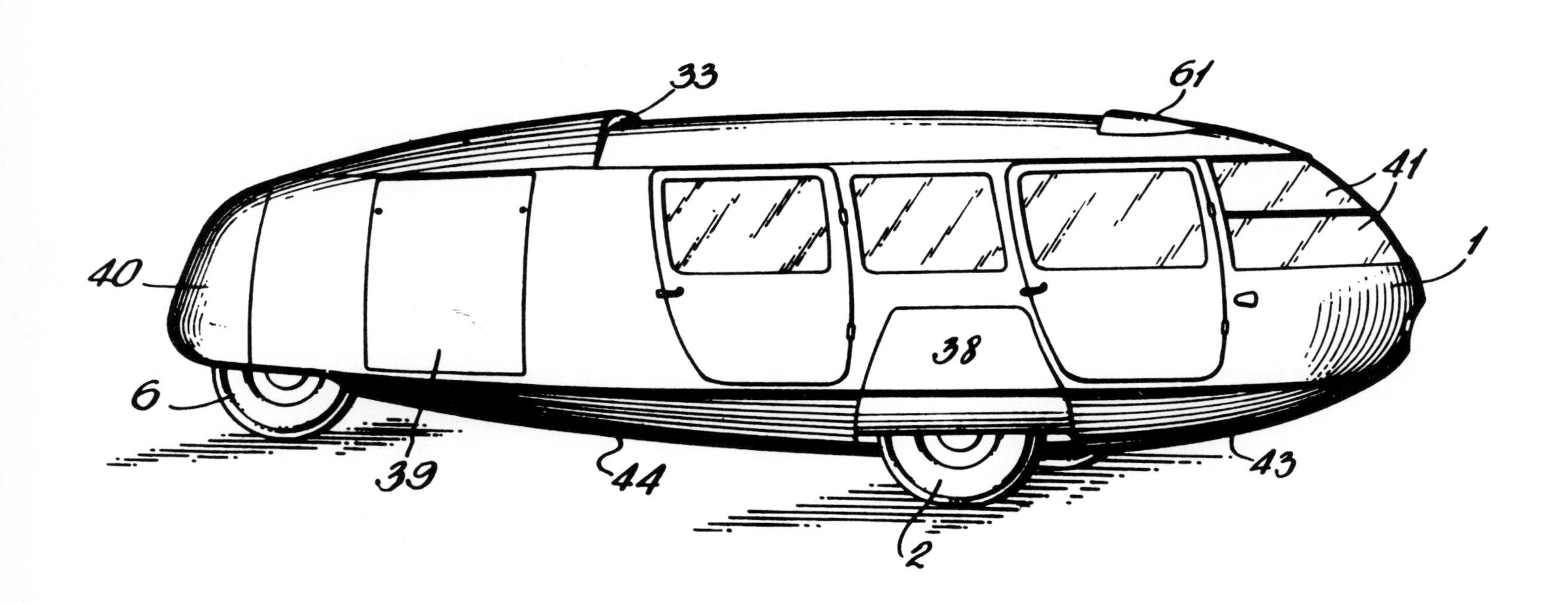

FIGURE 2

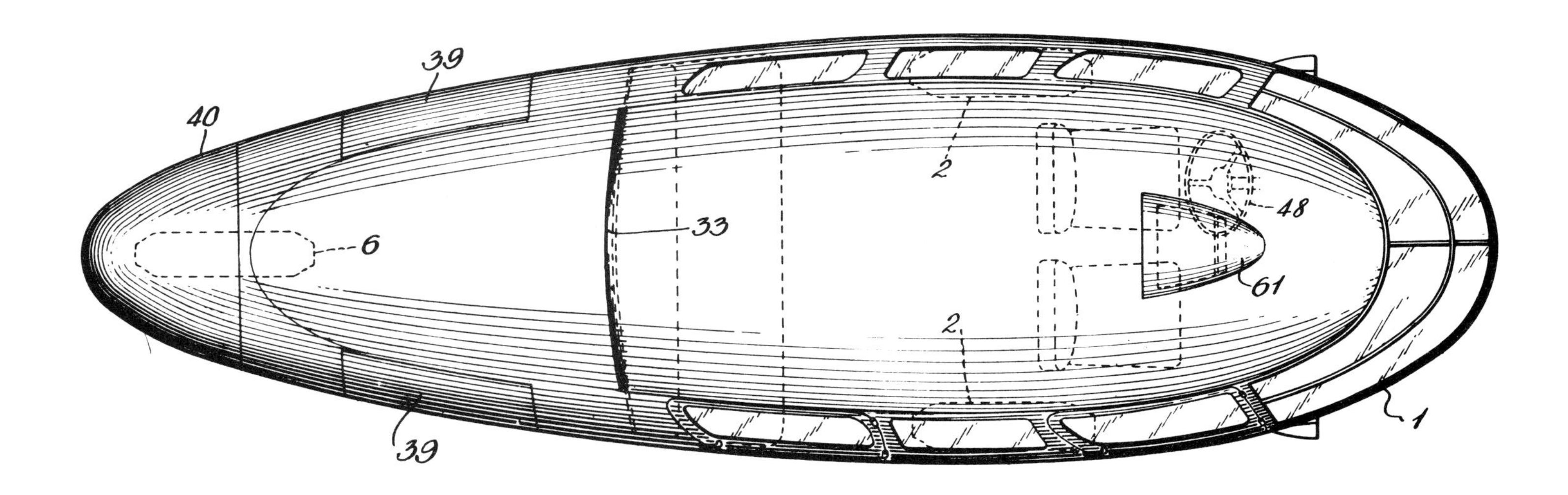

SINCE I WAS INTENT ON DEVELOPING a high-technology dwelling machine that could be air delivered to any remote, beautiful country site where there might be no roadways or landing fields for airplanes, I decided to try to develop an omni-medium transport vehicle to function in the sky, on negotiable terrain, or on water—to be securely landable anywhere, like an eagle.

There are two kinds of birds—soaring birds like gulls and nonsoaring birds like ducks, whose wings are too small to soar. Birds with wings too small to soar can flap their small wings very powerfully against their body, producing air jets, and can do so in such a manner as to make the jets orientable forwardly, backwardly, inwardly, and outwardly. All our heavier-than-air aeronautical development has been of the soaring-wing type. The structural weights of the "main spars" and of the wing-routes, etc., are great. I saw that with the duck-type nonsoaring jet flying device, much weight could be saved.

The first full-scale experimental building of what I called the 4D (four-dimensional) transport was undertaken to discover the ground-taxiing controlling requirements of an omni-streamlined vehicle that would ultimately be developed into a twin orientable jet-stilts, flyable vehicle. Such a vehicle had to be superbly faired, i.e., streamlined.

Everything in Universe is in constant motion. Everything in Universe is always moving in the direction of least resistance. When what we call a light plane, one flown by an individual, lands crosswind, its fairing or streamlining makes it want to turn violently in the direction of the wind—the direction of least resistance. This is called ground looping. I realized that the most difficult conditions for my omni-medium jet-stilt superbly faired flying device would be when it was on the ground. What is popularly called the Dymaxion Car were the first three vehicles designed to test ground taxiing under transverse wind conditions.

I produced only these three prototypes, the first with financial help from a friend, the other two with all of the money I inherited from my mother.

UNITED STATES PATENT OFFICE

Buckminster Fuller, Bridgeport, Conn., assignor to the Dymaxion Corporation, Bridgeport, Conn., a corporation of Connecticut

MOTOR VEHICLE

The invention relates to the construction of motor road vehicles whereby they are adapted to the economical operation resulting from full streamline formation and whereby other and independent advantages are obtained as will be apparent to those skilled in this art from this disclosure. The principles of the invention are exemplified by the vehicle illustrated in the accompanying drawings, but without limitation to such particular form.

Fig. 1 is a side view of the vehicle.

Fig. 2, a top plan.

Fig. 3, a longitudinal vertical section.

Fig. 4, a horizontal section on line IV—IV.

Fig. 5, a cross section on line V—V.

Fig. 6, a cross section on line VI—VI.

Fig. 7, a cross section on line VII—VII.

Fig. 8, an end view at line VIII—VIII with bussle removed.

Fig. 9, a detail section on line IX—IX.

The streamline body covers or encloses all of the chassis including all the wheels. For best economy it should be so designed that every axial section has a full streamline contour, which is to say that the body should be continuously curved from a round or blunt front to a tapered tail and that all its transverse maximum diameters should occur at a point about one-third of the length from the front end with no substantial interruption to the curvature and with no more excrescencies exposed to the relative wind than necessary for operation. The front wheels 2 are the driven or traction wheels and are located at the widest part of the streamline body, that is, at a point about one-third of its length from the front end. They are journalled at the ends of an axle structure or housing 3 and driven through differential gearing indicated at 4 by a propeller shaft 5 or in any equivalent differential manner. The axle structure may be the same as the rear axle structure of standard automobiles.

The forward wheels can be organized as the steering wheels within the broader aspect of this invention, but it is preferred that the steering is done by a rear wheel or wheels such as indicated at 6 which is central of the two forward wheels, being journalled on a stub shaft 7 rigidly fixed in the end of a single-tined steering fork 8, the head 9 of which is swiveled to turn on an upright axis. This wheel is preferably of the same size as the forward wheels and interchangeable therewith as in standard automobiles, being readily removed from its stub shaft on the single-tined fork. It may however be dual-tired if desired, or may consist of twin wheels turning together as a unit or like a single wheel and such variants are to be understood as included within the term single steering wheel as used herein.

The steering head 9 is journalled on vertically spaced bearings in a deep barrel socket 10 formed in the rear apical end of a generally triangular or A-shaped frame 11 herein termed the sub-frame, and is slightly castored therein as shown in Fig. 3, to facilitate steering. The forwardly extending legs of this frame 11 are supported on the forward axle housing 3 close to the front wheels 2, thus to provide a wide bearing for the sub-frame on the

forward wheels and it has deep web sections with liberal flanging and gusseting and is reinforced by an arched cross-brace 12, all for the purpose of producing a maximum degree of rigidity in the torsional sense between the steering head and axle housing. Such rigidity is important in three-wheeled vehicles intended for passenger car speeds because, if the steering axis is not kept to a plane parallel with the planes of the front wheels (or if these are canted, then to an intermediate plane bisecting the angle between them) the steering becomes unsteady and dangerous. On this account the sub-frame 11 is specially stiffened as stated and no spring intervenes between it and the wheels such as might permit the steering axis to change its lateral position in relation to the forward wheels. In this sense the sub-frame is an unsprung frame. It may however be connected to the axle by a joint, if the joint axis is horizontal and such joint is preferably used and appears at 13, where the ends of the frame legs are attached to the axle housing. It does not impair the rigidity of the frame against torsion. Preferably also the sub-frame is dropped or formed with an angle at or near the cross brace so that its forwardly extending leg members are substantially horizontal and at about the level of the wheel hubs and only the pointed rear part rises above the hub level.

The propelling engine 14 occupies the space within and below the narrow part of the sub-frame and under the arched cross-brace 12. It may be mounted on that frame with appropriate cushionings, if desired, but is preferably mounted on a second frame 15, herein called the main frame, which carries the body 1 and is spring-supported. This frame extends from a rear point just forward of the rear wheel to a forward point well beyond the forward wheels and has a kickup over the forward axle. Its rear part lies in substantially the same level as the legs of the sub-frame and between them and about one-third of its length overhangs the forward axle. Cross bracing, not shown, may be provided to give it requisite stiffness. Its forward point of support is by a cross bolster 16 and a transverse spring 17 which is shackled at its ends to appropriate brackets on the axle structure; see Fig. 5. At its rear end it is flexibly connected to the sub-frame in such manner as to accommodate the action of the forward spring and preferably the connection includes a spring such as cross-spring 18 which is centrally fastened to the cross-bolster 19 of the main frame and suspended at its ends by a pair of hanger-links 20 depending from the high part of the sub-frame. These links include turn-buckles, as indicated, which can be adjusted to raise or lower the main frame. By reason of their substantial parallelism they permit a certain amount of sidesway to the main frame relatively to the sub-frame but tending at the same time to restrain careening of the body.

The chassis of the illustrated vehicle thus includes the sub-frame which as stated is unsprung, and the main frame which is sprung both front and rear. It is desirable that the normal amplitude of the rear spring action be relatively less than that of the front spring. This can be done by loading it with a resistance of some sort, such as provided by connecting ordinary hydraulic shock absorbers 21 between the two frames at this point. If absorbers are also associated with the front spring, as indicated at

FIGURE 3

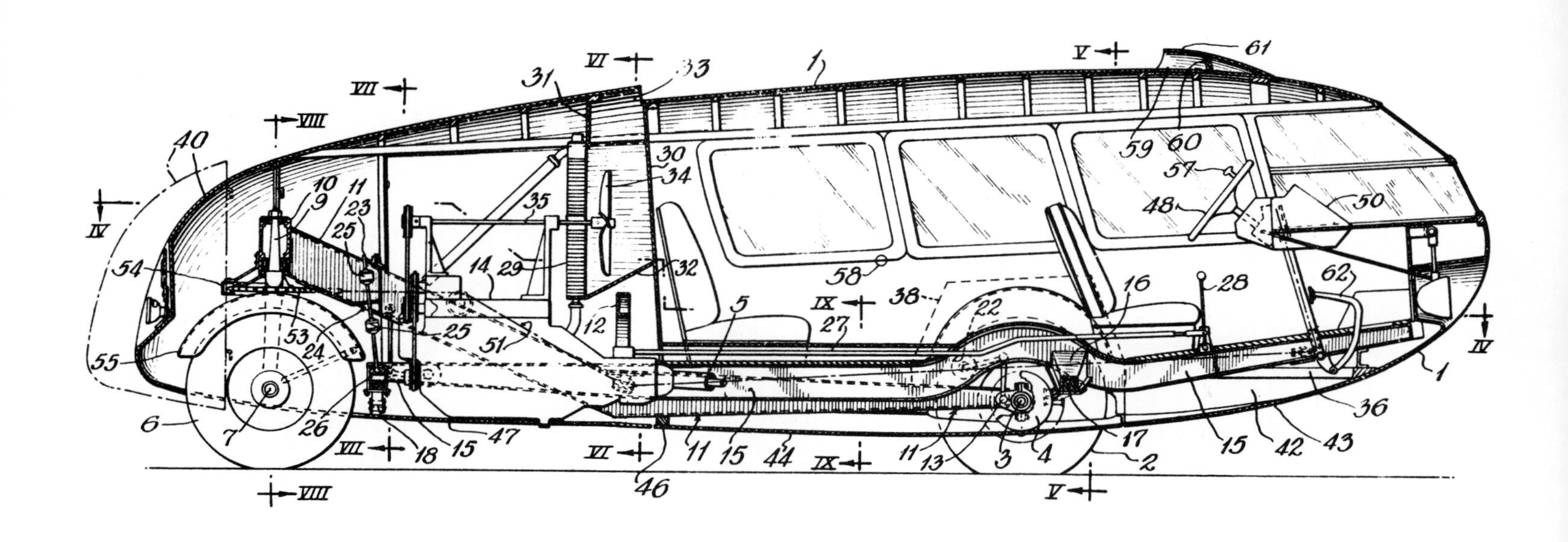

22, the resistance of the rear absorbers is made to exceed materially that of the front absorbers, so that the action of the rear spring is relatively stiff or sluggish. The throw of the rear spring is limited by a check rod 23 which is connected at its foot with the end of the main frame or with the cross bolster 19 thereof or otherwise and plays in a hole in a cross flange 24 (Fig. 3) of the sub-frame 11 with rubber-backed collars 25 fixed on the rod above and below such flange. These collars coact with the flange as spring-bumpers, in both directions, and in addition the upper collar serves also as a safety support to hold the main frame in the event of failure of its rear spring or support. The engine is mounted with the end of its crankshaft accessible to the rear through a hole marked 26 and the check rod is attached to the frame by a connection above this hole, as indicated in Figs. 3 and 7 so that by turning the steering wheel to a transverse position room is available to introduce a hand crank in the hole when the engine requires to be hand cranked.

The transmission case is on the forward end of the engine and connected to drive the forwardly extending propeller shaft 5 through an appropriate universal joint or joints and with or without a torque tube as preferred. The transmission mechanism is controlled by a selector rod 27 extending forwardly to the gear shift lever 28 at the operator's station. The usual engine controls, though not shown in the drawings, will be understood to be arranged in any suitable way.

In the case of a watercooled engine, the radiator 29 is preferably located directly over the transmission or flywheel case and just abaft the after bulkhead wall 30 of the cabin compartment, and suitable partitions 31 and 32 (Fig. 6) are provided to form an air channel for conducting air to it from an air scoop slot 33 which extends across the roof part of the body. A fan 34 is located in the air channel, being driven by the engine in any suitable manner, as for example, by the belt-driven shaft 35, which is journalled on the engine and extends through the radiator. The cooling air passing the radiator flows over the engine and out around the rear wheel.

While the body 1 can be variously constructed and wholly of metal, if desired, it is shown as built of wood framing with a light metal covering. Its main sills 36 are carried on brackets 37 which project laterally from the main frame, some of them extending over and some under the legs of the sub-frame and all shaped or located to afford the necessary clearance for the relative movement of the unsprung sub-frame and the sprung main frame. These sills extend aft of the main frame as cantilevers to support the tail part of the body. Doors and windows are provided and also a number of removable panels, those marked 38 being for providing access to the forward wheels and those marked 39 to afford access to the engine while the rear end or bussle 40 is removable to afford access not only to the rear wheel but also to the crankshaft of the engine for hand-cranking it. The forward windows 41 are either curved to the streamline contour or composed of smaller flat sections collectively approximating such contour.

The bottom of the body is preferably closed by a belly wall in one or more sections which are longitudinally and transversely curved to conform to the streamline contour.

FIGURE 4

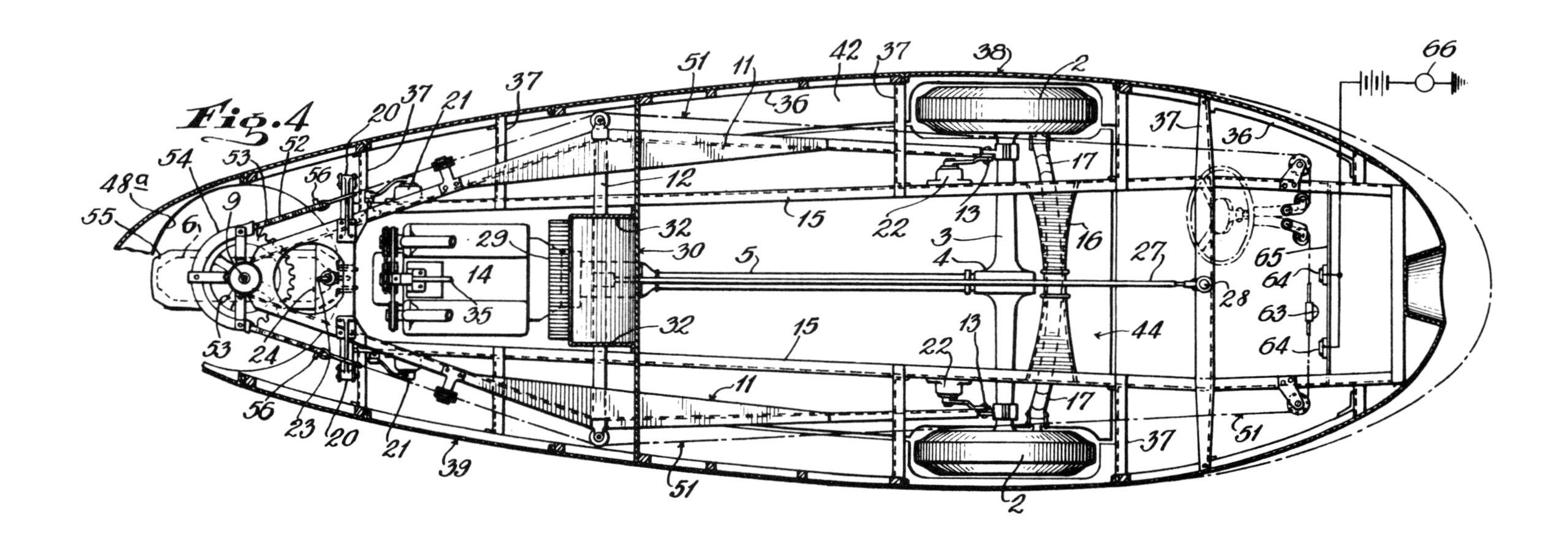

FIGURE 5

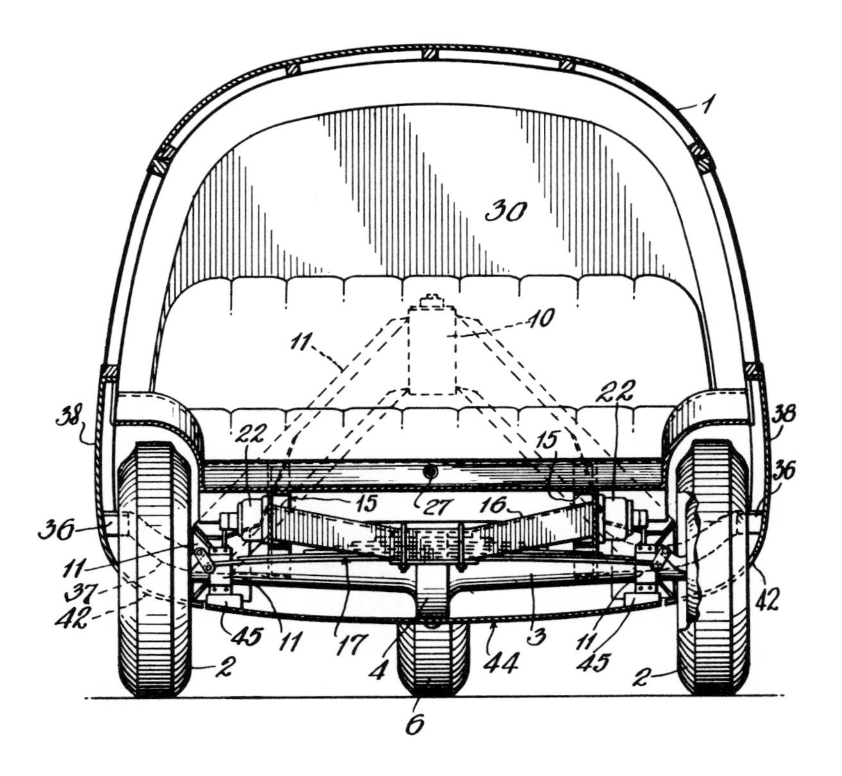

To this end the metal body covering below the sills 36 is inwardly curved at the sides, as indicated at 42 and the middle space is closed in by a curved belly wall section marked 43 in Figs. 1 and 3. This section 43 of the belly wall is a part of the body proper. Next in the rear of the belly wall section 43 comes a continuation section 44 which is removably fastened to the unsprung parts of the chassis, that is, to the differential casing and the legs of the sub-frame, the forward attachment points being marked 45 and the rear points 46 (Fig. 3). Aft of this section and extending to the end of the tail, the belly wall section 47 is fastened to the incurved sides 42 of the body in some removable manner and this section is cut with a circular hole 48A (Fig. 4) to accommodate the rear steering wheel. As thus organized the belly wall is formed of three sections of which two are carried by the body proper and the other intermediate section by the running gear. The sections meet without contact in normally flush relation so as to provide a substantially smooth belly from front to rear but it will be apparent that the edges will play past each other according to the action of the springs. The gap may be covered or faired over if desired to exclude entrance of air. While shown as made of metal the belly wall may be made of fabric, if desired, in which case it may be continuous from end to end.

By thus enclosing the whole running gear, including as much of the wheels as consistent with road clearance in a properly streamline external contour, the advantage is gained that the rate of fuel consumption, as compared with conventional cars of equivalent size and weight, falls off rapidly as the speed is increased above about 10 m.p.h. being some 30% less at 30 miles and 50% less at about 50 miles, while within the overall dimensions of such conventional cars the volume of useful cabin space inside the streamline body is much increased, being practically doubled. All of the interior of the body forward of the drop-angle or bulkhead wall 30 constitutes the useful space for passenger or cargo, and due to the drop-angle the rear seat can extend the full width of the body over the sub-frame 11, as well as over the main frame 15 and with cars of standard tread gauge this provides a seat some 6 feet wide, long enough to serve as a bunk for sleeping purposes.

The forward overhang of the main frame 15 pitches upwards from the forward wheels and terminates at about the bumper level of conventional cars or slightly higher, the purpose of which, among other things, is to take any collision impact in the event of accident at a point well in advance of the front seat and to receive it on the main frame, so that the inertia of the engine fixed on the rear of that frame will be available to absorb the impact, as is the case in conventional cars having the engine in front.

Steering is done by a hand-wheel 48 mounted at a convenient angle in front of one of the forward seats and according to this invention its connections to the steering head 9 provide for a maximum variation of steering angle of at least about 160° and in any event over 100°. With the traction wheels located at one third the body length abaft the front end, such range of steering angle affords a degree of maneuverability not heretofore attained in automobiles. In the present case the steering system includes a windlass contained in a case 50 with cables 51 trained over sheaves on the chassis or sub-frame and attached to the ends of the sprocket chain 52 of a full circular sprocket wheel 53 which is fixed to and below the steering head 9. By the use of a full circular sprocket wheel the same constant degree of tautness is kept in the cables for every steering angle, without which the steering would be erratic and unsatisfactory. A keeper or guard 54 is provided about the sprocket wheel rim to guide and retain the chain thereon. This keeper is fixed by rigid bracket arms to the steering head barrel 10 of the sub-frame 11, directly over the steering-wheel mudguard 55 which turns with the steering head 9. The lugs 56 (Fig. 4) on the ends of the sprocket chain serve to limit the steering angle by their abutment against the ends of the keeper 54. They limit the steering range to something less than 180° of arc. The gear ratio of the steering system is about 30 to 1 and in order to make quick changes through large angles, the hand-wheel 48 is provided with a crank knob 57 by which it may be easily spun.

While rear-steering greatly improves maneuverability as compared to conventional cars, and particularly with the traction wheels in the position described, it is apt to give rise to a tendency to skid when braking or rounding corners. This however is eliminated according to this invention by the distribution of the weight and the location of the center of gravity of the vehicle. It is found that such center should be forward of the mid-point of the wheel base and must not in fact be located further aft from the forward wheel axis than a distance equal to about 40% of the wheel base length. The importance of the pronounced forward body overhang will now be apparent, since even with the engine in the rear it brings the center of gravity to the position of maximum safety against skidding. In the car taken for illustration, the center of gravity is about 20% aft of the front axle, some 75% of the total weight being on the two forward wheels, and this location of the gravity center is preferred. The normal loading of the vehicle will not appreciably shift it. Also specially contributing to the maneuverability and ease of handling generally is the fact that the traction center as well as the gravity center are both located in the same general position, forward of the center point of the wheel base and that this position also substantially coincides with what may be called the streamline center of the body which may be taken as its center of volume or the center of area of its axial section. This center is indicated roughly in Fig. 3 by the small circle 58; the gravity center is lower down and the traction center of course coincides with the axis of the front wheels. The consequences of the grouping of these important centers in the same general forward location are reflected in the structural economy of the vehicle and become obvious on comparison with the action of conventional cars and especially those which have their fraction center rearward of the mid-point of the wheel base.

A view to the rear is afforded to the driver through a water-tight roof window 59 and an exterior inclined mirror 60 mounted on the roof at its highest point and within a rearward open hood or fairing 61 to avoid wind resistance and also shelter the mirror from the weather. The mirror may be viewed through the window and by reason of its position at the highest point gives unobstructed vision to the rear through a wide arc. This makes it easy for the driver to avoid swinging the tail of the vehicle so far to the outside when turning a corner, as to collide with adjacent cars or objects. To the same end the invention contemplates as an additional safeguard, useful in the case of drivers unaccustomed to rear-steered vehicles, a warning device of some kind which will announce the fact whenever the driver turns the rear wheel to such an angle as might be likely to result in a sideswipe. This may take any suitable form and as shown herein consists

FIGURE 6

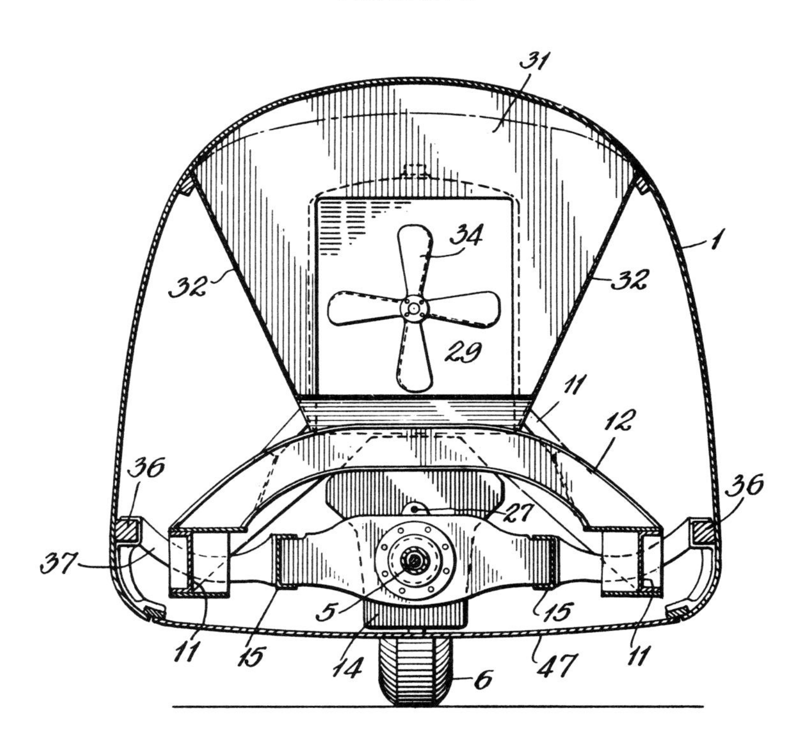

FIGURE 7

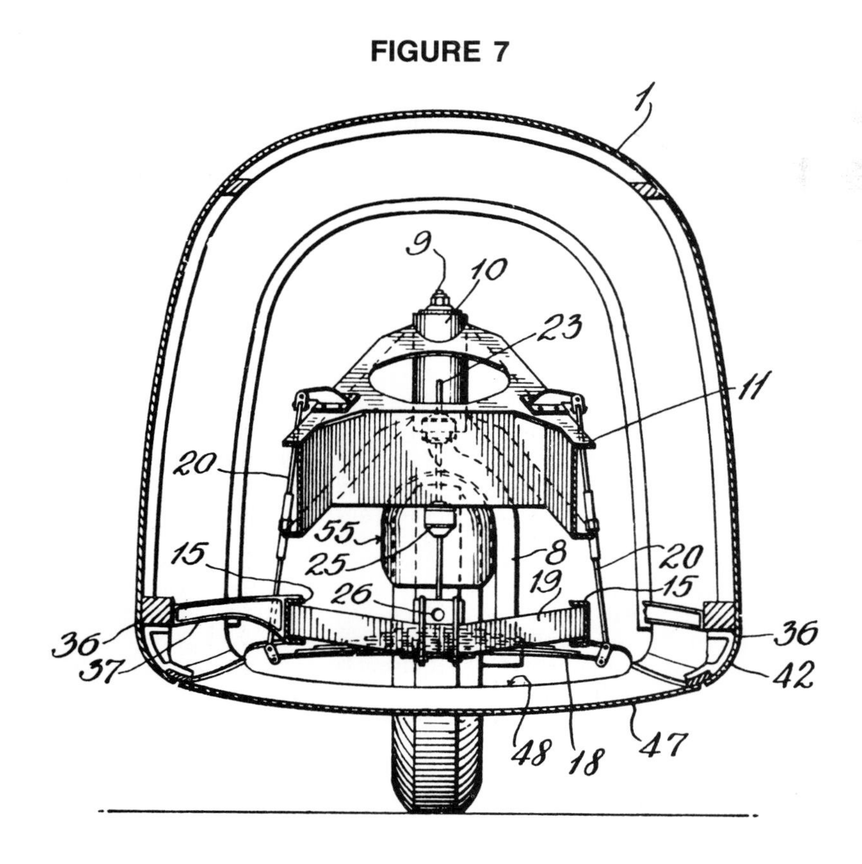

FIGURE 8

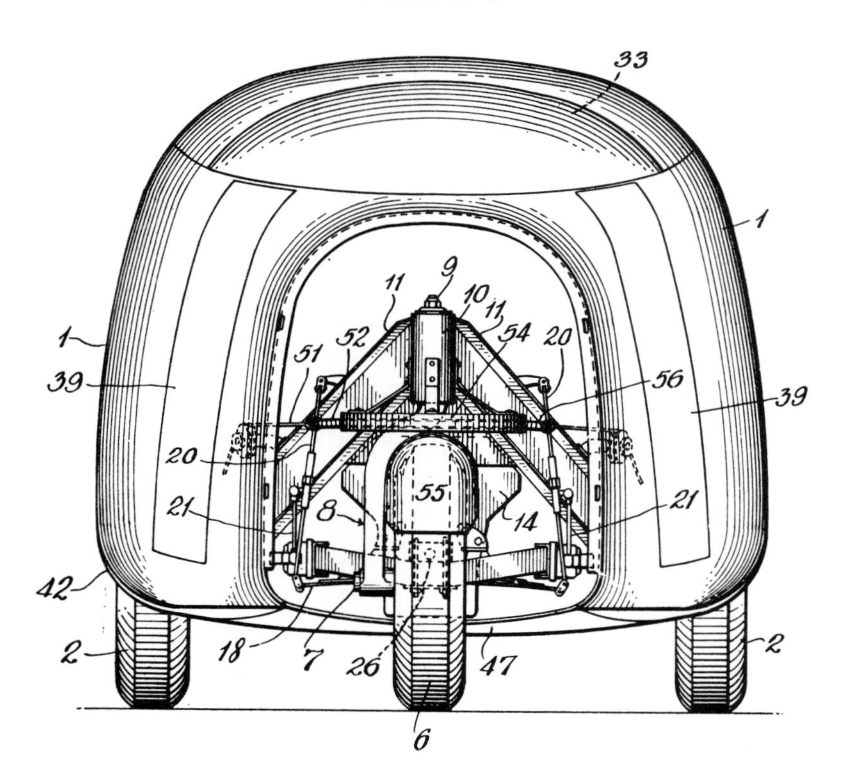

of a wiper button 63 (Fig. 4) fastened to the steel steering cable 51 and adapted to contact with either of two electrically insulated terminal plates 64 mounted on a cross bar 65 and connected in circuit with a buzzer or the like 66. Whenever the rear wheel reaches or passes the angle which will run it outside of the tracks of the forward wheels, the signal is given and the driver's attention is thereby called to the need of caution in the event there should be an adjacent object in position to be sideswiped. When operating within the limits represented by the two terminal plates, the driver may handle the car without concern for side collisions more than with ordinary automobiles. Instead of an audible signal any other device may be employed which will guard against involuntarily exceeding the normal range of steering angle.

A brake pedal is indicated at 62 but the braking system has been omitted; it may be applied to all three wheels, if desired, but braking on the two forward wheels alone has been found sufficient with the weight distribution as described.

While the various features of this invention have been above described as mutually combined and cooperating in a single structure which is rear-steered, it is to be understood and will be apparent that there is no intention to limit this patent to such single combination inasmuch as certain subcombinations set forth in the claims obviously have important uses in independent relations.

FIGURE 9

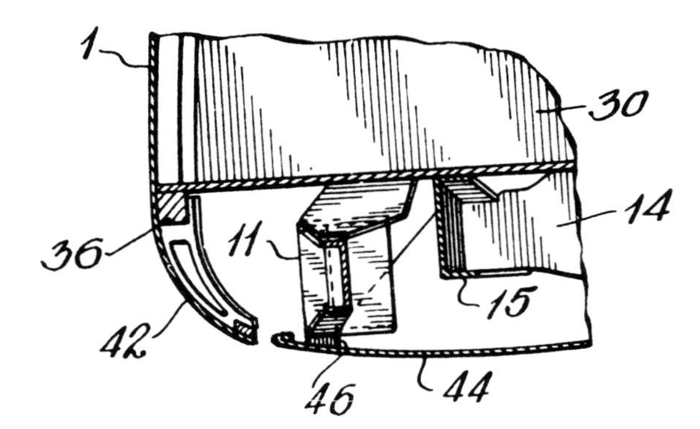

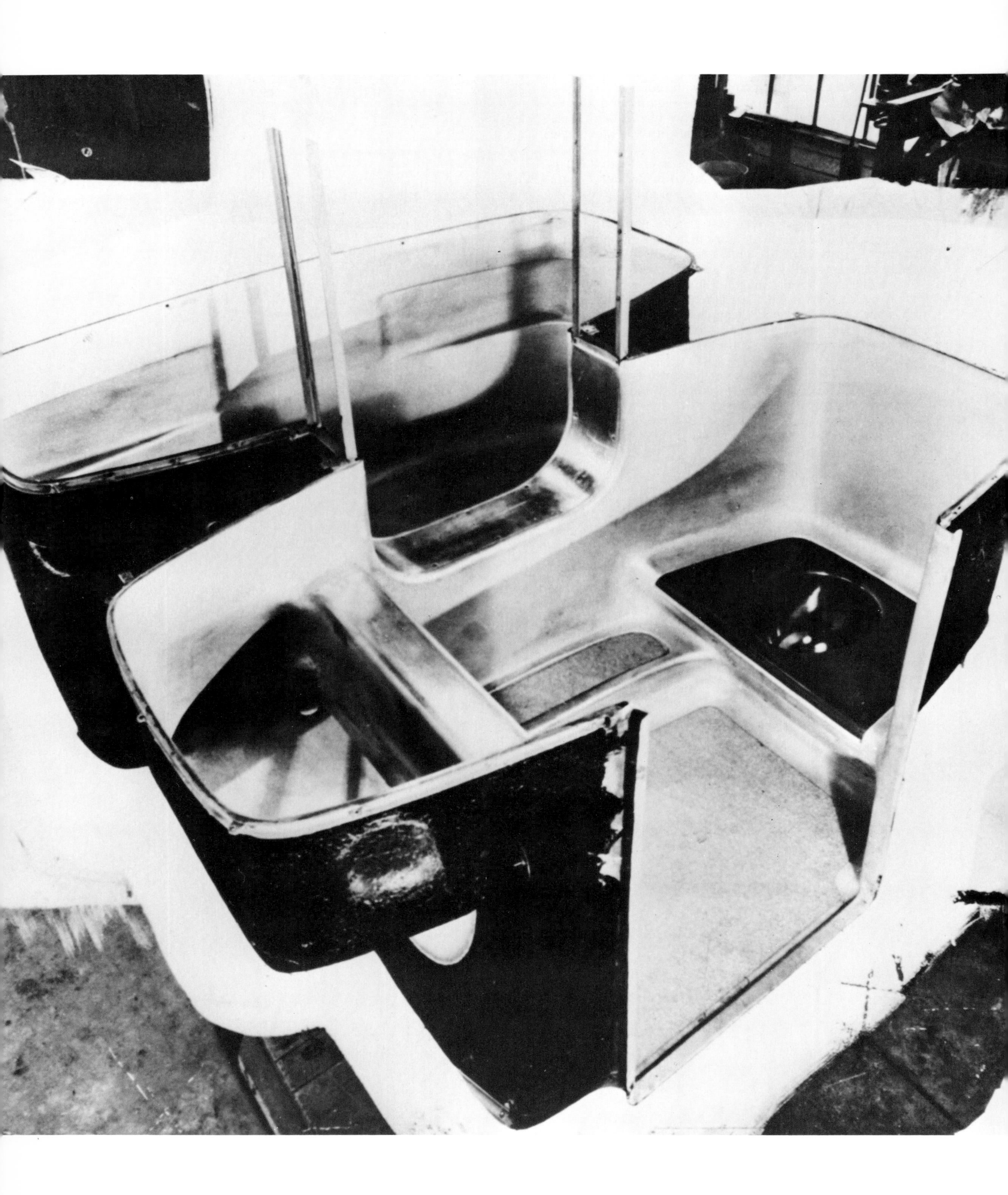

5▲DYMAXION BATHROOM (1940)

U.S. PATENT—2,220,482

APPLICATION—MAY 12, 1938

SERIAL NO.—207,518

PATENTED—NOVEMBER 5, 1940

IN THE ORIGINAL 4D HOUSE, I planned to introduce all accessories such as bathroom functions in one prefabricated unit, this unit to be hung-swung into position as we would hang a corner cabinet, with gravity holding it firmly in place. I developed the first full-scale working prototype of the bathroom for American Standard Sanitary Heating and Plumbing Company in 1930 and developed twelve new prototypes six years later for the Phelps Dodge Copper Company, then the third-largest copper company in the world. One of these twelve units was set up at the U.S. Bureau of Standards where it was tested for two years, and it was found to conform satisfactorily to all U.S. building codes, but Phelps Dodge, as a copper company, decided not to go into production of the unit.

This bathroom unit was designed to be produced not in metal but in glass-reinforced plastic. This production took place in Germany years later, apparently following my drawings, as the unit produced was exactly as I designed it in 1937, although my patent had long since expired.

UNITED STATES PATENT OFFICE

Richard Buckminster Fuller, New York, N.Y., assignor to Phelps Dodge Corporation, New York, N.Y., a corporation of New York

PREFABRICATED BATHROOM

This invention relates to a prefabricated building unit suitable for use as a bathroom.

Attempts have been made heretofore to provide prefabricated bathrooms with the object of lowering the cost of building a bathroom into a dwelling. Such bathrooms, however, by reason of their great weight and more or less conventional construction have involved relatively high costs by the time they have been shipped and installed ready for use. Furthermore, such bathroom units as heretofore known have been largely designed for introduction into a new building under construction and have not been particularly practical for installation in a dwelling already built without involving too great an expense.

It is an object of my invention to provide a compact, light, prefabricated bathroom which may be readily installed either in a dwelling under construction or in a dwelling that is already built.

It is another object of this invention to provide a prefabricated bathroom of such a compact construction that it can be separated into a few sections which may be readily carried by hand through a doorway and up a staircase of the average house.

It is another object of this invention to provide a prefabricated bathroom fashioned from a relatively few units made of sheet material to provide an integral structure light in weight but having the requisite strength and rigidity when assembled.

A further object of my invention is to provide prefabricated bathroom sections of sheet metal having bathroom fixtures formed integrally therewith.

Another object of this invention is to provide a two chamber prefabricated bathroom, one chamber being suitable for use as a lavatory and water closet, and the other chamber being useful as a combined tub and shower, and either of these chambers being useful alone.

Further objects and advantages of my invention will be apparent from the following detailed description of the embodiments thereof illustrated in the accompanying drawings, in which:

Figure 1 is a front elevation of my preferred bathroom as assembled with the lower portions of the outer front decorative panel and the doors broken away;

Figure 2 is a horizontal cross section of my completed bathroom in installed position, taken about on the offset line 2—2 of Figure 1;

Figure 3 is a vertical cross section through approximately the center of the lavatory and water closet, or outer chamber of the bathroom, taken about on the line 3—3 of Figure 2;

Figure 4 is a vertical cross section of my bathroom,

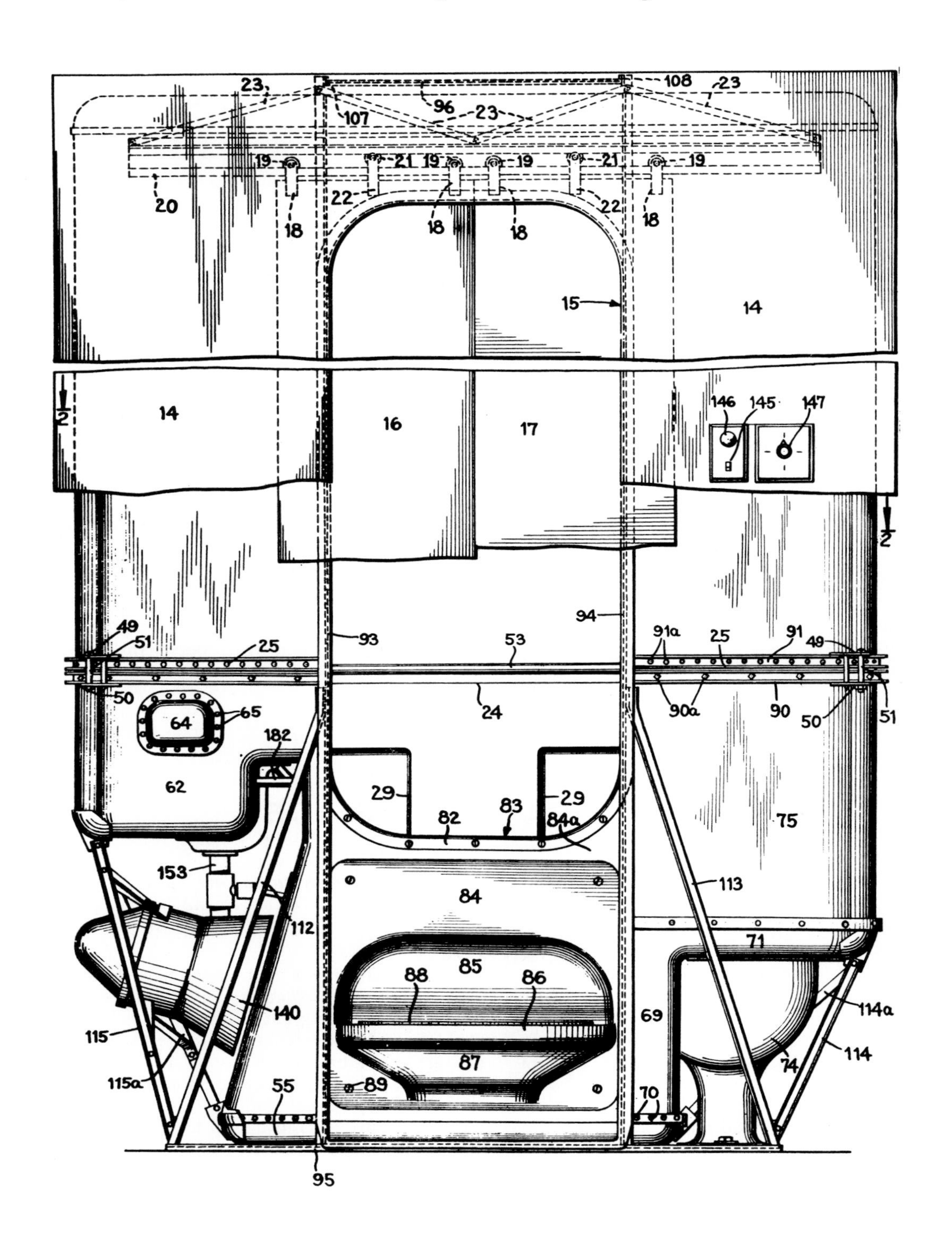

FIGURE 1

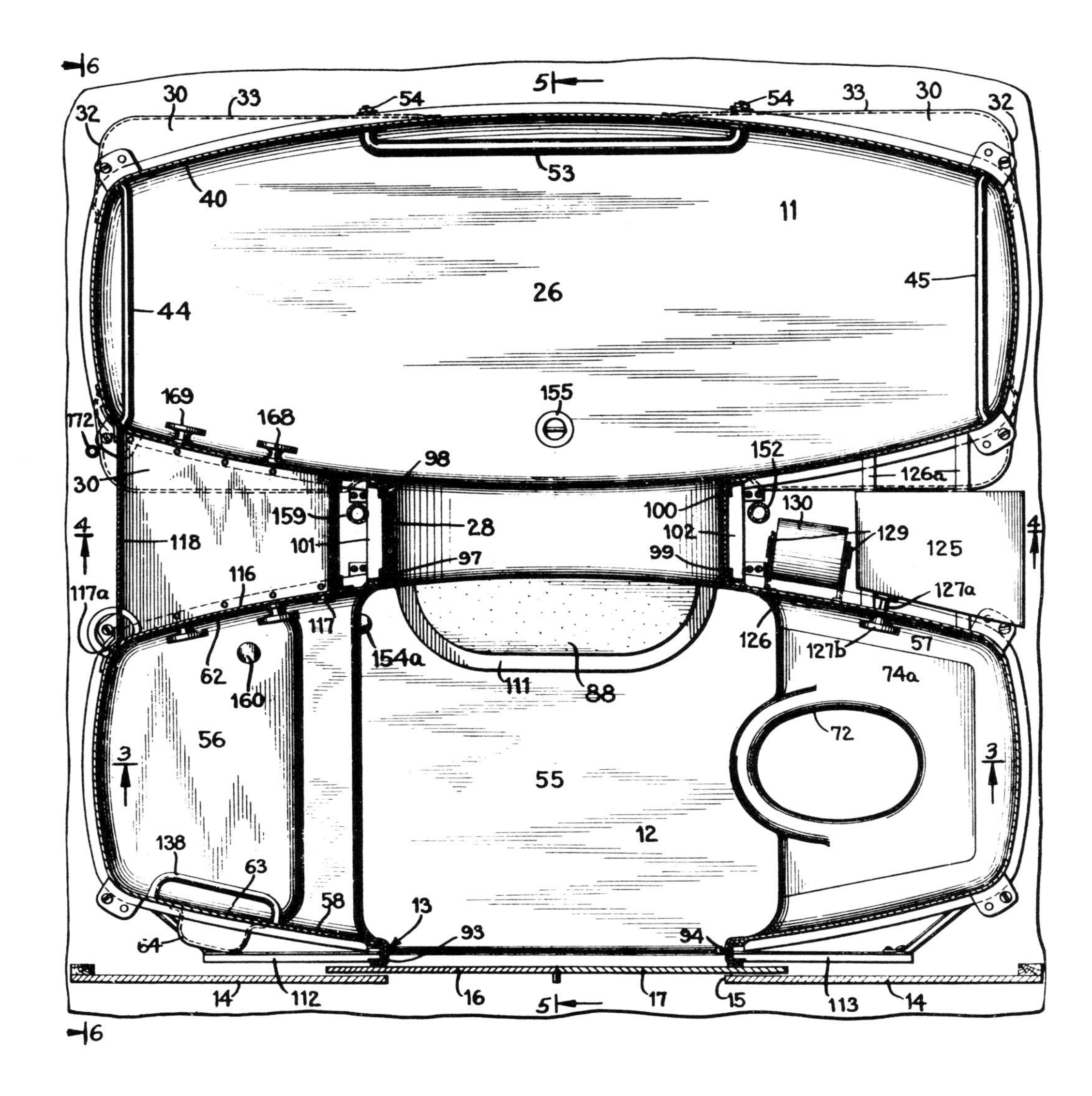

FIGURE 2

taken approximately on the line dividing the two chambers thereof, such as the line 4—4 of Figure 2;

Figure 5 is a vertical cross section through approximately the center of my bathroom and at right angle to the section line 2—3 and 4—4, taken approximately on the line of 5—5 of Figure 2;

Figure 6 is an end elevation of the two chambers of my bathroom as it appears from the left-hand side of Figure 1;

Figure 7 is a detail view of a ventilating grill in the doorway of the bathroom connecting the two chambers thereof and taken on the line 7—7 of Figure 3.

Figure 8 is a vertical section of a detail of the bathroom constructed on an enlarged scale, and taken through the junction of the upper and lower sections of the bathroom, showing the manner of clamping the upper and lower sections of each chamber together;

Figure 9 is a perspective view partly broken away illustrating a modification of my bathroom with the upper portion thereof comprising the walls of the room in which the unit is installed; and

Figure 10 is a vertical sectional view of a detail on an enlarged scale showing the way the upper edge of the bathroom in Figure 9 is joined to the walls of the room and taken about on the line 10—10 of Figure 9.

With reference, more particularly, to Figures 1 to 8, inclusive, of the drawings, my bathroom is preferably constructed with two chambers of similar shape in horizontal cross section. These two chambers are indicated generally by the numerals 11 and 12. The outer chamber 12 is designed as a lavatory and water closet, and is provided with a doorway 13 permitting entrance into the chamber from an adjoining room of the dwelling. The side of the chamber 12 containing doorway 13 may be concealed from the adjoining room by the fixed vertical panel 14 provided with an opening 15, which may be constructed of metal, wood, Bakelite, or other plastic or composition material. The panel 14 may be rigidly attached to the chamber 12 of the bathroom, although it is preferably provided as a separate unit which may be affixed to and supported by the adjacent portions of the dwelling. Between the panel 14 and the doorway 13 of chamber 12, sliding doors 16 and 17 have been illustrated, it being understood that other forms of doors could be employed if desired. Sliding doors have the advantage, however, of space economy that is not obtained with a hinged door, and they may be used where a hinged door would not be practical. The sliding doors 16 and 17 are designed so that they may be moved outwardly away from each other, sliding into the space between the outer wall of chamber 12 and the panel 14 to leave an opening registering with doorway 13 and the panel opening 15. The sliding doors 16 and 17 may be hung by means of suitable straps 18 fixed to them and carrying the rollers 19 which rest on the track 20 and support the weight of the

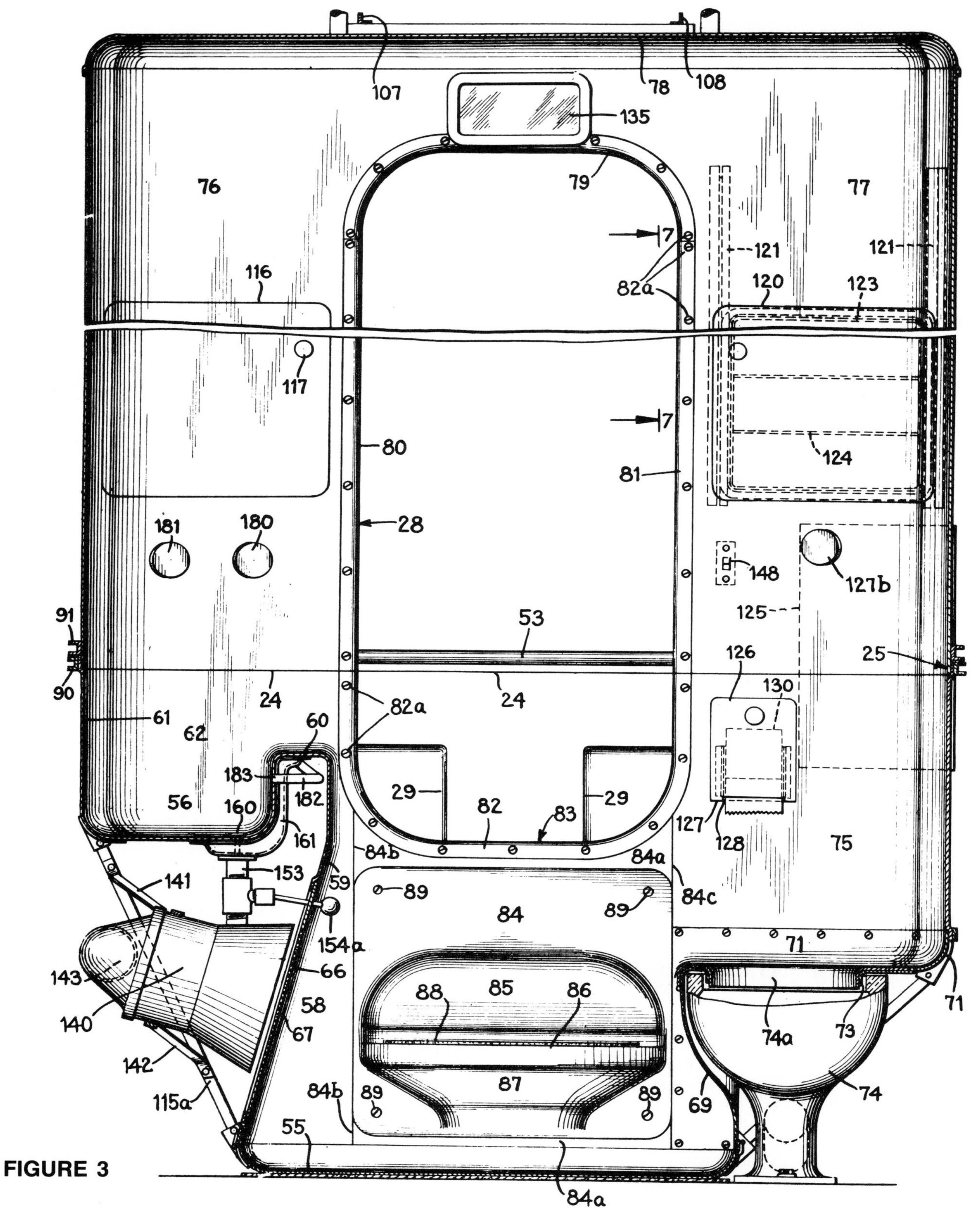

FIGURE 3

sliding doors. Track 20 may be formed with the shape of a block C in cross section and the rollers 21 carried on straps 22 intermediate the straps 18 on the doors bear against the upper inside portion of the track 20 to prevent dislodgement of the doors from the track by accident. The track 20 may be conveniently suspended from a framework for the chamber 12 described below and rigidly supported by the diagonal braces 23.

The bathroom chambers 11 and 12 may be conveniently constructed with their lower portions or sections made of sheet material such as sheet metal or thin-walled plastic material and with their upper portions formed from a lightweight metal sheet or from suitable plastic or composition material to provide as light a weight as possible. In order to provide as much strength as possible both the upper and lower sections of the chambers may be conveniently made of sheet metal such as copper, aluminum, or steel or a combination of two of them. These upper and lower portions of chambers 11 and 12 are illustrated as joined along the horizontal line 24 and held in assembled position by a suitable clamping arrangement 25 to be described in detail hereinafter.

The floor of the bathing chamber 11 may be formed from a single sheet 26 of thin metal such as copper, stamped or drawn into the shape illustrated in Figures 2 and 5, with its edges curving upwardly to form the lowest portions of the chamber walls. The remainder of the lower section of this chamber 11 may be formed from a single sheet 27 of the same metal bent or stamped to expand from below the center of the doorway 28 entirely around the chamber 11 and back to the center of the doorway 28 where the ends of the sheet are joined in the vertical seam 29a. This sheet 27 is preferably shaped to provide all four walls of the chamber 11 curved outwardly about relatively long radii, these walls being joined at the corners by coved portions curved about comparatively short radii.

Intermediate the top and bottom of the metal sheet 27 forming the side walls, at a suitable height above the floor, four openings 29 of rectangular shape are formed in the sheet 27. These openings 29 are covered by pocket members 30 stamped out of sheet metal and secured to the outer surface of sheet 27 by riveting, welding or the like. The pocket members 30 are preferably formed with relatively flat lower surfaces 31 and relatively straight side and end walls 32 and 33. When these pockets are sufficiently shallow, they may be stamped or otherwise formed integrally with sheet 27. However, these outwardly extending pockets are preferably formed of separate sheets of metal attached to the outside of the main sheet 27 by the rivets 34, with suitable provision being made at the juncture of the layers of metal to form a water-tight joint having a flush surface inside the chamber 11. Similarly, the bottom metal sheet 26 of chamber 11 may be provided with an offset upper edge 35 so that the lower edge of sheet 27 will be flush therewith on the inside of the chamber, and these two sheets may be secured together in a similar manner by the rivets 36 or by welding or the like. The pockets 30 are constructed to provide the lower wall portions of the chamber 11 with increased rigidity and strength, even though very light-gauge metal may be used in making the walls and floor. In addition, these pockets serve as convenient supports for soap and other bathing accessories, and also as arm rests or supports to assist one in sitting down or rising in the chamber 11 when using it as a tub.

The floor of the bathing chamber 11, that is the lower portion of sheet 26, is preferably supported on a suitable wooden platform 37 raised on legs 38 made of angle iron or other material, so that the floor of this chamber is about nine inches above the level of the floor of the chamber 12. The metal sheet 26 may be separated from the wooden platform 37 by a layer of insulation 39, if desired. This arrangement provides adequate room for plumbing and also enables the bottom of the bathing chamber 11 to be cleaned more easily by a person standing in chamber 12.

The upper portion of the chamber 11 comprising the walls 40 and the ceiling 41, may be constructed of the same or different material. This upper portion of the bathroom unit is preferably made of a material as light in weight as possible, while still providing the desired strength, rigidity, and pleasing appearance to the entire structure. Various substances may be suitable for this purpose, although I have found that alloys of aluminum or thin sheet steel are particularly satisfactory when the lower portion of the bathroom unit is constructed of sheet copper. The upper portion of chamber 11 may be formed, for example, from a sheet 40 of light-gauge steel or a strong aluminum alloy providing curved end and side walls which are extensions of the lower walls formed from sheet 27. Sheet 40 may extend from the horizontal juncture line 24 to the juncture of this sheet with the ceiling 41 at 42. The ceiling member 41 may similarly be constructed from a single stamped sheet of metal or other material and attached to the side wall 40 by the riveted joint 43 in the same manner that the base sheet 26 is attached to the lower side wall 27.

To lend the structure increased rigidity, I prefer to attach firmly inside the sheet 40 at a suitable height above the floor, the rods 44 and 45 extending across the ends of this section. The ends of these rods may be welded or otherwise rigidly secured to the sheet 40 and serve to brace it inside, as well as to provide towel racks and hand grips.

The upper portion of chamber 11 comprising the wall sheet 40 and the ceiling member 41 form one prefabricated unit section which is fitted to the lower section as a unit when the chamber 11 is assembled. For this purpose, an outwardly opening channel member 47 may be shaped and riveted to the outer surface of 40 spaced a small distance from its lower edge. A similar rigid channel member 48 may be riveted to the outer surface of sheet 27 adjacent its upper edge. The joining of these walls in assembled position of the units forming chamber 11 is best illustrated in Fig. 8, from which it will be seen that the upper edge of the metal sheet 27 is provided with an offset portion 46 so that the lower edge of sheet 40 below the channel member 47 overlaps therewith to provide a substantially flush interior surface for the chamber 11.

In this manner, the sections of the chambers 11 and 12 above the line 24 may be prefabricated as integral units and the sections below the line 24 may likewise be prefabricated as integral units. These unit sections may then be readily transported and assembled in the desired location by clamping the two channel members 47 and 48 together. This might be accomplished by simply bolting the adjacent flanges of the channel members together, although I prefer to employ a clamp consisting of the upper and lower plates 49 and 50 held together by a suitable bolt or bolts 51. These clamping plates 49 and 50 are simply applied to the uppermost and lowermost surfaces of the channel members 47 and 48, respectively, and held in position by tightening the bolts 51. To deaden reverberations within the chambers 11 and 12 and to avoid any hollow metallic sounds, a dope material such as a mixture of asphaltum and asbestos 52 may be applied to the exterior surfaces of the various units either prior to assembly or after assembly. Other materials could be employed for this purpose.

Along the solid or innermost wall of the chamber 11, a further bracing rod 53, which may be used to provide additional rigidity to the side wall structure, is preferably attached inside the lower edge of metal sheet 48. The rod 53 may be secured in place by bolts extending through the channel members 47 and 48 and held in place by suitable nuts 54 outside of the channel members. The rod 53 may also be used as a towel rack or as a support to assist one in sitting down or getting up when using the chamber 11 as a tub. When the rod 53 is held in place as illustrated, it is secured to the chamber wall after the upper and lower units have been assembled.

The outer chamber 12 serves as a lavatory and water closet, and communicates with the inner chamber 11 by means of the door 28. The lower portion of this chamber 12 may be constructed with a metal sheet 55 of copper or the like forming a floor, stamped with upwardly curving edges and having a shape corresponding to the central portion of sheet 26 in chamber 11. In one end of the chamber 12, a wash basin 56 is provided and in the other

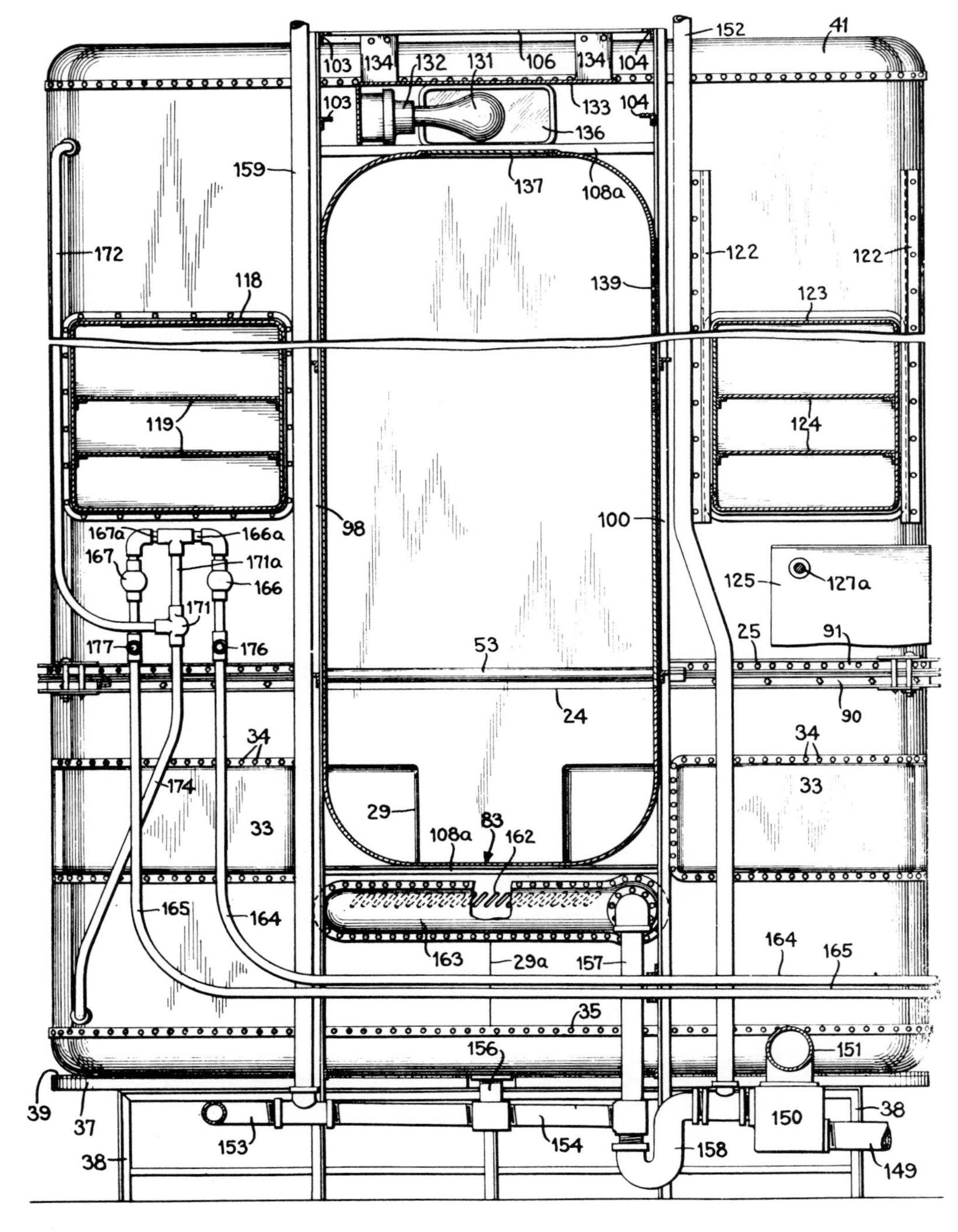

FIGURE 4

end of chamber 12, a water closet seat 57 is provided formed integrally with the adjoining walls. The wash basin 56 may be constructed from a single sheet of metal 58 stamped out to provide an inclined wall 59 extending up from the floor sheet 55 and suitably curved to provide the wash basin with an upstanding forward portion 60 and a vertical curved back wall 61 together with suitably curved side walls 62 and 63. The inclined portion 59 provides adequate foot space for one using the wash basin 56. An opening may be provided in the side wall 63 well above the bottom of the wash basin 56 which is closed by a cup-shaped metal member 64 providing a soap receptacle and affixed to the outer surface of wall 62 by rivets 65, welding or the like. Sheet 58 is also provided with a suitable opening located centrally of the inclined wall portion 59 and adapted to be closed by the flat plate 66, having suitable perforations 67 therein for a purpose to be described. This plate may be removably attached to the wall portion 59 by the bolts 68.

At the other end of chamber 12, a suitably curved sheet of metal 69 may be attached to the edges of the floor sheet 55 by rivets 70 or the like, and provided with upwardly extending curved edges 71 and shaped to provide a horizontal surface having an oval-shaped opening 72. The metal around this opening 72 is preferably bent downwardly as indicated at 73. A porcelain or other suitable water closet bowl 74 may thus be supported immediately below the oval opening 72. I prefer to provide a separate removable sheet of metal 74a which may be coated with porcelain or other suitable metal and which is provided with an oval opening conforming to the opening 72 and fitting thereover. This cover member may be then easily removed for cleaning the water closet bowl or other purposes. The extension of the side walls above the upwardly curved edges 71 of sheet 69 may be formed from a single sheet of metal 75, suitably curved and extending from the line 24 to the upper edge of the metal sheet 65. A portion of sheet 69 on the inner side of chamber 12 may be removably secured in place, if desired, by the bolts 69a, thereby providing access to the plumbing after the chamber 12 has been assembled and installed.

The upper portion of the side walls of the chamber 12 may be formed from lightweight metal, such as aluminum alloy, sheet steel, or plastics or other material, in a similar manner to the upper portion of chamber 11. The sheet 76 may be curved to provide suitable side walls extending above the line 24 as extensions of the walls 61, 62, 63. Another sheet 11 may be suitably curved to provide upward wall extensions from the sheet 75. Openings would thus be left on each side of chamber 12 between the sheets 78 and 71 and the sheets 56 and 75 above panel 84a in the shape of the doorways 13 and 28. The ceiling of chamber 12 may be formed from a single sheet 78 of stamped metal or molded plastics and permanently secured to the upper edges of sheets 76 and 77 as by rivets 78a.

When the two chambers 11 and 12 of the bathroom are assembled, the upper and lower sections of each of these chambers may be held together by an upper U-shaped member 79, side wall plates 80 and 81, and the lower U-shaped member 82 forming a frame for doorway 28. These members and plates may be provided with suitable flanges which may be removably attached to the edges of the various sheets defining doorway 28 by the bolts 82a. The lower member 82 thus provides a flat horizontal surface 83 which can be used as a step for stepping from the lavatory and water closet chamber 12 into the bathing chamber 11, or for a seat when bathing.

Inside chamber 12 and immediately below the horizontal surface 83, a panel 84 is removably attached to the side wall of chamber 12. This portion of the chamber side wall may be a single sheet 84a having a suitably opening normally covered by panel 24. Sheet 84a may be joined to sheet 58 along the seam 84b, and to sheet 75 along the seam 84c. Panel 84 is preferably stamped or drawn to provide a depression 85 on the inside of the panel, a flat

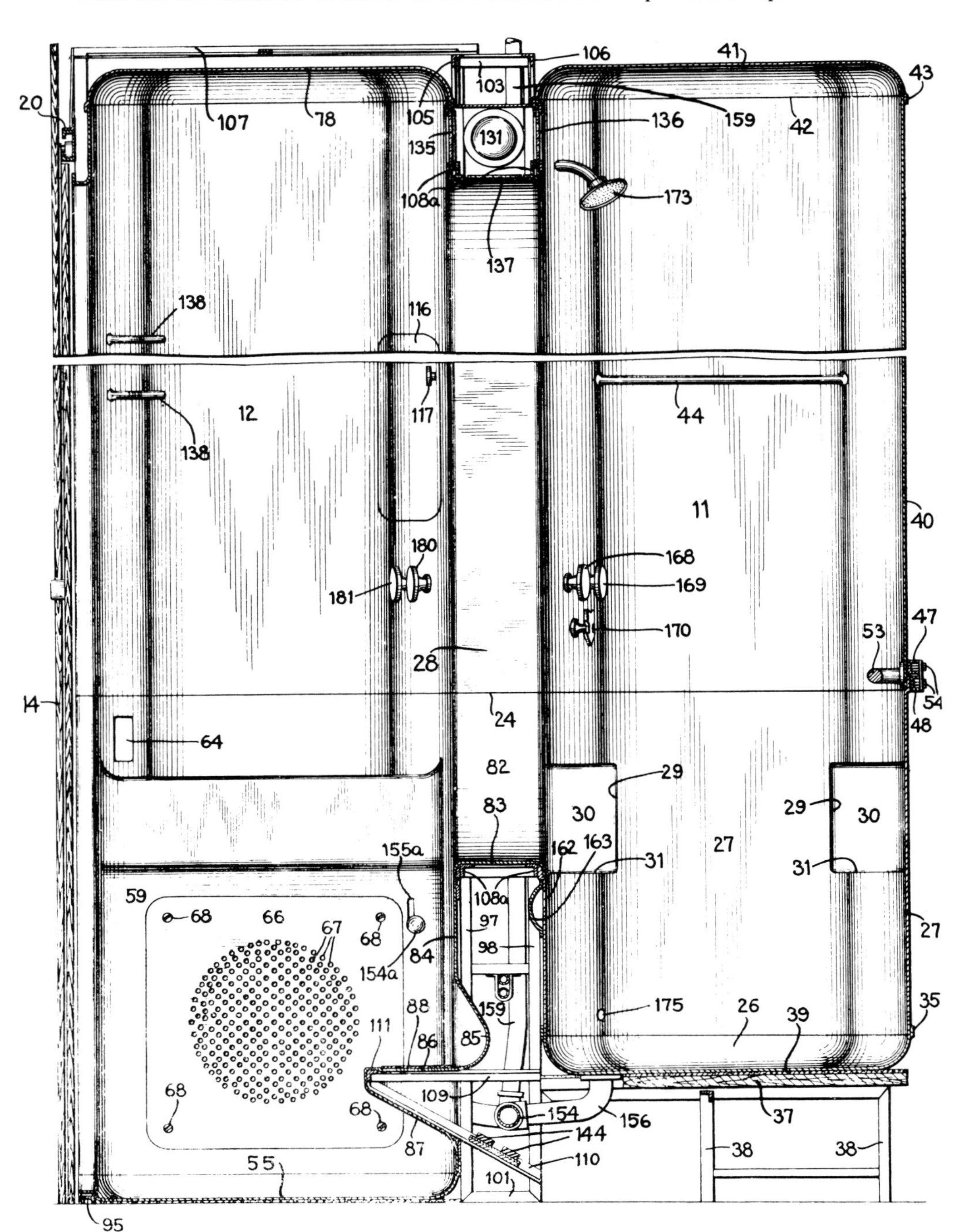

FIGURE 5

horizontal surface 86 of substantial width extending into the chamber below depression 85, and a curved inclined supporting surface 87 immediately below the horizontal surface 86. A mat 88 of pressed cork or the like may be employed on the surface 86, thereby providing a step to facilitate stepping from chamber 12 through the doorway 28 into the bathing chamber 11. Panel 84 may be removably attached to the structure by the bolts 89. The upper side wall and ceiling sheets 76, 77 and 78 of chamber 12 are preferably constructed as a single unit section which may be attached to the lower section of chamber 12 also constructed as a single unit in the same manner as described above in connection with the upper and lower sections of chamber 11. Thus, an outwardly opening channel-shaped member 90 may be affixed to the upper edges of sheets 59 and 75 and the channel member 91 may be affixed near the lower edges of sheets 76 and 77 as by the rivets 90a and 91a respectively, or by welding or

FIGURE 6

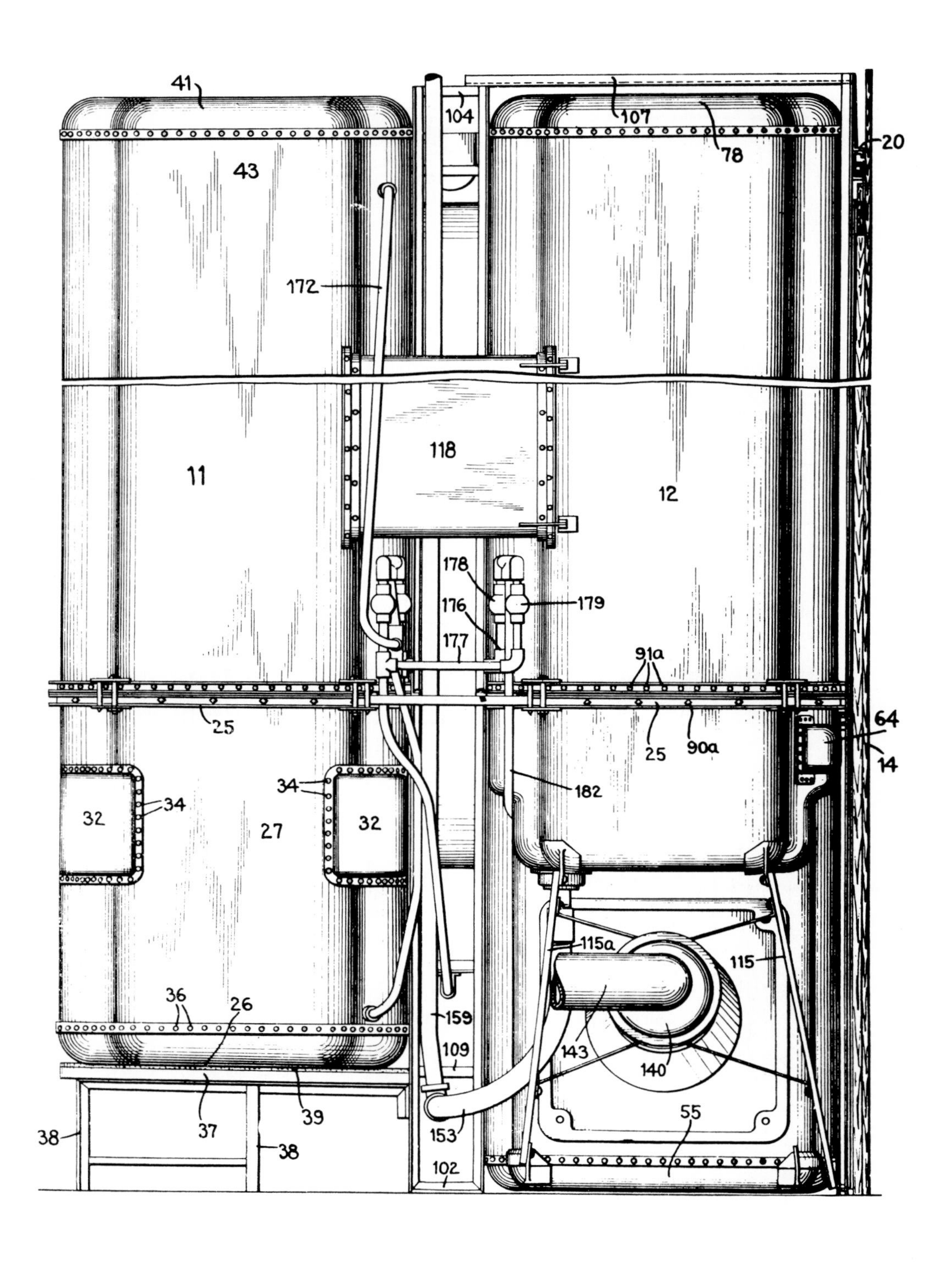

the like. These channel members 90 and 91 may be clamped together in any suitable manner when the two sections are assembled, as by the use of the same clamping plates 49 and 50 and bolts 51 as heretofore described for the unit sections of chamber 11.

A framework is preferably constructed around the center of chamber 12 and around doorway 28 to provide additional strength. This framework may conveniently be made of lengths of angle iron bolted, riveted, welded or otherwise secured together. For example, the vertical angle irons 93 and 94 joined together at the bottom by a suitable horizontal member 95 provide in effect a frame for the main door 13 to the bathroom. The upper ends of these frame members 93 and 94 may be connected together by a suitable cross member 96. The track 20 for the sliding doors 16 and 17 is preferably separate from this framework but suspended from the top of members 93 and 94 by the diagonal straps 23.

Disposed between the two chambers 11 and 12 is a box-shaped frame constructed around the doorway 28 and made up of the four upright frame members 97, 98, 99 and 100. The frame members 97 and 98 are joined together at their foot by a short cross member 101, and the frame members 99 and 100 are similarly joined together at their foot by a cross member 102. At their top these upright members are connected together by corresponding short angle iron pieces 103, 104 and by the longer angle iron pieces 105, 106. The upper portion of this box frame is preferably connected to the front frame members 93, 94 by the angle iron pieces 107 and 108. The portions of the box frame adjacent the upper and lower ends of doorway 28 may be provided with four horizontal members 108a connecting the upright members 97, 98, 99 and 100 to reinforce the doorway. In the lower portion of this box frame, the horizontal angled pieces 109 braced by the inclined angle pieces 110 and connected by the cross member 111 serve to reinforce the step 86 in the panel 84. This box frame may also advantageously be connected to and form some of the legs for the platform 37 which supports the floor of bathing chamber 11.

The front frame made up of vertical members 93 and 94 may also be braced by the inclined members 112 and 113 connected to the ends of cross member 95 and to intermediate portions of members 93 and 94. The inclined braces 114 and 114a may also advantageously be arranged to brace the sheet metal forming the water closet seat. Brace 114 extends from the end of member 95 to the nearest upper corner of sheet 57 while brace 114a extends from the inside corner of the floor sheet 55 to the other upper corner of sheet 57. At the other end of

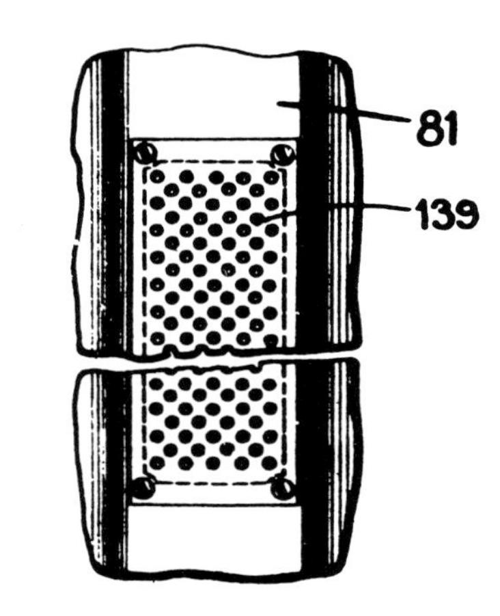

FIGURE 7

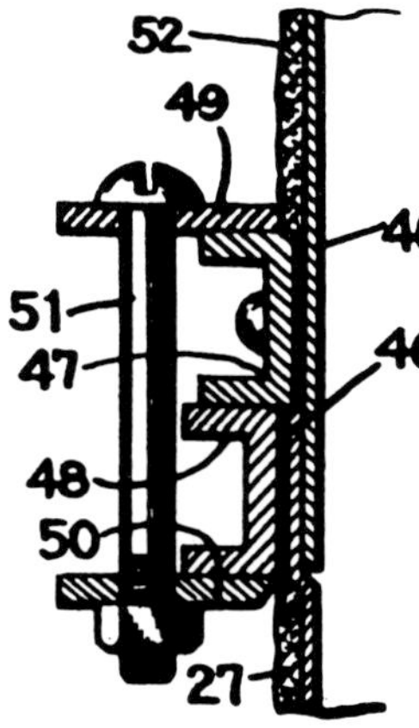

FIGURE 8

the chamber 12, the braces 115 and 115a are preferably employed to support the wash basin. Brace 115 may extend from the end of cross member 95 to one upper corner of the wash basin 56 while the other brace 115a extends from the other lower inside corner of sheet 55 to the other corner of wash basin 56.

Braces 114 and 115 are preferably pivotally connected at their upper ends so that during transportation of the lower section of chamber 12, the lower ends of these braces may be detached and swung inwardly. These braces are constructed of such a length that their lower ends in this latter position will bear against the lower outside corners of sheet 55 to which they may be attached during transportation, if desired. When the chamber 12 is assembled, the lower ends of braces 114 and 115 may then be swung outwardly and bolted or otherwise attached to the ends of frame member 95.

Inside chamber 12, a suitable door 116, provided inside with a handle 117, may be vertically hinged at 117a in an opening in sheet 76. By opening this door 116 inward into the chamber 12, access to a cabinet 113 having suitable shelves 119 is provided. The cabinet 118 is secured to the outside of the sheet 76 between the two chambers 11 and 12. Where the two chambers are to be employed together, this cabinet 118 may advantageously be attached also to the exterior of sheet 40 of chamber 11 as illustrated, thereby serving to hold the two chambers together more rigidly. A mirror may be affixed to the inner surface of door 116 and a light suitably connected to the inner surface of the door above the mirror may be used. Thus the lighted mirror is in the best position for use when the cabinet door 116 is open, and is concealed when this door is closed.

At the other end of chamber 12, a similar door 120, suitably hinged to sheet 77, may be provided. Parallel tracks 121 may be secured to the exterior surface of sheet 77 and another pair of tracks 122 may be attached to the outside of sheet 40 of chamber 11 to provide a vertical runway for the cabinet 123 having suitable shelves 124 therein. The advantage of this vertically movable cabinet 123, exposed by opening the door 120, is to provide access to the water tank 125 located immediately below it in case repairs should be necessary. This water tank 125 may be of a suitable shape to fit between the two chambers 11 and 12 and may be supported by the cross members 126a. A rod 127a extending into tank 125 and through sheet 75 and provided inside chamber 12 with a handle 127b may be used for controlling the water in tank 125 to flush the water closet. At one side of this water tank 125, a panel 126 may be provided hinged to sheet 75

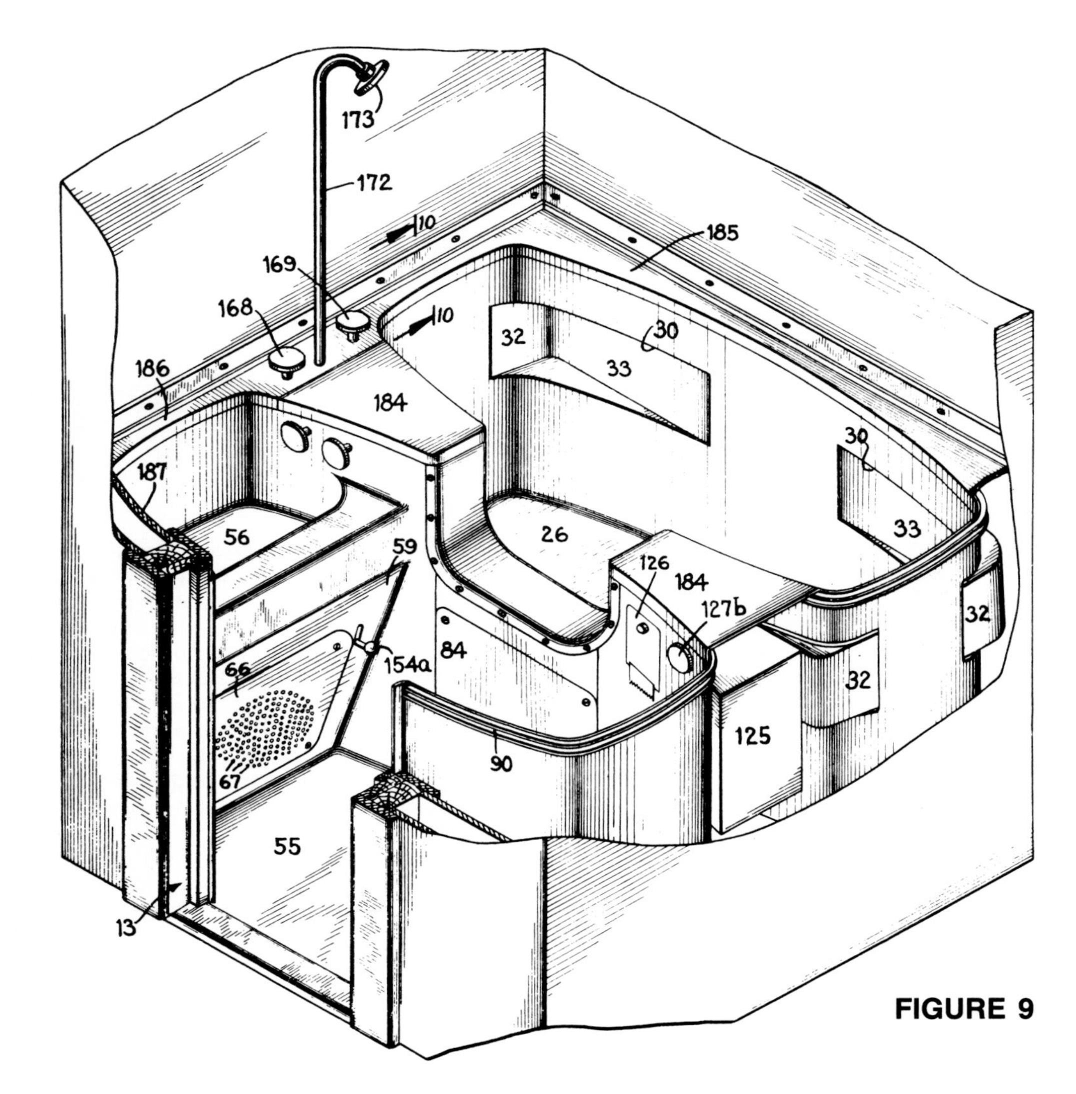

FIGURE 9

at its lower edge 127 and leaving a narrow horizontal opening 128. On the exterior surface of this panel 126, a suitable bracket 129 may be provided for supporting a roll of paper 130. The paper is thus adapted to be threaded through the opening 128 into chamber 12 and roll 130 may be replaced when needed by opening the panel 126.

Light for both chambers 11 and 12 may be provided from a bulb 131. A suitable fixture 132 supported in the frame 133 is suspended between the upper portions of chambers 11 and 12 by suitable strap members 134 from the cross bar 108 of the external angle iron framework. Transparent or translucent panels 135, 136 and 137 may be provided surrounding this bulb so that indirect lighting is supplied from a single bulb to both chambers 11 and 12. Towel racks 138 may be provided in chamber 12 above the wash basin 56 if desired.

Ventilation is obtained by drawing air in through the grill 139 removably set into the upper end of frame member 81 of doorway 28. Air and steam may be forcibly drawn downwards and exhausted from the chambers through the perforations 67 in panel 66 by means of a suction fan associated with the hood 140. This suction fan and hood may be supported by suitable straps 141 and braces 142 from the frame members 115 and 115a. The air withdrawn from the chamber 12 may be carried away through conduit 143 and discharged to the outer atmosphere in any suitable way. By this provision for lighting and ventilation, the necessity of using a special ventilator shaft or window for the bathroom is avoided. This is particularly advantageous for large buildings such as apartment houses where a bathroom may be near an air shaft, or when a bathroom is needed in a location of the building where a window would not be practical. Cutting new windows is also obviated with my bathroom when it is used for remodeling a dwelling.

Heating of the chambers 11 and 12 may be taken care of by drawing in warm air through the grill 139 from a room in which the bathroom may be installed. I prefer, however, to provide a heating unit under the step 86 such, for example, as the simple resistance heater 144 mounted on the diagonal brace 110. This resistance unit 144 heats the space below the step 86 and the step 83, and when a heat conductive metal, such as copper, is employed for the lower portions of the chambers, heat is readily and rapidly conducted throughout the entire lower section of chambers 11 and 12, thereby providing surfaces lukewarm and pleasant to the touch, as well as serving to heat the entire chambers.

The lighting, ventilation and heating may be controlled from inside or outside the bathroom as desired. For example, a switch 145 may be provided on the exterior of the panel 14 outside of the bathroom, together with a signal light 146 to operate simultaneously the heating unit 144 and the light bulb 131. The ventilation fan in hood 140 may be controlled by a three speed switch 147 adjacent the switch 145. Alternatively, the heating unit 144 and/or the ventilating fan may be independently operated by a switch 148, or switches, mounted on an inner wall or chamber 12 of the bathroom. As will be readily understood, the electrical connections for the ventilating fan, the light, and the heating unit may be taken from a single cable, which cable is provided for connection to the house lighting system in any suitable manner when the bathroom is assembled and installed.

The plumbing connections are preferably arranged in conjunction with a manifold to facilitate connection to the main supply and waste lines of the dwelling. The main waste pipe 149 for the outlet of all waste from the bathroom may be connected through a trap 150 of suitable construction to a pipe 151 which is in turn connected to a side outlet from the water closet bowl 14. This trap 150 may be ventilated through pipe 152 if necessary. Waste water from the wash basin 56 may be withdrawn through the outlet pipe 153. Pipe 153 may be connected to the main pipe 154 which also serves to withdraw waste water from the bottom of the bathing chamber through outlet 155 and pipe 156. The bathing chamber 11 may also be provided with an overflow outlet connected to pipe 157 which joins pipe 154 near the trap 158 connected to trap 150. Another ventilation pipe 159 may be provided for this portion of the waste manifold, if necessary.

The handle 154a extending into chamber 12 through the slot 155a may operate a suitable closure 160 in the wash basin 56, which allows water to flow into the auxiliary chamber 161 arranged below the wash basin outlet and connected to pipe 153 to carry away waste water from the wash basin outlet proper and from the overflow outlet for the waste basin. The overflow outlet for the bathing chamber 11 may comprise a series of diagonal openings 162 in the chamber wall immediately below the threshold 83 of the doorway 28. Covering these openings 162 is a stamped metal fitting 163 riveted or otherwise secured in watertight fashion to the pipe 157 and to the outer surface of the wall of chamber 11 between chamber 11 and the chamber 12.

Hot and cold water may be supplied for the entire unit through the main supply pipes 164 and 165 suitably bracketed together and supported by the angle iron framework between the two chambers 11 and 12. These pipes 164 and 165 are connected to the valves 166 and 167, operated by the handles 163 and 169, respectively, inside the bathing chamber 11. Immediately below and disposed between these handles 168 and 169 is a third handle 170 which may be in the form of an indicator, and which operates a three-way valve 171. Pipe connections 166a, 167a, and 171a are provided between these three valves. Valve 171 determines whether the hot and/or cold water from valves 166 and 167 flow through pipe 172 to the shower spray 173 in chamber 11, or through the pipe 174 to the opening 175 near the floor of the bathing chamber 11. The hot and cold water inlet pipes 164 and 165 are also connected by branch lines 176 and 177, respectively, to the valves 178 and 179. These valves are controlled by handles 180 and 181 inside the chamber section 12 immediately above the wash basin 56. Hot or cold water or a mixture thereof from valves 178 and 179 flows through pipe 182 to the inlet opening 183 in the side of the wash basin 56 nearest the user. In filling the basin, therefore, the water spurts away from the user, avoiding splashing.

It will thus be seen that the prefabricated bathroom of my invention is constructed with the three primary bathroom fixtures, tub, wash basin and water closet seat as integral parts of the floor and walls of the lower one-third of the bathroom. This lower one-third of the bathroom is divided into two sections, one for each of the chambers 11 and 12 of the bathroom and each of these chambers 11 and 12 may be constructed as two separate prefabricated units including the necessary fixtures. The wall extension and ceiling units, or the upper portions of the two chambers 11 and 12 of the bathroom may similarly be constructed as two separate prefabricated units. The frame members 79, 80, 81 and 82 for doorway 28, the angle iron framework, and the plumbing manifold constitute separate units, together with any accessories of an optional nature, such as cabinets and the other relatively few pieces that require attachment to the various units when the bathroom is assembled.

In assembling and installing my new bathroom, the lower unit sections of chambers 11 and 12 are disposed side by side with the lower section of chamber 11 supported on the platform 37, the water closet bowl 74 having been previously bolted or otherwise secured in the proper position. The box frame of angle iron members is placed in correct position between chambers 11 and 12 and the flanged U-shaped doorway frame member 82 is then bolted to the two lower sections of chambers 11 and 12 to hold them in place. The upper unit section of chamber 11 may now be fitted in place on the lower section and the clamps 49 and 50 applied and secured by bolts 51. The plumbing manifold may now be connected in place to chamber 11 and the main supply and waste lines of the dwelling.

The accessories, such as the water tank 125, light bracket 132, cabinets, heating unit, rod 53 and ventilating unit 140 may conveniently be secured in the proper position at this time and connected by any piping or wiring necessary. Finally, the upper section of chamber 12 may be fitted and clamped in place, and the remainder of the angle iron framework may be adjusted. Before securing the doorway frame members 79, 80 and 81 in place, the plumbing connections should be completed and the cabinets and other accessories suitably attached to the walls of chamber 12 as may be necessary.

As soon as the doors 16 and 17 are hung, the panel 14 secured in place and the switches carried by it connected, the bathroom is ready for immediate use.

I have found that this bathroom can be readily made from integral metal stampings. When the units are constructed with the curved side walls as illustrated in Fig. 2, the various units may be made from stampings of a very light gauge sheet metal, such as copper, while still providing adequate strength. By the curved wall and integral

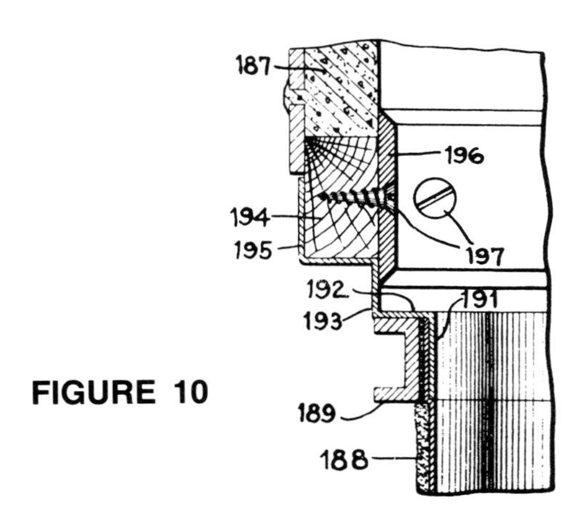

FIGURE 10

metal stamping construction illustrated, no corners are left which are difficult to clean.

Accessibility to the plumbing and other appliances of the bathroom after assembly is provided through the panels 86, the U-shaped door frame member 82, the panel 84, grill 139 and the vertically movable cabinet 120. Other removable panels may be provided at any points desired.

After assembly, or before assembly, the copper sheets of the lower portion of the bathroom unit may be finished as by spraying a metal coating of a corrosion resistant alloy on the surfaces inside the chambers 11 and 12, followed by buffing or polishing the surface to give it a slightly hammered texture which will prevent slipping and provide a surface with a minimum of dirt adherence. An alloy of about 98% tin and 2% silver is suitable for this coating, although other alloys could, of course, be used. If the upper portions of the chambers 11 and 12 are made of some sheet metal such as steel, it may be desirable to employ an insulating material such as a lacquered masking tape at all joints between the steel and copper to avoid the formation of galvanic couples which might promote corrosion. The electrical connections are also preferably grounded to the lower portions of the chambers.

The outer chamber 12 of the bathroom may be used alone without the bathing chamber 11 by simply providing a panel for closing off the doorway 27. This chamber 12 may be very suitable for use alone where a "powder room" is desired, or with the bathing chamber 11 in places where space is at a premium, such as in boats, trains and the like. A curtain may be provided for the doorway 28 such as a metal Venetian blind to separate the two chambers when they are used together. Such a curtain as a Venetian blind is particularly suitable because it allows the ventilating system to exhaust steam from the bathing chamber 11, thereby ventilating both chambers at the same time even though both are being used simultaneously. Chamber 11 may also be used alone when the facilities of chamber 12 are not required.

The complete prefabricated bathroom with both chambers 11 and 12 is so devised that it may be installed in a room of a completed house as a unit cabinet with or without partitioning off, and without the necessity of providing a window. The whole assembly may be constructed of sheet metal such as aluminum with a total weight of approximately 250 lbs., and made up of separate integrated sections, each of which can be readily carried by two men through the doorway of the ordinary house and assembled ready for use in a few hours. The bathroom may also be installed as easily during the erection of a dwelling.

A modification of my bathroom is illustrated in Figs. 9 and 10 and comprises substantially the same construction as described above, with the exception that the portions of the chambers 11 and 12 above the line 24 in Figs. 1–8 are left off and the lower portions of the chambers are capped to conceal the plumbing. The walls of plaster, wood, tile or the like of the house in which the lower portions are installed provide the upward extensions for the walls of the chambers. Similar reference numerals indicate similar parts in Fig. 9 to those described in connection with Figs. 1–8. In this modification, suitable metal capping plates 184, 185 and 186 may be provided to cover the openings between the chamber walls and are preferably removable to get at the plumbing.

The shower spray 173 may be attached to the pipe 172 which extends up through the cover plate 186 instead of being between the walls of the chambers 11 and 12 as in Figs. 1–8. The control handles for the shower 168, 169 may also be disposed in the cover plate 186 instead of on the side wall. For this modification, some form of window or ventilator is necessary. The cabinets and other accessories for the room above the line 24 in Figs. 1–8 must also be supplied separate from the prefabricated metal units of the bathroom.

For joining the metal of the lower portion of this bathroom unit to a plaster wall 187, such as illustrated in Figs. 9 and 10, the wall 188 of the bathroom unit may have suitable channel members 189 attached to the outer surface adjacent its upper edge. The cover members 190 have flanges 191 depending inside the chambers and lying against the offset upper edge 188 of the chambers to provide a flush interior surface. Horizontal portions 192 of the cover members 190 overlie the channel members 189 extending to the vertical portions 193 in the plane of the plaster wall 187. Wooden members 194 support the lower edge of the plaster and lie flush with it. A further flange 195 of the cover members 190 extends underneath and up along the back of these wooden members 194. These joints and the wooden members 194 are concealed by the strip of beveled metal plate 196 held in place by the wood screws 197. A relatively smooth exterior is thus presented to the room containing the bathroom chamber units. The prefabricated portions of this modification of my invention may, of course, be constructed at less cost and of less weight than the preferred modification although more time may be required under certain conditions for installation.

The terms and expressions which I have employed are used as terms of description and not of limitation, and I have no intention, in the use of such terms and expressions, of excluding any equivalents of the features shown and described or portions thereof, but recognize that various modifications are possible within the scope of the invention claimed.

6▲DYMAXION DEPLOYMENT UNIT (SHEET) (1944)

U.S. PATENT—2,343,764

APPLICATION—MARCH 21, 1941

SERIAL NO.—384,509

PATENTED—MARCH 7, 1944

FIGURE 1

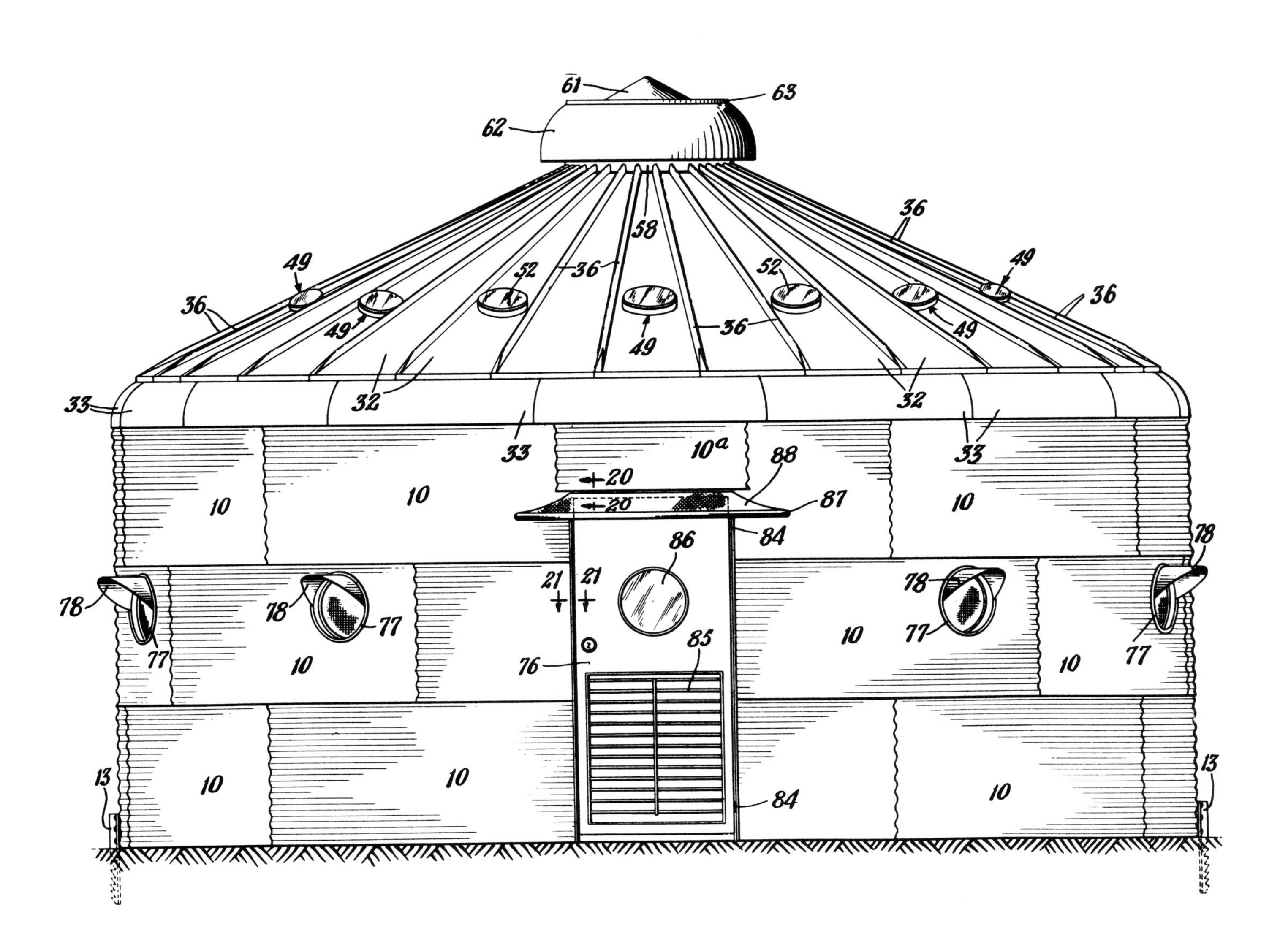

LOUISIANA STREET
CALIFORNIA STREET

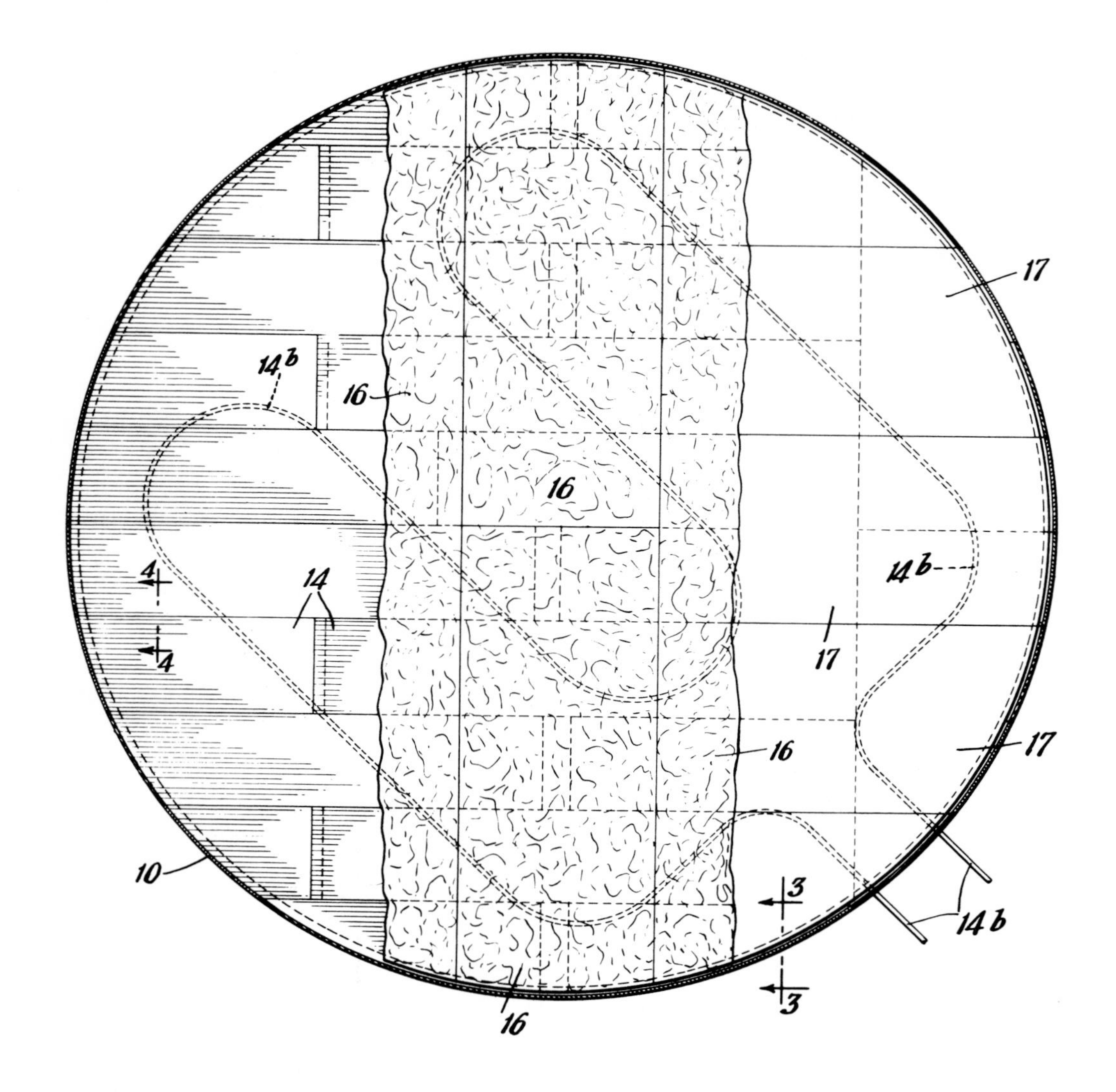

FIGURE 2

WHEN THE NEW DEAL SET ABOUT TO SOLVE the problems of a paralyzed economy, amongst the greatest of these problems were those of the farmers. The U.S. Department of Agriculture developed a program entitled the Normal Grainery Program. This called for the grain producers to grow bumper crops and store their grain in galvanized corrugated sheet-steel grain bins with flat conical tops. The bins were of standard size. They were rodent- and rainproof. When a farmer had filled a grain bin, the local bank would seal the bin and pay the farmer for it at a fixed price. The government would then sell the grain wherever it found it expedient, so that the price paid to the farmer was not depressed by a national abundance. I found that these eighteen-foot-diameter mass-produced grain bins could be lined inside with fireproof insulation and could be floored with plywood. With windows and other modifications plus Sears Roebuck furnishings (kerosene flame icebox, stove, and other items) a bin could be turned into a humble but adequate dwelling unit.

At this time, 1940, World War II was clearly approaching. It was assumed by the British that their major industrial cities would be bombed, and leading Scottish landowners offered their lands for the encampment of people from the bombed-out British cities. The Scottish government asked me to develop the Dymaxion Deployment Units with the converted grain bins. When the war broke out, the British found that the steel requirements of their armaments would have to take priority over their people's livingry requirements, and the Dymaxion Deployment Unit orders of the Scots were canceled. The United States later bought the Deployment Units and installed them with air-conditioning units at the head of the Persian Gulf to house the American aircraft-ascending teams who delivered the "flyaway" U.S. aircraft to our then allies, the Russians.

UNITED STATES PATENT OFFICE

Richard Buckminster Fuller, New York, N.Y., assignor to The Dymaxion Company, Inc., Wilmington, Del., a corporation of Delaware

BUILDING CONSTRUCTION

This invention relates to prefabricated buildings, and more particularly to prefabricated shelter units capable of quick erection at low cost.

A considerable effort is being made and has been made in the past to develop a truly low cost housing unit that would be capable of mass production. Most of such housing units, however, have conformed more or less strictly to conventional architectural concepts of housing units, the principles of which have made it extremely difficult to produce prefabricated houses capable of rapid mass production and easy erection in a few hours time.

The problem is not simplified by the fact that such buildings must afford not only good protection against the weather, but should also provide comfortable as well as conveniently arranged living quarters.

One object of my invention is to provide a shelter unit or building having curved walls that answers these requirements and does not require the use of the ordinary internal framework or stress members.

A further object is the provision of a shelter unit of generally curved shape composed of a laminated shell, the different layers of which strengthen the other layers by reason of their manner of assembly and the curvature of the shell.

Another object is the provision of a shelter unit in which an outer curved shell of sheet material is strengthened and held more rigidly in place by sheets of resilient material sprung and held in the sprung position against the inside of the sheet material.

A further object of the invention is the provision of a novel floor structure for a building that is weather proof and capable of quick assembly at low cost from standard sheet materials.

Another object is the provision of a building having a curved side wall, a roof, and a curved eave portion connecting the roof and side wall that reinforces the entire structure.

A further object of the invention is the provision of low cost housing units capable of prefabrication on a large scale, quick assembly, and of being demountable for re-assembly at a different location.

FIGURE 4

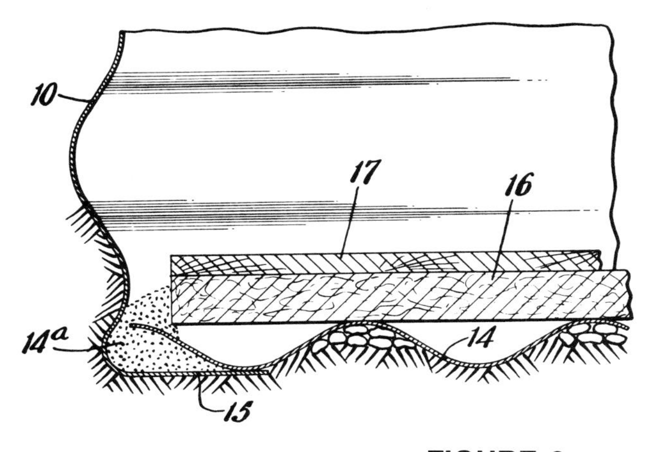

FIGURE 3

Further objects of the invention are the provision of an adjustable ventilator system and movable partitions for such a shelter unit or building.

Other objects and advantages of the invention will be described and will be apparent from a description of the embodiments of my invention shown in the accompanying drawings, in which

Figure 1 is a front elevation of a housing unit constructed in accordance with my invention.

Figure 2 is a plan view partly broken away to show a built-up floor for such a housing unit.

Figure 3 is a vertical sectional view on an enlarged scale of a detail, taken on the line 3—3 of Figure 2, to show the junction of the floor with the side wall.

Figure 4 is a vertical sectional view of the base layer of the floor, taken on the line 4—4 of Figure 2.

Figure 5 is a fractional horizontal sectional view showing the laminated structure of the curved side walls.

Figures 6, 7 and 8 are fractional vertical sectional views of the wall structure showing the bottom, intermediate and top of the side walls, respectively, said views being taken approximately on the line *x—x* of Figure 5.

Figure 9 is a fractional vertical sectional view, taken on about the line 9—9 of Figure 5.

Figure 10 is an elevation of a section of a slotted channel bar that may be attached to the side walls as shown in Figures 5–8.

Figure 11 is a top plan view of a wall bracket that may be supported by the channel bar shown in Figure 10.

Figure 12 is a vertical sectional view showing a shelf supported by one of these brackets fixed in the slotted channel bar.

Figure 13 is a rear elevation of one of the brackets that may be used for supporting shelves, cabinets or other wall fixtures.

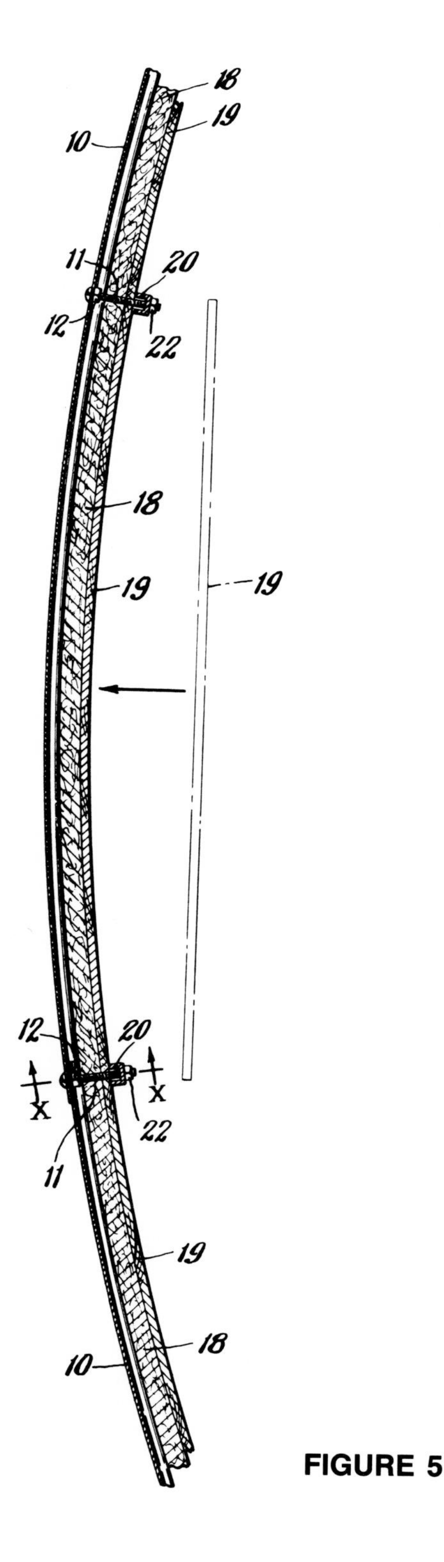

FIGURE 5

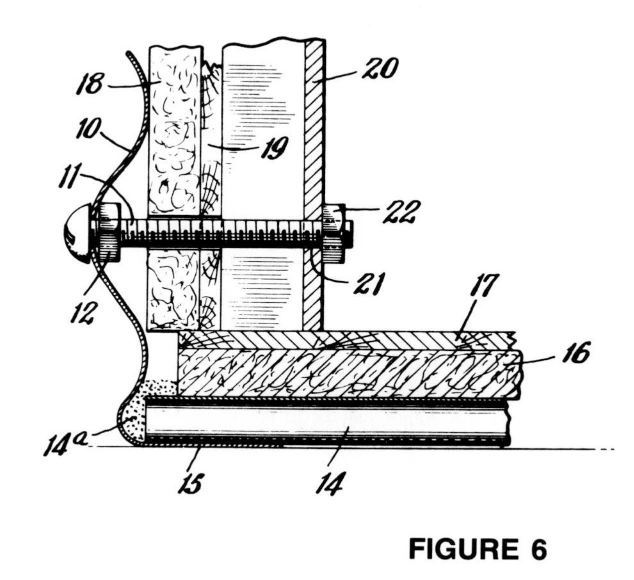

FIGURE 6

Figure 14 is a vertical sectional view through the eave portion of the building shell or wall.

Figure 15 is a vertical sectional view through approximately the center of one of the roof panels.

Figure 16 is a fractional sectional view, taken on the line 16—16 of Figure 15, to show the manner in which the roof panels and insulation are joined.

Figure 17 is a plan view of one of the roof panels.

Figure 18 is a fractional sectional view through one of the roof panel apertures for light, taken on the line 18—18 of Figure 17.

Figure 19 is a vertical sectional view through a centrally located ventilator at the top of the roof.

Figure 20 is a fractional vertical sectional view of a door frame, taken on the line 16—16 of Figure 1.

Figure 21 is a fractional horizontal sectional view of the door frame, taken on the line 17—17 of Figure 1.

Figure 22 is a horizontal sectional view through the

building to illustrate one possible arrangement of partitions and furniture.

Figure 23 is a perspective of an end of a movable partition that may be used to divide the interior of the building into two or more rooms.

Figure 24 is a fractional elevational view of a modification of a building constructed according to my invention.

Figure 25 is a horizontal sectional view through the side wall of the building shown in Figure 24.

Figure 26 is a fractional vertical sectional view of the side wall, taken approximately on the line 26—26 of Figure 25.

Figure 27 is a fractional vertical sectional view of the side wall, taken approximately on the line 27—27 of Figure 25.

Figure 28 is a vertical sectional view of a detail of the wall structure, taken on the line 28—28 of Figure 26.

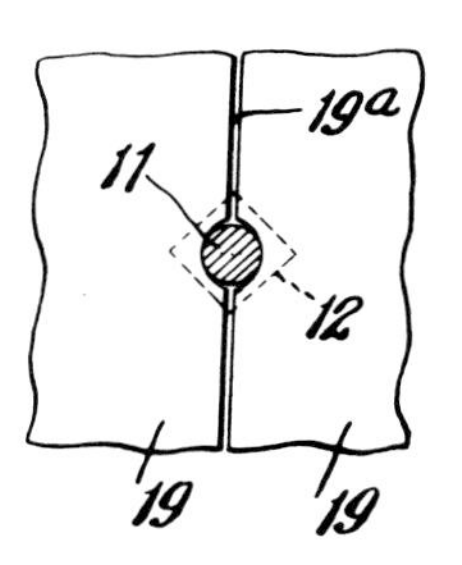

FIGURE 9

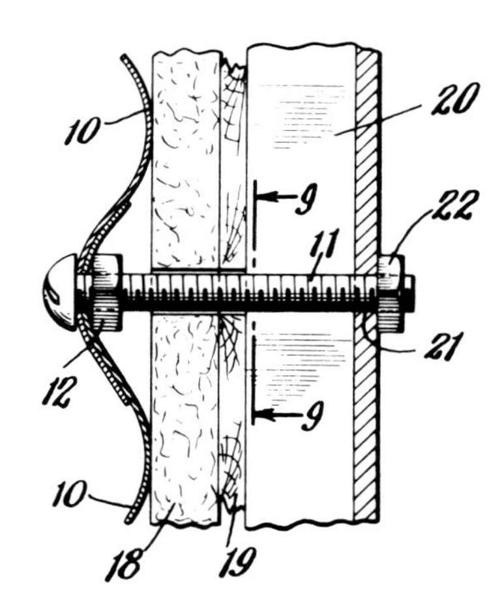

FIGURE 7

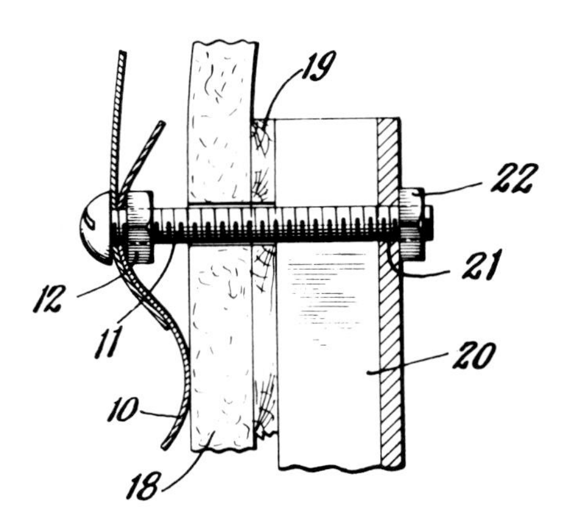

FIGURE 8

Figure 29 is a vertical sectional view through the eave and roof portion of the building shown in Figure 24.

Figure 30 is a plan view of a portion of one of the roof sections.

Figure 31 is a perspective view of the lower portion of one of the roof sections.

Figure 32 is a vertical sectional view through a modified ventilator for the building shown in Figure 24.

Figure 33 is a partially diagrammatic view of the interior of this building to illustrate a modified movable partition arrangement.

The embodiment of my invention illustrated in Figures 1 to 21 of the drawings is composed of a number of rectangular sheets 10 of suitable sheet material, such as a lightweight plain or corrugated galvanized steel or other metal, curved to form, when assembled, a substantially cylindrical shell. Three horizontal rows of such sheets are shown in Figure 1, although a smaller or larger number of rows may be used, and these sheets preferably overlap each other, with the uppermost sheet overlapping the sheet below it, and the vertical edges overlapping each other. The sheets 10 may be bolted together where they overlap by means of the bolts 11 and nuts 12 shown in Figures 5 to 8. A suitable anchoring means, such as the angle iron posts 13, may be driven into the ground around the lower edge of the lower row of sheets 10 and bolted to the sheets to hold them securely in place during and after erection of the unit. The sheets 10 are also preferably staggered with respect to the adjacent horizontal rows of sheets.

Referring now more particularly to Figures 2, 3 and 4, a concrete or other permanent type of flooring may be provided for the building, although a more portable form of flooring may be used consisting of a lower layer of sheet metal or other strong sheet material such as corrugated sheets of galvanized steel 14. These sheets may be generally rectangular in shape except for their outer edges which are curved to produce a generally circular floor, and the sheets 14 are preferably laid with their sides and ends overlapping each other. A suitable asphaltic mastic or other water-proofing material may be applied to these joints if desired. The sheets 14 may be laid directly on the ground or, if desired, a concrete or other suitable foundation may be provided for them.

The lower edges of the lower row of said sheets 10 are preferably bent inwardly at right angles to form the inwardly projecting flange 15 and the outer curved edges of the base layer floor sheets 14 may be laid on top of this inwardly projecting flange and sealed with a suitable asphaltic mastic 14a or other suitable material.

On top of sheets 14 may be placed a layer of flat sheet material. This second layer may be composed, for example, of a number of rectangular sheets 16 of a suitable insulating board of more or less porous or solid construction, with these rectangular sheets extending in a general direction at right angles to the direction of the sheets 14. These sheets 16 may be laid with their edges abutting each other or, if desired, the edges may be grooved and overlapped. In order to provide as weatherproof a floor as possible, I prefer to coat, as by spraying or dipping, the sheets 16 with a suitable asphaltum composition before they are laid in place.

A third layer may be applied on top of the sheets 16 composed of a plurality of rectangular sheets 17 of pressed wood, plywood or other suitable finishing material. These sheets 17 may be of generally rectangular shape, except for their curved outer edges, and laid in a general direction extending at right angles to the direction of the sheets 16 or, in other words, in the same general direction as the bottom layer of sheets 14. The sheets 16 and 17 may thus be arranged so that the joints of the different layers do not coincide in any portion of the floor.

This top layer of pressed wood, plywood or similar material may, if desired, be held in place by means of a suitable adhesive between the sheets 16 and 17, and if desired a water-proof paper, canvas or layer of similar material may be interposed between the sheets 16 and the sheets 17. This provides a floor with a suitable finish for direct use, although if desired it may be covered, of course, by linoleum or other suitable floor coverings.

The use of corrugated sheet metal for one layer of the floor furnishes a certain amount of resiliency to the floor, and, in addition, the sheets of metal are held more securely in place relative to each by reason of the corrugations. Such sheets are also easier to lay flat on the earth because the earth may be readily conformed to their under surface.

While the outer edges of the sheets 16 and 17 may extend into contact with the side sheets 10, it is preferable to leave a small clearance at this point to prevent subsequent buckling of the sheets 16 and 17 after the floor is laid.

Provision may also be made under the floor for heating by the use of a coil of pipe as shown in dotted lines at 14b

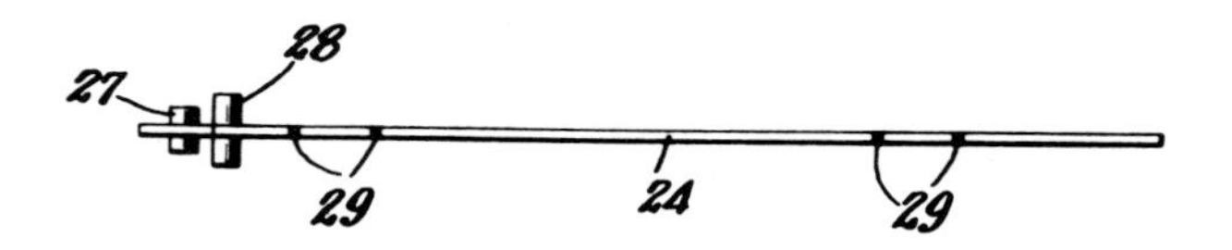

FIGURE 11

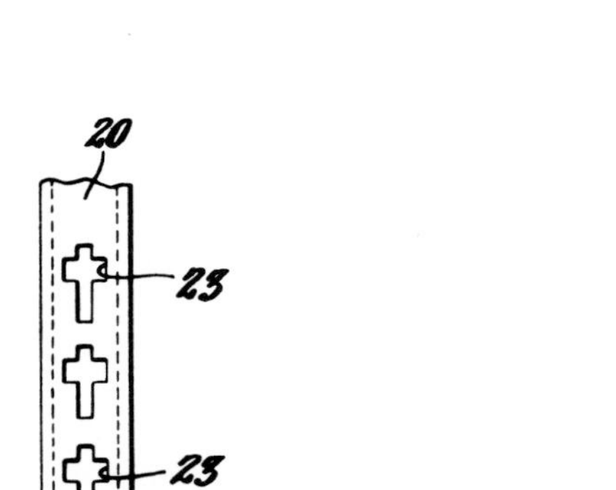

FIGURE 10

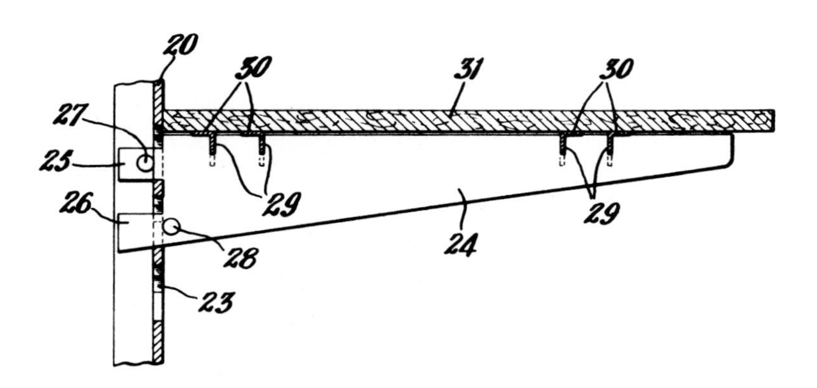

FIGURE 12

FIGURE 13

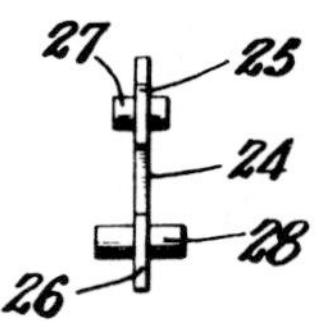

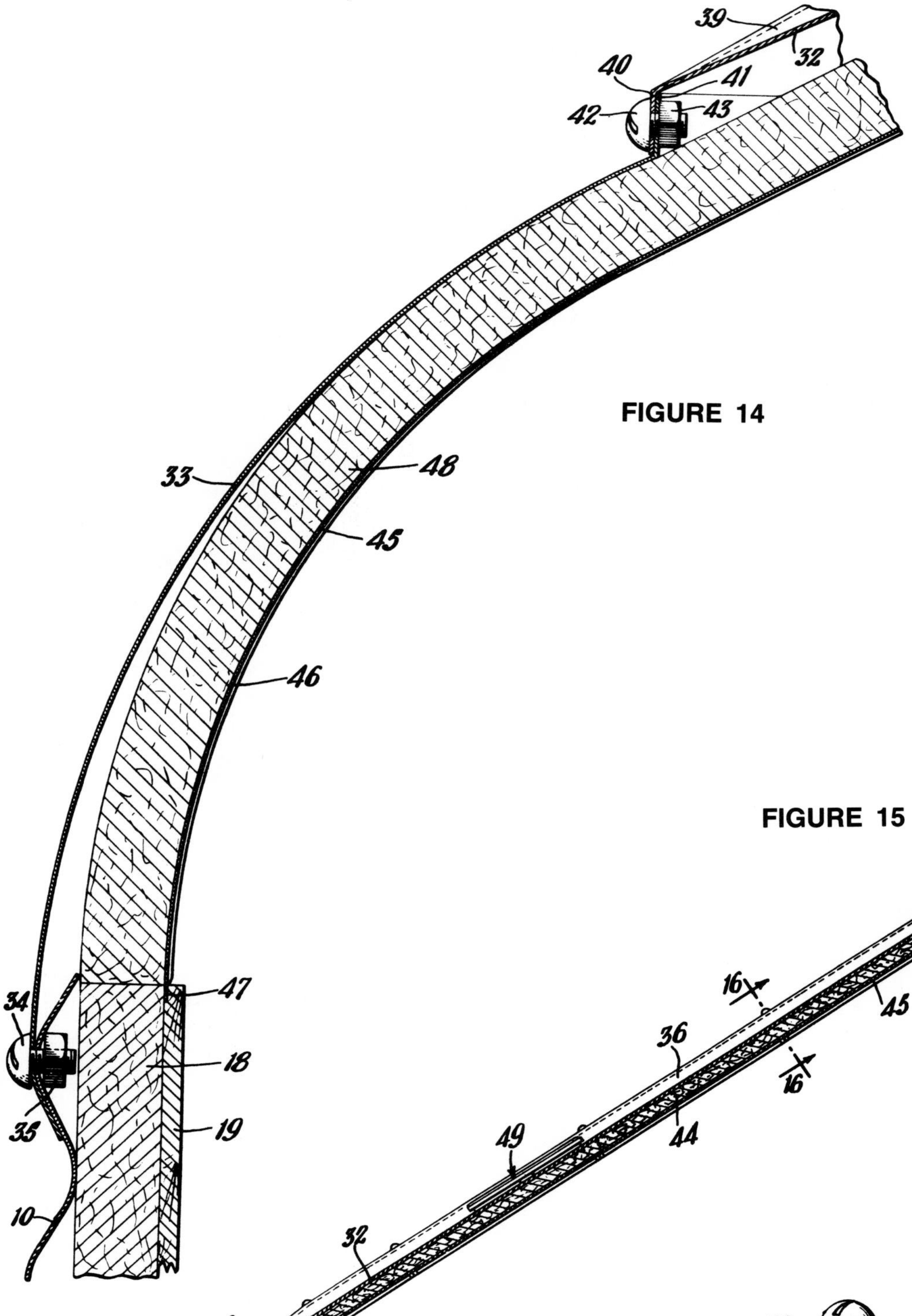

FIGURE 14

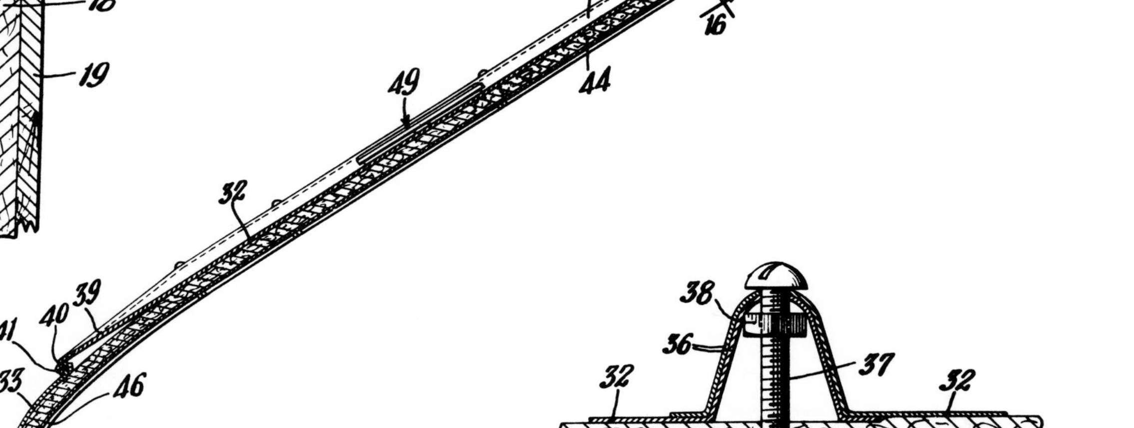

FIGURE 15

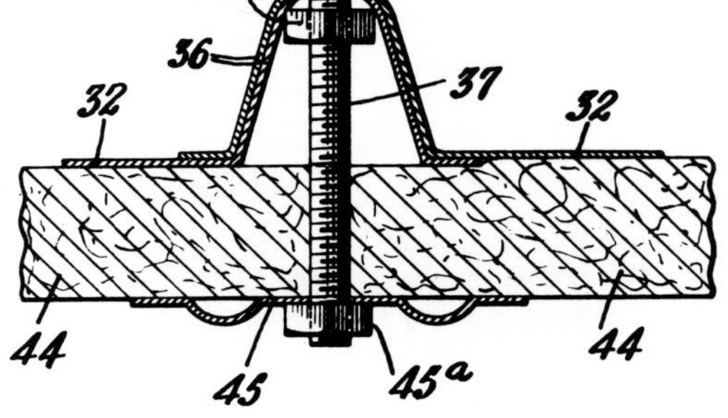

FIGURE 16

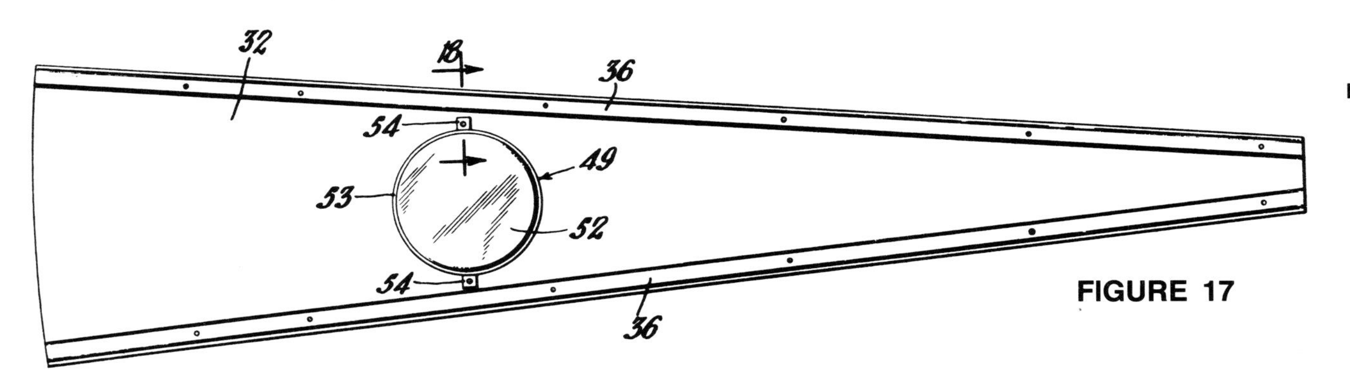

FIGURE 17

in Figure 2. Such a coil may be conveniently connected to the hot water supply so that hot water passes through the coil before reaching faucets or other outlets in the building. By placing such a heating coil under the metal floor sheets 14, the heat is uniformly distributed by the high conductivity of these sheets over the entire floor. If such a heating arrangement is employed, the various layers of the floor should not have too great an insulating effect but should be constructed of materials that will allow the heat to pass through into the room.

The side wall structure may be composed of the curved sheets 10 of suitable material such as corrugated galvanized sheet steel as described above, these sheets being held together where they overlap each other by means of the bolts 11 and the nuts 12. Suitable water-proofing or sealing means may be applied at these joints if desired.

The bolts 11 preferably project a substantial distance into the building, and when the sheets 10 are of generally rectangular shape and arranged in staggered fashion, additional inwardly projecting bolts may be affixed to the centers of these sheets to provide evenly spaced vertical rows of inwardly projecting bolts around the side walls.

Suitable insulating material in the shape of rectangular sheets, preferably of a height corresponding to the height of the vertical walls and of a width just allowing them to fit in a curved position between the vertical rows of inwardly projecting bolts 11, may now be applied inside the sheet metal walls. For example, sheets 18 of a suitable insulating material, preferably having some inherent resilience, may be provided with semi-circular notches along each side at points corresponding to the position of the inwardly extending bolts 11. The sheets are then held at the sides and the center portions of each sheet are pressed against the sheet metal wall to spring the sheets of insulating material into place. Natural resilience of the sheet insulating material will tend to press the side edges against the bolts 11.

By applying sheets of insulating material all around the interior of the side walls in this fashion, a structure is provided in which a curved wall of sheet insulating material is formed, held securely in place by its own resilience and stiffening the entire outer sheet metal wall.

Another layer comprising sheets 19 of plywood, pressed wood, or similar material may now be applied to the wall in a manner similar to the application of the sheet insulating material. In other words, sheets 19 of a height extending substantially the entire distance of the vertical wall and of a width allowing just sufficient tolerance to fit between the rows of bolts 11 in a curved position, may be similarly notched along their sides at points corresponding to the positions of the bolts 11, and sprung inwardly into place between the vertical rows of bolts. These sheets, referring to Figure 5, may be naturally flat as shown in dot-dash line, but when sprung into place

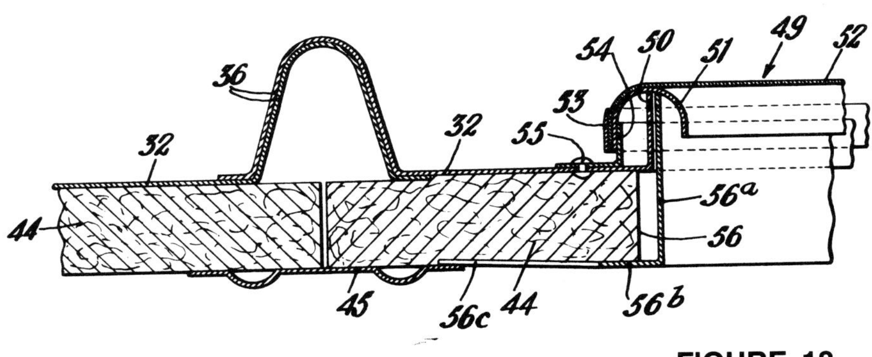

FIGURE 18

between the rows of bolts assume more or less of the curvature of the outer sheet metal wall made up of the sheets 10 and, of course, the similar curvature of the sheets 18 of insulating material.

The lower edges of the sheets 18 and 19 rest on the outer upper surface of the floor sheets 17, thus giving a finished appearance to the interior and assisting in maintaining a tight joint at the base of the side walls.

As shown in Figure 9, the notched edges of the sheet insulating material 18 and of the finish sheets 19 fit around the shanks of the bolts 11 so that the edges of the sheets 19 abut each other or practically meet, as may be

thus cover the vertical seams 19a between the side edges of sheets 19 and, in addition, hold these edges securely in place against any accidental displacement.

Referring now to Figures 10, 11, 12 and 13, the channel sections 20 or other cover members may, if desired, be utilized to support shelves, cabinets and various other types of fixtures. For example, the faces of these channel strips 20 may be provided with a number of spaced keyhole or cross-shaped slots 23 and various fixtures may thus be secured in these slots. For example, wall brackets 24 may be utilized having a pair of rearwardly projecting ears 25, 26 with suitable pins 27, 28 pressed through

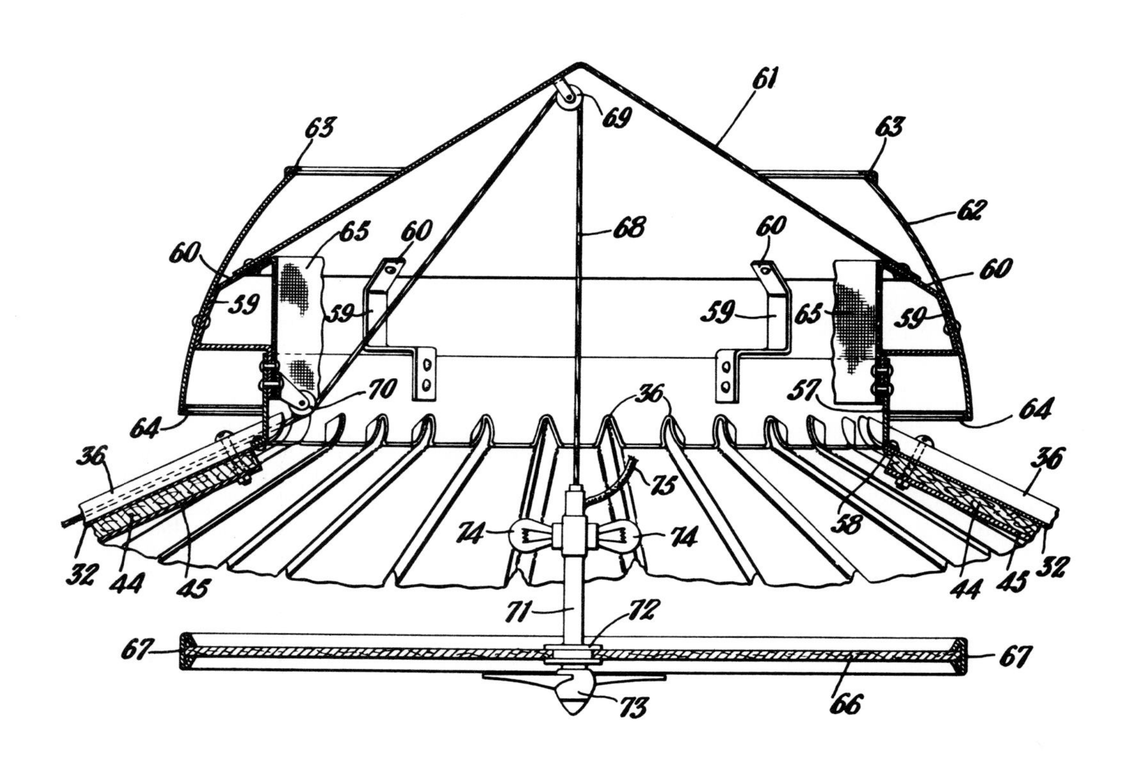

FIGURE 19

desired, presenting a smooth, pleasant finish to the walls inside disturbed only by the ends of the inwardly projecting bolts 11 and the vertical seams where these sheets come together. If desired, however, the sheets 18 and 19 may be sprung into place without notches in their sides, thus leaving wider seams between the sheets.

To cover the seams between the side wall sheets, vertical channel members 20, or other suitable cover members, having openings 21 to receive the ends of the bolts 11 may be fitted in place over the ends of the bolts 11 and held in place securely by the nuts 22. The nuts 22 may, if desired, be special round headed finish nuts that conceal entirely the ends of the bolts 11. These channel strips 20

openings in the brackets, or welded in place at right angles to the plane of the brackets 24.

By providing the pins 27 and 28 in a staggered position relative to each other as illustrated, the ears 25, 26 and the pin 27 may be inserted in two adjacent cross-shaped slots 23 of the channel member 20 and then allowed to drop down into a position in which the pin 27 presses outwardly against the inside of the channel strip 20 and the pin 28 is forced inwardly against the exterior thereof, thus holding the bracket 24 securely in place.

These brackets may be used for supporting shelves in a relatively fixed position. For example, the brackets 24 may be provided with narrow vertical notches 29 to re-

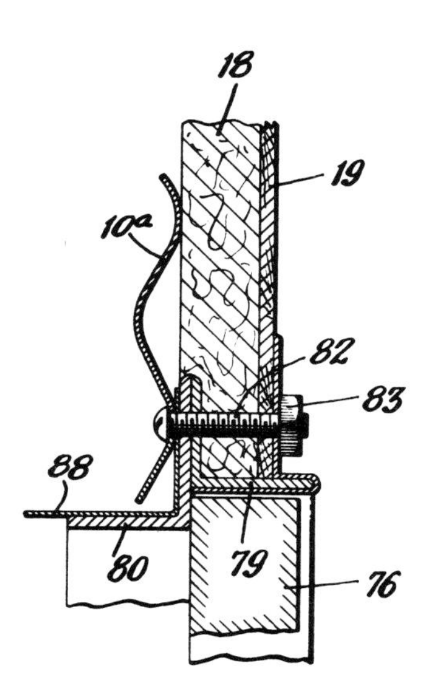

FIGURE 20

ceive the downwardly projecting portions of angle members 30 fixed to the lower surface of shelves 31. The shelves 31 may thus be secured in place, being rigidly supported and held against sliding movement toward or away from the wall structure.

Referring now to Figures 14 to 18, inclusive, the roof of the housing unit may be composed of a number of pie-shaped sections 32 of galvanized sheet steel or other strong sheet material, connected at their lower edges to the wall sheets 10 by means of the compound curved eave sheets 33. These eave sheets 33 may conveniently be made as single stampings of galvanized sheet steel or other material curved to conform to the general cylindrical shape of the sidewall sheets 10 and also curved inwardly about a horizontal axis toward the center of the structure. The sheets 33 may overlap the uppermost edges of the upper row of sheets 10 and be bolted together, with a suitable sealing material therebetween if desired, by means of the relatively short bolts 34 and the nuts 35.

The roof sections 32 are preferably formed along their side edges with raised ridges 36 so that they may be bolted together along these side edges by means of the bolts 37 and the nuts 38. Since these bolts 37 project through the roof sections 32 only at the top of the ridges 36, a relatively water-proof joint is provided, particularly if a sealing means such as asphaltum is applied between the overlapped portions. The lower portion of the roof sections 32 may have the ridges 36 tapered off to a flat surface, as shown at 39, and the lower edges of the roof sections 32 may be bent downwardly to provide flanges 40. The upper edges of the eave sheets 33 may be formed with corresponding flanges 41, so that these two flanges can be bolted together in overlapping position by means of the bolts 42 and nuts 43.

In order to insulate the roof, pie-shaped segments 44 of a suitable sheet insulating material may be utilized corresponding in shape and size to the individual roof sections 32 or to two or more of such sections when bolted together. The side edges of these sheets 44 of roof insulating material may be notched similar to the insulating sheets 18 to fit around the shanks of the inwardly projecting bolts 37, or if larger sheets 44 are used they may have holes to enable them to be pushed flat against the roof with the bolts 37 projecting through the sheets. These sheets 44 may be held in place by suitable cover strips 45 fitting over the ends of the bolts 37 and secured by the nuts 45a. The cover strips 45 preferably extend downwardly and are curved, as at 46, so that their lower edges fit in between the insulating sheets 18 and the finish sheets 19 of the side walls, as shown at 47. This provides a convenient means for holding the insulating material 48 in place inside the curved eave sheets 33.

In order to provide the interior of the building with sufficient light, a number of the roof sections 32, for example every alternate section, may be provided with a suitable skylight 49. A simple circular opening may be formed in these sheets 32 with the edges thereof bent outwardly to form a cylindrical flange 50. An annular channel frame 51 may be made to fit over the flange 50 and have stretched over its exterior a suitable transparent or translucent material, such as a sheet 52 of one of the transparent plastics known as "Celoglass." The edge of this sheet of material is preferably curved down over the frame 51 and held in place by an annular band of metal 53 pressed on over the outside of the light transmitting material 52 and frame 51.

Angular tabs 54 may be provided with their vertical sides welded or otherwise secured to the frame 51, for removably securing these skylights to the roof by means of the rivets or bolts 55. If desired, the inside of the frame 51 may be filled with a suitable asphaltic waterproofing material before the skylight is secured in place.

The insulating sheets 44, of course, have suitable openings 56 cut out to coincide with these skylights, and frames may be applied to these openings inside the building to conceal the edges of the openings in sheets 44. For example, an annular frame may be formed with the flange 56a extending up into the frame 51 and with the flanges 56b covering the edges of the opening in insulating sheets 44. These frames may be simply pressed into place and rotated until the tabs 56c are caught under the edges of the cover strips 45.

The roof of the building may be provided at its top with a suitable ventilator. This ventilator may include a vertical cylindrical ring 57 notched along its lower edge so that the ridges 36 of the roof sections 32 may project therethrough. This cylindrical ring 57 may be secured to

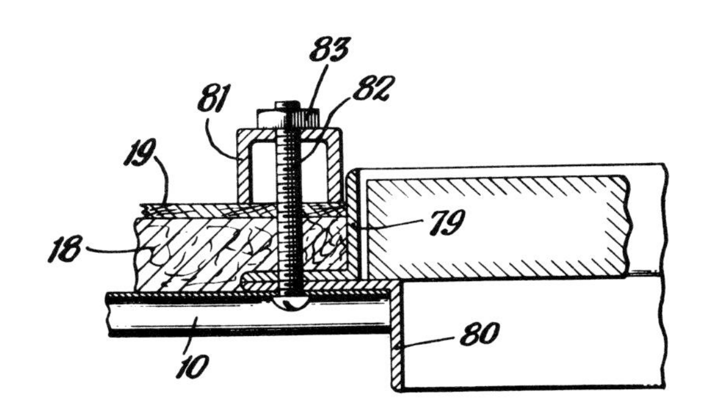

FIGURE 21

the upper ends of the roof sections 32 by means of suitable bolts or rivets 58.

Projecting upwardly and outwardly from the ring 57 are a number of brackets 59 which have secured to their inwardly inclined upper ends 60 the lower edges of a one-piece conical sheet metal top 61. The outer edges of the brackets 59 may have secured thereto a curved cowling 62 extending a substantial distance above the lower edge of the conical top 61 and also extending down sufficiently far to protect the upper ends of the roof sections 32 from the weather. The cowling 62 may be strengthened by a suitable rigid circular frame 63 at its top and a similar circular frame 64 at its bottom. A space is thus provided between the lower edge of the conical sheet 61 and cowling 62, so that circulation of air, either upwardly or downwardly, may take place while rain or snow is intercepted by the conical top 61 and deflected to drop down on the roof sections 32 at a point below the upper edges.

Inside the ring 57 and extending between the ring and the conical member 61 is a cylindrical screen 65 which may be used to keep out insects.

With such a ventilator, it is desirable to provide special means for controlling the circulation of air. To prevent too strong a draft upwardly or downwardly through the ventilator, a circular sheet of solid material, such as plywood, pressed wood, or a suitable transparent or translucent material, forming a valve 66 may be held in the circular frame 67 and suspended by means of a cable 68 from the top of the conical ventilator top 61. For example, the cable 68 may pass over a pulley 69 affixed near the undersurface of the conical ventilator member 61, and another pulley 70 may be provided to lead the cable down to a point where it can be grasped by a person inside the building to raise or lower the valve 66.

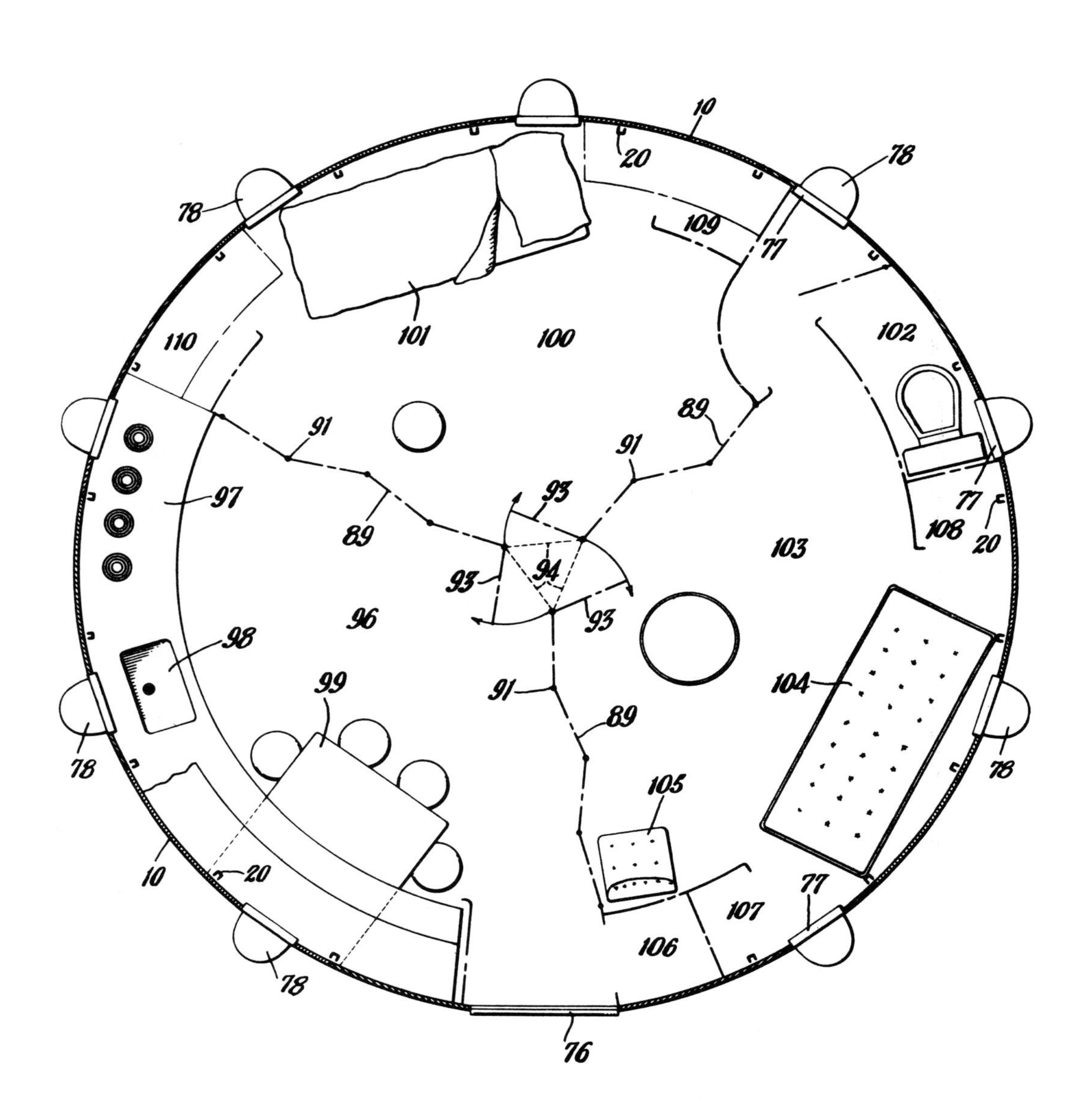

FIGURE 22

Raising or lowering this ventilator valve, of course, controls the amount of air that is allowed to flow in or out through the ventilator.

If desired, one end of cable 68 may be attached to a frame 71 having a suitable collar arrangement 72 for attachment to valve 66. An electric fan 73 may thus be supported directly below the valve 66. Suitable lights 74 may also be provided in this suspension, if desired, to furnish

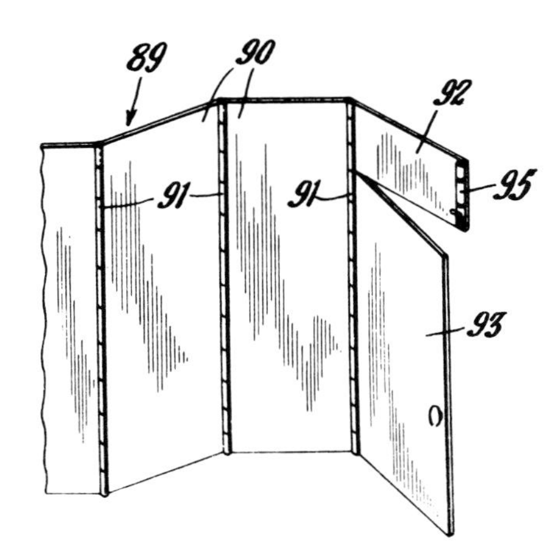

FIGURE 23

indirect lighting or a direct lighting if the valve 66 is made of transparent or translucent material. Current for the fan and lights may be supplied through a flexible cable 75 from any suitable source (not shown).

The housing unit may also be provided with a suitable door 76 and a number of windows 77 in the side. Because

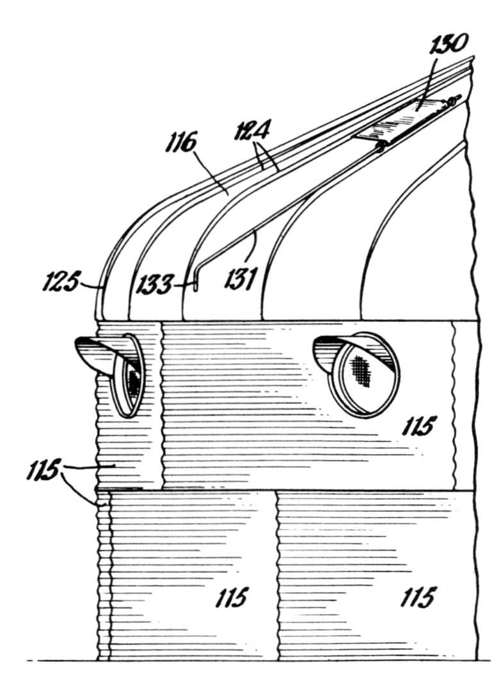

FIGURE 24

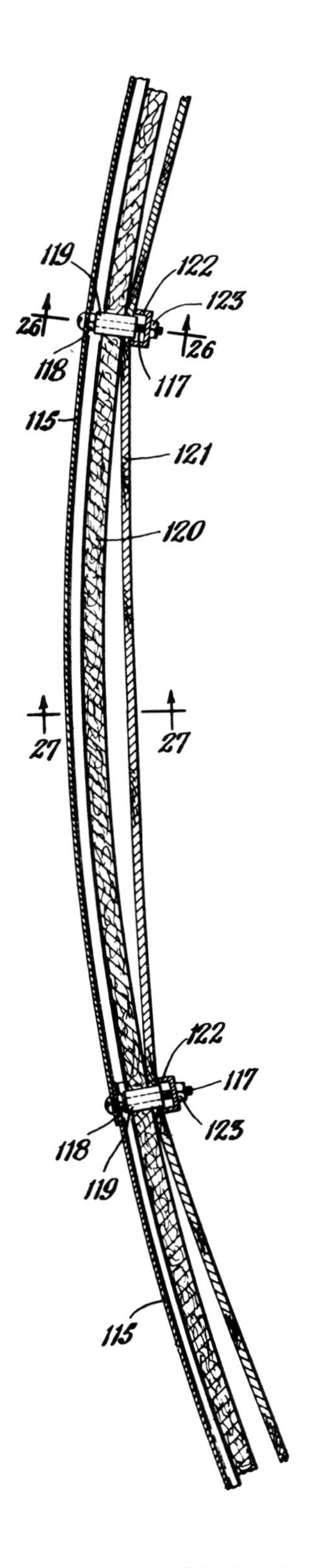

FIGURE 25

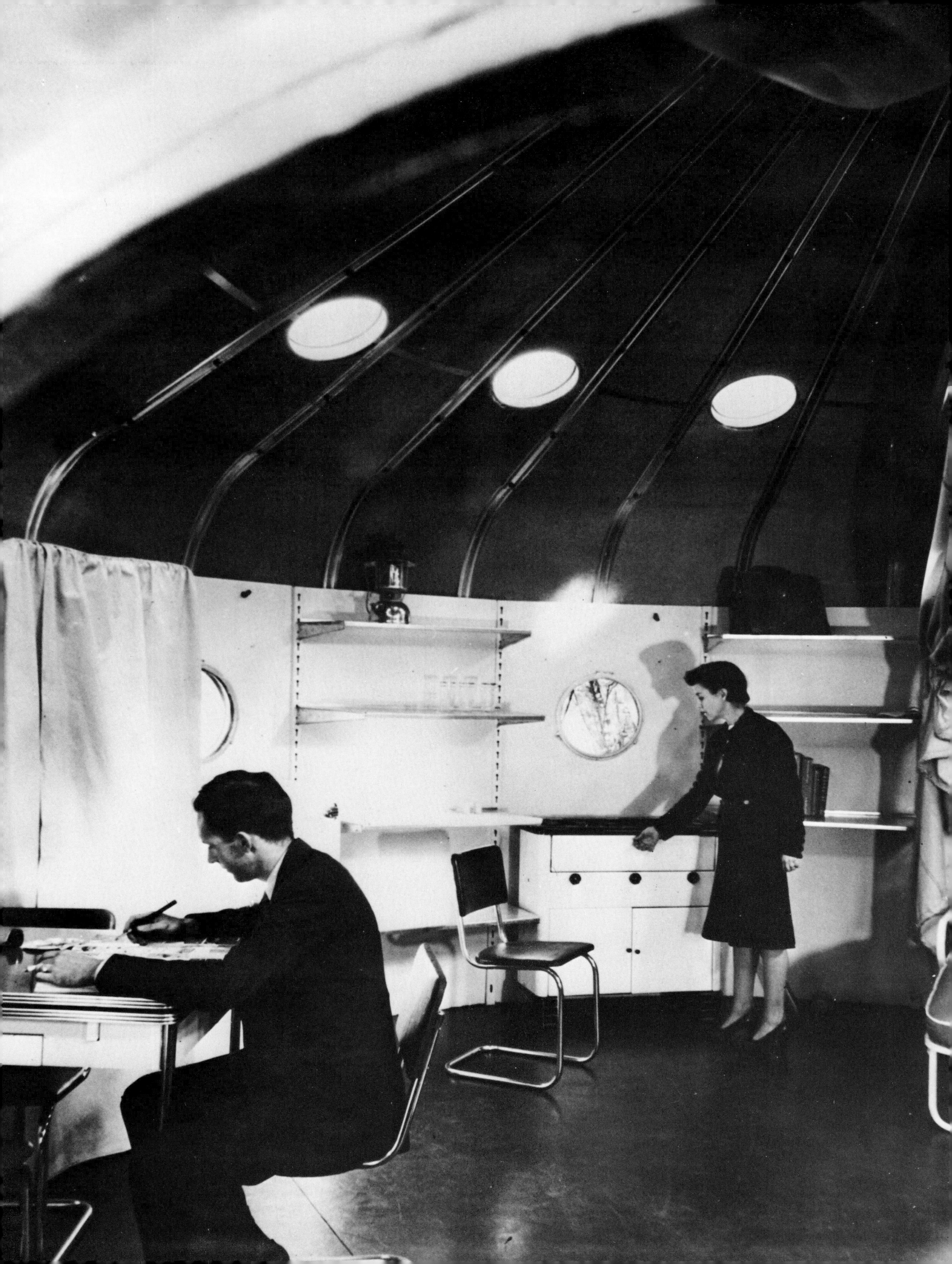

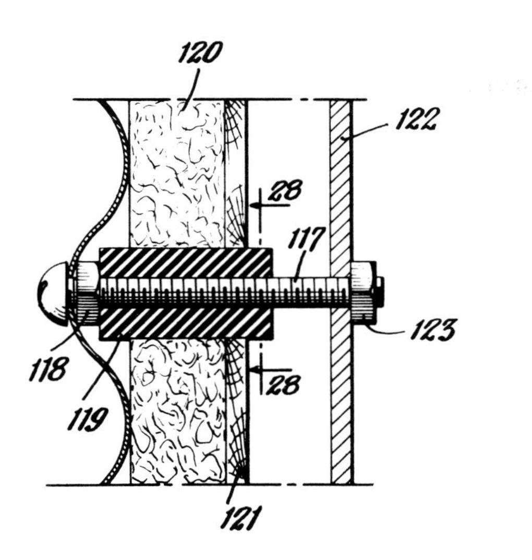

FIGURE 26

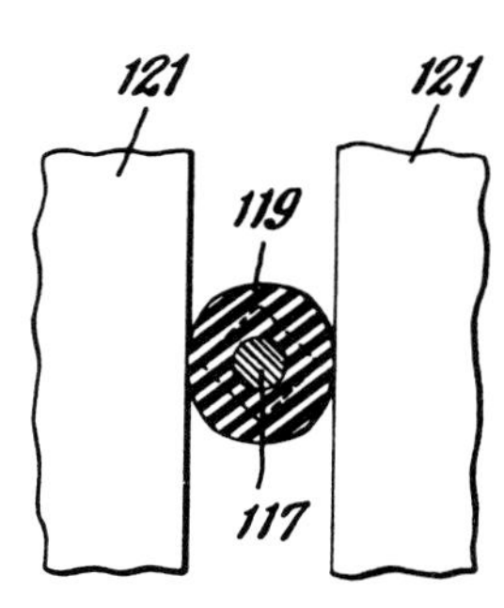

FIGURE 28

framework to receive the door 76, which is hinged in the usual manner at one side as shown at 84 (Fig. 1). If desired, the door may be provided with adjustable louvres 85 for ventilation without opening the door, and a suitable window 86.

It may also be desirable to provide a canopy for the door which may be supported by a partially circular frame 87 having its ends suitably secured to the vertical walls of the building above the door frame. A layer of canvas, sheet metal or other suitable material 88 may be secured to the frame 87 around its periphery and secured to the wall of the building, preferably inside and under the lower edge of the sheet 10a just above the door frame.

Referring more particularly to Figures 22 and 23, the interior of the building may be divided into a number of rooms by means of movable partitions 89, three of which are illustrated. These movable partitions may, if desired, be affixed at one end directly to the channel bars 20 by

of the curvature of the side wall, it is convenient to make the windows 77 circular in shape. For this purpose, the middle row of sheets 10 may have openings and flanges formed in them in a manner similar to the roof sections 32 that are provided with skylights. Suitable window frames may be mounted in these openings and provided, if desired, with external shields 78. Suitably fixed or removable screens may be fitted into the frames, and windows arranged for opening may be provided. The shields 78 aid in rendering the windows tight and proof against

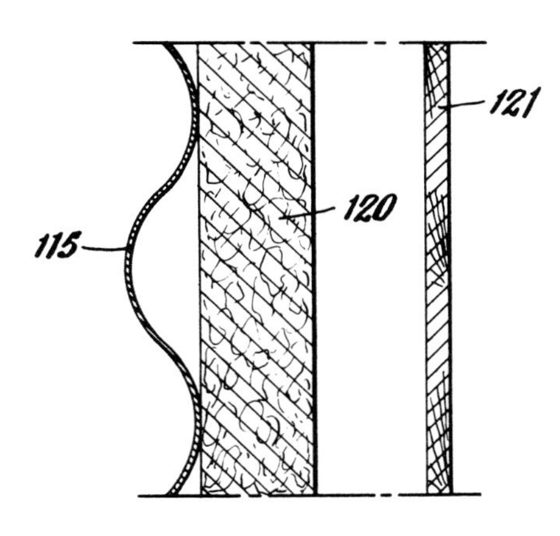

FIGURE 27

the weather, and at the same time serve as wind scoops when the windows are open to catch the cooler air rising outside the building so that it will be drawn in and then carried up through the ventilator.

A suitable door frame may be formed by the angle bars 79 and 80 bolted around the top, bottom and sides of the door opening to the edges of the sheets 10 by means of the bolts 82 and nuts 83. At the sides of the door frame, channel bars 81 lining the inside of the door may also be held in place by the bolts 82 and nuts 83. This provides a

means of brackets similar to the brackets 24. The partitions may comprise rectangular sheets 90 of plywood or pressed wood hinged together as at 91.

The free ends of the partitions can be formed with small upper sections 92 and lower sections 93 in the nature of doors, both hinged to the same sheet 90. Thus, when these three partitions are all extended to the center of the building, the three upper sections 92 may be connected together, as shown in dotted lines 94 in Figure 22, to form a triangle in the center of the building. Suitable sliding bolts 95 and corresponding sockets may be provided for this purpose.

A small triangular hallway is thus furnished, with doors from each of the three rooms thus formed leading into this small hallway. This provides a convenient way of passing from any one room to either of the other two rooms.

Various interior arrangements may be employed, and the building is well adapted for other uses than as a dwelling. In the layout shown in Figure 22, however, the room 96 is a kitchen with a stove or electric burners at 97, a suitable sink at 98, and a table and chair arrangement at 99. These fixtures may be readily attached to the channel bars 20 and supported thereby. The room 100 is arranged as a bedroom with a suitable day bed or cot 101, which may be converted to a couch in the day time. A lavatory is shown at 102, which may utilize a chemical hopper or be provided with suitable plumbing as is expe-

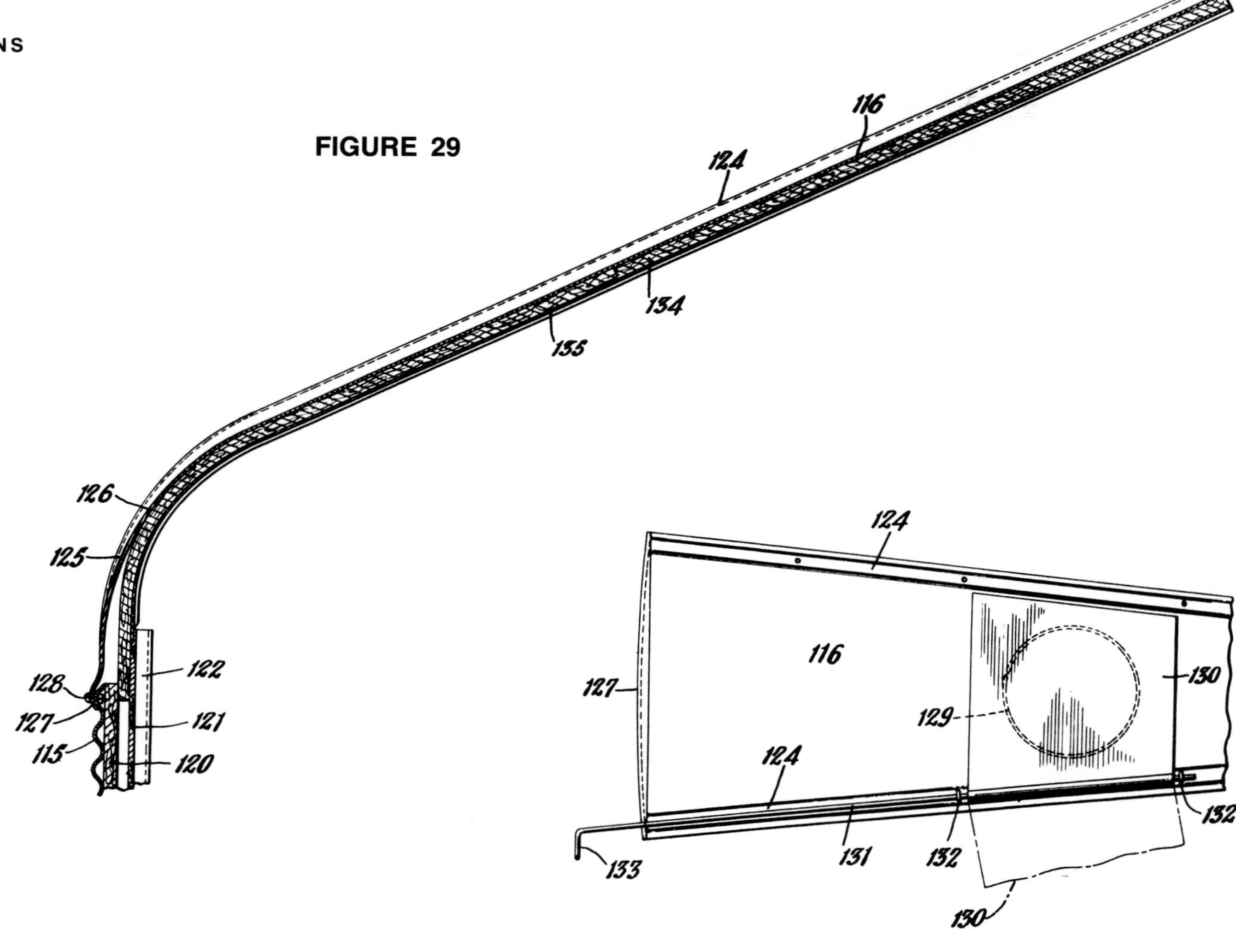

FIGURE 29

FIGURE 30

dient. The living room 103 may be fitted with a suitable divan 104 and chair 105. If desired, the partitions may be so arranged that convenient closets are formed at 106, 107, 108, 109 and 110.

It will be apparent that the external sheets 10 forming the outer shell of the building will be cut out to provide suitable openings for the windows and door. Similarly, the sheet insulating material 18 and the internal finish sheets 19 may be cut out to provide for the door and windows.

In erecting a building of the construction illustrated, it may be desirable to make a shallow excavation although, if the ground is reasonably level, no excavation at all may be needed. The lower tier of sheets 10 is first assembled, being secured together at their ends and held in place by suitable anchoring means such as the posts 13. The floor can then be laid directly on the ground or on any suitable foundation by simply laying the different layers of sheets 14, 16, and 17, and applying asphaltic or other sealing material to the joints as needed.

While the floor is being laid, or thereafter, the second and third tiers of sheets 10 and the eave sheets 33 may be assembled. The roof sections or sheets 32 are next bolted in place, and for this operation it is desirable to provide some temporary support for the inner ends of the roof sections 32 while their outer edges are being secured and before the ventilator is assembled.

This temporary support may be easily made by bolting together inside the house a scaffolding composed of the channel bars 20. These bars are provided with numerous holes for bolts and are sufficiently strong to support the weight of the center of the roof. After the ventilator has been assembled in place, this scaffolding may be taken down and the channel bars 20 used for holding the side wall sheets 18 and 19 in place.

A building constructed in accordance with my invention may be used as a dwelling, but is also well suited for other uses. For example, a building 20 feet in diameter would have about 60 feet of wall space for shelves if the building is used for a store. Various other uses include schools, churches, wayside establishments, camps, boat-houses, etc.

The building is also easily adapted to military use, because it may readily be surrounded by a suitable wall of concrete poured after the building is partially or completely erected and using the building wall as a part of the form.

Another modification of my invention is illustrated in Figures 24 to 33, inclusive, in which numeral 115 indicates side wall sheets similar to the sheets 10 forming an external layer of the side walls of the building. In this case, however, the lower edges of the combined roof and eave sheets or sections 116 are directly attached to the upper edges of the upper row of sheets 115. Two horizontal rows of the side wall sheets 115 are illustrated, although a single row of such sheets or any suitable number of rows may be employed, as desired. These sheets are preferably made of a relatively strong material, such as a galvanized sheet steel, and may have horizontal corrugations for reinforcement. The sheets preferably overlap and are sealed at their joints with a suitable water-proof material such as an asphaltic mastic, as described in connection with the building shown in Figure 1.

The side sheets 115 are also preferably bolted together in the position illustrated in Figure 24 to form an outer shell for the house. Bolts 117 holding these sheets 115 together may be secured in place by means of the nuts 118, leaving a substantial length of the bolt shanks projecting inwardly from the sheets 115. To support the inner layers of the side walls and permit easy installation of these inner layers after the outer shell of the house has been erected, collars 119 of suitable resilient material, such as rubber, may be slipped over the protruding shanks of the bolts 117. Vertical rows of rubber coated bolts are thus provided extending into the interior of the building between which the insulating sheets 120 and the sheets 121 of suitable finishing material may be fitted.

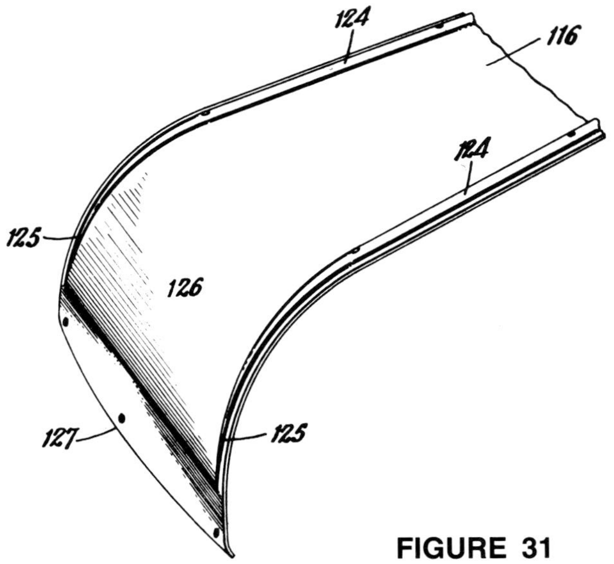

FIGURE 31

For example, the sheeting insulating material in the form of rectangular sheets 120 may be simply sprung into place and forced in between the rows of bolts 117 carrying the collars 119. The edges of the sheets 120 may be notched, if desired, to fit around the collars 119, although notching is not necessary to obtain a good fit of the sheets 120 with the extra tolerances provided by the resilient collars 119. The sheets of finishing material 121, such as plywood, pressed wood or other suitable material, may be similarly sprung into place but are illustrated as sprung only to a slight degree instead of lying substantially flat against the insulating sheets 120.

In this manner, the resiliency of the sheets 120 and 121, whether they are sprung completely or partially, exerts a pressure against the sides of the bolts 117 with a resultant outward pressure on the outer shell composed of the sheets 115 that serves to increase the rigidity and strengthen this outer shell. In addition, of course, this stressed condition serves to hold the sheets 120 and 121 securely in place even if shrinkage or swelling of the sheets 120 and 121 takes place.

To cover up vertical seams between these side wall sheets, suitable channel members 122 similar to the channel members 20, or other vertical extending cover members, may be fitted over the ends of the bolts 117 and secured in place by means of the nuts 123.

The roof sections 116 can be formed from single sheets of suitable material, such as galvanized sheet steel, with raised ridges 124 along the side edges of these pie-shaped segments, tapering off to a flat sheet near the lower edges of the sections, as indicated at 125. The lower portions of these roof sections or sheets are preferably curved, as illustrated at 126, about a substantially straight horizontal axis to form the building eaves. The lower edges of the sections 116 may terminate at the lower end of this curve or be curved down in a section as shown at 127 that is not curved about a horizontal axis. In either case, this lower edge should be curved about a vertical axis to conform to the curvature of the sheets 115. Thus, the lower portions of the roof sections 116 are either formed with a definite compound curve or are curved first in one direction and then in another to produce substantially the same effect. This compound curve arrangement at the eaves serves to stiffen and strengthen the side wall sheets 115, and also simplifies the erection of the building, providing an external construction of pleasing effect with fewer horizontal seams or joints.

A suitable number of roof sections 116 are secured at their lower edges in overlapping position to the upper edge of the sheets 115 by means of suitable bolts 128, and these individual roof sections 116 are bolted together along their sides with the ridges 124 overlapping, as previously described. Suitable asphaltic water-proof material may be used to seal these joints, if desired, and skylights 129 may be provided in the roof sections, as previously mentioned.

Under some conditions, it may be desirable to black-out the interior of the building or shut off the passage of light through one or more of the skylights 129. One convenient way of accomplishing this is to provide cover plates 130 fastened to rods 131 extending along the tops of the ridges 124 and passing through the eye-bolts 132. These rods may extend down to the eave of the house and be formed at their ends with suitable handles 133 to enable a person standing outside the house to rotate the cover plates 130 from a position covering the skylights 129 to a position in which the cover plates 130 lie flat on the next roof section 116, as shown in dot-dash lines in Figure 30.

Suitable insulating material may be provided underneath the roof sections 116, as illustrated at 134. This insulating material may be held in place by cover strips 135 bolted to the roof sections 116 and overlapping the edges of adjacent insulating sheets 134, as previously described.

The insulating sheets 134 may extend downwardly inside the curved eave portion of the roof sections 116 and abut the upper edges of the vertical insulating sheets 120. If a relatively compressible sheet insulating material is employed, however, these roof insulating sheets 134 may extend down far enough to overlap with the vertical sheets 120, and these sheets 120 and 134 where they overlap are simply compressed together at the points where necessary between the outside sheets 115 and the sheets 121 of finishing material.

Referring now to Figure 32, the upper ends of the roof sections 116 may project into an outwardly facing annular channel bar 136. Suitable sealing means, such as asphaltic material 137, may be applied around the channel of this annular bar or ring to make a water tight construction. A number of brackets 138 may be bolted to the annular bar 136 to support at their upper ends a conical ventilator top 139. This conical top preferably has a downwardly extending flange 140 around its outer edge. A suitable cylindrical screen 141 may be provided between the upper edge of the annular channel bar 136 and the lower surface of the conical top 139.

Suspended from the top of the conical piece 139 is a rod 142, and an annular ring 143 surrounding this rod may be suspended from the roof by means of brackets 144 bolted to the roof sections 116. This provides a simple support for movable fabric partitions 145 (Figures 32 and 33). Each of these fabric partitions or curtains may be supported at their inner end by a ring 146 supported for rotation about the rod 142 by means of the nut and washer 146a. A second ring 147 may be provided to slide around on the ring 143 through an angle of approximately 120° between two of the brackets 144.

At the side walls, the curtains 145 may be supported by suitable pulleys 148 removably attached to one of the vertical channel bars 122. These pulleys also carry a suitable cord 149 running through a number of rings or loops 150 sewed or otherwise secured on the curtains 145 and extending down to the lower central corners 151 of the curtains where the ends of these cords are attached to the curtains.

The curtain partitions 145 may thus be drawn back into a draped position by simply pulling the cords 149 until the curtains lie along the inside of the roof and against the side wall of the building. The pulleys 148 may be moved around and attached to any suitable channel bar 122 to divide the room up into different sizes of rooms as may be required.

A valve or ventilator regulator 152 may be provided underneath the conical ventilator top 139 with a collar 153 arranged to slide up and down on the rod 142. This regulator or valve 152 may be raised or lowered by means of a suitable cable or cord 154 running over the pulleys 155 and 156 and extending down to the side wall within reach of a person in the building. This form of sliding ventilator regulator is simple to adjust and is not readily tipped or otherwise moved out of place by sudden up or down drafts through the ventilator.

Various other modifications of my invention will be readily apparent to those skilled in the art.

A particular advantage of a building shaped like the

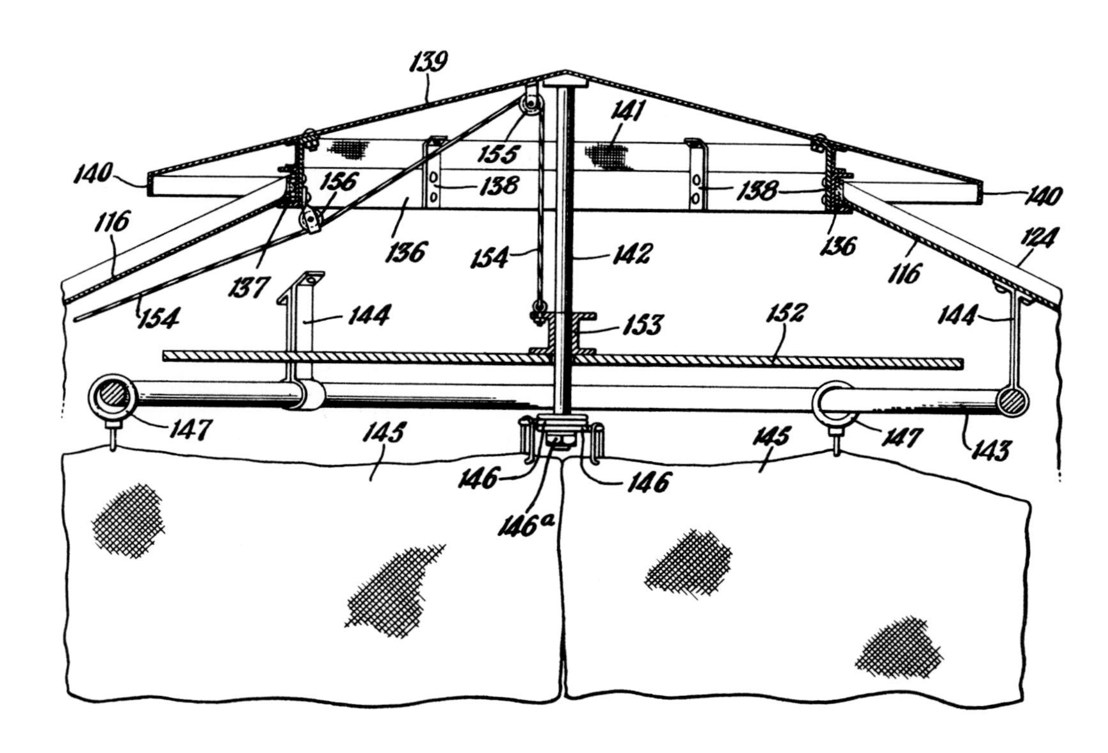

FIGURE 32

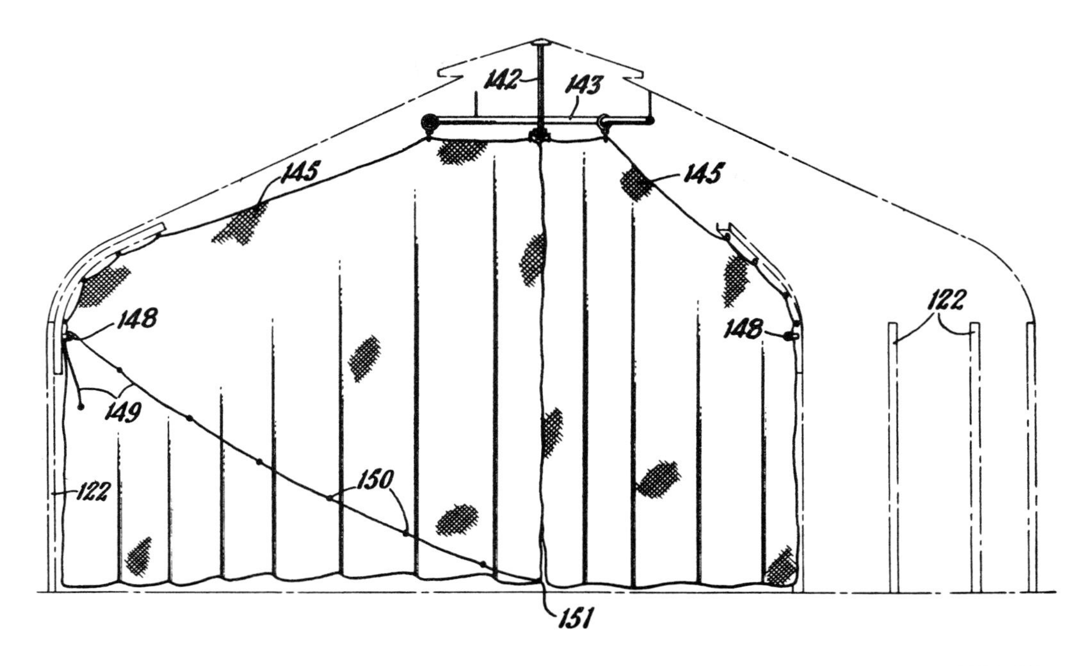

FIGURE 33

embodiment shown in the drawings is the marked efficiency in heating. By reason of its shape, wind currents do not burble but flow smoothly around the outside of the building and do not carry away as much heat by convection as is carried away from the ordinary building of angular shape. In addition, the heating system may be centrally located so that the heat is uniformly distributed throughout the building interior.

The construction illustrated and described also provides a building that can be readily taken apart and moved to a new location without substantial loss of materials. At the same time, the use of curved wall segments makes this possible using relatively light weight construction materials.

The construction and shape of the building lends itself to the use of many duplicate parts that nest together and may be packed into a relatively small space for shipment.

The terms and expressions which I have employed are used as terms of description and not of limitation, and I have no intention, in the use of such terms and expressions, of excluding any equivalents of the features shown and described or portions thereof, but recognize that various modifications are possible within the scope of the invention claimed.

7▲DYMAXION DEPLOYMENT UNIT (FRAME) (1944)

U.S. PATENT—2,351,419

APPLICATION—APRIL 9, 1941

SERIAL NO.—387,589

PATENTED—JUNE 13, 1944

THE DYMAXION DEPLOYMENT UNIT IN THE FRAME VERSION was a logical evolvement from my experience in making the standard grain bin acceptable as a dwelling unit. I could see ways of producing the dwelling units in the same fireproof materials but more efficiently than by converting a grain bin into a dwelling unit. This latter method taught me much that became useful in the Beech Aircraft House and in later geodesic dome development.

UNITED STATES PATENT OFFICE

Richard Buckminster Fuller, New York, N.Y., assignor to The Dymaxion Company, Inc., Wilmington, Del., a corporation of Delaware

BUILDING CONSTRUCTION

The invention relates to the fabrication of structures suitable for small houses, or for storage bins and the like, and more particularly to structures of this class which have walls of sheet metal or other sheet material arranged to form a shell or enclosure which is of substantially cylindrical form.

It is an object of my invention to fabricate structures of the class described in such a manner that the walls of sheet metal or other material can be of extremely light gauge by reason of distribution of stresses throughout the entire structure.

A further object of my invention is to provide a stressed-wall structure in combination with a semi-flexible supporting framework so as to segregate tensile and compression forces in such a manner that tensile forces are confined substantially entirely to the wall covering whereas compression forces are confined substantially entirely to the supporting framework.

Another object is to make possible the rapid erection of metal structures at low cost for both material and labor.

Another object is to provide a structure of the class described which can be readily taken down and transported to another location for erection there.

Another object is to provide improved means for utilizing the strength characteristics of both the inner and outer wall coverings in structures of the type referred to.

A still further object is to provide means inherent in the walls and roof of structures of the class described for controlling or modifying convection currents within the structure.

Other objects and advantages will appear as the description proceeds.

In the drawings:

Fig. 1 is a diagrammatic side elevational view of the compression frame used in a preferred embodiment of my invention; and Fig. 2 is a diagrammatic top plan view of the same frame, but with the ventilator removed.

FIGURE 1

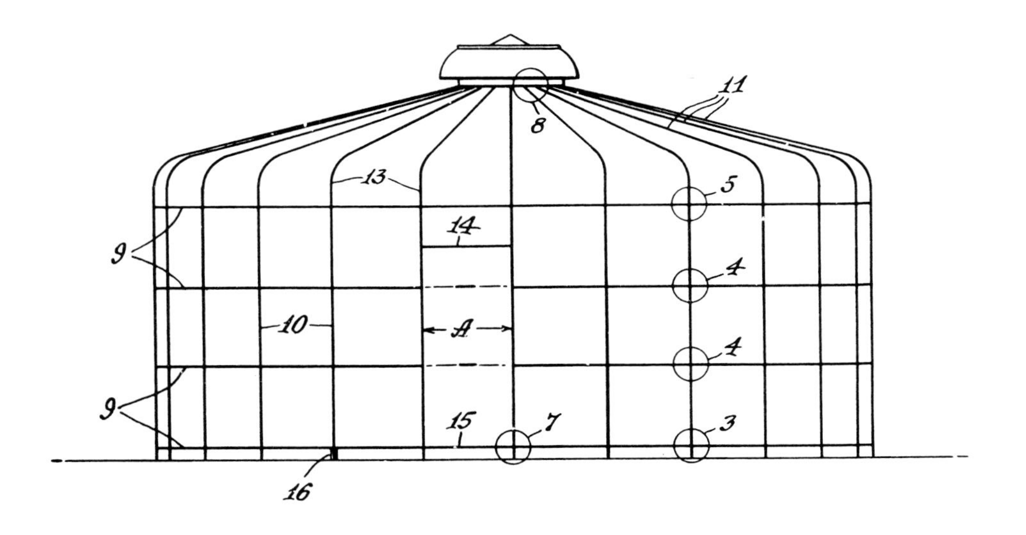

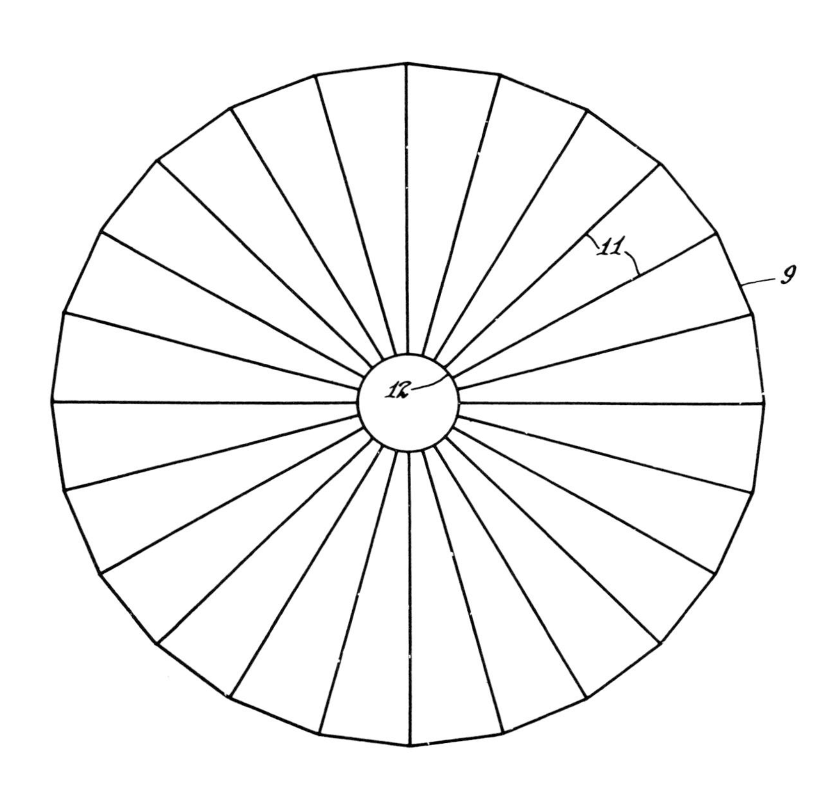

FIGURE 2

FIGURE 3

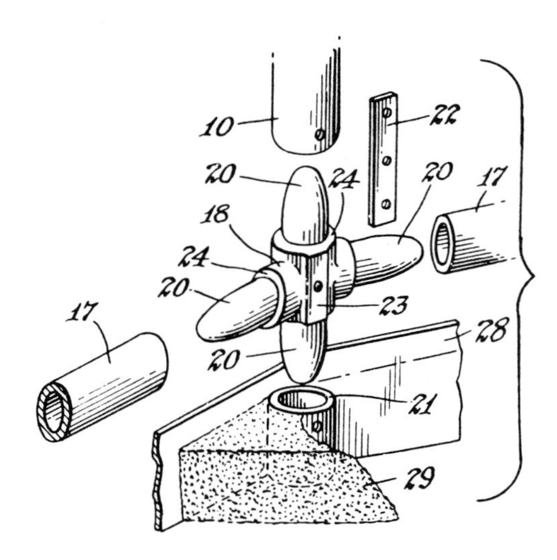

FIGURE 4

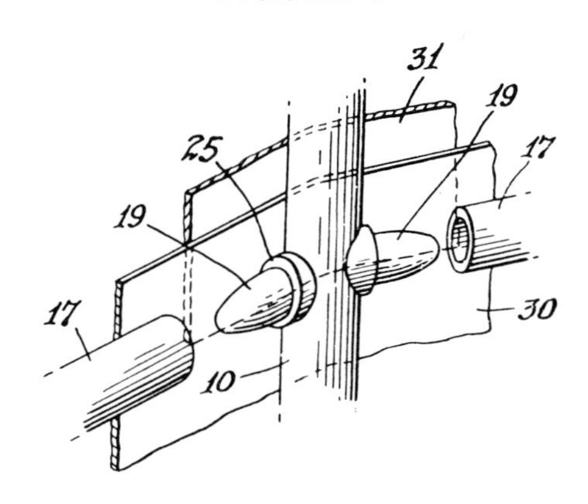

FIGURE 5

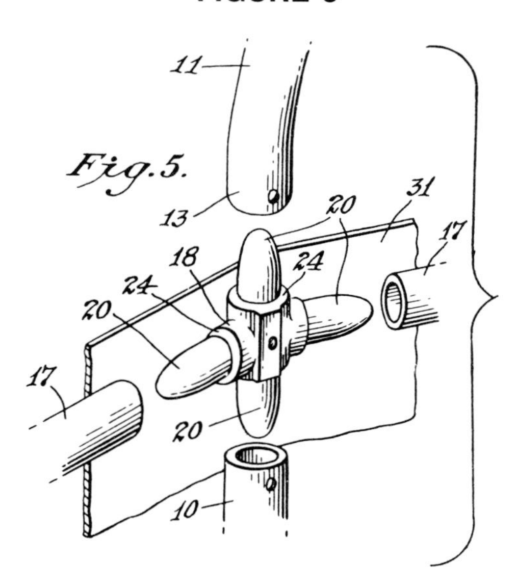

Figs. 3 to 8 , inclusive, are detail views showing the manner in which the structure of Figs. 1 and 2 is assembled, and how the outer stressed covering is draped thereon.

Fig. 3 is an exploded view of the joint indicated within the area of circle 3 on Fig. 1, looking from within the structure. Fig. 4 is a similar view of the joints indicated within the circles 4 on Fig. 1, and Fig. 5 is a similar view of the joint indicated within the circle 5 on Fig. 1.

Fig. 6 is a detail view covering that portion of the structure indicated within either of the circles 4 shown on Fig. 1, being a fragmentary elevational view looking at the exterior of the structure, and indicating the manner of draping the outer stressed covering on the framework.

Fig. 7 is a detail elevational view covering that portion of the structure indicated within the circle 7 shown on Fig. 1, and shows the door jamb and sill; also a portion of the flooring.

Fig. 8 is a detail view covering that portion of the structure indicated within the circle 8 shown on Fig. 1, and shows one manner of applying the roof sections to the framework.

Fig. 9 is a perspective view of a portion of the structure illustrated in the preceding views, showing one manner of applying and securing the stressed covering sheets to the framework.

Fig. 10 is a perspective view similar to Fig. 9, but showing another embodiment of the invention in which the means for stressing the cover sheets are so arranged as to leave a doorway free from obstruction.

Figs. 11 and 12 are detail horizontal sectional views of the means for stressing the covering sheets employed in the embodiment illustrated in Fig. 10. Fig. 11 shows the relative position of the parts before the covering has been stressed, and Fig. 12 shows their relative position after the covering has been stressed.

Fig. 13 is a detail perspective view showing one form of roof construction.

Figs. 14 and 15 are detail transverse vertical sectional views illustrating a modified form of roof construction in which the roof covering sheets are stressed. Fig. 14 shows the arrangement of the parts before the covering sheets have been stressed, and Fig. 15 shows the completed roof section.

Fig. 16 is a detail elevational view of a modified construction looking from the inside of the structure at the doorway and adjacent panels.

Fig. 17 is a detail perspective view looking from the outside of the same structure at a lower corner of the doorway.

Fig. 18 is a horizontal sectional view through a portion of the wall structure illustrated in Fig. 16, showing the manner of applying the insulation and interior wall covering thereto.

Fig. 19 is a detail horizontal sectional view taken as indicated at 19—19 in Fig. 16.

Fig. 20 is a detail view showing the manner of attaching the reinforcing strut members shown in Fig. 16.

One of the features of my invention resides in providing a more or less flexible supporting frame over which covering sheets are draped and then stressed under considerable tension. With such a construction the tensile and compression forces are segregated in such a manner that the tensile forces are confined substantially entirely to the covering sheets, whereas compression forces are confined substantially entirely to the supporting frame. In its simplest embodiment, such a segregated compression and tension structure may consist of an annular ring with a sheet member adjacent said ring and wrapped around it, together with means for tensioning the sheet member to set up compression forces peripherally of the ring. The annular ring preferably is formed of a plurality of sections arranged end to end. The ring may be substantially circular or of other generally curved form, or it may be polygonal in form. In the several embodiments illustrated in the accompanying drawings, a plurality of such rings are arranged in substantial parallelism, and assembled with vertical members to form a cage-like structure over which the covering is stretched.

Referring to Figs. 1 to 9, inclusive, I shall describe one of the preferred embodiments which have been selected for the purpose of illustrating certain features of my invention. The construction shown in these views has been designed primarily for use as a storage unit, such as a grain bin or the like. The construction shown, however, is also applicable to the fabrication of houses, and my invention is not to be understood as being restricted to any particular field of use.

In Figs. 1 and 2, the supporting framework is illustrated diagrammatically, the constructional details of the various portions of this framework being shown in Figs. 3 to 9, inclusive. In its general arrangement, this framework consists of the polygonal or substantially circular horizontal ring members 9, intersecting vertical members 10, and inclined radial roof supporting members 11. The roof supporting members 11 at their inner ends abut a compression ring member 12 (Fig. 2), and at their outer ends are preferably curved downwardly, terminating in vertical sections 13 for alignment with the vertical members 10. The top and bottom ring members 9 are continuous, but the two intermediate ring members 9 may have an open section to provide a doorway or the like, as at A. The top of the doorway is indicated at 14 and the sill at 15. A jack 16 may be provided under the end of each vertical member 10 for the purpose of leveling the structure.

The annular rings 9 preferably are formed of a plurality of sections arranged end to end with suitable means for holding them in alignment. Figs. 3, 4 and 5 illustrate my preferred form of aligning means. These views should be considered in conjunction with Fig. 1 wherein the circles 3, 4 and 5 indicate the respective portions of the structure which are illustrated in more detail in Figs. 3, 4 and 5. In the detail views the parts have been shown separated, or "exploded," in order to more clearly illustrate the manner in which they are assembled. In these three views, the horizontal ring members 9 are made up of a series of straight tubular sections 17 joined together by cross-shaped connecting members 18 (Figs. 3 and 5), or by studs 19 on the vertical member 10 (Fig. 4). The cross-shaped connecting members 18 also serve to connect the horizontal sections 17 to the ends of the vertical members 10 and to the vertical portions 13 of the roof members 11. The intermediate joints 4 may, if desired, be constructed in the same manner as the lower and upper joints 3 and 5, but in the construction shown in Fig. 4 the vertical member 10 is shown as being constructed in a single piece.

The cross-shaped connecting members 18 have projecting dowels 20 which fit within the ends of the members 10, 11 and 17, and also within the end of a tubular member 21, shown in Fig. 3, which may be driven into the ground to provide a support beneath the column formed by the vertical member 10. There is one of these supporting members 21 for each of the vertical members 10, and the first step in the erection of the structure consists in driving the tubular supporting members 21 into the ground, preferably with the use of a centering plug placed within the upper end of the member 21. After the members 21 have been driven so that their top faces are

FIGURE 6

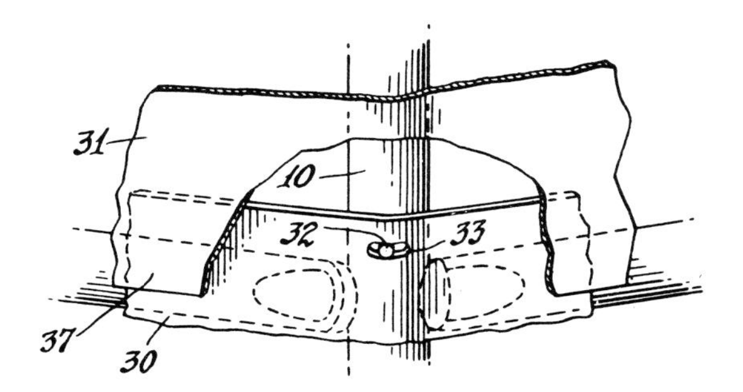

FIGURE 7

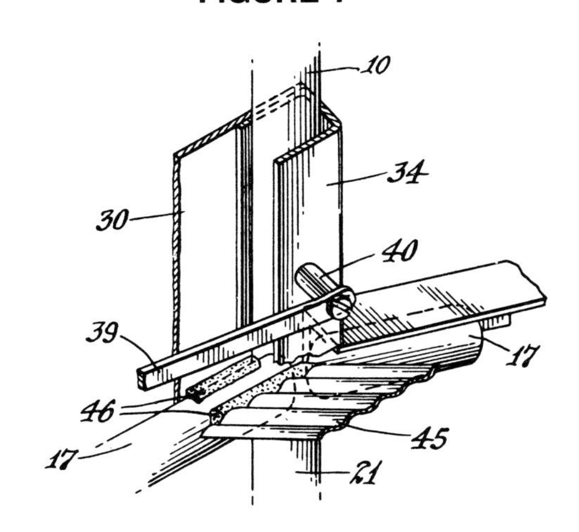

FIGURE 8

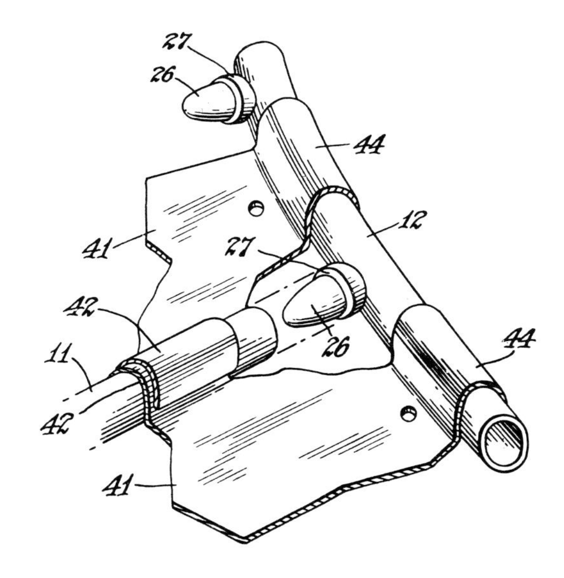

brought to a common level, the elements 17 and 18 of the lowermost ring member 9 are assembled on the foundation thus provided. In a similar manner, the vertical members 10 and the elements of the intermediate and upper ring members 9 are assembled to complete the formation of a substantially cylindrical or prismatic framework. The ends of the projections 20 of the cross members 18 are rounded, as shown, to facilitate assembly, and also to provide a certain amount of flexibility in the completed structure. If desired, the members 10, 11, and 21 may be held in assembled relationship with the crosses 18 by any suitable fastening means such as the strap 22 shown in Fig. 3, which has three holes arranged for alignment with tapped holes in the cross member 18, and in the members 10 and 21 (or 10 and 13). The cross member 18 may be provided with a flat inner face 23 against which the plate 22 is brought to bear by the machine screw which passes through the center hole of the strap 22. Machine screws passing through the end holes of the strap 22 engage the tapped holes in the ends of the member 10, and the end of the member 13 or 21, as the case may be. The cross members 18 preferably are constructed with shoulders 24 so as to bring the body portion of the member 18 into flush alignment with the outer surfaces of the members 10, 11, 17 and 21.

The studs 19 shown in Fig. 4 may be welded or otherwise secured to the vertical tubular member 10 and preferably are tapered similar to the studs 20 of the cross members 18 so as to provide a certain amount of flexibility in the completed structure, and also to assist in its assembly. The studs 19 likewise are formed with shoulders 25 so that the attaching portions of these studs come into flush alignment with the outer surfaces of the sections 17.

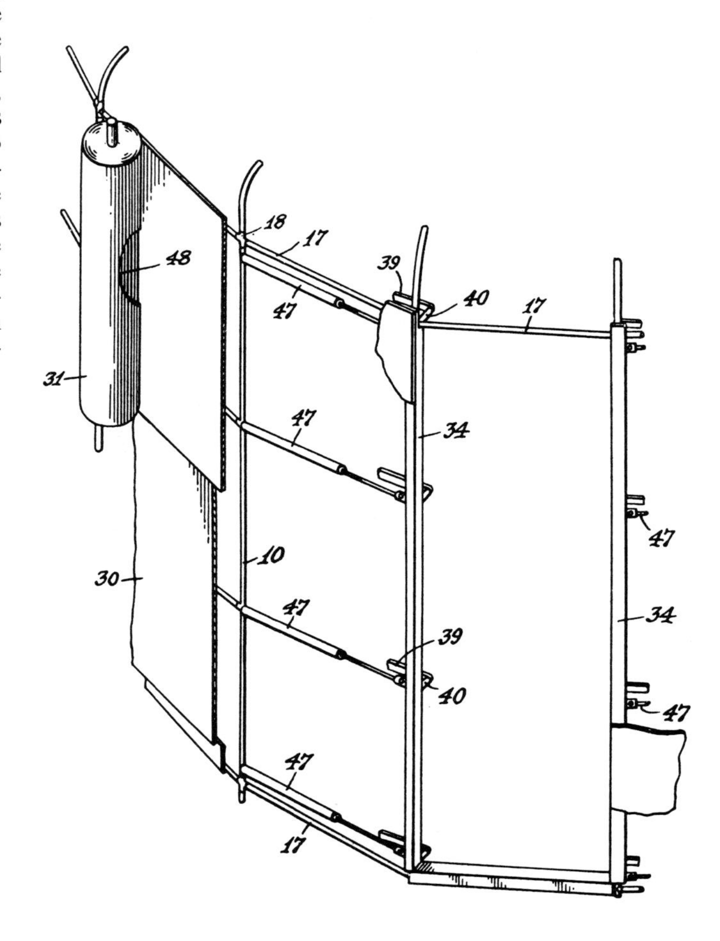

FIGURE 10

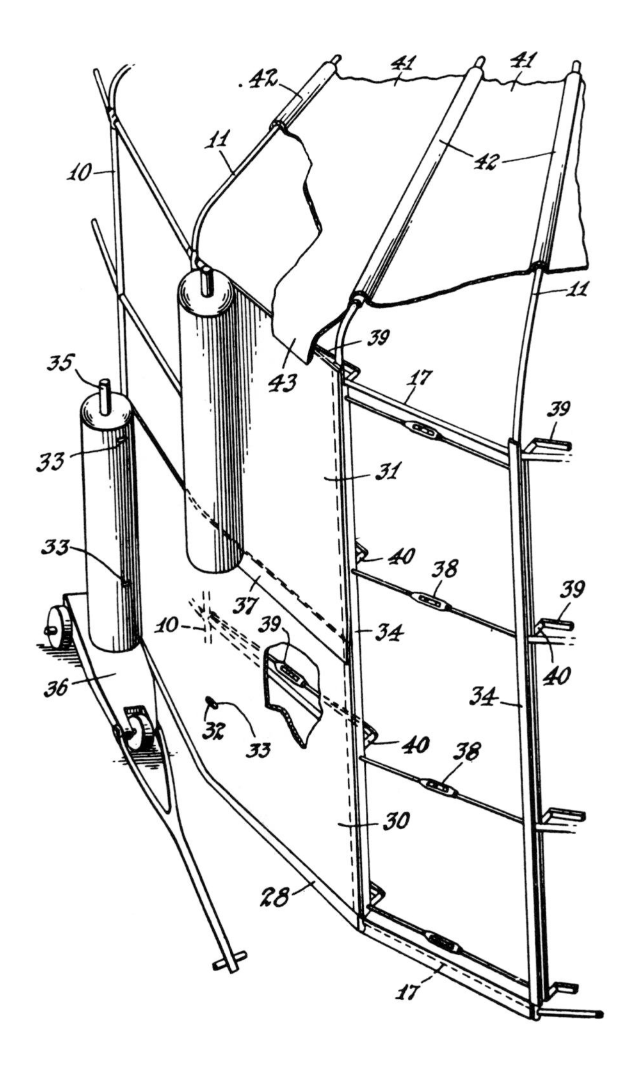

FIGURE 9

The connection between the roof members 11 and the compression ring 12 is illustrated in Fig. 8. The ring 12 is provided with studs 26 similar in construction to the studs 19 which have been described with reference to Fig. 4. These studs may be welded or otherwise secured to the ring 12. The inner ends of the tubular roof members 11 fit around the ends of these studs, and are forced against the shoulders 27 thereof when the outer wall covering is stretched over the cylindrical walls of the frame. The ring 12 in conjunction with the roof members 11 thus serve to true up the completed structure.

After the supporting foundation members 21 have been driven and leveled, a band or strip 28 of metal or other suitable material is wrapped around the outside of the members 21 as shown in Fig. 3. This may be done

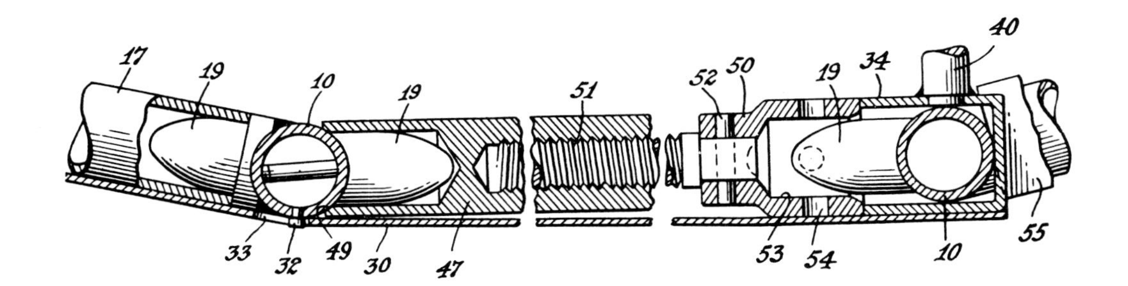

FIGURE 11

after the rest of the framework has been assembled. With the strip 28 in place, the foundation of the structure is filled to the desired level, as indicated at 29. The fill 29 may be of earth, or of any other material. Then, after the framework has been assembled in the manner previously described, the wall covering sheets 30 and 31 are draped on the structure. For this purpose, the vertical members 10 are provided with outwardly projecting studs 32 (Fig. 6) for engagement with slots 33 formed in the sheets 30 and 31. The engagement of the slots in the sheets with the studs 32 serves to support the sheets in the proper position during their application, while allowing relative movement between the sheets and the vertical members 10 during the operation of stressing the covering sheets. The ends of the sheet members are secured to vertical members such as the channels 34 (Figs. 7 and 9), as by welding or bolting the sheets thereto. The channel 34 hooks over one of the vertical members 10, holding one end of the sheet as it is wrapped around the structure to have its opposite end secured to another channel 34 at the adjacent vertical member 10. Application of the covering sheets 30 and 31 may be facilitated by having the sheets delivered in roll form and, if desired, the slots 33 may be punched prior to application of the sheet. The rolled sheet may be mounted on a spindle 35 carried by a hand truck 36 on which it is rotatably mounted. As the truck is pushed or pulled around the structure, the sheet unrolls and the slots 33 brought into proper registration with the studs 32 in the framework. The lower sheet 30 is applied first, and the upper sheet 31 overlaps the lower sheet, as at 37, to provide proper protection against the weather.

The covering sheets 30 and 31 may be made of light gauge metal, reinforced plastic, plywood, canvas, or any sheet material capable of withstanding moderate tensile stresses.

After the covering sheets have been loosely draped on the structure, the vertical members 34 are drawn toward one another by means of turnbuckle rods 38, the ends of which are suitably secured to the channels 34. As will be seen in Fig. 9, the sections 17 of the uppermost and lowermost compression rings 9 extend across the opening between the ends of the covering sheets. The back and one flange of the channels 34 are cut away so as to avoid

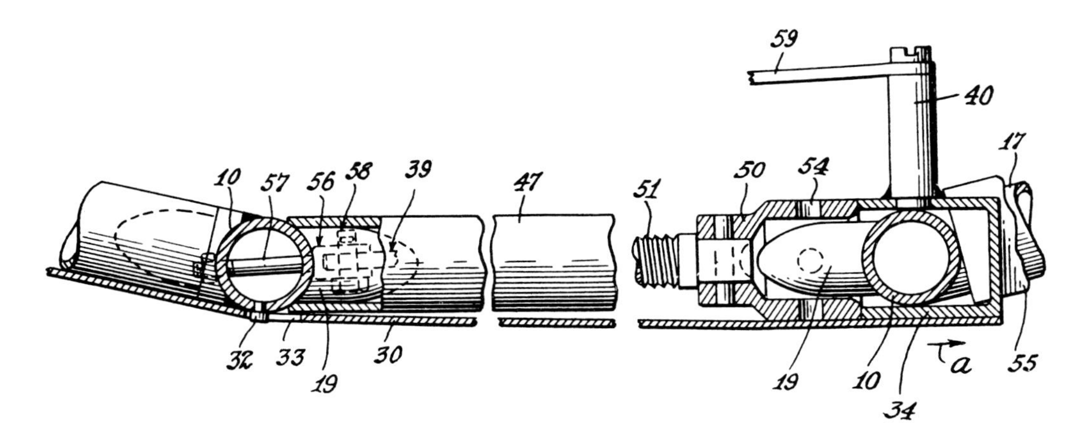

FIGURE 12

interference with these sections 17. As the turnbuckle rods 38 are tightened to draw the channels 34 toward one another, the rings 9 are placed under compression, holding the sections 17 and 18 thereof in assembled relationship. The vertical members 10 adjacent the channels 34 are prevented from being drawn toward one another by reason of the fact that the uppermost and lowermost compression rings are continuous. Therefore, as the channel members 34 are drawn toward one another, the vertical members 10, which are surrounded by the flanges of these channel members (Fig. 7), are held apart by the sections 17 of the compression rings which extend across the same space that is occupied by the turnbuckle rods. Since the pull applied by the turnbuckle rods is not directly in line with the plane of the covering sheets, a turning moment is applied to the channels 34 in a direction about an axis which parallels the length thereof. This turning moment is resisted by a second set of turnbuckle rods 39 which are secured at one end to the vertical member 10 at the end of the adjacent panel, and at their other ends to studs 40 (Figs. 7 and 9) rigidly secured to one side of the channel 34. It will be seen that by properly adjusting the turnbuckles 38 and 39, proper alignment of the channels 34 may be secured.

The roof covering may consist of a plurality of segments 41 having curved flanges 42 adapted to extend over the roof supporting members 11. The roof segments 41 are provided with downwardly extending flanges at their outer ends arranged to overlap the top of the wall as at 43. At their inner edges, the roof segments are provided with flanges 44 adapted to extend over the compression ring 12.

In Fig. 7 I have shown an illustrative section of corrugated flooring 45. Asphalt mastic 46 or other suitable material may be used to seal the joints between the flooring and the ring sections 17 and between the ring sections 17 and the covering sheet 30.

In Fig. 10 I have shown a modified form of my invention which is particularly suited for the construction of houses, where an opening for a full height door must be left free from obstruction by turnbuckles or the like. The construction of this embodiment is the same as that which has been described insofar as concerns the arrangement of the supporting framework and the wall covering sheets 30 and 31. Also, the ends of the wall covering sheets are attached to channels 34 in the manner which has been described with reference to Fig. 9. However, in place of the turnbuckle rods 38 I employ a series of jacks 47 arranged in the panel adjacent the doorway. As shown, there is a set of these jacks 47 at each side of the doorway

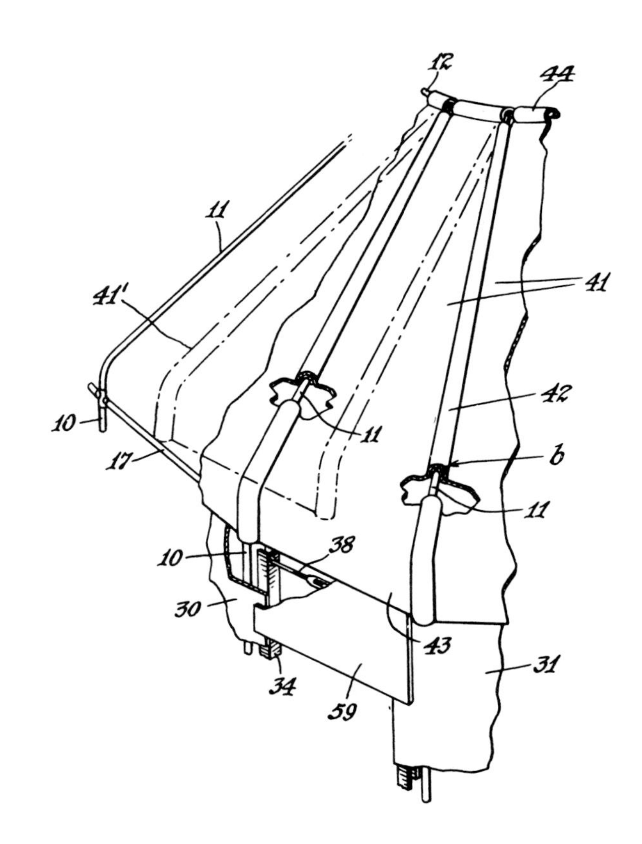

FIGURE 13

FIGURE 14

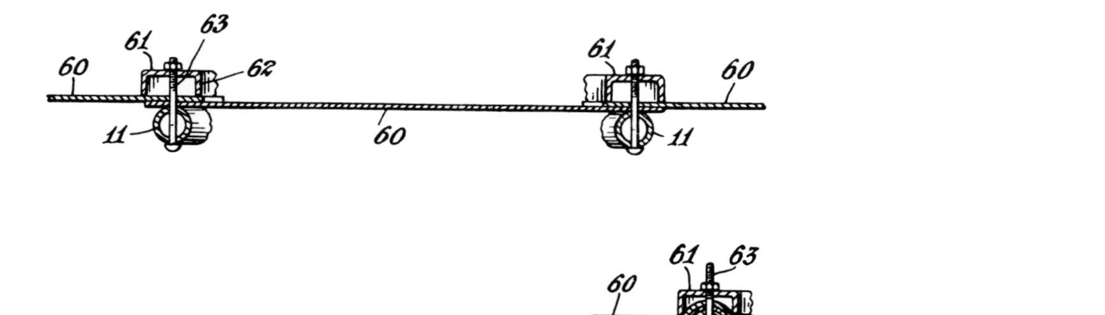

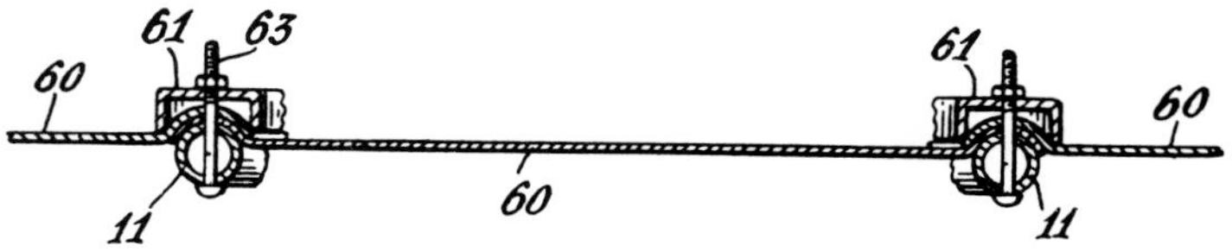

FIGURE 15

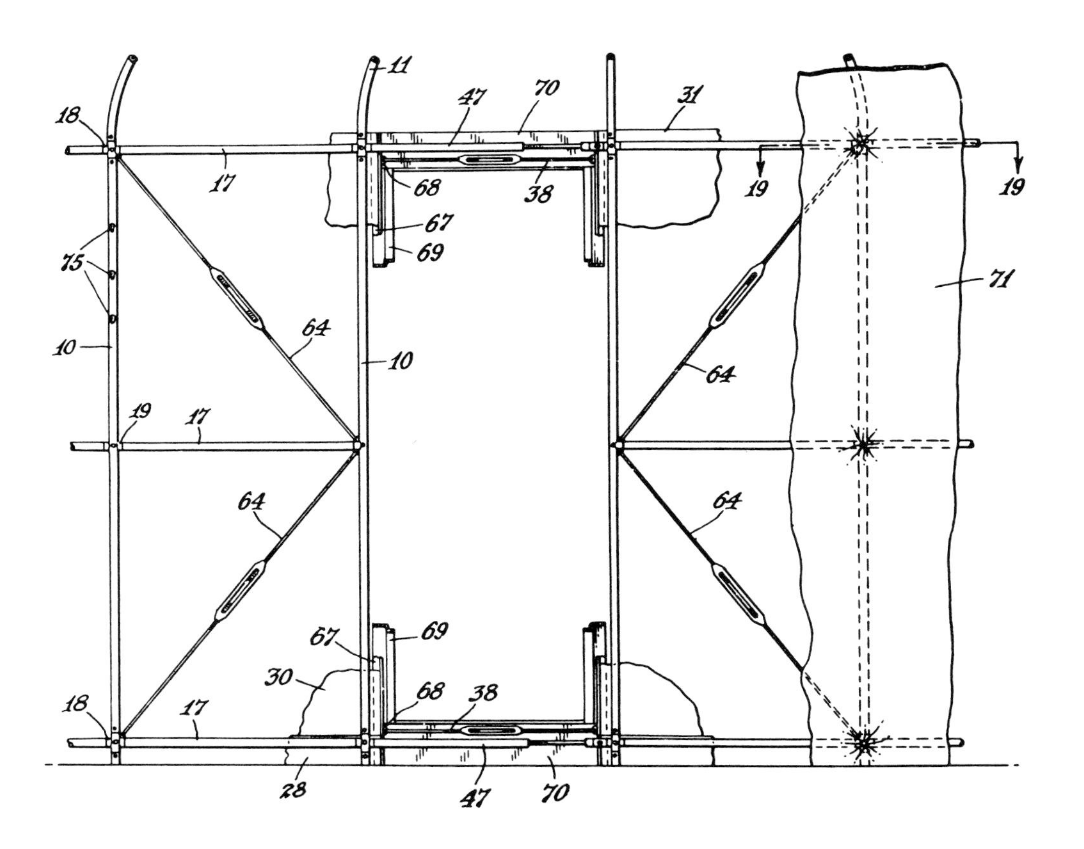

FIGURE 16

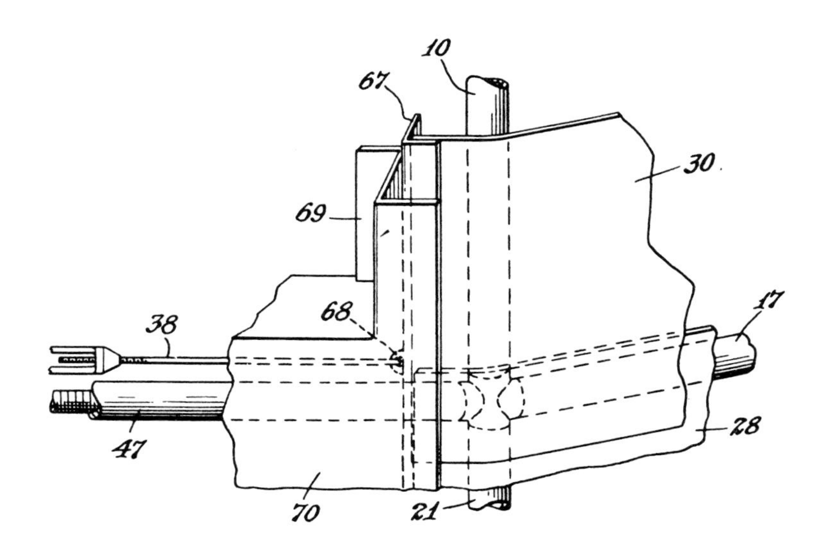

FIGURE 17

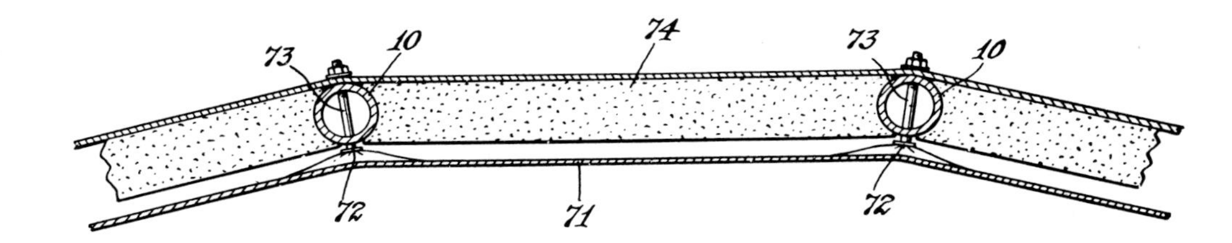

FIGURE 18

so that the take-up in stretching the covering sheets may be distributed evenly around the structure. However, if desired, the jacks may be omitted at one side of the doorway, in which case the channel 34 at that side will simply be drawn up tightly against the vertical member 10 so that all of the stretch may be taken up at the other side. The two intermediate jacks 47 bear at one end against the vertical members 10 at the studs 19, and at the other end bear against the flanges of the channel 34. The lowermost and uppermost jacks 47 are arranged adjacent the compression ring sections 17. Thus, just as has been described with reference to the embodiment of Fig. 9, the uppermost and lowermost horizontal compression rings 9 are continuous, i.e., they extend completely around the structure and prevent it from collapsing when the sheets are stretched thereover by means of the jacks 47. Turnbuckle rods 39 attached at one end to one of the vertical members 10, and at the other end to a stud 40 on the channel 34, cooperate with the jacks 47 and, when properly adjusted, balance the turning moment on the channel 34, as has been described with reference to Fig. 9. If desired, window openings 48 may be preformed in the covering sheet 31.

It will be understood that the tensioning of the covering sheets is accomplished with very little relative movement between the two channel members 34. The action of the jacks 47 in tightening the covering sheet 30 is illustrated in Figs. 11 and 12, Fig. 11 showing the relative position of the parts before tightening, and Fig. 12 their relative position after tightening. The jack 47 is recessed at one end to receive the dowel 19 secured to the vertical member 10. If desired, the base of the jack may be made arcuate in form, as at 49, for engagement with the vertical member 10 to prevent rotation of the base portion of the jack during its operation. The head 50 of the jack is keyed to the screw 51 thereof as by means of the pin 52, and is provided with a recess 53 arranged to surround the stud 19 of the adjacent vertical member 10. Apertures 54 are provided in the head 50 of the jack for engagement with a suitable tool for operating the jack. As the jacks are operated to draw the sheets 30 and 31 taut over the frame, the channel 34 moves to the right as indicated by the arrow *a* in Fig. 12, but the position of the adjacent vertical member 10 remains substantially unchanged for it is held by the continuous compression rings at the top and bottom of the structure. As the channel 34 moves to the right as indicated in Fig. 12, its lower end slides over the top of the sill 55 without interference therewith.

The connection for the inner end of the turnbuckle rod 39 is shown in Fig. 12, and may comprise a shackle 56 secured to the vertical member 10 by means of a bolt 57, the end of the turnbuckle rod being received between the bifurcations of the shackle 56 and secured thereto by a pin and cotter 58.

Fig. 13 shows a detail of the construction of the roof for a grain bin or the like to provide an opening for loading and unloading or other purposes. The construction of the supporting framework may be the same as has been described with reference to Figs. 1 to 9, and as shown, comprises the tubular roof-supporting members 11 which extend radially off the structure, and the compression ring 12 to which the inner ends of the members 11 are secured. Each roof panel 41 spans the distance between two of the members 11, and is provided with marginal flanges 42 arranged for engagement with the members 11 or with the flanges of adjacent panels. The flanges 42 preferably are curved or bent over in such a manner as to provide an interlocking engagement between the flanges of adjacent panels, and an interlocking engagement with the members 11 as indicated at *b* in Fig. 13. The flanges 42 of the central panel 41 shown in Fig. 13 overlie the flanges of the adjacent panels so as to permit the central panel to be raised in the manner indicated by the dot-dash lines 41′, swinging about the compression ring 12 as a pivot. The engagement of the roof panels 41 with the ring 12 has been described with reference to Fig. 8. The dot-dash lines 41′ in Fig. 13 show the central panel in only partially elevated position, and it will be understood that it may be swung upwardly to the extent which may be desired for loading operations. The panel 59 which completes the wall enclosure between the ends of the wall covering sheets 30 and 31 likewise may be removed to facilitate loading or unloading of the bin. If desired, the panel 59 may be connected to the downwardly extending portion 43 of the central roof panel so that when this panel is raised, the section 59 will be raised with it.

In Figs. 14 and 15 I have illustrated a modified form of roof construction to provide for stressing of the roof covering sheets. These are cross-sectional views taken transversely of one of the roof panels. The construction of the supporting framework may be similar to that which has been described comprising the tubular supporting members 11. In this embodiment of the invention, the covering sheets 60 may be flat sheets devoid of marginal flanges, as indicated in Fig. 14. Adjacent edges of the sheets 60 are arranged in overlapping relationship above the supporting members 11. A clamping element or batten 61 is placed above the overlapping portions of the sheets. The clamping element may be in the form of a channel in cross section so as to provide downwardly extending flanges 62 bearing against the overlapping edges of the sheets at each side of the supporting member 11. A series of clamping bolts 63 extend through the member 11, through the overlapping portions of the sheets 60, and the center of the clamping element 61. When the clamping bolts 63 are tightened, the overlapping portions of the sheets 60 are drawn down around the member 11, as indicated in Fig. 15. This action stresses the covering sheets because of the material taken up by wrapping the marginal edges of the sheets around the surface of the member 11. If desired, the bolts 63 which are nearest the center of the structure, may be tightened to a lesser extent than bolts which are toward the outside, so as to properly distribute the stress throughout the roof. Also, the tightening of the bolts 63 may be performed in such a manner as to regulate the clamping action of the elements 61 to stress the roof to approximately the same tension as the wall covering. The resulting even distribution of stresses throughout the entire structure makes it possible to employ lighter sections than would otherwise be possible.

In Figs. 16 to 20, inclusive, I have illustrated another embodiment of my invention which is more particularly adapted to the construction of houses or other structures in which it is desirable to provide a full height door opening. In this embodiment, the construction of the supporting framework may be substantially the same as has been described with reference to the preceding views, consisting of vertical wall members 10, horizontal compression rings made up of tubular sections 17, and roof members 11. However, only one intermediate compression ring is employed in place of the two rings 9 shown in Fig. 1. In this connection, I wish to make it clear that the number of intermediate rings is purely a matter of choice, and that, in the embodiment about to be described, one or more intermediate rings may be employed, as may be desired. The intermediate compression ring (or rings) does not extend completely around the structure because of the necessity of providing a full-height door opening. The top and bottom compression rings do extend completely around the structure. However, at the doorway a jack 47 takes the place of one of the sections 17. Compression forces which are set up in the supporting framework upon stressing of the covering sheets are transmitted from the upper and lower compression rings to the intermediate compression ring by means of struts 64. These struts 64 preferably consist of turnbuckle rods so as to provide a suitable adjustment. The rods 64 may conveniently be secured at their ends to the cross-shaped connecting members 18 and studs 19 previously described. For this purpose, the members 18 and 19 may be tapped to receive eye-bolts 65 (Fig. 20) for engagement with eye-bolts 66 formed in the ends of the rods. The provision of the struts 64 makes it possible to utilize the compressive strength of the discontinuous intermediate compression ring.

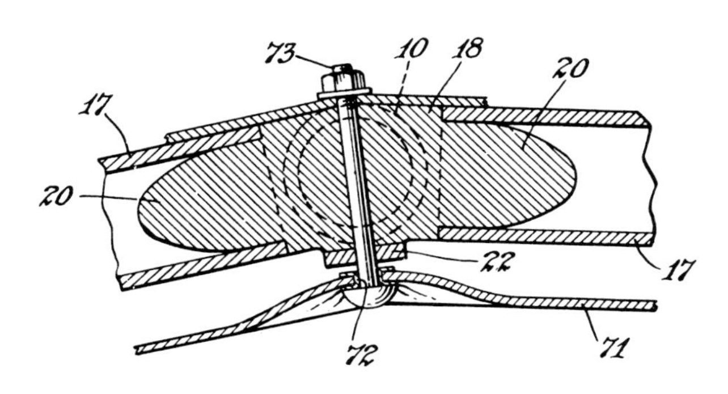

FIGURE 19

The wall covering sheets 30, 31 are applied in the manner which has been described with reference to Fig. 9. In this embodiment the ends of the sheets 30, 31 are welded or otherwise secured to angle members 67. Eye-bolts 68 are secured to the corners of the angle members 67 in line with the plane of the covering sheets, and turnbuckle rods 38 adjacent the jacks 47 at the top and bottom of the doorway are secured to the eye-bolts 68. A door frame member 69 of suitable form and construction may be provided. As shown in Fig. 17, the door frame 69 has a face flange 70 covering the turnbuckle rods 38 and jacks 47 and also overlying the angle members 67 at the ends of the covering sheets.

In the construction just described, stressing of the covering sheets 30 and 31 may be accomplished by tightening the turnbuckle rods 38, or by extending the jacks 47, or by conjoint manipulation of the jacks and turnbuckles. This makes it possible to obtain the desired stress in the covering while at the same time avoiding interference between the angle members 67 and the door frame. If, as the rods 38 are tightened, the members 67 come up against the door frame, the jacks 47 may be operated to complete the stressing operations; or, if desired, the turnbuckles 38 and jacks 47 may be so regulated as to bring the members 67 tightly against the door frame 69 just as the desired stress has been secured, thus providing a tight seal between the wall and door frame. Again, the jacks and turnbuckles may be so adjusted as to leave just a small clearance between the members 67 and door frame sufficient to permit the introduction of the desired quantity of mastic sealing compound therebetween. It will be seen that with the construction described, operation of the jacks 47 serves to enlarge the entire supporting framework slightly so as to stress the covering sheets, whereas operation of the turnbuckle rods 38 likewise serves to stress the covering sheets, but by direct action.

My invention also contemplates the stressing of the interior wall covering so as to utilize its strength characteristics in the finished structure. This feature is illustrated in Figs. 16, 18 and 19 which show an interior wall sheet 71 provided with grommeted apertures 72 at intervals which may be spaced in conformity with the spacing of the intersections between the frame members 10 and 17. Bolts 73 or other suitable clamping means pass through the grommets 72 and through apertures in the cross-members 18 or vertical members 10. In the joint shown in Fig. 19, the bolt 73 also passes through the plate 22, which was described with reference to Fig. 3. When the bolts 73 are tightened, the wall sheet 71 is drawn outwardly toward the framework of the structure so as to place this sheet under tension and impose additional compressive forces on the frame.

Before the interior wall covering 71 is applied, insulating batts 74 may be placed against the outer wall covering between the uprights 10 as shown in Fig. 18. In some cases it may be preferred to fill the space between the inner and outer wall coverings with insulation in loose form after both coverings are in place.

When the structure is to be employed as a tool supply

house, store, or for other purposes, it may be desired to arrange shelving around the inside, and for this purpose, the vertical members 10 may be provided with slots 75 (Fig. 16) for engagement by projections on the shelf supports.

In its preferred form, my invention contemplates the application to the finished structure of coatings of different heat absorptive capacities. For example, in the case of a grain bin or other storage unit where it is desired to reduce the magnitude of convection currents

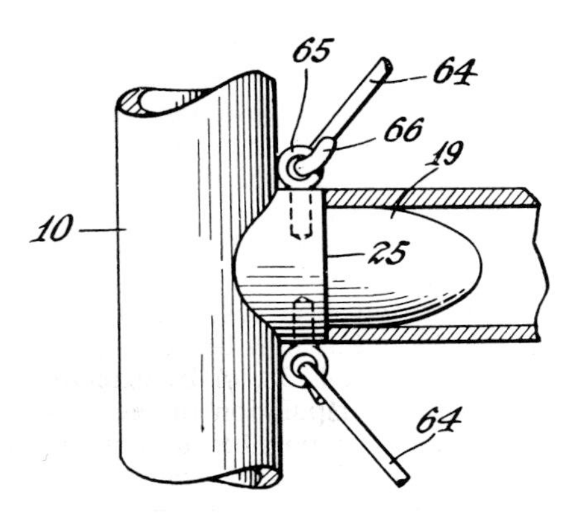

FIGURE 20

within the structure notwithstanding the provision of a ventilator at the apex of the roof, I paint the roof with a light-colored paint such as aluminum or white, and the side walls with a dark paint which has a greater heat absorptive capacity. Preferably, a plurality of coatings are used such that the color becomes gradually darker toward the bottom of the bin to create a low temperature gradient within the structure dependent upon the differential in heat absorptiveness of the pigment used in the coating. The diminished temperature gradient within the bin results in a more nearly static air condition while at the same time permitting proper ventilation.

In a structure used for houses, the roof will ordinarily be provided with a dark, heat absorptive coating, and the walls will be made progressively lighter in color so as to be more heat reflective or heat insulating at the bottom. This will create convection currents moving upwardly along the interior walls and ceiling to the ventilator, while cool fresh air will be drawn in around the bottom. Depending upon climatic conditions, it may be desired, however, to utilize in a house the same means as has been described above in connection with storage bins and the like, viz., to provide a structure with roof and walls of heat absorptivity which increases toward the bottom of the structure.

I also contemplate the use of interior coated surfaces having a differential in heat emissivity such that the interior heating effect of radiant heat at the outer surfaces of the structure varies from the top to the bottom thereof whereby convection currents within the structure are modified accordingly. For example, the ceiling may be painted a dark hue to provide high heat emissivity, and the interior walls coated with varying hues becoming lighter toward the bottom so as to gradually decrease the heat emissivity factor. This differential in heat emissivity of the interior walls and ceiling may be used either alone, or in conjunction with the differential in heat absorptiveness of the roof and exterior walls. In every case, the coating or coatings are such as to create a differential in the interior heating effect of radiant heat reaching the outer surfaces of the structure, as by direct sunlight or infrared rays, such that this heating effect varies from the top to the bottom of the structure whereby convection currents within are modified accordingly.

It will be understood that the features which I have described in connection with certain embodiments of my invention may be applied to others of the embodiments. For example, the manner of attaching the turnbuckle rods 38 to the ends of the covering sheets, as described with Figs. 16 and 17, may be employed in conjunction with the other structure illustrated in Fig. 9, in which case the turnbuckle rods 39 can be eliminated because there will be no turning moment on the channels 34. The terms and expressions which I have employed are used in a descriptive and not a limiting sense, and I have no intention of excluding such equivalents of the invention described, or of portions thereof, as fall within the purview of the claims.

8▲DYMAXION MAP (1946)

U.S. PATENT—2,393,676

APPLICATION—FEBRUARY 25, 1944

SERIAL NO.—523,842

PATENTED—JANUARY 29, 1946

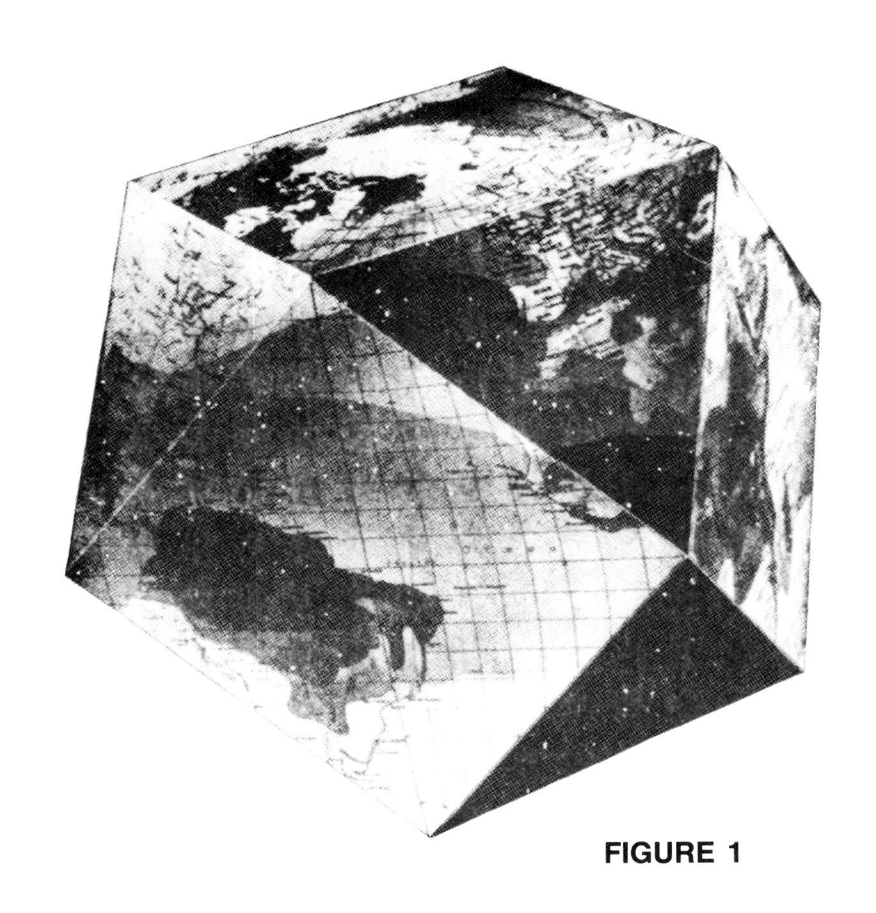

FIGURE 1

DYMAXION A
THE RALEIGH EDIT
R. BUCKMINSTER FULLER
PUBLISHED BY THE STUDENT PUBLICATIONS OF THE SCHO
COPYRIGHTED 1954
MEAN LOW ANNUAL TEMPERATURE
LAND
WATER
50° C
40° C
30° C
20° C

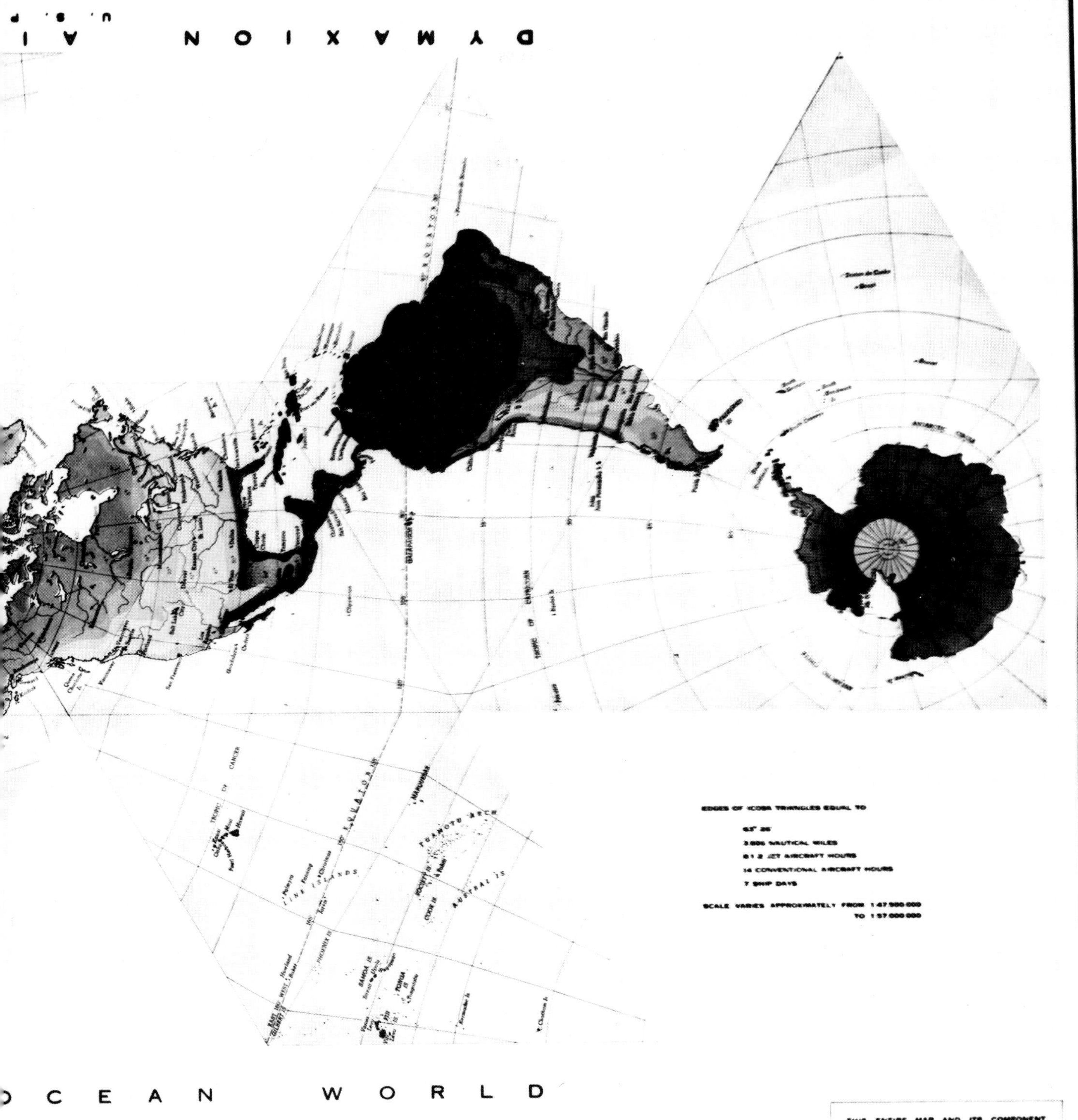
DYMAXION
EDGES OF ICOSA TRIANGLES EQUAL TO
63° 26'
3.906 NAUTICAL MILES
8.1.2 JET AIRCRAFT HOURS
14 CONVENTIONAL AIRCRAFT HOURS
7 SHIP DAYS
SCALE VARIES APPROXIMATELY FROM 1:47 500 000
TO 1:57 000 000
CEAN WORLD
FULLER PROJECTION
JI SADAO. CARTOGRAPHERS
SIGN. NORTH CAROLINA STATE COLLEGE. RALEIGH. N. C.. U.S.A.
U. S. PAT. 2.393.676
THIS ENTIRE MAP AND ITS COMPONENT PARTS AND METHOD OF PROJECTION ARE FULLY PROTECTED BY INTERNATIONAL COPYRIGHT CONVENTION AND PATENTS AND MUST NOT BE REPRODUCED WITHOUT PERMISSION OF R BUCKMINSTER FULLER
THIS EDITION OF FULLER PROJECTION LICENSED BY HIM UNDER HIS U.S. PATENT 2 393 676 CANADIAN PATENT 448 064 AND HIS COPYRIGHTS OF 1943 1944 1953 THIS EDITION COPYRIGHTED BY HIM 1954
ADDITIONAL COPIES AVAILABLE FROM
STUDENT PUBLICATIONS. SCHOOL OF DESIGN
BOX 5273
RALEIGH NORTH CAROLINA
UNITED STATES OF AMERICA
10° C
0° C
10° C
20 C

MY 1927 COMMITMENT TO DEAL HENCEFORTH only with total planetary physical and metaphysical resources employed only in technology useful for all people around the surface of Spaceship Earth called for a non-distorted map of the world upon which to identify the resources and the people. I first tried making a Mercator-type map using the over-the-Poles 90th meridian of longitude instead of the equator as the point of "true" reference. I used this North-South world island map as the endpapers in my first published book, *Nine Chains to the Moon* (1938). I then found I could develop a projection with what I called unbroken uniform boundary scale symmetrical polyhedral facets. My first such projection was published in an eighteen-page, four-color edition in *Life* magazine in March 1943. My next improved version was published in *American Neptune* in 1944 and the first of my icosahedral versions in the Raleigh, North Carolina, 1954 edition of the map. In 1947, my patent was

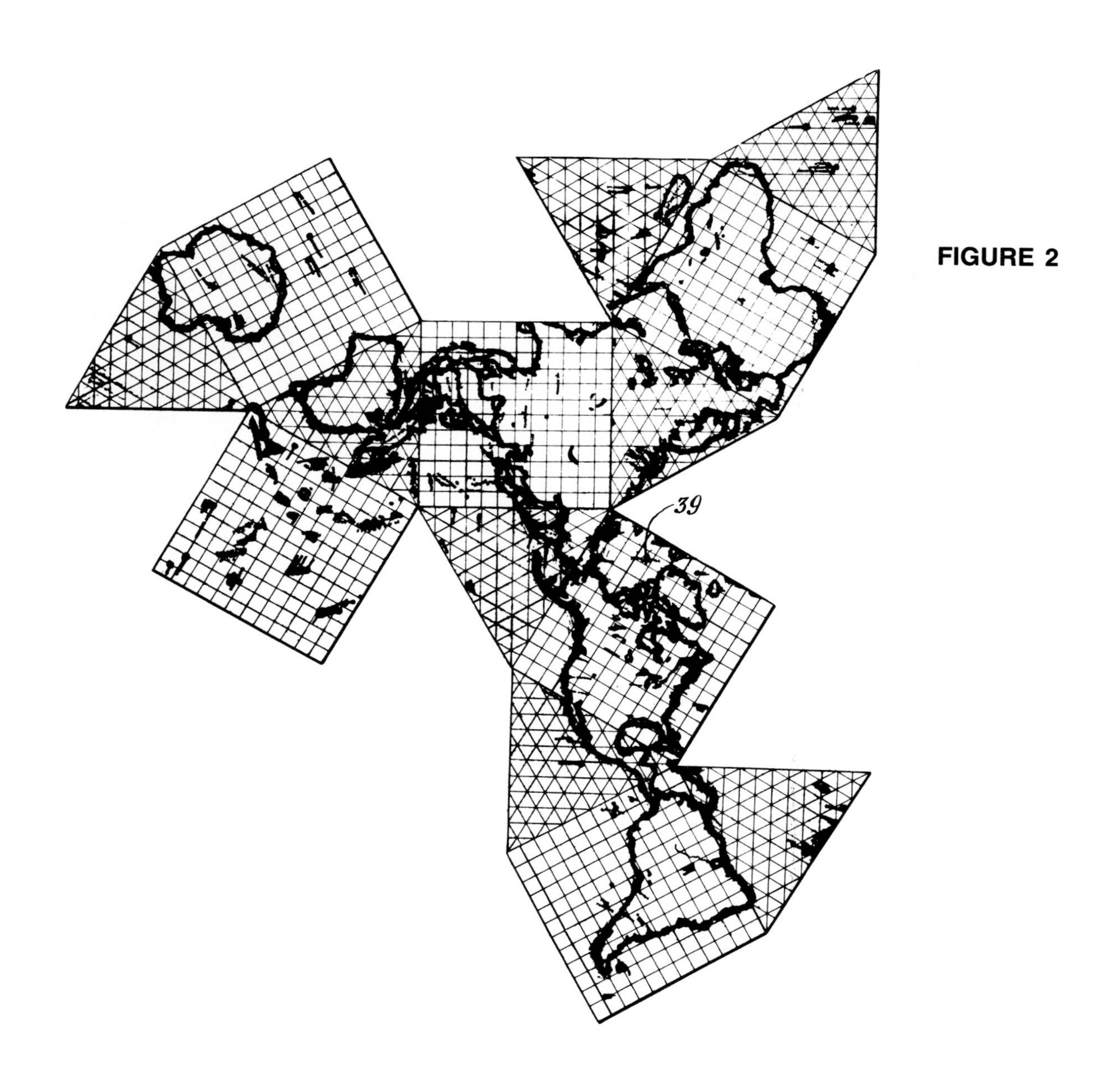

FIGURE 2

FIGURE 3

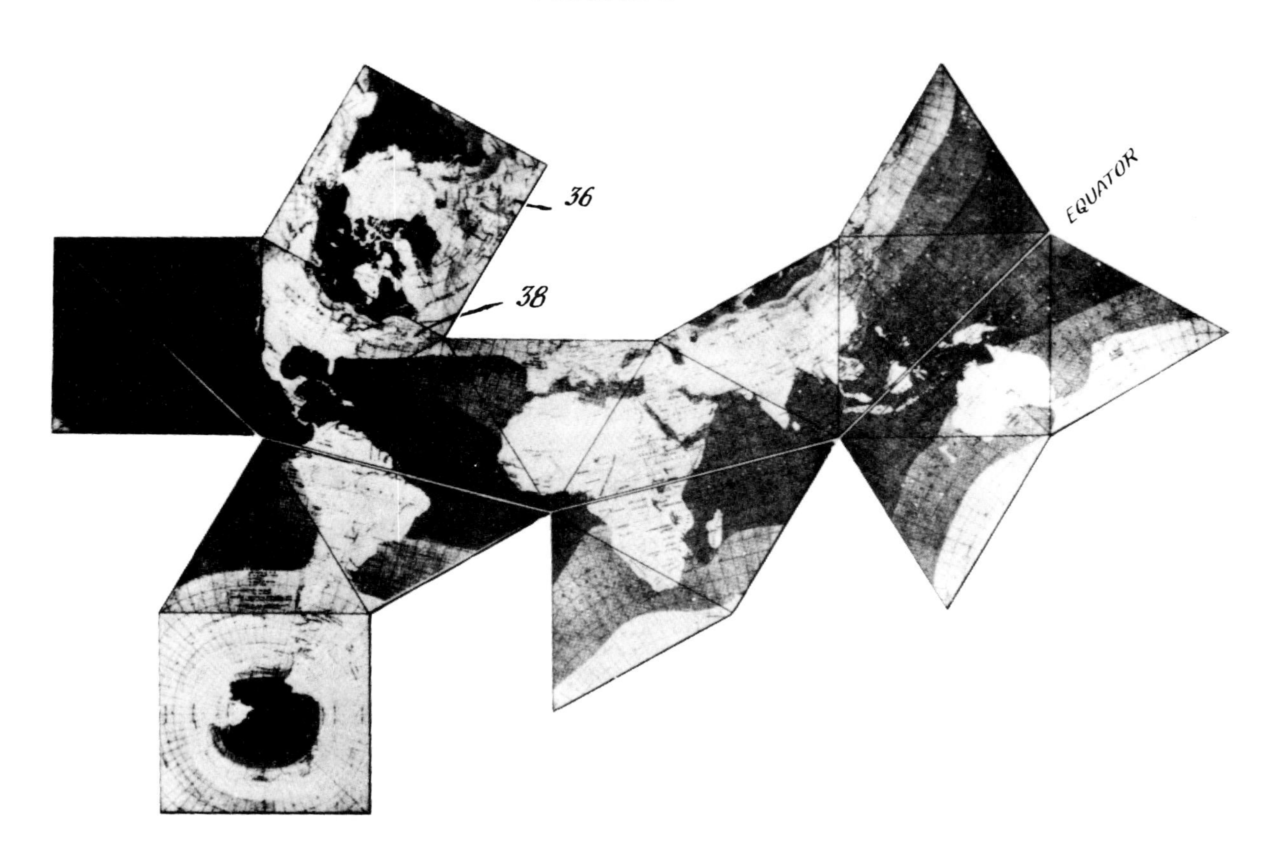

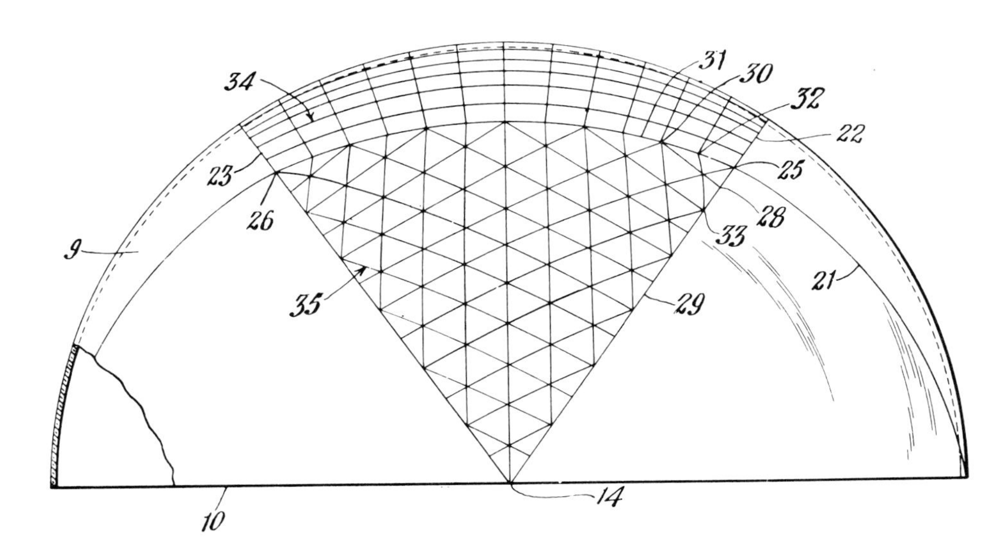

FIGURE 4

granted, and in that year *Science* magazine published a statement that my projection method was the first to be granted a patent by the U.S. Patent Office. *Science* magazine was in error. In 1900, the Patent Office had ruled that no further applications for patents for world maps would be considered as all possible methods had been exhausted. When *Life* magazine was considering publishing my Dymaxion Map, they asked several great experts to review my work, and the experts dismissed it as "pure invention." These statements were used by my patent attorney to persuade the Patent Office to consider my patent application, which was granted.

UNITED STATES PATENT OFFICE

Richard Buckminster Fuller, Washington, D.C.

CARTOGRAPHY

The invention relates to cartography.

As the earth is a spherical body, so the only true cartographic representation of its surface must be spherical. All flat surface maps are compromises with the truth. For example, Mercator's projection is true to scale only along the equator, and azimuthal projection is limited to convergence of the meridians at one pole at a time. Other known systems of projection can be made to give uniform scale along parallels, or to give equal areas albeit with exaggerated shape distortions.

Another expedient has been to resolve the earth's surface into a polyhedron, projecting gnomonically to the facets of the polyhedron, the idea being that the sections of the polyhedron can be assembled on a flat surface to give a truer picture of the earth's surface and of directions and distances. Such a system is fettered to the limitations and gross radial distortions which characterize gnomonic projection.

It is an object of my invention to provide a sectional map of the world, or of a portion of its surface, which is so constructed that its parts can be assembled to give a truer over-all picture of areas, boundaries, directions and distances than is attainable with any type of plane surface map heretofore known.

Another object has been to provide a subdivision of the earth's surface for cartographic purposes which will result in sections that can be assembled with fewer sinuses in land areas than is possible with sectional maps heretofore known.

Other objects and advantages will appear as the description proceeds.

I have found that by resolving the earth's surface into sections which are entirely bounded by straight line projections of great circles, and constructing a map on great circle grids, it is possible to maintain uniform scale peripheral cartographic delineations and to distribute all subsidence distortion from the periphery toward the center. I have discovered further that this system brings the subsidence distortion to an irreducible minimum which, without correction of any kind, is very considerably less than with any system of projection heretofore devised.

Another discovery which I have made is that if the earth's surface is resolved into six equilateral square sections and eight equilateral triangular sections whose edges match throughout, there is formed a polyhedron all of the vertexes of which lie in great circles of a sphere. This figure I call a "dymaxion." As a consequence, all of the sides of all of the sections are true projections of great circles, and uniform scale peripheral cartographic delineations can be constructed.

With reference to the accompanying drawings, I shall now describe a preferred form of my improved map and the method of constructing it.

Fig. 1 is a perspective view of a "dymaxion," in which the earth's surface is resolved into that form of polyhedron which has six equilateral square sections and eight equilateral triangular sections whose edges match throughout and all of the vertexes of which lie in great circles of a sphere.

Fig. 2 is a map of the world made up of a plurality of square and triangular sections, the cartographic delineations being constructed on great circle grids. In this embodiment of the invention the location of the pole and the orientation of the map relative to the "dymaxion" are such that the land areas can be joined without land sinuses.

Fig. 3 is a view similar to Fig. 2, and shows an arrangement of the sections of the polyhedron of Fig. 1. In this embodiment of the invention the poles are located arbitrarily at the centers of two of the square sections or facets of the polyhedron. The sections are laid in a pattern that approximates the familiar appearance of the Mercator projection. The equator is a continuous line, orienting the world east to west.

Fig. 4 is an elevational view of a cartographic device having a spherical grid composed of intersecting great circles. This is the form of device which I prefer to use in transferring cartographic delineations from a spherical to a plane surface.

Fig. 5 is a plan view of the cartographic device of Fig. 4.

Figs. 6 and 7 depict two arrangements of selected map sections illustrating the relationship of sinus to arc. The sections shown have the great circle grids which I will describe, but for the sake of simplicity the cartographic delineations have been omitted in these views.

Fig. 8 is a detail view of one of the triangular sections showing a three-way great circle grid and, superimposed, meridians and parallels.

An essential feature of my invention resides in constructing the map on great circle grids. In the case of the square section, a two-way grid is employed. In the case of the triangular sections, a three-way great circle grid is employed. This will be understood in part from Figs. 4 and 5 which show one form of cartographic device used in constructing my polyhedral map from a globe.

Construction of the cartographic device

The invention, and its distinguishing attributes and advantages, may best be understood by considering first my preferred method of translating true spherical cartographic delineations to the flat sections which are the facets of the particular form of polyhedron shown in Fig. 1 heretofore referred to as the "dymaxion." Let us assume that we start with a standard globe. First, we construct a member in the form of a hemisphere 9, illustrated by Figs. 4 and 5. The size of this hemisphere is such that it will fit closely over the surface of the globe that has been selected—i.e., the inside diameter of the member 9 will be approximately equal to the diameter of the globe. The member 9 may conveniently be formed of a transparent plastic, although this is a matter of choice and other materials may be employed. Preferably, it is made as thin as will permit convenient handling so as to avoid undue parallax in reading the great circle coordinates off the globe.

We have seen that all of the vertexes of the polyhedron represented in Fig. 1 lie on great circles of a sphere. The first step in laying out the great circle grids on the member 9 is to locate these vertex points for at least one triangular section and for at least one rectangular section. In other words, we are going to construct on the member 9 a spherical triangle and a spherical rectangle the vertexes of which coincide with selected vertexes of a polyhedron which has six equilateral square facets and eight equilateral triangular facets. The locating of the vertexes on the member 9 may be accomplished by any desired means and it is not essential that the grids formed on the member 9 be disposed in the exact positions shown in Figs. 4 and 5 relative to the edge 10 of this member. However, it is convenient to have the spherical rectangle centered on the member, as shown in Fig. 5.

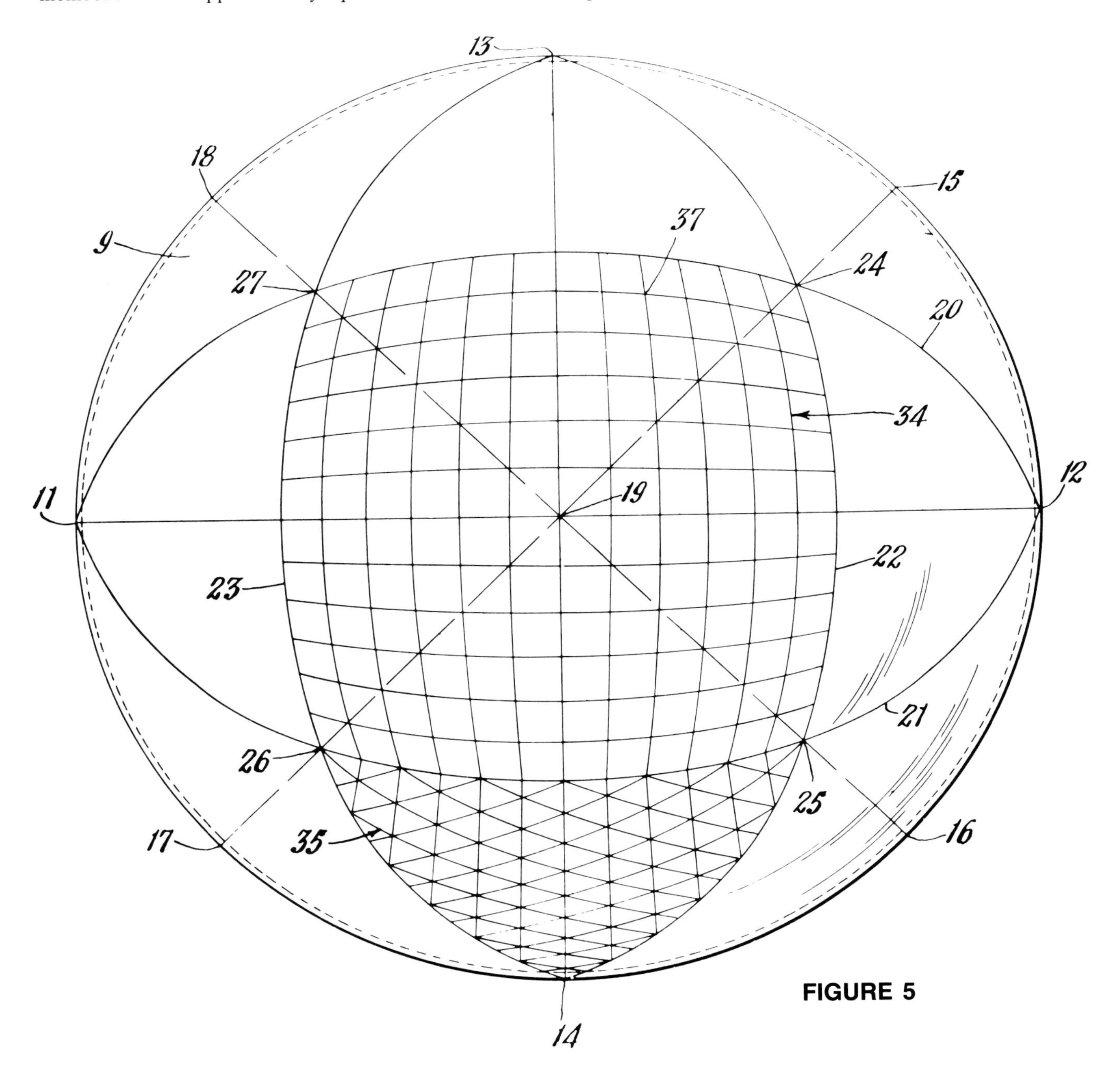

FIGURE 5

One method of locating the vertexes on the member 9

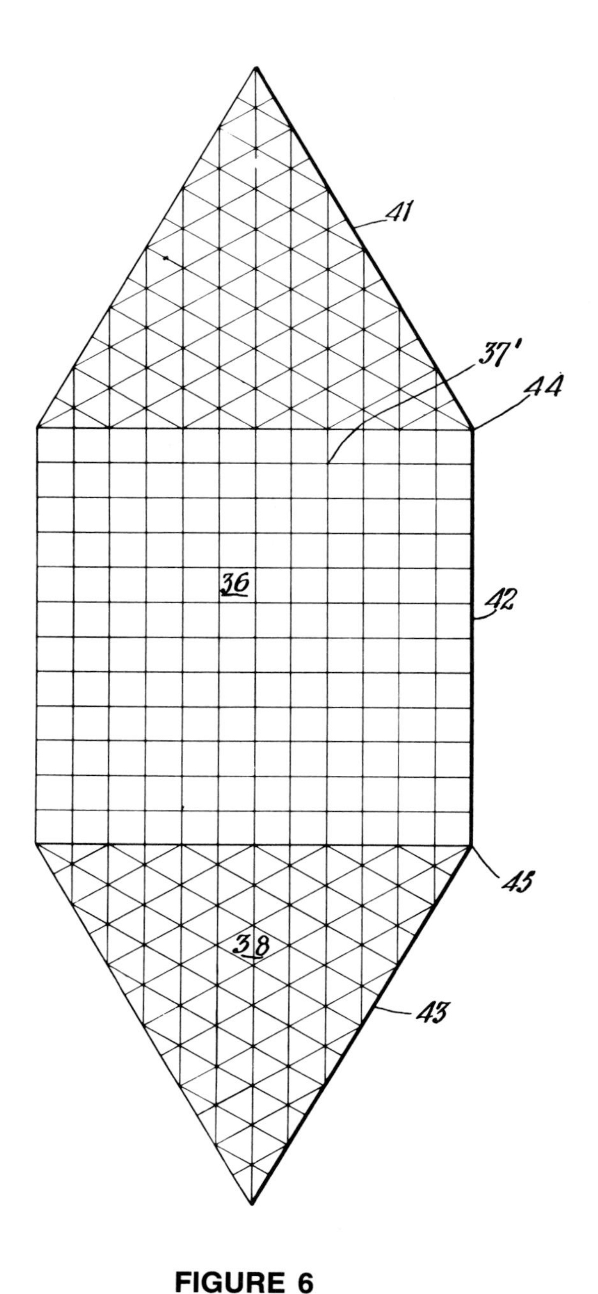

FIGURE 6

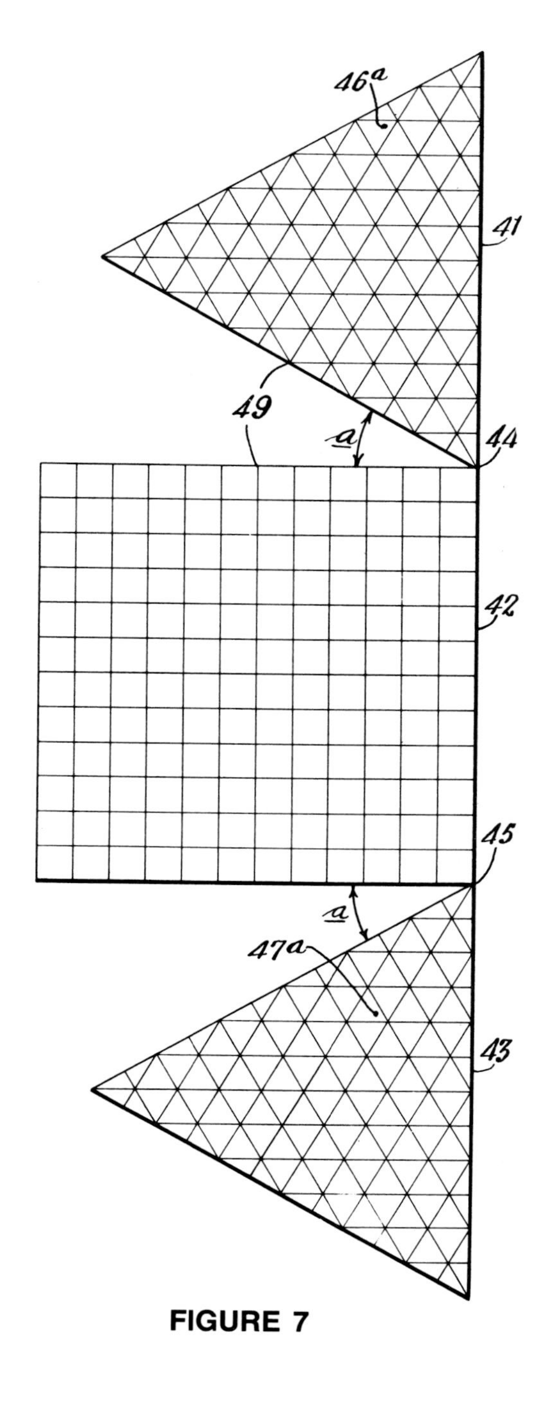

FIGURE 7

is as follows: locate the points 11 and 12 (Fig. 5) along the edge 10 of the member 9 at diametrically opposite positions. Locate the points 13 and 14 along the edge 10 midway between the points 11 and 12. Locate the points 15, 16, 17 and 18 along the edge 10 midway between the points 12—13, 12—14, 11—14, and 11—13, respectively. What this amounts to is dividing the great circle edge 10 of the member 9 into eight parts. Scribe four great circle arcs (spherical straight lines) between the points 13—14, 15—17, 11—12, and 16—18. These great circle arcs can conveniently be drawn by placing the member 9 on the globe with the diametrically opposite points on one of the great circles of the globe with the center point 19 at the pole or somewhere along the same great circle, and then tracing the great circle onto the surface of the member 9. The points 24, 25, 26 and 27 are located midway between the point 19 and the points 15, 16, 17 and 18 respectively, i.e., 45° from the point 19. The great circle arcs 20, 21, 22 and 23 are then constructed so as to intersect at points 24, 25, 26 and 27. This divides each of the great circle arcs 20, 21, 22 and 23 into three equal parts of 60°. We now have a spherical rectangle 24—25—26—27 whose vertexes coincide with the vertexes of any one of the six equilateral square facets of Fig. 1, and a spherical triangle 14—25—26 whose vertexes coincide with the vertexes of any one of the eight equilateral triangular facets of Fig. 1.

Each side of the spherical rectangle is now divided into any desired number of equal parts. In the embodiment shown the division is into twelve parts, each of which is 5° of arc. By joining all of these points with great circle arcs, we obtain the grid shown in Fig. 5 consisting of great circle arcs. These can conveniently be drawn by placing the member 9 on the globe and locating each pair of corresponding points along any great circle, then tracing the great circle onto the member 9.

To construct the three-way grid on the spherical triangle, the sides of the triangle are similarly divided into any desired number of parts (twelve, as shown) and the great circle arcs are scribed as shown in Fig. 4, connecting the first division point 28 of the side 29 to the second division point 30 of the side 31 (and so on), and connecting the first point 32 of the side 31 with the second point 33 of the side 29, and so on. In each case, the great circle arcs can be drawn by tracing from a globe as previously described.

We now have on the member 9 a two-way great circle grid indicated generally at 34 (Fig. 5) and a three-way great circle grid indicated generally at 35 (Fig. 4). A peculiarity of a three-way great circle grid constructed on an equilateral spherical triangle whose vertexes correspond to the polyhedron shown in Fig. 1 (consisting of six equilateral rectangular facets and eight equilateral triangular facets) is that all of the great circle arcs intersect—that is, the entire pattern of the grid shows intersections of three great circle arcs.

It will be understood, of course, that the provision of both the rectangular and triangular grids on the one cartographic device 9 is largely a matter of convenience, and I contemplate that if desired, the triangular grid could be placed upon a separate device apart from the rectangular grid; also, that they could be located in different positions on the hemispherical member 9. Furthermore, it is not necessary that the rectangular and triangular grids be conjoined in the manner shown in Figs. 4 and 5 as they are susceptible of use entirely independently of one another.

Translation from the spherical to the plane surface

The flat sections or tiles shown in Figs. 2, 6 and 7 are the facets of the polyhedron, and the corners or vertexes of each tile are the vertexes of the polyhedron. Each tile is provided with a great circle grid, those on the square tiles corresponding to the spherical grid 34 and those on the triangular tiles corresponding to the grid 35 previously described. The grids are constructed as before by dividing the sides into the desired number of equal parts. Assuming 5° intervals, we will divide into twelve parts to gain correspondence with the spherical grids. In this case, the points along the edges of the tiles are joined by straight lines. These straight lines are true representations of a projection of a great circle since the projection of a great circle is a straight line. All of the edges of each tile are projections of great circles.

Having the device of Figs. 4 and 5 and the tiles with their great circle grids as shown in Figs. 6 and 7, we now proceed to translate cartographic delineations from the spherical to the plane surface. For example, let us suppose that we are mapping on one of the square tiles. We will place the member 9 on the globe with the spherical grid 34 overlying that portion of the earth's surface which is to be translated to the tile (Fig. 6). If it is desired to have the poles at the centers of the square tiles as in the case of the map represented by Fig. 3, the spherical grid 34 will be centered on one of the poles and oriented as shown (or otherwise, as may be desired). The coordinates of the particular city, coast-line point, or other cartographic feature are read on the grid 34 and plotted on the grid of the tile 36. Thus, if the particular point being plotted lies at the point 37 on the grid 34, it will be plotted at the point 37′ on the grid of the tile 36. This process is repeated for each point that is being translated from the spherical to the plane surface. It will, of course, be understood that the grids may be subdivided as finely as may be desired. That is, the grid may be carried down to 1° intervals or to fractions of a degree, depending upon the accuracy desired.

The same procedure is followed in translating from the globe to the triangular tile 38. First, the member 9 is so located on the globe that the corners 25 and 26 of the triangular grid 35 overlie the desired points, that is, they will overlie the same points as did the corners 25 and 26 of the rectangular grid 34 during the use of the latter. Coordinates are read on the grid 35 and plotted on the grid of the tile 38.

It is not necessary that the poles be located at the centers of square tiles, and in Fig. 2 I have illustrated a modification in which the north pole is located in an arbitrary position 39 which was selected with view to having the land areas so placed as to eliminate land sinuses. It will be observed that in this embodiment, the land areas have been joined without sinuses.

The embodiment illustrated in Fig. 3 is useful because of its approximation of the Mercator projection which brings the character of the Renaissance world into bold relief.

A distinguishing feature of the maps of both Figs. 2 and 3, and of other maps which can be constructed in accordance with my invention, is that each plane section or tile has uniform scale peripheral cartographic delineation. This is possible because the sections match along edges which are representations of projected great circles. This means that distances measured along the edges of any section are true to scale, and that scale is uniform throughout. Moreover, by reason of employment of the particular method of translating from the spherical to the plane surface which I described, subsidence errors are distributed interiorly of the periphery. This is accomplished by plotting on the great circle grid and no corrections are required. The tiles may be arranged in any manner which may be desired for the study of particular land or water features, directions and distances. With three tiles arranged as shown in Fig. 6, we have along the edges 41, 42, 43 the arc of a great circle which may extend from pole to pole, or which may extend from any point on the earth to any point on the opposite side of the earth. If we shift the upper and lower triangular tiles into the position shown in Fig. 7 by simply turning them through an angle of 30° about the points 44 and 45, respectively, we see at a glance the straight line distances and directions between any two points along the line 41, 42, 43.

With reference to Fig. 7, it will be observed that points 44 and 45 are 30° removed in arc from the center axis 49 of the sections, and that the triangular tiles have been opened away from the square section by a corresponding 30° sinus angle *a*.

Fig. 8 shows one of the triangular tiles in which the three-way great circle grid is shown in light lines at 46, and superimposed thereon are the meridians 47 and parallels 48 which may be plotted on the tile in the same manner as the other cartographic delineations are plotted. If desired, the map may be employed without adding the superimposed meridians and parallels. Where the meridians and parallels are used, the great circle grid may, if desired, be removed either before or after the map has been plotted. In this case the great circle grid will have been used purely as a construction device.

When the great circle grid is employed merely as a construction device for translating meridians and parallels to a plane surface, it may be desirable to plot these down to single degrees, or even to fractions of a degree. The map itself may then be plotted directly on the coordinates of latitude and longitude, with the result of reducing the map to imaginary great circle grids, and producing a map which possesses the various advantages I have described. This system makes it possible to construct my novel map with the use of available cartographic data based on the system of latitudes and longitudes.

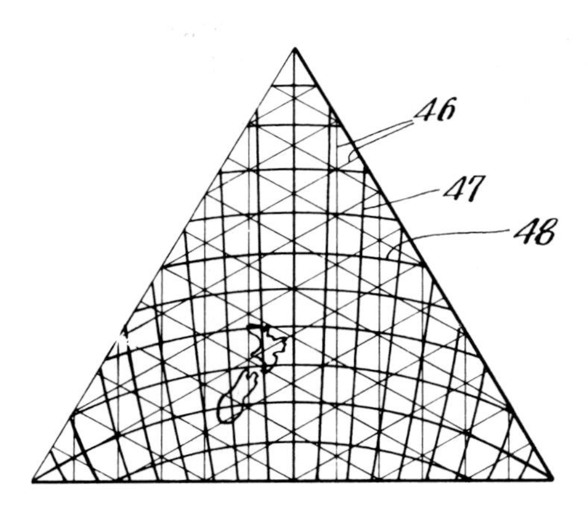

FIGURE 8

Among the advantages of my invention may be cited the provision of uniform scale along the periphery of all of the sections, the provision of a sectional map which can be assembled in a manner which eliminates land sinuses, and the fact that by having uniform peripheral scale with subsidence errors distributed interiorly of the periphery by plotting on a great circle grid, distortion is less than with any form of projection heretofore known. With gnomonic projection, the scale is true only at the exact center of a section, and subsidence errors build up in a radially outward direction at an alarming rate. Some systems of cartography resort to correction of areas on what is known as the "equal area" basis, which only serves to enormously distort shapes. Careful study of maps constructed in accordance with my invention will show that it gives a truer overall picture of areas, boundaries, directions and distances than is attainable with any type of plane surface map heretofore known.

The terms and expressions which I have employed are used in a descriptive and not a limiting sense, and I have no intention of excluding such equivalents of the invention described, or of portions thereof, as fall within the purview of the claims.

9▲DYMAXION HOUSE (WICHITA) (1946)

IN 1944 THE AERONAUTICAL PRODUCTION DIVISION OF THE U.S. War Production Board and the U.S. Department of Labor became alarmed over the fact that they were being completely frustrated in their bomber, fighter, and auxiliary aircraft production because labor did not like the tour of Wichita, Kansas, and other longtime aircraft-production company towns of the West. These locations were proving so unsatisfactory to labor that labor was deserting these towns and was turning to other war production activities. In Wichita, Kansas, the vitally essential B-29 bomber was being produced, together with many other fighter and auxiliary craft. In 1944 the Wichita population doubled from 100,000 to 200,000. People were sleeping in three shifts in the same beds. Life was intolerable. The workers could see no future for aircraft production after the war (international flights and jets had not yet been developed). Workers by the thousands were quitting Wichita daily to find more comfortable war work elsewhere where postwar employment conditions would be more favorable. Labor leaders in Washington, D.C., recalled that I had been developing the Dymaxion House to be produced only by the most advanced aeronautical technology. The War Production Board and U.S. labor leaders asked me to go to Wichita and meet with the chief executive officer of Beech Aircraft, where they had the best labor relations. I did so and we produced my Dymaxion dwelling there. Finished, it weighed the three tons that I had estimated it would in 1927, in contrast to the 150-ton-volume, floor area, and techno-facilities equivalent one-family house produced by conventional building methods. In my book *Grunch of Giants* I relate what happened subsequently.

UNITED STATES PATENT OFFICE

This invention relates to houses, this term being comprehensively used to include shelters in the forms of dwellings for people, commercial establishments, schools, manufacturing plants, and, in general, for any purpose or things requiring shelter from the weather.

The primary object is to provide a house assembled

from parts having a total weight so light and capable of being grouped in a unit package so compactly as to make it economically practicable to manufacture and package the house in a factory and then to ship the packaged house from this factory to anywhere in this or foreign countries, together with the requirement that the assembled house must be capable of functioning adequately as a home for the average small family or, in general, as necessary to meet the requirements of sheltering constructions of comparable size. However, the principles of the invention are applicable to larger structures. These principles provide for a house that is not only economically competitive with conventional houses but which has many advantageous features unobtainable by any prior art construction methods.

A house intended as a home for a small family, which embodies specific examples of the principles of the present invention, is illustrated by the accompanying drawings in which:

Fig. 1 is a perspective view having broken-away portions and showing the general construction of the house in its entirety;

Fig. 2 is a side view of the compression mast and footing;

Fig. 3 is a cross section taken from the line 3—3 in Fig. 2;

Fig. 4 is a top view of the mast;

Fig. 5 is a partly sectioned side view of the top of the compression mast and including the rotative ventilator mounting;

Fig. 6 is a top view of one of the compression rings;

Fig. 6a is an enlargement from Fig. 6;

Fig. 7 is a side view of the inside of one of the outer deck ring sections;

Fig. 8 is an end view of Fig. 7 showing how the ends of adjacent sections internest where they connect together;

Fig. 9 is a top view of Fig. 7;

Fig. 10 is a detail showing how the tension strands connect with the upper flange of the outer deck ring;

Fig. 11 is a side view of the inside of one of the inner deck ring sections;

Fig. 12 is a view like Fig. 8 excepting that it applies to the inner deck ring;

Fig. 13 is a top view of Fig. 11;

Fig. 14 is a sectioned view showing how the flooring beams connect with the deck rings;

Fig. 15 is a cross section showing the connection between a flooring beam and the inner deck ring;

Fig. 16 is an enlargement taken from Fig. 14;

Fig. 17 is a transverse section through the deck showing how it is trussed;

Fig. 18 is a plan layout of Fig. 17;

Fig. 19 is a side view showing the inter-fastened ends of two truss ring sections;

Fig. 20 is a detail showing how the outer deck ring is anchored to the ground;

Fig. 21 shows how the carling supporting ring is spaced above the upper and inner compression ring;

Fig. 22 shows the same as Fig. 21 excepting that it shows the inner joined ends of the carling ring sections;

Fig. 23 is a top view of Figs. 21 and 22;

Fig. 24 is a cross section showing how the carling supporting ring is fastened to the uppermost compression ring;

Figs. 25 and 26 are cross sections showing how the cowling carling clips are fastened to the intermediate and lowermost compression rings;

Fig. 27 is a top view showing how the clips of Fig. 26 appear;

Fig. 28 is a side view of one of the cowling carlings;

Fig. 29 is a cross section taken from the line 29—29 in Fig. 28;

Fig. 30 is a top view of the bottom end of a cowling carling;

Fig. 31 is a top view of an upper end of a cowling carling;

Fig. 32 shows the overlapping joint between two cowling carling sections;

Fig. 43 is a side view of the lowermost end of one of the ceiling carlings;

Fig. 44 is a cross section taken from the line 44—44 in Fig. 43;

Fig. 45 is a top view of Fig. 43;

Fig. 46 is a cross section showing the various adjacent parts around the uppermost compression ring;

Fig. 47 is a cross section through the side portion of the house around and beneath the uppermost compression ring;

FIGURE 1

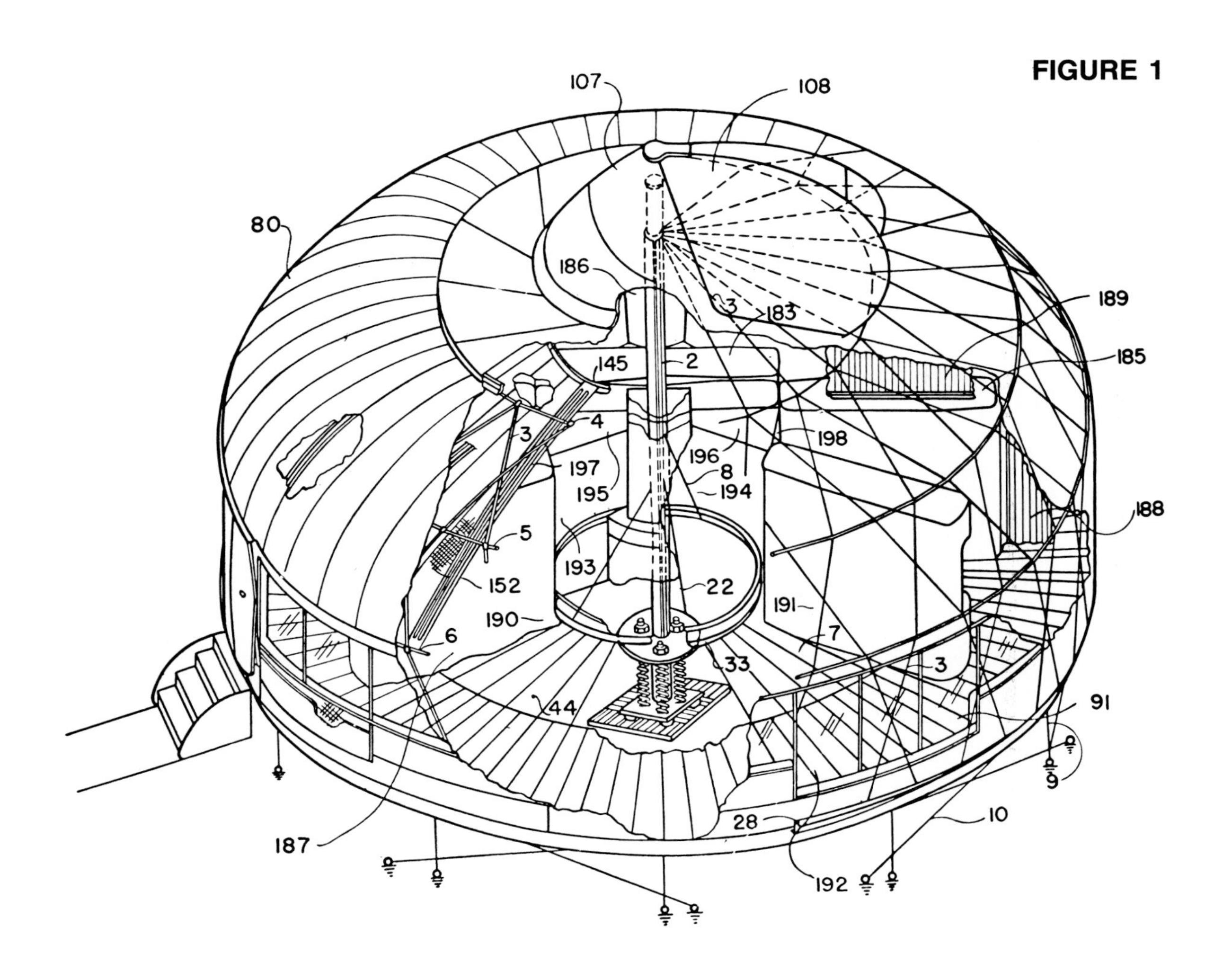

Fig. 33 is a cross section showing how the cowling carling clips work with one of these carlings;

Fig. 34 is a top view of one of the cowling gores;

Fig. 35 is a top end portion of one of these gores;

Fig. 36 is a side view of one of the carling supporting ring sections showing how the cowling carlings connect therewith;

Fig. 37 is a top view of Fig. 36;

Fig. 38 is a cross section taken from the line 38—38 in Fig. 36;

Fig. 39 is an edge view of one of the clips for connecting the cowling carling through the intermediate compression ring;

Fig. 40 is a side view of Fig. 39;

Figs. 41 and 42 are similar to Figs. 39 and 40 excepting that the clip is the one connecting the cowling carling with the lowermost compression ring;

Fig. 48 is a broken-away perspective of the rotating ventilator shown on top of the house of Fig. 1;

Fig. 49 is a top view of this ventilator;

Fig. 50 is a cross section taken from the line 50—50 in Fig. 49;

Figs. 51 and 52 are top and side views of the ventilator's outer cone only;

Figs. 53 and 54 are top and side views of the ventilator's inner cone only;

Fig. 55 is a cross section through the house generally indicating the association between details shown by the previous figures;

Figs. 56 and 57 are cross sections showing details illustrated by previous figures;

Fig. 58 is a top view of the gutter used by the house;

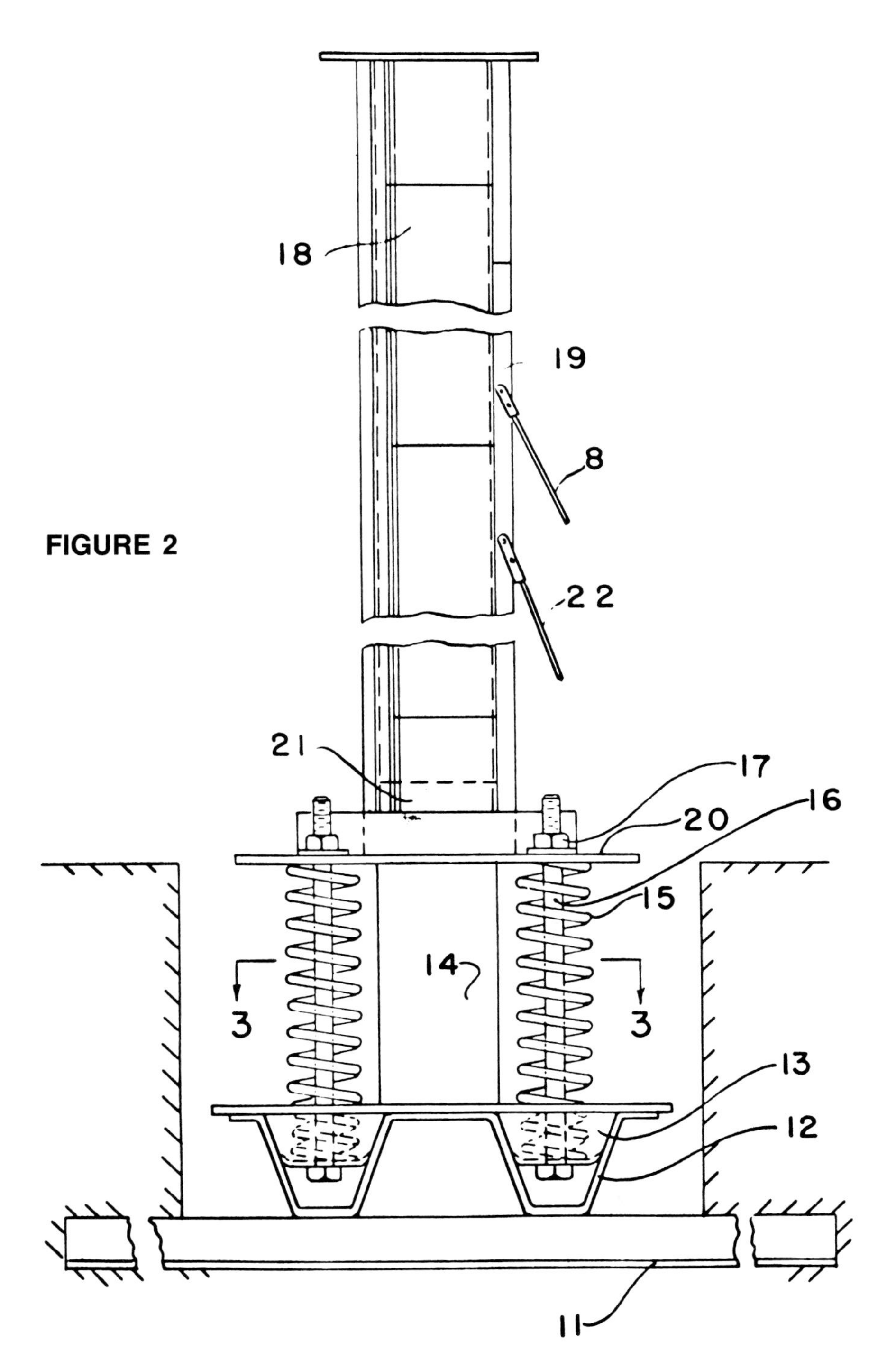

FIGURE 2

Fig. 59 is a cross section taken from the line 59—59 in Fig. 58;

Figs. 60 and 61 are top and side views of a window sill section;

Fig. 62 is a cross section taken from the line 62—62 in Fig. 60;

Fig. 63 is a top view of one of the flooring panels;

Figs. 64 and 65 are sections taken from the lines 64—64 and 65—65 in Fig. 3;

Figs. 66 and 67 are side and end views of one of the saddle and dowel units used to connect the flooring beams with the deck rings;

Fig. 68 is a perspective looking from the inside of the house toward the outer edge of the deck and the inside of the house side;

Fig. 69 is an elevation of details shown in Fig. 68;

Fig. 70 is an enlargement showing a detail in Figs. 68 and 69;

Fig. 71 shows the front of the door of the house and adjacent portions of the house side;

Fig. 72 is a cross section taken from the line 72—72 in Fig. 71;

Fig. 73 is a partially broken-away plan of the deck; and

Fig. 74 is an end elevation of the mechanism operating the endless conveyer system used in at least one of the wall units of the house.

This illustrated house includes two bedrooms, two bathrooms, an entrance hall, living room and kitchen, but all of the parts required to assemble this house and its foundation can be manufactured in a factory and grouped into a unit package approximating the size and weight of a crated large automobile of current manufacture. The

FIGURE 5

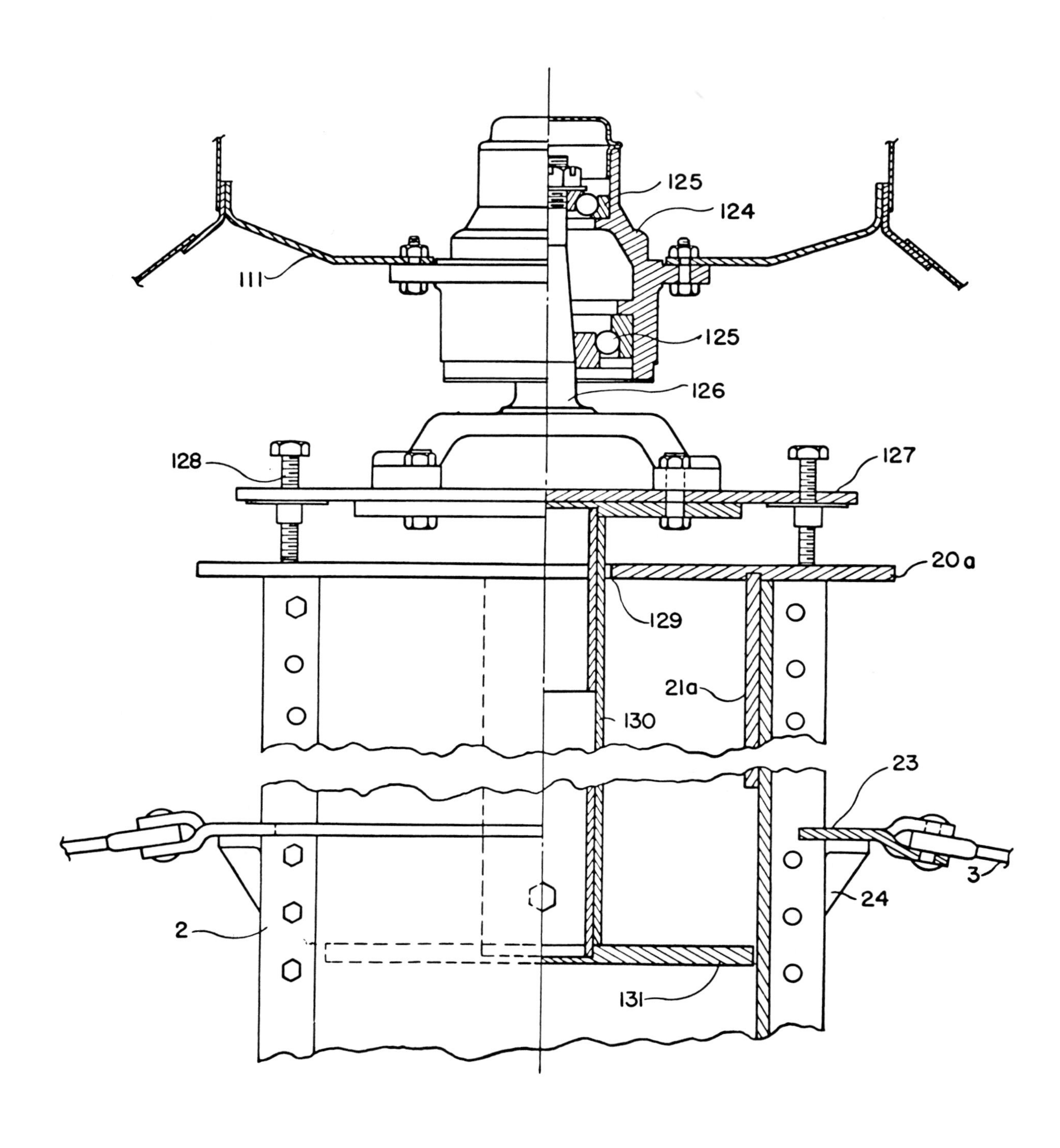

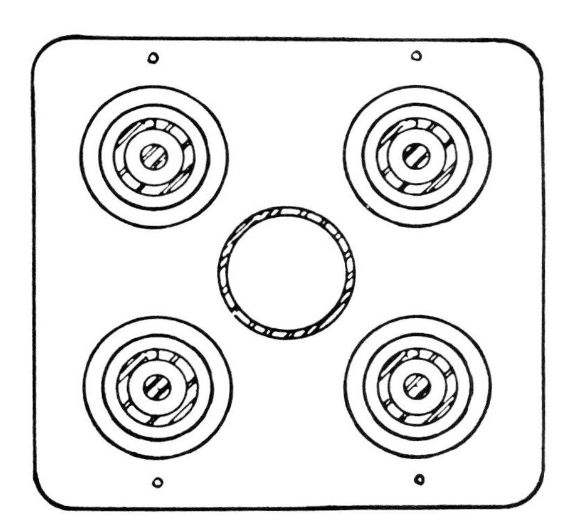

FIGURE 3

parts, and therefore the assembled house, are capable of mass production as this term is used in the automobile industry. Each part is adequately light and small to permit its being handled manually by one or two men during the assembly of the house. The assembled house provides for its own foundation, no large excavations being required for this.

Since the house may be manufactured by mass production methods, its total manufacturing costs are low by comparison with the total cost of making the components of a conventionally constructed house of comparable size. The cost of shipping the house is radically lower, to a critical degree, than the cost of shipping the components of a comparably sized house of conventional construction, and the house may be assembled in the field by a few skilled mechanics, as contrasted to the large number of laborers and craftsmen required to construct a conventional house of comparable size. Therefore, it is economically practicable to make the parts of this house from the best of the structural metal alloys, such as aluminum alloys and the stainless steels.

Broadly speaking, the house is made almost entirely of aluminum alloy and stainless steel and it includes a footing 1 placed in a comparatively small-diametered hole dug in the ground to just safely below the frost level, a compression mast 2 supported by this footing and extending vertically for approximately the height of the house, tension strands 3 spreading outwardly from the upper portion of this compression mast and bearing against compression rings 4, 5 and 6, spreading these tension strands in a shape approximating a dome, and a floor or deck 7 in the form of a flat annulus suspended around its outer periphery by portions of the strands 3 which depend vertically from the outer and lower compression ring 6 and around its inner periphery by strands 8 supported by the compression mast between its upper and lower portions. The outer periphery of the deck is anchored by vertical tension strands 9 and is held against rotation by diagonal stabilizing strands 10, the lower ends of the strands 9 and 10 being anchored to the ground. Stressed sheet-metal skins provide a roof or cowling and a cylindrical side, all supported by the mast working through the tension strands 3.

A house constructed as described above provides an extremely stable structure with all the compression stresses carried by the compact compression mast and compression rings and with all the other major parts carrying tension only so that they may be made very thin in cross section and take advantage of the great tensile strengths of presently available metal alloys. The provision of a frame and skin made according to these principles is responsible for the production of a comparatively large house assembled from parts very light in weight and which may be mass produced largely as thin-gauged sheet-metal pressings and drawn shapes, and packaged very compactly. The circular shape contributes to the lightness too, because this shape provides for the maximum enclosed volume with the least amount of materials.

The footing 1 provides a broad base in the subsoil, for supporting the compression stress of the mast, and it also functions as a spring for continuously biasing the mast upwardly in an elastic manner, whereby thermal expansion and contraction of the house components may be accommodated by permitting vertical movement of the mast bottom while keeping tension on the tension-carrying parts of the house. Under some circumstances this spring action is unnecessary.

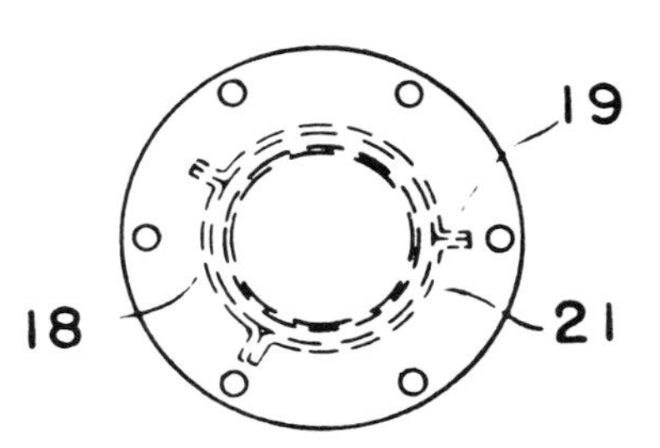

FIGURE 4

The mast footing includes a lower level of channel bars 11 and an upper level of upwardly facing channel bars 12 transverse to the bars 11. These bars may be of the same size throughout but in all events it is preferred that they have shapes permitting compact internesting when packaged. A spring cluster basket 13 is carried by the upper level of channels 12 with its baskets depending into these channel bars, and a mast guide 14 extends upwardly from this basket to the mast and is constructed or associated with the mast to permit guided vertical movement of the latter. Compression springs 15 extend upwardly from the spring baskets, as a geometrical cluster around the guide 14, and are compressed by tension bolts 16 having nuts 17 above a mast base 20, to be described presently, the bolts projecting through this base and the latter being guided by the mast guide 14.

All of the above parts, excepting the mast base, are preferably made of stainless steel, to resist corrosion, and the footing is placed in a hole in the ground, in the manner usual in the case of a tower footing, but preferably without filling the removed dirt back into the hole. The idea is to provide an elastically restrained yielding effect somewhere between the mast beneath its connection with the strands 3 and the ground, although this effect could be introduced into the tensioned strands but with greater difficulty in obtaining geometrical uniformity. This elastically restrained yielding is theoretically necessary to take up thermal expansion and contraction but, due to the novel stress pattern of the house and the opposite thermal effects between the compression and tension members, such yielding may be unnecessary excepting to the extent that it is inherent to the house.

The mast 2 is made from a plurality of partially cylindrical sections 18 with their edges having outwardly radiating flanges 19 which are fastened together by fasteners such as rivets or bolts, the various sections together forming a cylindrical mass of great compression strength and stiffness to bending, the inter-fastened flanges contributing greatly to the latter characteristic. Preferably these mast sections are aluminum alloy extrusions. For greater convenience in packaging, these mast sections are made

FIGURE 6

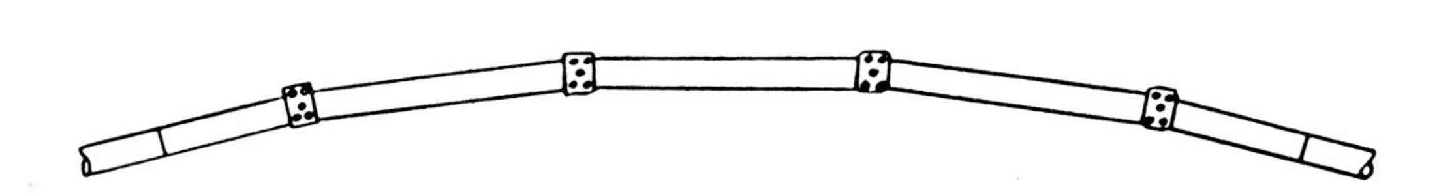

shorter than the total length of the mast, some being longer than others so that the flanges of the longer upper ones may be fastened to the flanges of the shorter lower ones and vice versa, whereby to effect a strong joint. The individual sections are light enough to be manually handled by one or two men and their shape permits them to be internested for compact packaging.

FIGURE 6a

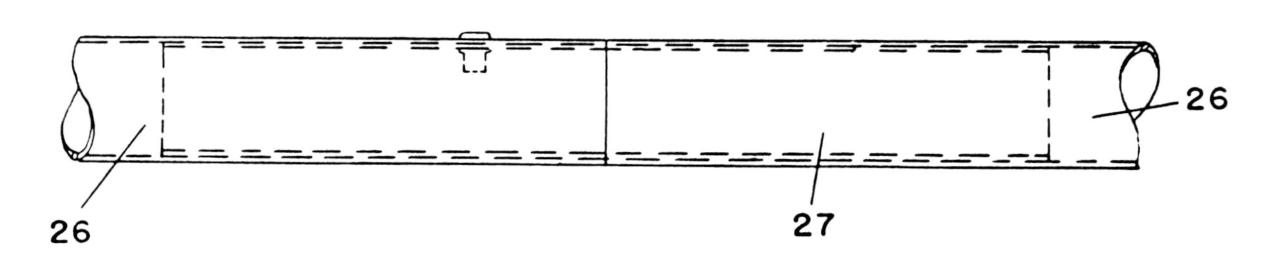

The mast is centered respecting the mast base by a centering plate 20, this centering plate being an aluminum alloy casting and having a centering stud 21 fitting inside the mast. Tension strands 22 are fastened to the mast partway up its height and extend down to anchorages embedded in the earth at spaced locations around the footing. These strands function as guys which steady the mast immediately after its erection and prior to further assembly of the house, and they also provide greater rigidity in the case of the completed assembly.

The tension strands 3 are fixed to the mast near its top by a ring member 23 which encircles the mast and is held against falling down around the mast by a series of steps 24 fixed to the outwardly radiating flanges of the mast sections 19. The tension strands 3 may be wire ropes but solid constructions are preferred, such as solid rods. When rods are used, the connections between the tension strands 3 and the ring member 23 may be by way of pin and clevis connections.

From the upper part of the mast, the tension strands 3 extend outwardly and downwardly over the various compression rings 4, 5, and 6 and from the ring 6 vertically downwardly, and these strands are graduated in cross-sectional areas and arranged geometrically, as crisscrossing diagonals, as required to carry the maximum tension and torque stresses with the least amount of materials. The tension strands are of the larger cross-sectional areas when they first leave the mast and gradually reduce in these dimensions as they progress outwardly and downwardly, this being because the tension stress on them decreases to the extent that the weight carried by these strands decreases and because their number is geometrically doubled at the ring 4. Where the strands pass over the various compression rings they may be fixed against relative slipping by fastenings 25, which also hold the rings against the strands, and the various lengths of strands, required to effect most conveniently the reductions in the cross sections of the strands outwardly and downwardly, are inter-fastened in any adequate strong manner that is flexible or pivotal, so that the strands may be folded for packaging or capable of easy field assembly. These strands and their connections are preferably made of high tensile strength stainless steel.

The various compression rings 4, 5 and 6 are each

made from a plurality of extruded aluminum alloy tubular sections 26 of convenient lengths for packaging, the various sections being joined together in the field around the mast, during the erection of the house, by means of dowels 27 retained in one end of each section by fastenings to restrain them from slipping longitudinally, into any one ring tube, which would permit the joint to come apart. These various rings are not truly circular but are made up of straight sections extending transversely between the points of their connections with the various tension strands, this being because a straight section is better able to carry compression than is a curved section. That is to say, the compression rings are made up of a plurality of straight sections between the various points of connections with the various tension strands, there being enough of these points for the rings to closely approximate the circular contour of the outer periphery of the deck below. These various rings may be made with different cross-sectional areas as required by the different compression stresses they must carry.

The vertical portions of the strands 3, which depend from the ring 6 so as to support the outer periphery of the deck 7, are fastened to an outer deck ring 28 which is made of a plurality of partially circular Z-sections 29 stamped or extruded from aluminum alloy and arranged so that when inter-fastened with their webs vertical they provide the circular contour needed. These sections are made in appropriate lengths for easy handling by one or two men and are Z-sections so that they may be nested compactly together for packaging. Their upper flanges are provided with holes 30 from which narrow slots 31 extend, the lower ends of the strands 3 being provided with enlargements 32 on their ends which will pass through the holes 30 but not through the slots 31, whereby the enlargements may be dropped through the holes 30, during the assembly of the house, and moved over into the slots 31 so as to transmit the tension stress on the strand to the outer deck ring, and the deck weight to the strands. The slots should extend in the diagonal directions of the various strands so as to assure a constant tendency to pull the enlargements 32 away from the holes 30 and under the metal surrounding the slots 31. Each of the sections 29 is provided with fastening holes arranged so that when one of the Z-section ends overlaps another these holes register, so that fastenings may be passed through them to inter-fasten the various sections. Also, the webs of the various sections are preferably provided with lightening holes.

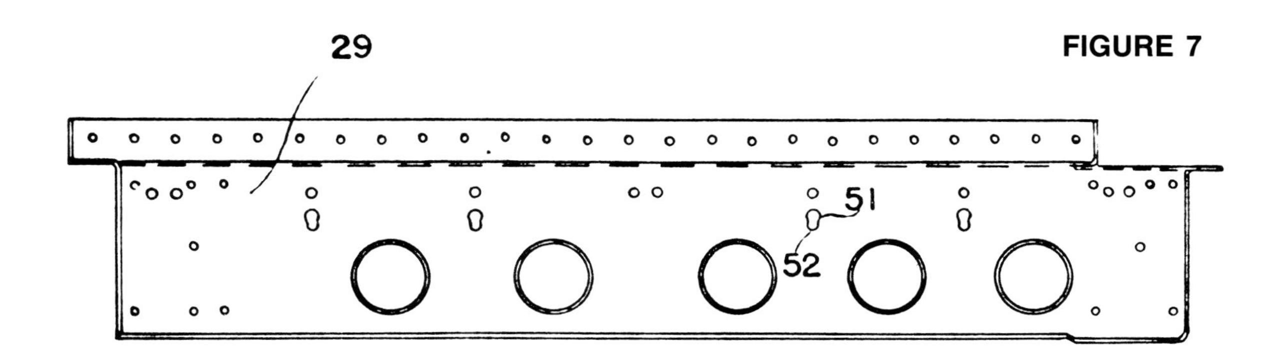

FIGURE 7

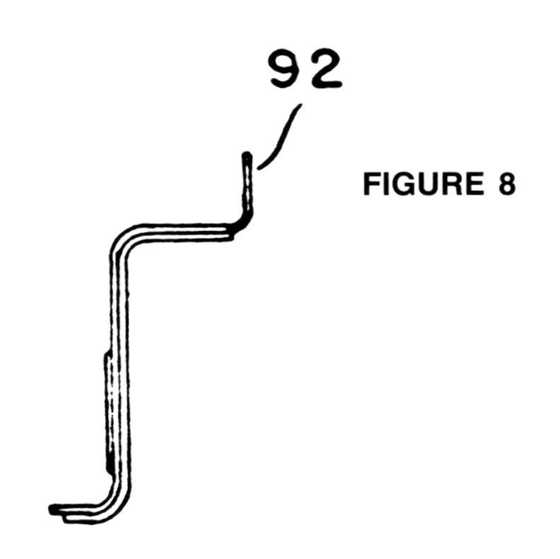

FIGURE 8

It might be mentioned at this point that wherever possible the parts of the house are constructed in the manner of aircraft parts so as to obtain great lightness in the case of the completed structure without any loss of adequate structural strength. Also, they are made to internest, during packaging, in all practicable instances.

The deck includes an inner deck ring 33 which is also made of stamped or extruded aluminum alloy in the form of sections substantially similar to those of the outer deck ring, excepting for curvature, the inner deck ring being much smaller in diameter than the outer and encircling the mast fairly closely. Thus, the various sections 34 of the inner deck ring are also partially circular Z-sections, which may be assembled with their ends overlapping and fastened together by way of suitable fastening holes which may be registered appropriately, but there is one difference in that the tension strands 8, from which the inner deck ring hangs from the mast, are formed in the webs of the inner deck ring sections, the holes being shown at 35 with their slots at 36, the slots 36 extending vertically so that the enlargements 37, on the end of the tension strands 8, will be drawn upwardly and away from the holes 35 and snugly into the slots 36.

The deck beams radiate from the inner deck ring to the outer deck ring and are in the form of W-sections 38 which are narrowest at the inner deck ring and gradually

FIGURE 9

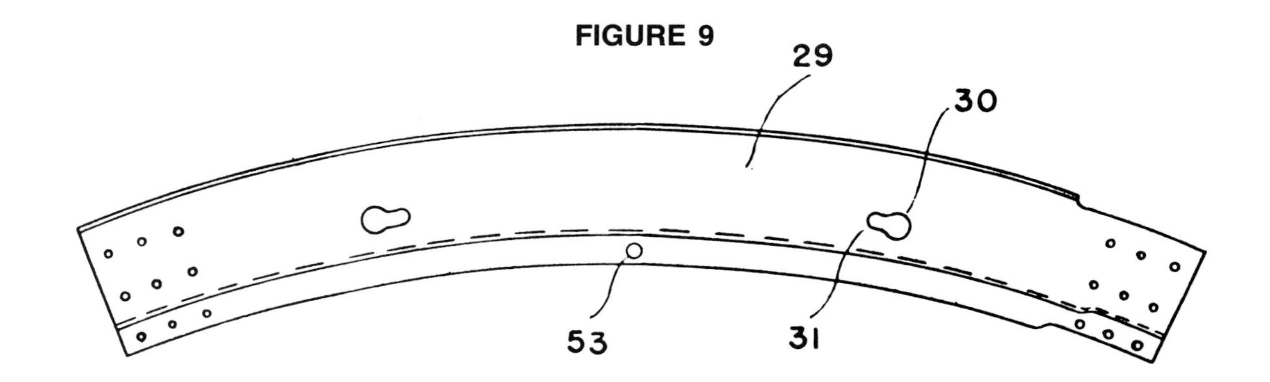

FIGURE 10

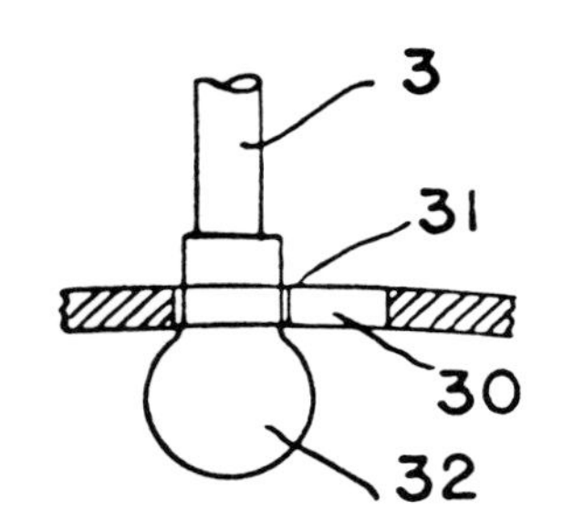

FIGURE 11

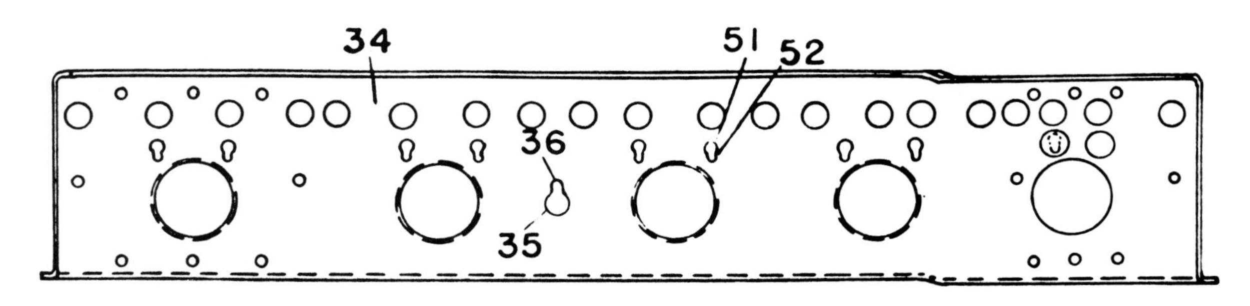

FIGURE 12

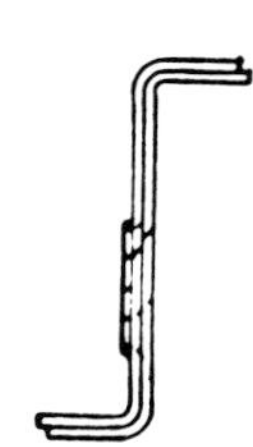

flare, with the increasing diameter, as required to make a deck structure, in the form of an annulus, the W-sections being arranged transversely horizontal and having depending flanges 39 on their longitudinally extending side edges. Connection between the W-sections and the inner and outer deck rings is effected by cast aluminum alloy units including saddles 40 having dowels 41 which fit properly positioned holes in the webs of the inner and outer deck rings, the W-sections 38 having their upwardly looped portions resting on the saddles 40, and top edge flanges 42 also resting on such saddles, and connected therewith by fastenings 43 during the assembly of the deck in the field.

The flooring for the deck comprises radial flooring panels 44 that radiate from the inner to the outer deck rings and which also appropriately flare to provide a complete flooring, the edges of the panels 44 being downwardly beveled and being fastened to the deck beams by clip bars stamped from aluminum alloy to provide a triangular cross-section 45 and depending flanges 46 which terminate with upwardly extending clip edges 47 that clip beneath the flanges 39 of the W-section floor beams, whereby to fasten the latter laterally together with the triangular cross section portions keying the flooring panels 44 in place. The clip bars may be placed in position prior to the flooring panels 44, during the assembly of the deck, the tops of the W-sections being substantially flush with the tops of the upper flanges of the outer deck Z-sections, whereby the flooring panels may be slightly spaced outwardly, laid down so as to clear the triangular section 45 of the clip bars, and then shoved slightly inwardly so as to key beneath these triangular sections. The flooring panels are preferably made of either wood or plastic, for aesthetic reasons, and they may be made in sections of less length than the full length required by the use of a single inner panel having an outer end edge against which two other panel sections abut and from which they continue to radiate, this requiring long and short clip bars, all as illustrated.

To assure adequate deck stiffness with least weight, the W-sections 38 gradually become deeper from the inner deck ring toward the outer deck ring, their ends at the latter location being deepest. Furthermore, a plurality of truss rings 48 and 49, which may each be made of aluminum alloy Z-sections constructed and arranged along the lines described in connection with the inner and outer deck rings, are arranged concentrically beneath the deck beams so as to bear against their bottoms, and tension strands 50 are fastened to the webs of the inner and outer deck rings so as to radiate therebetween and alternately bear against the bottoms of the alternate truss rings, these strands being highly tensioned so that the deck becomes a stressed truss assembly of great rigidity. The inner and outer deck rings are provided with more of the holes 51 and slots 52, in their webs, so that by providing enlargements 53 on the ends of the strands 50, the strands may be fastened to the deck ring webs. The strands 50, forming the truss sling, may pass through the truss rings via slots 54 formed in these rings where required for this purpose, and the truss rings may advantageously be a plurality of straight sections, rather than partially circular sections as are the deck rings, with all

FIGURE 13

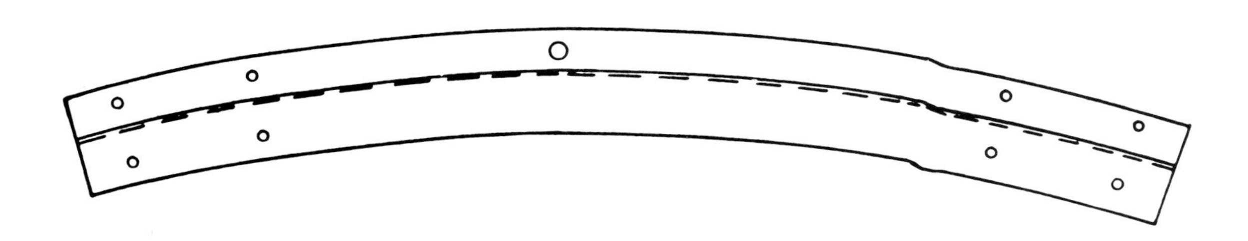

FIGURE 14

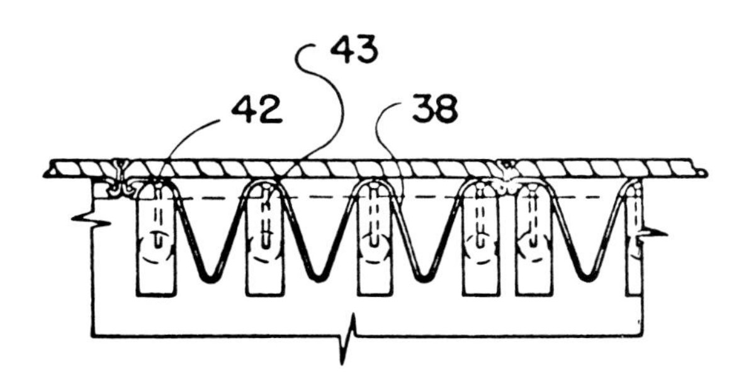

the parts arranged as required to obtain the maximum structural stability. These truss sling tension strands pull the inner and outer deck rings in the direction of each other while these rings are simultaneously strutted apart by the W-section floor beams, this action forcing the dowels 41 and the adjacent portions of the saddle 40, fixed to the inner and outer ends of the deck beams, in the directions of the rings so that the entire assembly is rigid and free from rattles.

The bottom flange of the Z-section outer deck ring 29 extends radially inwardly and is provided with holes for receiving fastenings for the anchoring strands 9 and the diagonal stabilizing strands 10. These strands are firmly fastened to the ground by tension rods 54 that extend downwardly through small-diametered holes dug well down into the ground, as by means of a posthole digger, with their bottom ends anchored by means of anchors 55 such as are used for ground anchorage purposes gener-

FIGURE 15

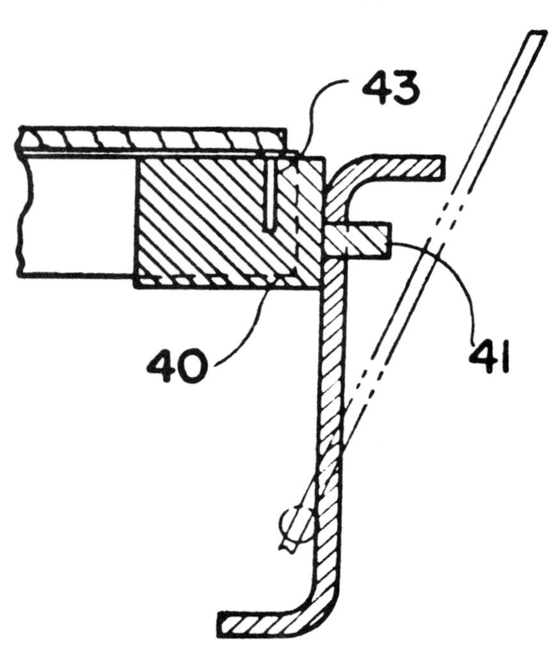

ally, the holes being filled over these anchors in the usual manner. Preferably, the various anchors and stabilizing strands are provided with turnbuckles 56 for the purpose of leveling the deck, and the house generally, when the ground is uneven.

With the house framework erected as described, the roof may now be assembled. This roof, in part, is in the form of a dome adapted to function aerodynamically somewhat like the cowling on an airplane, so it is called a cowling.

The cowling assembly includes a plurality of W-sections 57, permitting internesting for packaging, that radiate from the ring 4 down over the top of the ring 5 and to the ring 6, these three rings supporting the weight of the cowl. These W-sections constitute radial carlings which are circumferentially spaced sufficiently closely to provide the strength required to resist all the stresses to which the cowling will be subjected. Each carling or W-section is made up of a plurality of lengths with the upper lengths overlapping the lower or outer lengths to provide a joint which is completed by watertight fastenings.

A carling supporting ring is formed by partially circular pieces of Z-sections 58, which permit internesting, assembled to form a circle above the ring 4 with their bottom flanges extending outwardly, these flanges being formed to angle downwardly from the horizontal. All or a number of the Z-sections have these lower flanges pressed downwardly below the level of their remainders, as at 59, and these portions 59 have holes 60 to provide for their fastening to the ring 4 by means of the same fastenings fixing the tension strands 3 to this ring. As explained before, the various strands 3 are made with different cross-sectional areas, this requiring their joining, and this is done by flat straps 61 which lie on top of the various rings and have holes 62 which register with the holes 60 in the carling supporting ring sections so that the fastenings 25 may be in the form of bolts or rivets fastening all these parts together. Since the straps 61 are under heavy tension they remain flat at all times and provide firm support for the flange portions 59, thus holding the carling supporting ring firmly in place with the webs of its various sections 58 vertical.

The W-sections 57, forming the carlings, have the bottoms of their depending loops formed as flanges 63 and these flanges rest on the outwardly extending flanges of the Z-sections 58 of the carling supporting ring and are fastened there by rivets or bolts 64 passed through the flanges 63 and the flanges of the sections 58 by way of registering holes 65, in the ring flanges, and 67, in the carling flanges. The arrangement is such that the carlings are fastened to the carling flanges away from the latter's depressed portions 59 so that the upper ends of the carlings are spaced above the ring 4 and terminate thereover.

Upstanding clips 68 are applied to ring 5 by the fastenings 25, the arrangement being generally similar in this respect as at the ring 4, each clip having a curved flange 69 fitting the curvature of the strap 61, where it passes over the compression ring 5, and which is provided with a

FIGURE 16

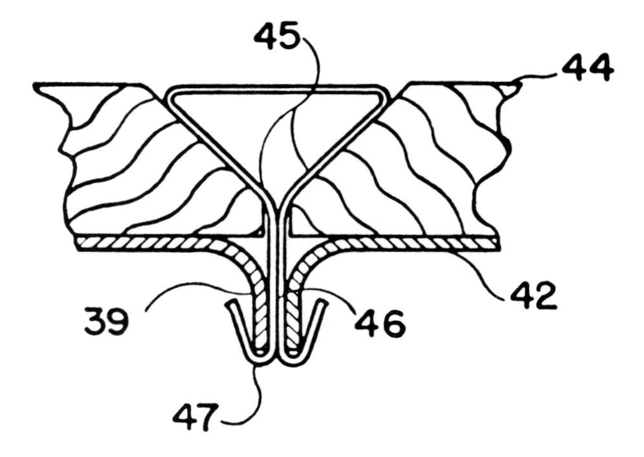

hole 70 for receiving the fastening 25. The upstanding portion of the clip is provided with a smoothly rounded head 71, which flares downwardly to lips 72, and the various W-sections, from which the carlings are made, are provided with holes 73 in the side walls of their middle loops so that by pushing a W-section downwardly, over the head 71, its middle loop sides spring outwardly, due to the camming action of the head 71, until the lips 72 of the clip register with the holes 73, whereupon the middle loop sides spring together again and the carling is held firmly by the clip. This requires that the clips be arranged transversely respecting the carlings. The clips should be sufficiently high so that the holes 73 may be formed well above the flanges 63 of the W-sections, forming the cowling carlings, so that water running down the flanges 63 of the W-sections will not leak through these holes.

The lower ends of the cowling carlings terminate at about the level of and outside of the compression ring 6 and at those locations their ends are connected with the ring 4 by clips somewhat similar to those already described, these clips each having a base 74 provided with a hole 75 through which the fastening 25 passes, which fixes the strap 61, interconnecting the tension strand lengths at the ring and which lies outside the ring, to the ring. Since the carlings are curved, so that together they provide a dome contour, their lowermost ends approach the vertical, so the clips at the ring 6 each have a horizontal extending portion provided with a head 76 and lips 77 for fitting the holes 78 formed as previously described in the cowling carling W-sections. Furthermore, these clips, for the fastening 25, are each provided with a gooseneck 79 so that the carlings are free to move longitudinally without causing trouble, the elasticity of the clips at the ring 5 also permitting such motion. This permits accommodation for thermal expansion and contraction and for strain resulting from stress.

The cowling, covering the domed framework provided by the carlings, is made up of a plurality of gores 80 made of sheet aluminum alloy with each gore free from joints its outside and is engaged in each instance by one of a series of tension screws 84 which extend horizontally through the upstanding wall sections and the webs of the Z-sections 58, forming the carling supporting ring, these screws 84 being provided with compression springs 85 so that there is a continuously exerted and upwardly directed elastic tension placed on all the cowling gores. At their lower ends, these gores are rigidly fastened to the flanges 81 at the lower ends of the cowling carlings 57. Therefore, the cowling gores 80 are continuously tensioned upwardly so that they fit tautly over the carlings provided for their support. Any leakage that might occur

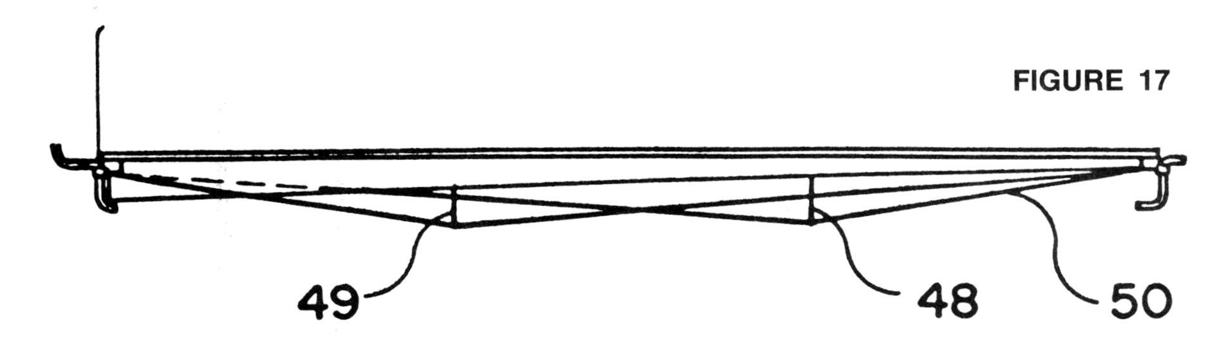

so that by arranging them radially on top of the cowling carlings with the edges of the various gores positioned so that they abut, in each instance, the outside of the upwardly extending loop 81A of the W-sections forming the carling underneath. This arrangement positively positions the gores and prevents their rotation when they are tensioned, as well as providing a tight joint, the gore edges wedging between the loops 81A. The upstanding edges of the outer legs of these W-sections are provided with longitudinal flanges 81 which provide good bearings for the cowling gores 80. There is a cowling gore filling the space between each two of these carlings. Each gore flares from its top to its bottom, as required by the increase in diameter of the domed shape, and each gore is bent upwardly at its inner end to provide a vertical wall section 82, these upstanding wall sections of all the gores providing a cylindrical wall structure at the tops of the carlings at their upper ends.

A nut 83 is provided for each upturned gore end 82 on between the radiating longitudinal edges of the gores and the carlings is caught in the troughs provided by the W-section carlings underneath, the water running down in the carlings and dropping from their lower and outer ends.

This rigid fastening of the bottom ends of the gores to the bottom ends of the carlings is effected by fastenings 86 such as rivets, and these extend outwardly through spacers 87 and fasten a cylindrical depending skirt 88 of sheet aluminum alloy around the periphery of the house, this being made of sheet segments so they may be packaged and handled conveniently. This depending skirt 88 comprises a tension member from which the cylindrical siding of the house depends with the weight transmitted through the carlings upwardly to the various compression rings in a distributed manner. This siding, in part, comprises a cylindrical wall 89 of transparent plastic material riveted or otherwise rigidly fastened to the lower edge of the skirt 88. Excepting for the door or doors, this trans-

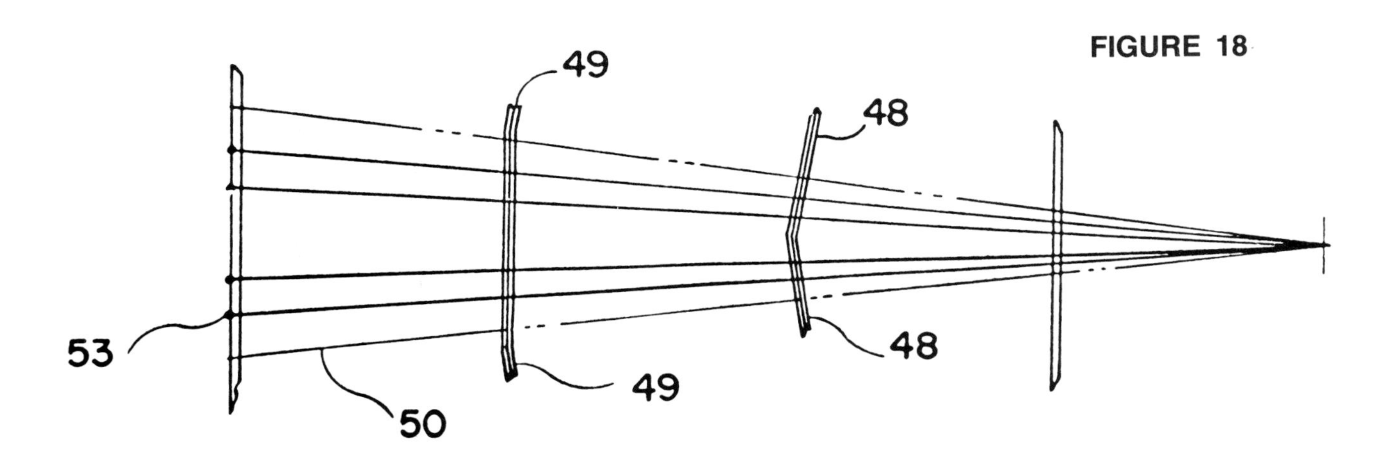

parent strip or wall section extends completely around the side periphery of the house. There are limitations on the available length of such material as it is now made but by using sections with overlapping ends which are rigidly fastened together, by rivets or the like, it may be made as a continuous piece. This strip of transparent plastic hangs in tension from the skirt 88, its lower edge terminating at about the height customary for windows in homes. This lower edge rigidly carries a short cylindrical skirt 90 and a circular window ledge 91 made up of assembled partly circular Z-sections with their webs horizontal, their outer flanges being upstanding and fastened rigidly to the lower window edge and their inner flanges depending. This window ledge also encircles the house beneath the window, with the strands 3 extending down through it and aiding in its support by having enlargements 91A secured to them beneath the windowsill, and is made of overlapped sections having lengths compatible with compact packaging and easy handling. In these parts again, aluminum alloy is used.

Going now to the bottom of the side, the outer deck ring 29 is made of partly circular Z-sections, as will be remembered, and the outwardly extending upper flanges of these sections are turned upwardly to provide a vertical flange 92 and a skirt 93, made of partly cylindrical sheet aluminum alloy sections that are inter-fastened, extends upwardly from this flange 92, this upstanding skirt being rigidly fastened to this flange. This skirt does not reach to the bottom of the windowsill 91 but terminates well below it, its upper edge being rigidly fixed to the vertical leg of an angle bar section ring 94 that encircles the house, excepting for the doorways, and which is made of inter-fastened lengths for reasons already described. The other leg of the section, comprising this ring, extends horizontally inwardly from the upstanding skirt 93 and interconnects near its outer periphery with the web of the windowsill 91 by way of tension bars 95 which draw the depending skirt 88 and the transparent plastic sheet window wall 89 downwardly, while drawing the angle ring 94, and therefore the skirt 93, upwardly.

The windowsill 91 and the parts above it depend from the carlings and cowling as previously described, so they naturally are tensioned and hang straight so that flexible materials may be used in their construction. The tension rods 95 transmit the weight of the skirt 93 to the windowsill and therefore upwardly to the other parts.

The above-described construction leaves a cylindrical opening completely around the house, and this is closed by partly cylindrical sheet aluminum alloy sections 96 with each section depending from its own piece of partly circular angle bar 97 having apertures through which the tension rods 95 pass, the bottom edges of these partly cylindrical sections 96 lying outside of the upstanding skirt 93 and the angle bars 97 riding the tension rods 95 so that the latter function as vertical guides. There are enough of the tension rods 95, distributed around the periphery of the house, to provide at least two of these rods for each of the partly cylindrical sections 96 so that the latter are adequately steadied when raised or lowered. This is one of the ways in which ventilation is provided, the angle bars 97 in each instance being provided with sheaves 98, near each of its ends, and cables 99 being arranged for each of the sheaves 98 so that each cable is fastened to the bottom of the windowsill 91, extends vertically downwardly and around a sheave 98, upwardly to a sheave 100, also fixed to the windowsill bottom and horizontally beneath the windowsill bottom, the cables from either end of each section approaching each other,

FIGURE 19

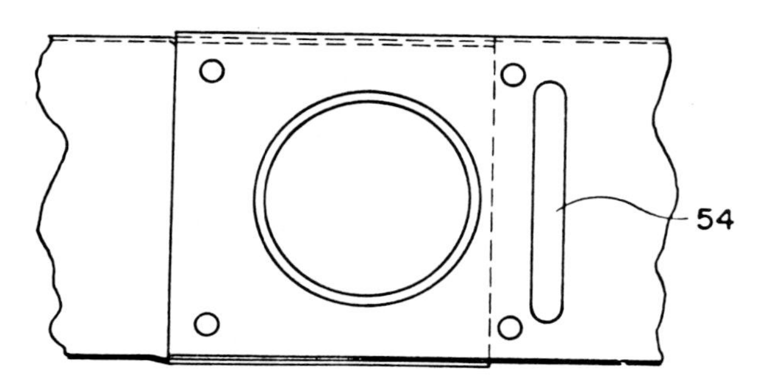

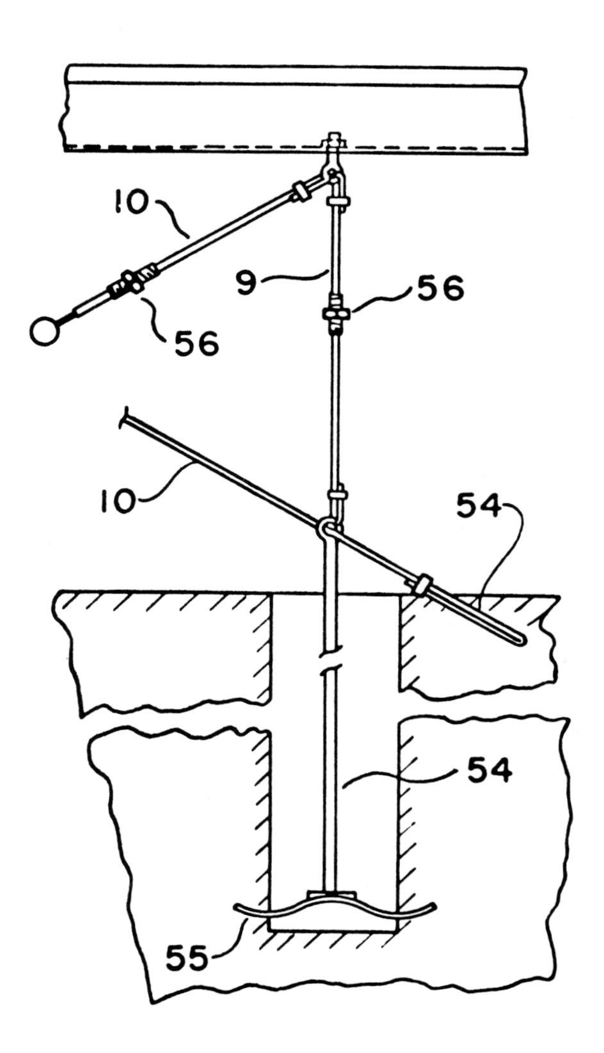

FIGURE 20

to a sheave 101 at about the center of each section and fixed to the windowsill bottom, the cables 99 then being fastened together and one of them going vertically upwardly, through a hole in the windowsill, to a sheave or ring 102, fixed well above the windowsill, through which it bends and starts downwardly, its end then being provided with a guide 103 which guides this end along itself in a parallel fashion downwardly. With this arrangement, when ventilation is not desired the guide 103 is pulled downwardly so as to lift the section 96 upwardly and close the opening between the bottom of the windowsill and the top of the upstanding skirt 93, the guide being fastened down in any manner desired. By extending the upper edge of the sheet section 96 a little above the angle bar 97, it fits beneath the little depending skirt 90 and forms a weatherproof seal. Vertical strips may be placed along the vertical edges of these sections 96 so that these strips cooperate with the edges of the sections to form tongue-in-groove sliding joints. Since the cylindrical sections 96 depend from the angle bars 97, they may be made of flexible sheet metal without causing difficulties in the way of their bending. When the sections 96 are raised their lower ends should overlap the outside of the upstanding skirt 93 sufficiently to provide a watertight effect.

It will be remembered that the depending skirt 88 is spaced from the outside of the cowling forming the roof, and that leaking water runs down the carlings. Therefore, the house is provided around its entire periphery with an annular gutter 104, preferably made of material having the properties of synthetic rubber, inside the skirt 88 and in line and just below the outside of the cowling and the

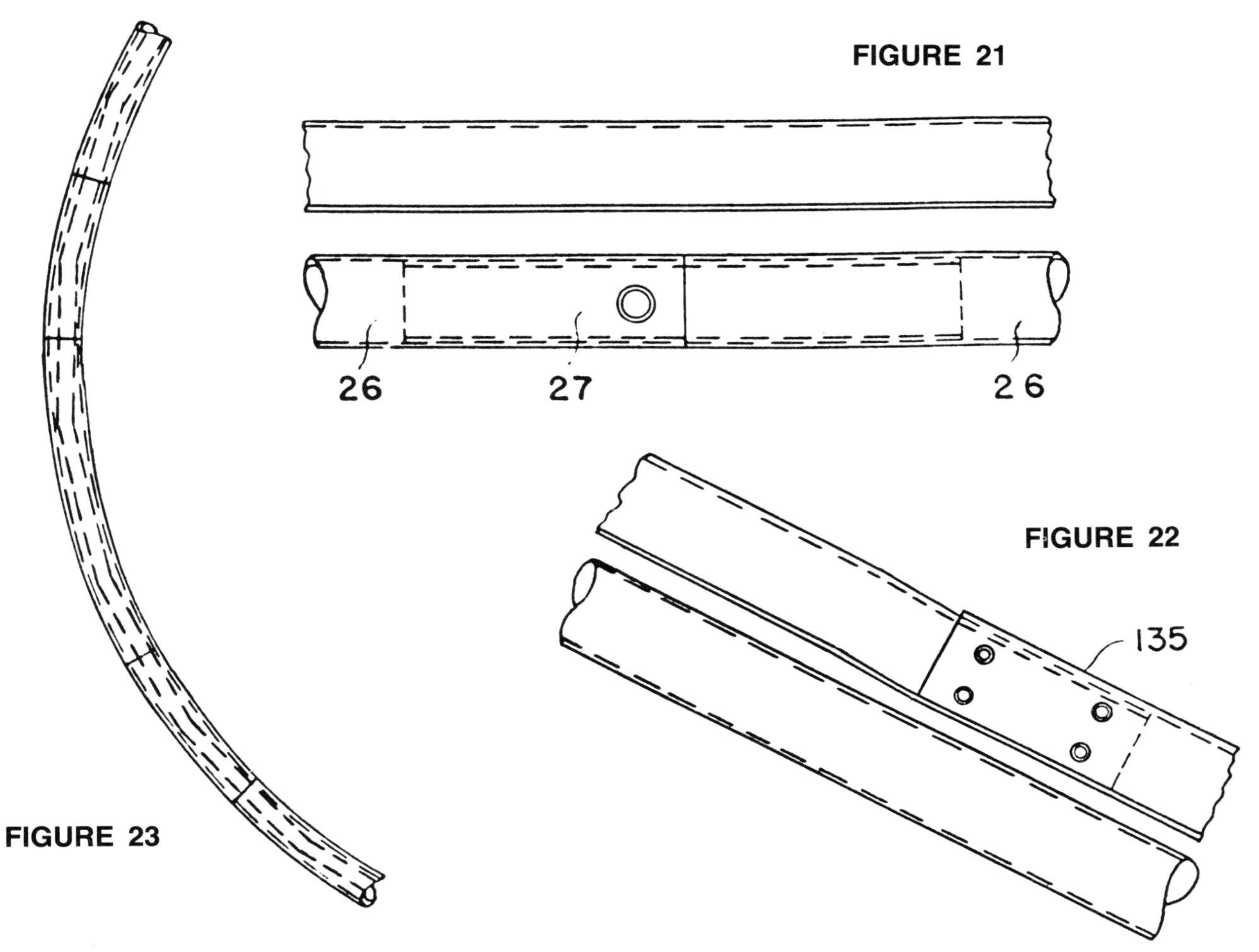

FIGURE 21

FIGURE 22

FIGURE 23

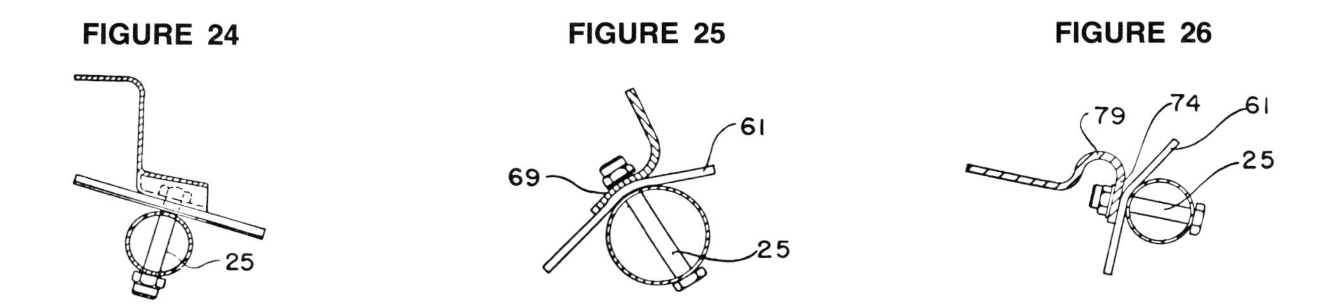

FIGURE 24

FIGURE 25

FIGURE 26

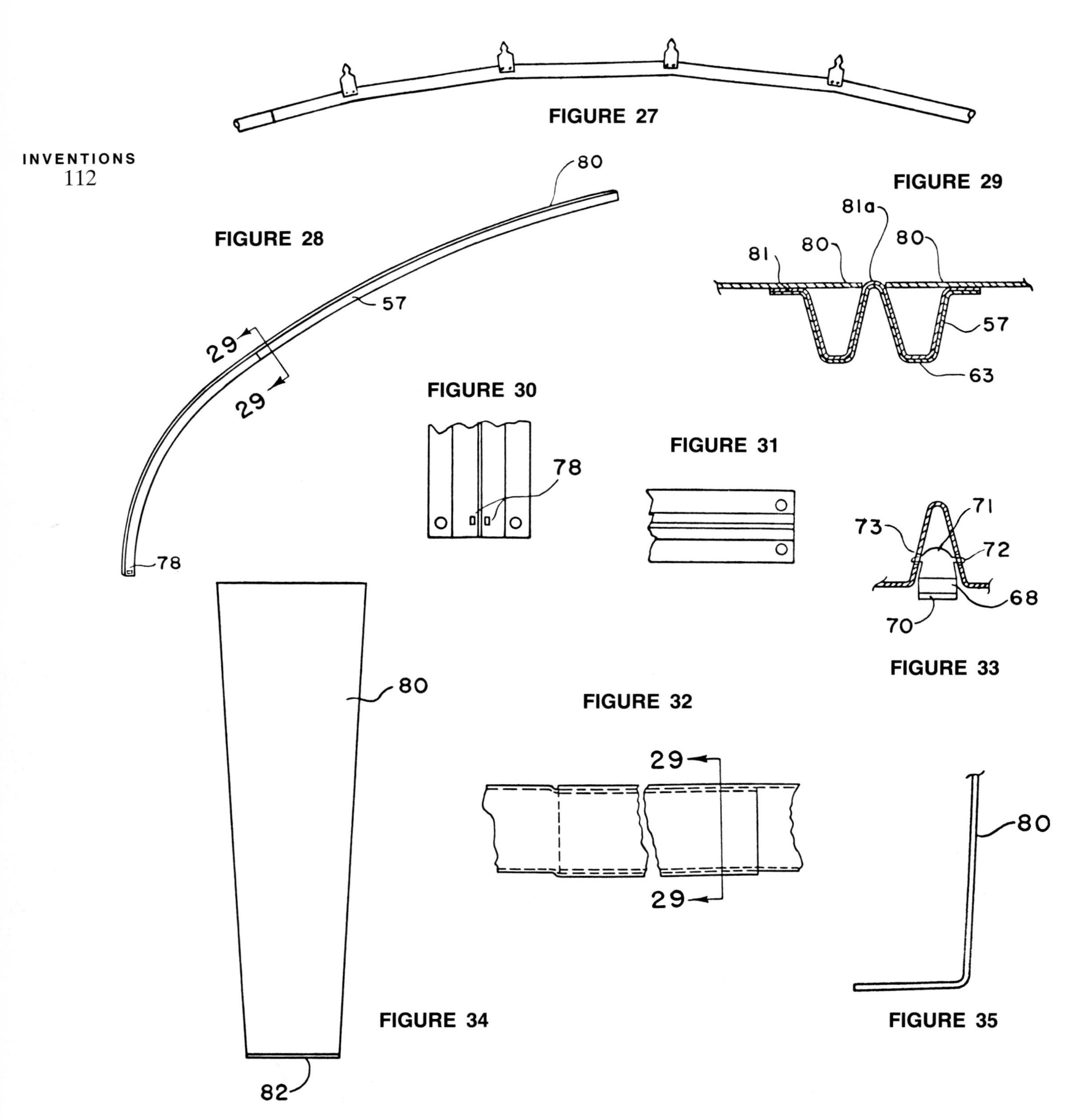

FIGURE 27

FIGURE 28

FIGURE 29

FIGURE 30

FIGURE 31

FIGURE 32

FIGURE 33

FIGURE 34

FIGURE 35

carling lower ends. This gutter may be semi-circular in cross section with one upstanding wall fixed watertightly to the inside of the depending skirt 88 and with its other upstanding wall fastened so as to keep the upper edges from spreading, this being illustrated as being done by fastening the inner wall of the gutter 104 to the connections for the tension rods 3 just below where their straps 61 pass over the ring 6. To assure permanent watertightness, the ring 6 is preferably an extruded synthetic rubber section having its ends permanently vulcanized together. This gutter 104 is completely enclosed and invisible from the outside of the house. In addition to catching rainwater, it performs the further very useful function of catching condensate which might form on and run down the inside of the cowling or the carlings.

Preferably, a header 105, made of pressed sheet aluminum alloy as partly circular sections, is installed above the window so as to completely enclose the gutter 104,

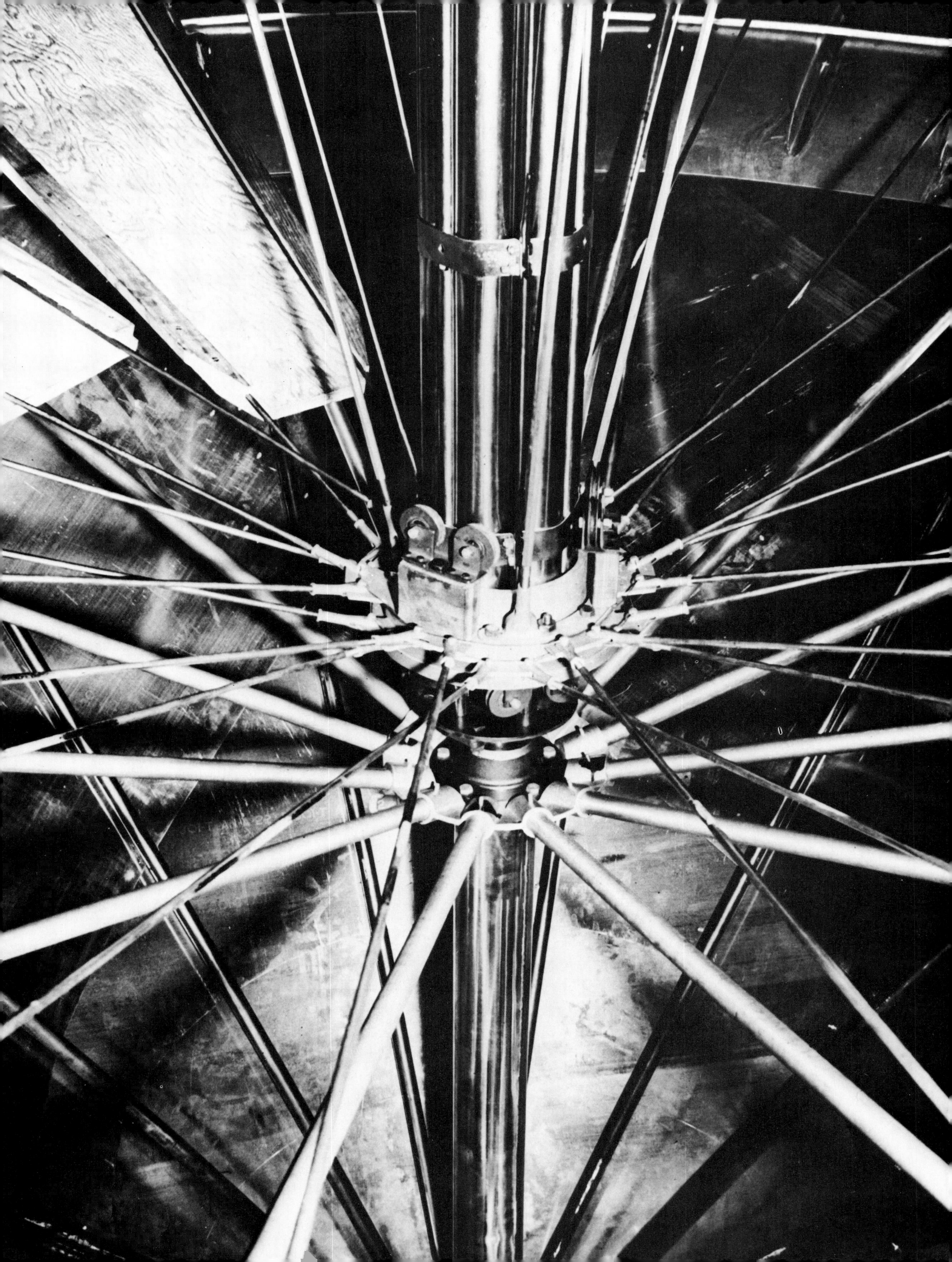

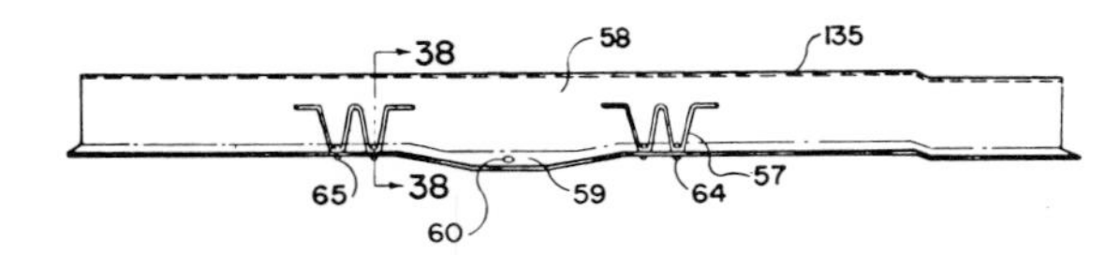

FIGURE 36

the connections for the tension rods 3 and the compression ring 6, this header being provided for aesthetic reasons. Its various sections should be capable of internesting.

The domed cowling goes upwardly and inwardly only to the compression ring 4, this leaving a large-diametered hole surrounding the mast. This hole is closed by a large rotating ventilator 107 having a wind vane 108, the trailing edge of this vane being open and communicating with

FIGURE 37

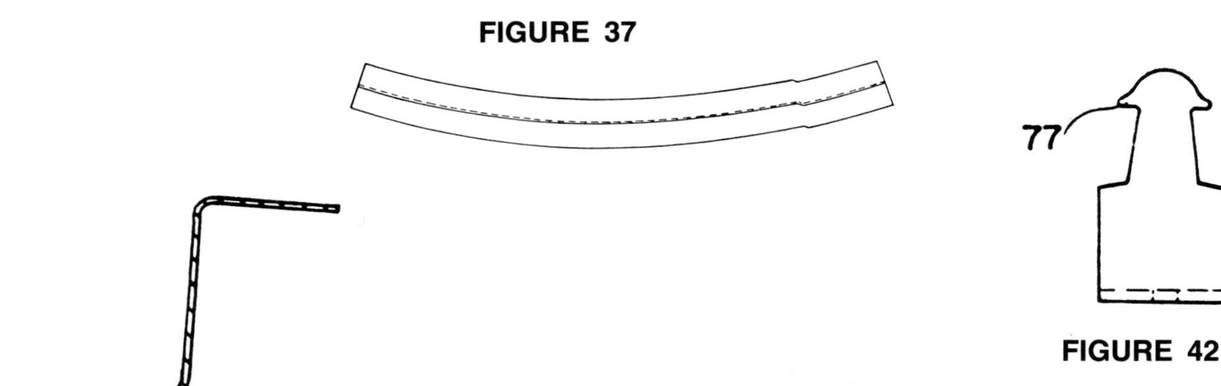

FIGURE 42

FIGURE 38

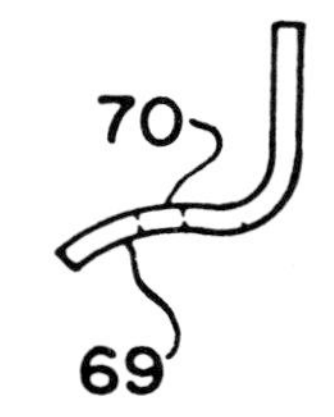

FIGURE 39

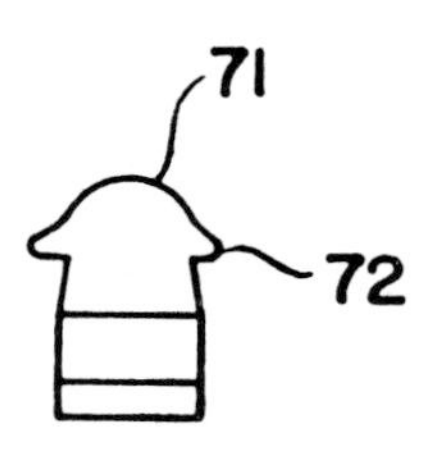

FIGURE 40

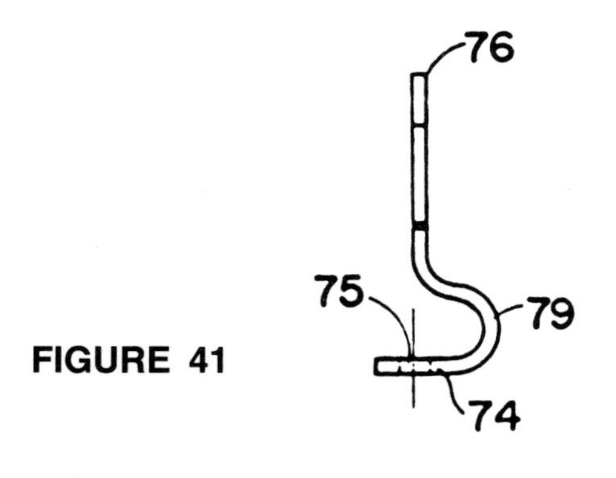

FIGURE 41

the interior of the house. The ventilator is designed to cooperate with the domed cowling according to aerodynamic principles, the wind flowing around the cylindrical house smoothly and that portion that flows around the domed cowling being deflected upwardly and circumferentially, the vane rotating the ventilator so that the trailing edge of the vane is always on the lee side of the ventilator where there is a natural reduction in the atmospheric pressure, whereby air may be continually sucked out from the house so as to obtain forced ventilation inside the house without the need for mechanical contrivances.

The rotating ventilator 107 is made of aluminum alloy and comprises an inner cone 109, made of sheet sections that are fastened together, provided around its side with large symmetrically arranged air passage holes 110. The top of this inner cone is open and peripherally connects with an inwardly extending flange 111 and also with a short upwardly extending sheet cylinder 112. The periphery of this cylinder 112 connects with an outer cone 113 which extends downwardly to the bottom periphery of the inner cone 109. Both inner and outer cones 109 and 113 connect at their bottom peripheries with an outwardly curving flared ring 114, the bottom periphery of which connects with a flat conical skirt 115 which extends outwardly so as to overlap the edge of the opening in the domed cowling.

The vane 108 is made from sheet side walls which extend away from the outer cone 113 toward the periphery of the lower edge of the conical skirt 115, these sides being close together at their tops, where they are closed by a flat metal plate 116 by joints including angle bars 117 extending along the upper edges of the vane. These sheet-metal side walls flare circumferentially of the ventilator downwardly, the lower edges of the side sheets being joined with the outer conical parts 113 and 115, and with the flaring ring section 114. A circular plate 118 closes the top of the cylinder 112 and has a tail 119 which joins with the plate 116 closing the top of the vane. The trailing or end edges of the side sheets, forming the vane

118, are curved away from the wind flow in accordance with good aerodynamic principles.

An opening 120 is formed in the outer conical skin 113 inside of the vane so that air can be sucked through the holes 110 in the inner cone, upwardly through this opening 120 and outwardly through the open end of the vane 118. A strengthening bar 121 extends across this opening 120 longitudinally of the outer cone to replace the strength lost by reason of this opening 120; and the side sheets of the vane 118 connect with vertical angle bars 122 which, in turn, connect with the longitudinally extending angle bars 117, and an angle bar 123 extends diagonally between the angle bars 122, all to provide further bracing.

It is to be understood that this ventilator is made of separate sheets of metal and may be assembled in the field at the site of the house, this assuring compact packaging. Furthermore, all the parts follow aircraft construction, whereby the total weight of the ventilator may be kept very low although, by reason of its construction, its strength is very great.

The mast 2 rises to an elevation high enough above the ring 4 to support the ventilator by way of the inwardly extending flange 111 previously described. This flange is rigidly fastened to a hub 124 provided with anti-friction thrust bearings 125 arranged to take thrust both upwardly and downwardly and which are carried by a vertical shaft 126 rigidly fastened to a horizontal plate 127 spaced just above a flange 20A, on top of the mast, and supported against downward movement thereby through the medium of tripod adjusting screws 128. The flange 20A is fixed to the mast top and centered there by a centering stud 21A. These tripod adjusting screws are adjusted so that the ventilator may freely rotate without its outer edges, which overlap the inner periphery of the cowling, contacting the latter. The flange or plate 20A has a hole 129 formed in its center through which a tube 130 depends vertically from the plate 127, the lower end of this tube being fixed to a disc 131 that rides inside the mast 2 and renders the plate 127 rigid to tilting action. This tube 130 slides in the hole 129 and the disc 131 slides inside the mast, the idea being that should there be a sudden reduction in pressure on the outside of the house, such as might occur during a hurricane, the entire ventilator may lift up until the disc 131 engages beneath the centering stud 21A depending from the plate 20, the parts 20 and 21 being fastened to the mast. This permits almost immediate pressure equalization between the inside and outside of the house so as to avoid the destruction that would occur in the case of a conventional house. Due to the light weight of the ventilator it may be necessary to bias the plate 127 downwardly by weight or springs so that the ventilator will not rise too easily in the case of minor pressure variations. Due to the aerodynamic construction of the ventilator a high wind will not lift it, but will instead tend to urge it downwardly. However, sudden pressure variations will tend to lift it, so the ventilator should be biased downwardly to an adjusted degree in the manner common to safety valves in general.

FIGURE 43

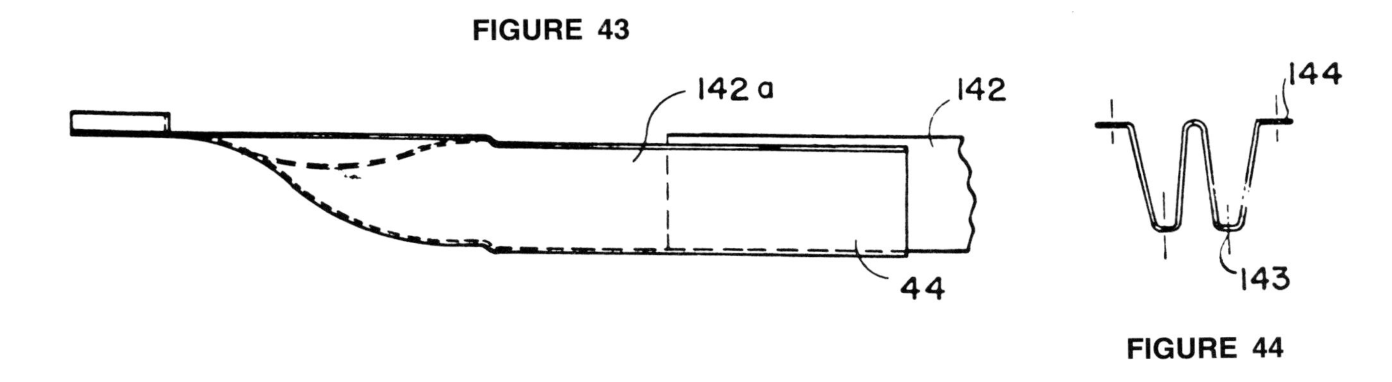

FIGURE 44

The outer periphery of the lowermost and most widely flaring conical skirt 115 carries a depending skirt 132 made of sheet sections rigidly fastened to the periphery of the skirt 115. This skirt depends outside of the wall formed by the upstanding wall sections 82 of the cowling gores 80, is radially spaced from the same and is made of sections which, when assembled, provide a skirt that is vertically concave-convex, this skirt being outwardly concave. The lower end of this skirt is fairly close to the cowling and is well below the upper edge of the wall formed by the wall sections 82 of the cowling gores, but its lower edge is spaced sufficiently above the cowling to clear it in the event of slight tipping action of the ventilator due to strain resulting from wind stresses. The ventilator tipping action is kept within limits preventing intercontact between the cowling and the lower edge of the skirt 132 by means of an annular series of rollers 133 mounted by strutting brackets 134 fixed to the bottom of the conical skirt 115. These rollers 133 contact the upward and inwardly extending flanges of the ring sections 58, forming the carling supporting ring, so that the ventilator can freely turn even should it tip somewhat while its tipping action is rigidly limited. The various ring sections 58 each have one end vertically depressed, as at 135, so that the overlapping end of the next section has its flange upwardly flush with the flange of the first one, whereby the annular surface provided by the tops of the upper flanges of the ring sections 58 provide a smooth trackway for the rollers 133. Due to the conical shape of the ventilator skirt 115, it is inherently rigid to bending even though made of thin-gauged sheets.

Since the cylindrical wall, provided by the upstanding end sections 82 of the cowling gores 80, is not circumferentially continuous, water blown against it by the wind through the space between the lower edge of the skirt 133 and the cowling might get inside the house. Therefore, a cylindrical ring 136, made in the form of an extruded section of elastic material having the properties of rubber, is provided. This ring 138 is elastically forced down over the wall formed by the upstanding end sections 82 and provides a watertight barrier. This ring 136 has an outwardly and downwardly flaring flange 137, on its lower edge, which clears the outer ends of the tension bolts or screws 84 and which terminates with a foot ring 138 which presses downwardly tightly against the upper surfaces of the cowling gores 80. Its upper edge has an inwardly, downwardly and reversing, looping portion 139, the edge of which grips the inside of the wall formed by the cowling gore sections 82, and it also includes an outwardly and downwardly extending flange 140 which positively reverses the flow of water blown upwardly along the outside of the ring 136. Weathertightness is further assured by a cylindrical skirt 141 made of flexible material like rubber and fixed by fastenings 142 to the ventilator skirt 132 above its lower edge, the lower edge of this flexible skirt 141 hanging well below the lower edge of the sheet metal skirt 132. This skirt is flexible enough so that the wind can blow it back against the ring 136 beneath its outwardly and downwardly extending flange 140, so it functions as an air seal on the windward side of the ventilator.

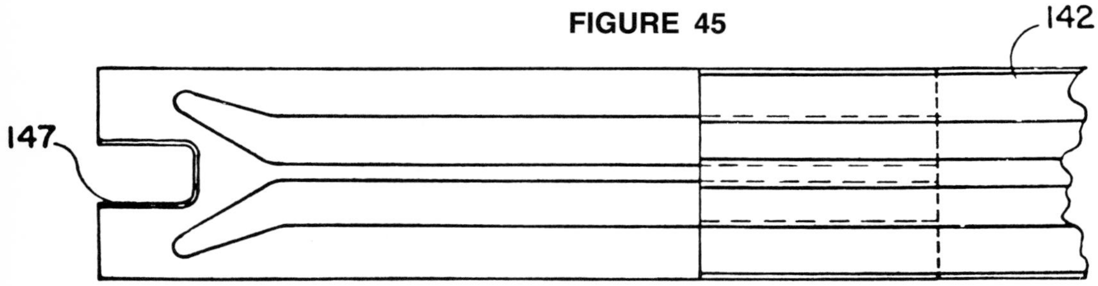

FIGURE 45

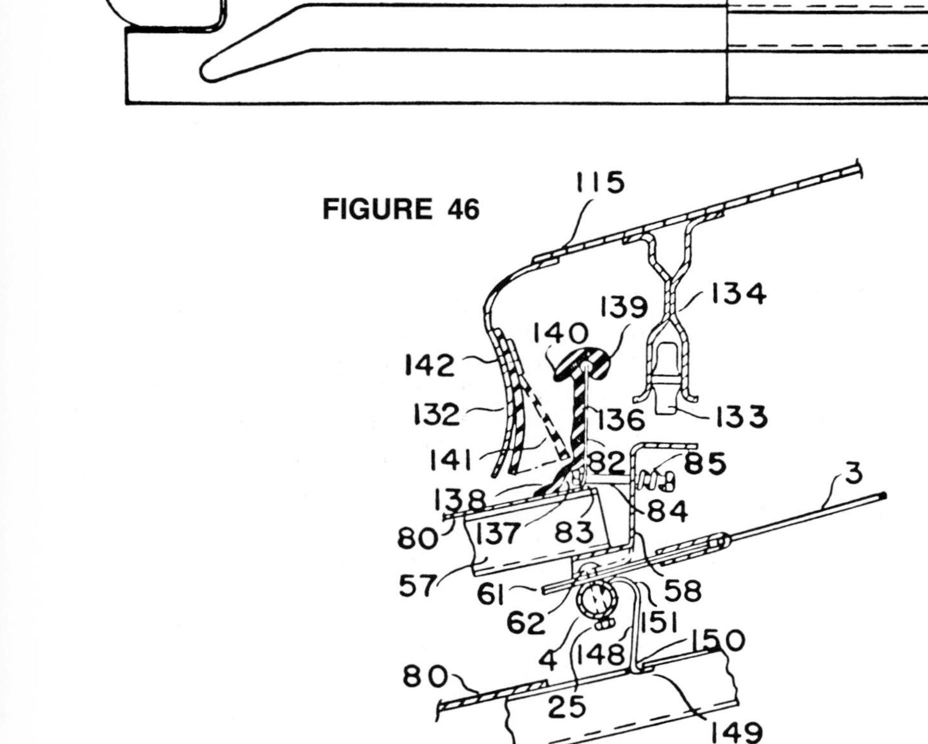

FIGURE 46

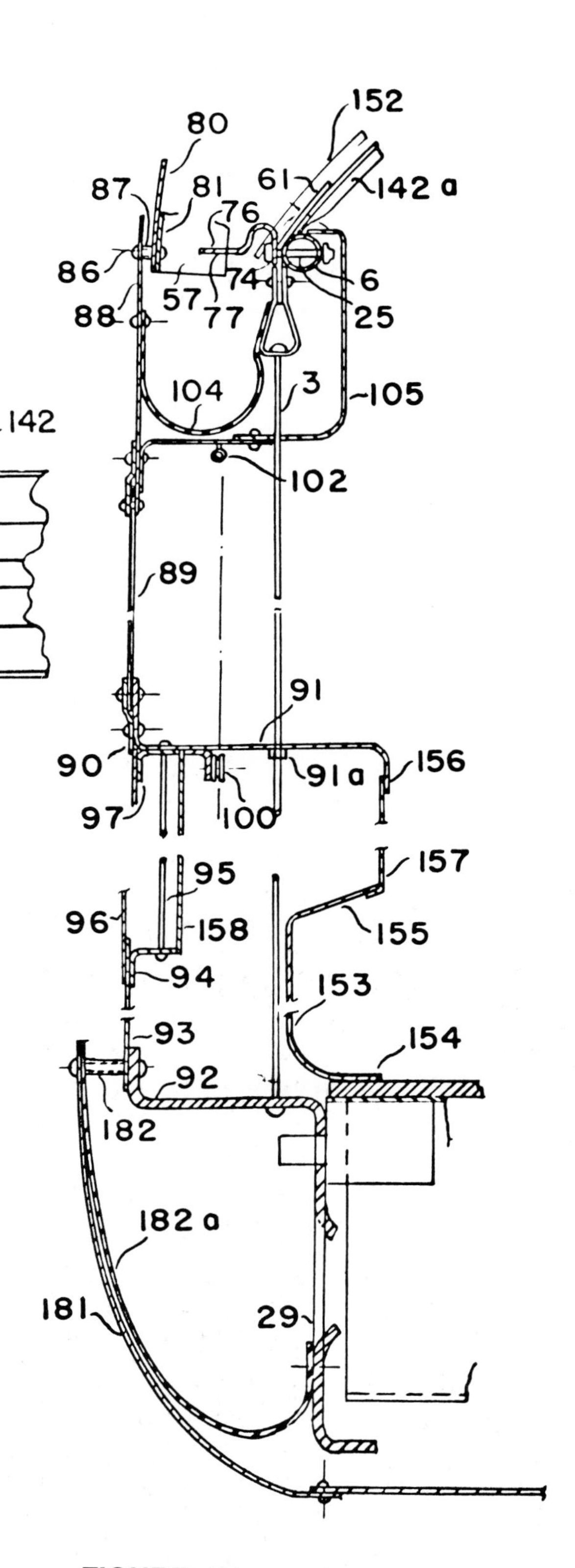

FIGURE 47

Provision for a double roof is made by having ceiling carlings 142 in the form of W-sections which gradually fade into flat sections at each end 142A, the bottoms of the two downwardly depending loops having flanges 143 and the upper edges of the outer sides having flanges 144. The ceiling carlings may be sections exactly like those of the cowling carlings 57 with the ends 142A separate and fixed thereto in water-shedding arrangement. These carlings 142 are fixed at their upper ends to a ceiling carling supporting ring 145, made of partly circular angle bar sections and supported from the strands 3 by depending brackets 146. This ring 145 is of considerably smaller diameter than the ring 4, it being of approximately the diameter of the inner cone 109 of the ventilator so as to encompass an area large enough to permit proper flue arrangements going up to the ventilator. The arrangement is such that the carlings radiate in a circumferentially spaced relation from this ring 145 downwardly until their lower ends reach the ring 6; these lower ends have slots 147. These lower ends are, therefore, forked and fit around the parts 74 of the brackets supporting the lower ends of the cowling carlings, this providing ends terminating over the gutter 104 encircling the house beneath the cowling and cowling carlings. Beneath the compression ring 4 these ceiling carlings 142 are supported by brackets 148 having ends 149 which may be slipped into holes 150 formed in the upper parts of the carlings and which have hooked upper ends 151 which may be hooked over the ring 4.

The arrangement is such that these ceiling carlings may have their lower ends slipped over the compression ring 6, by fitting their slots 147 around the brackets supporting the cowling carlings, and their inner ends then raised for application of the brackets 148 and fastened to the ring 145. Ceiling gores 152 made in the form of sections which flare from their inner to their outer ends, may be supported on top of the ceiling carlings with the gore longitudinal edges adjacent and overlying the centers of the ceiling carlings. The lower ends of these roofing gores are also slotted as required to clear the parts around the ring

FIGURE 48

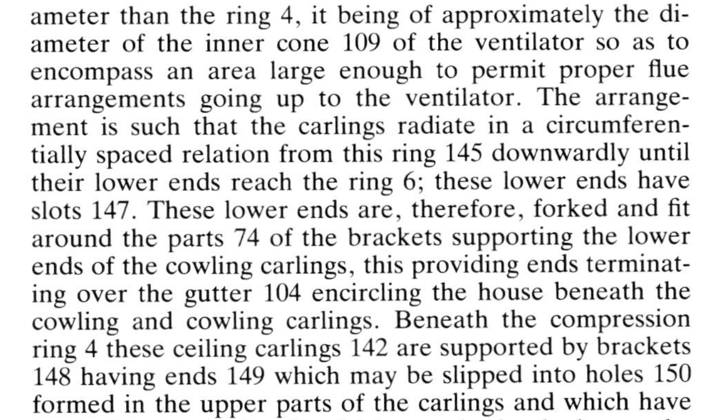

FIGURE 49

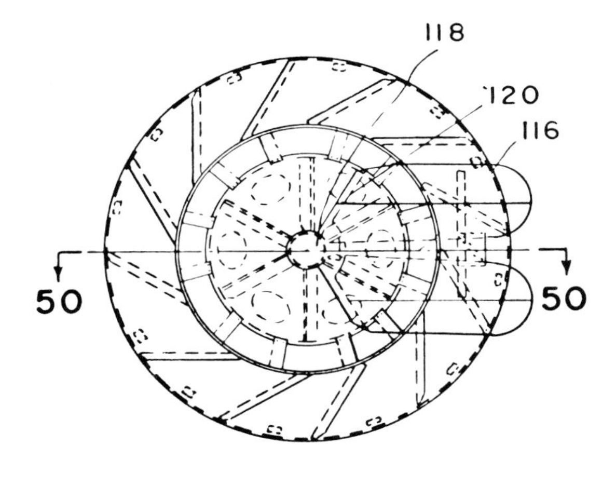

FIGURE 50

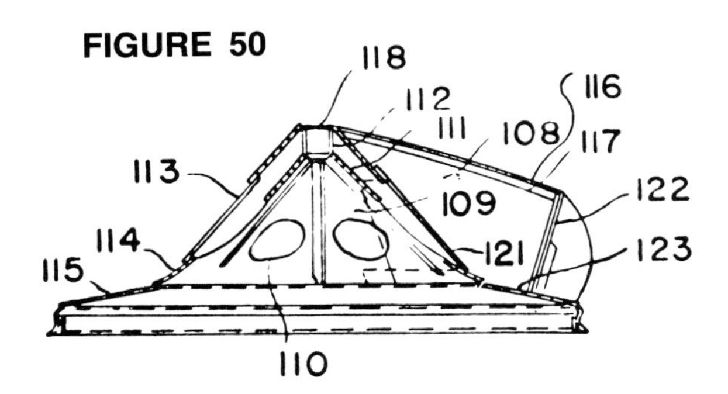

FIGURE 51

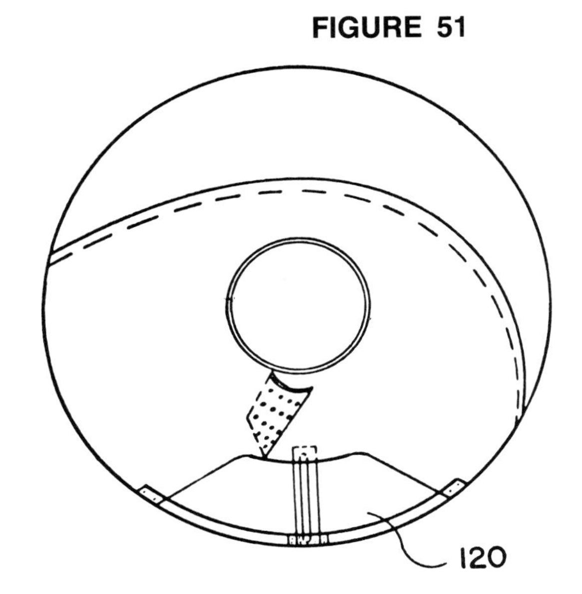

FIGURE 52

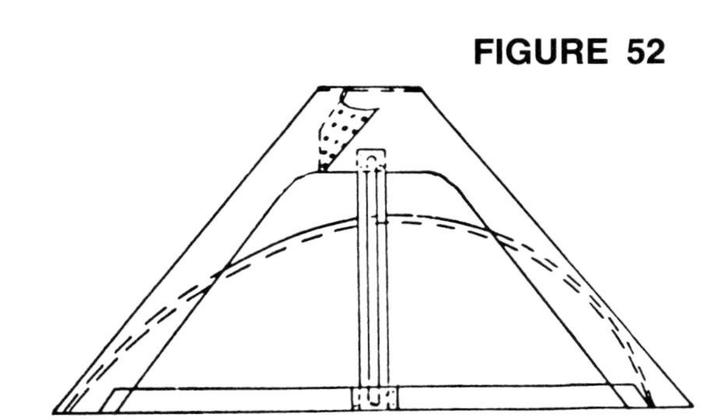

FIGURE 53

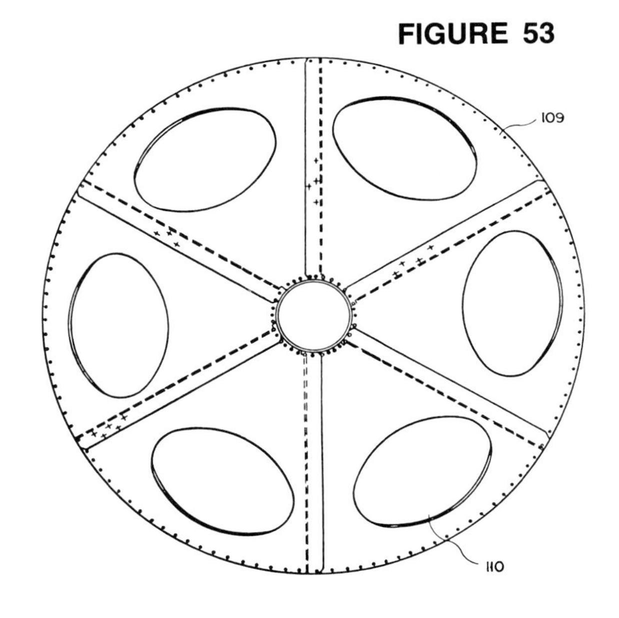

FIGURE 54

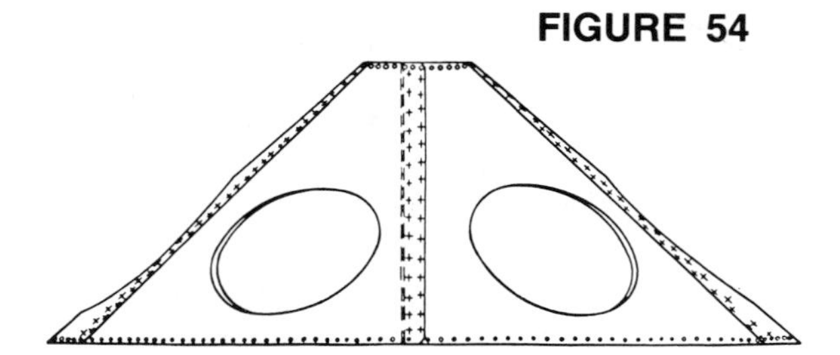

6 so that their lower ends may terminate over the gutter 104. The hooked brackets 148 are small and the holes through which their lower ends 149 are slipped, are preferably positioned in the top portions of the middle upward loops of the carling sections, thus permitting these hooks to pass upwardly through the spaces between the cowling gores. These gores are preferably made of hard plastic sections, or they might be made of plywood or the like. They may be provided with fastenings at their upper and bottom ends so that they may be stretched tautly over the ceiling carlings or they may be preformed as

FIGURE 55

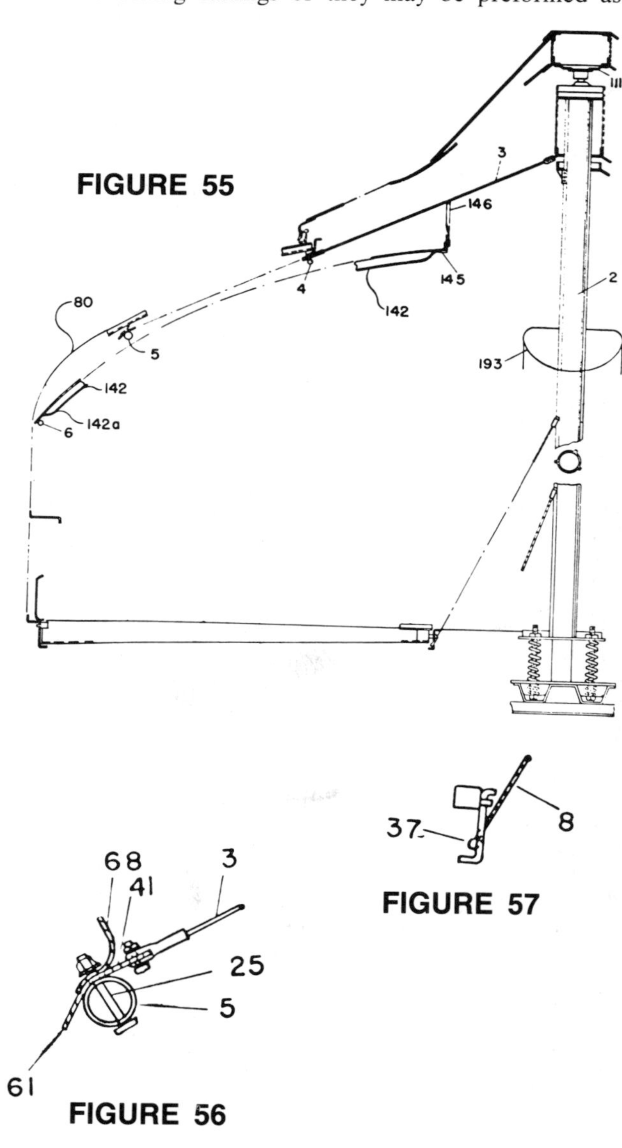

FIGURE 57

FIGURE 56

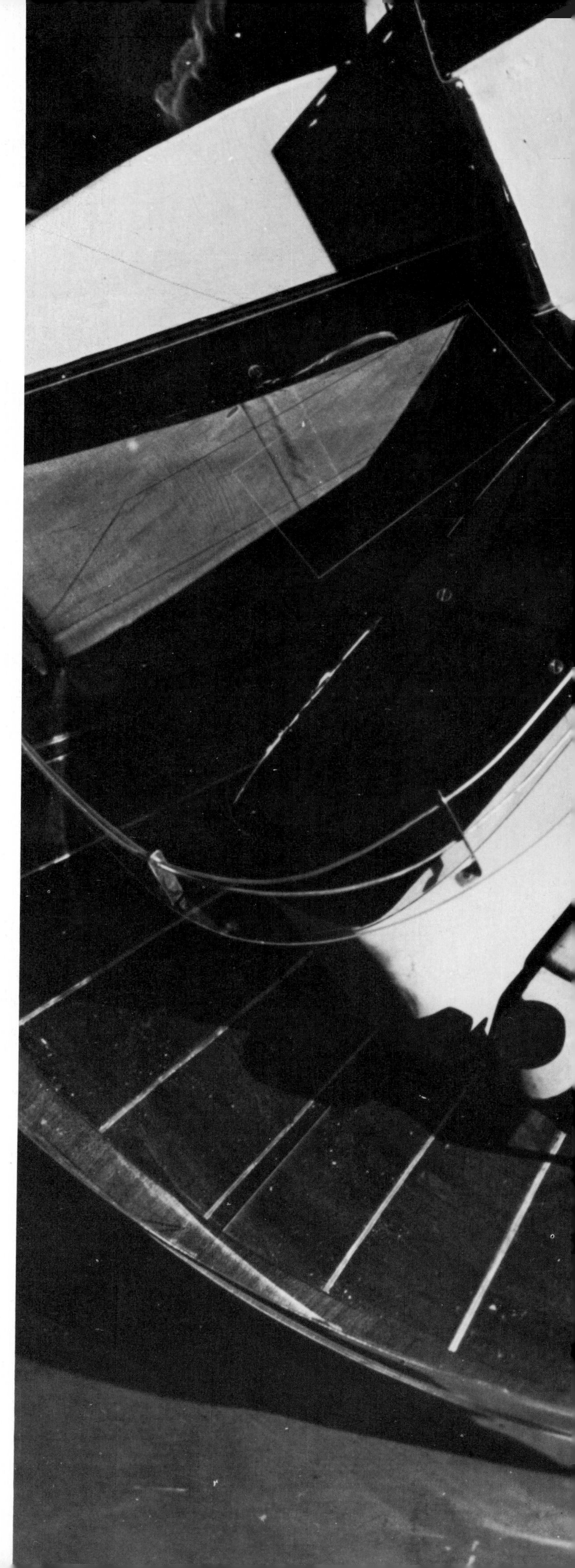

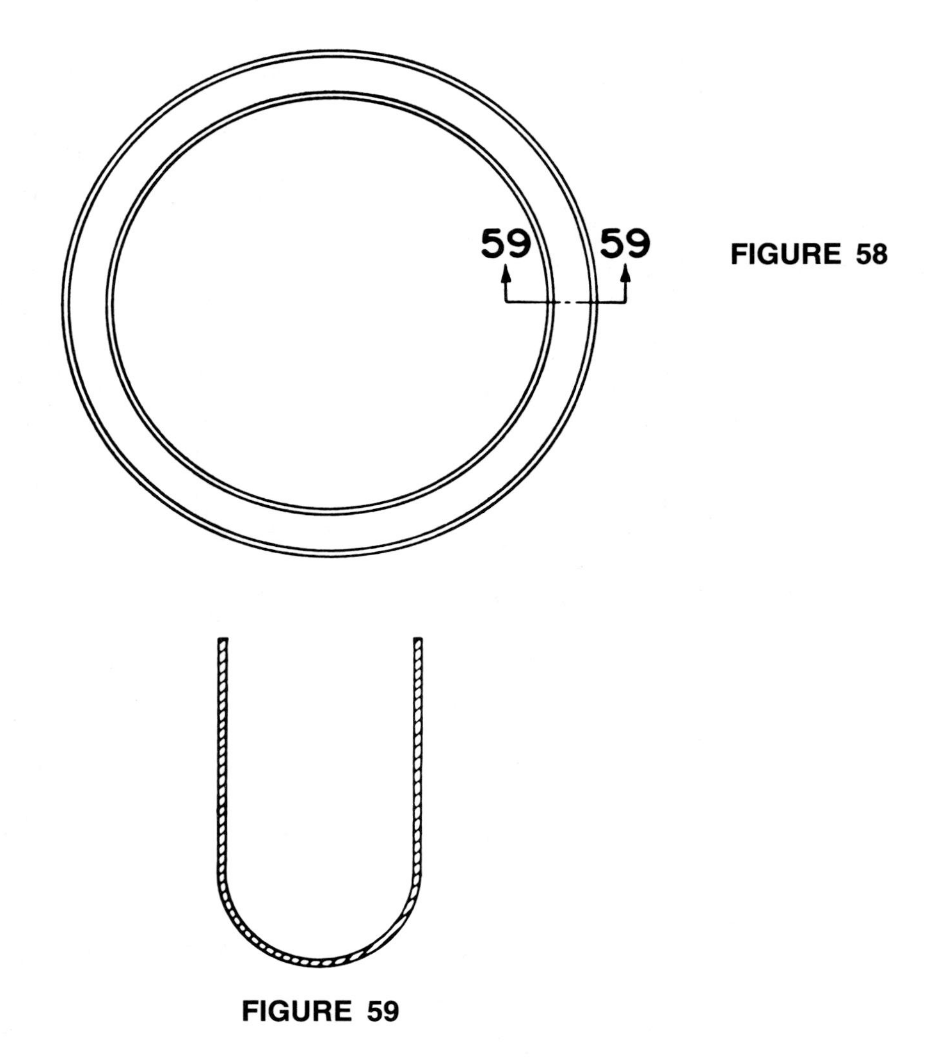

FIGURE 58

FIGURE 59

proper curves so that after assembly they provide a ceiling dome. They should be waterproof so that any leakage or condensate, dropping from the inside of the cowling, will be caught by them and guided down into the gutter 104 or into the ceiling carlings which then function as gutters carrying the water to the circular gutter 104.

The cowling gores are made of sheet aluminum alloy that is sufficiently thin to permit its being compactly rolled for packaging, and if they possess sufficient flexibility the ceiling gores may be similarly packaged. Otherwise, they may be made in sections which are fastened together during the assembly of the house, with the sections overlapping in a direction preventing water running through the joints. The ceiling carlings may also be made in sections which are fastened together during assembly of the house, like the cowling carlings, and they too are made of aluminum alloy and can internest for packaging most compactly.

Returning now to the side wall construction, partly circular sections 153 when assembled provide baseboards around the inside periphery of the house, these sections having their ends overlapped and inter-fastened and being pressed aluminum alloy channel sections with vertical webs and bottom flanges 154 which may be fastened to both the flooring panels 44 and to the deck sections, whereby the fastenings simultaneously position these baseboard sections and prevent outward movement of the flooring panels which would cause them to become loose. The bottom corner between the flange 154 and the upstanding web of the baseboard sections is provided by a fairly large radius curve so as to avoid the formation of corners inside of the house which might catch dirt. The upper flange 155 angles upwardly so that its upper edge terminates beneath the inner edge of the windowsill 91, the sections from which the latter are made being smoothly curved downwardly to provide a depending trim skirt 156. A series of air-impervious curtains, such as synthetic leather or rubber of attractive appearance, depend from this trim skirt 156 to the upper edge of the upper flange 155 of the baseboard 153, and may be fastened thereto by snap fasteners or the like so that these curtains may be released and raised when the previously described ventilating panels, forming part of the wall skin, are dropped for ventilation purposes. These curtains 157, in conjunction with the baseboard 153 and the header 105, function to provide a double-walled skin around the entire periphery of the house. If desired, the plastic windows 89 may also be of double thickness.

Fly screen panels 158, preferably made of plastic, may be cemented or otherwise fastened in place to close the openings resulting when the ventilator panels are dropped, and by arranging these well inside the double-walled construction beneath the windowsill 91, they are entirely concealed when the curtains 157 are dropped. Incidentally, these curtains may be rolled up and fastened when ventilation is desired.

The doorways are also made of aluminum alloy pressed parts and each includes a door frame 159 comprising an assembly of separate angle sections which are interfastened to form the door contour, which is preferably with a narrow top and bottom and more widely flaring central section. The webs of these sections are formed so that the sides of the door frame flare outwardly, they being formed by one of the legs of the angle sections. The door outline is formed vertically as a part of a cylinder so as to follow the cylindrical contour of the house side, this outline being provided by the other legs of the angle sections from which the door frame 159 is assembled. These legs forming this outline may be rigidly fastened to sheet-metal sections 160 filling the spaces between the terminating ends of the other wall sections and parts previously described.

The door closure comprises an assembly in the form of an outer frame 161 assembled from inwardly facing pressed sheet channel sections, so that their outer flanges extend inwardly, and a skin 162 is fixed to these outer flanges. The door, of course, follows the contour of the door frame. The door frame 161 is suspended by rollers 163 riding a trackway 164 fastened to the outside of the house and extending sufficiently far from the doorway to permit the door closure to ride along this trackway to a position where it is completely free from the door frame, the trackway 164 following the cylindrical contour of the house. The inner flanges of the door frame 161 are shorter than the outer flanges and are engaged outwardly of the house by equally short flanges of a second door closure frame 165 assembled from lengths of suitably shaped pressed sheet Z-sections, the contour of this frame also following that of the door frame and of the door. The flanges of these Z-sections inwardly of the house are fastened to an inner skin 166, this providing the door with spaced walls with the inner wall and its frame capable of telescoping inside the door closure frame 161, the arrangement being such that when the frame 165 and the skin 166 are pushed or pulled into the frame 161, the door can slide, and when these parts are pushed or pulled toward the door frame they close this door frame, a gasket 167 providing a weathertight seal in conjunction with the wedging action of the parts, and prevent sliding of the door.

Cylindrical sections 168 and 169 are respectively fixed to the inner and outer skins 162 and 166, these cylindrical parts being assemblies of sheets which can telescope inside one another so as to function as a guide during telescopic action of the door closure parts previously described. Compression springs 170 working in telescoped tubes 171, the telescoping sections being respectively fixed to the inner and outer door skins, bias the inner door closure skin and frame inwardly at all times, the center of the door inside the cylindrical sections 168 and 169 being provided with a latch bar 172 which interconnects with the inner skin 168 and projects slidably

FIGURE 60

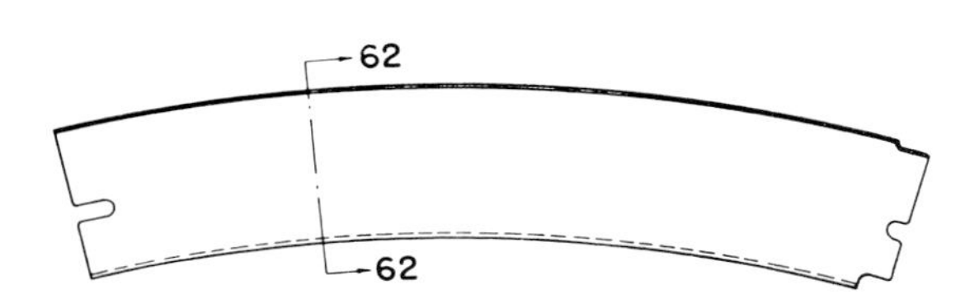

FIGURE 61

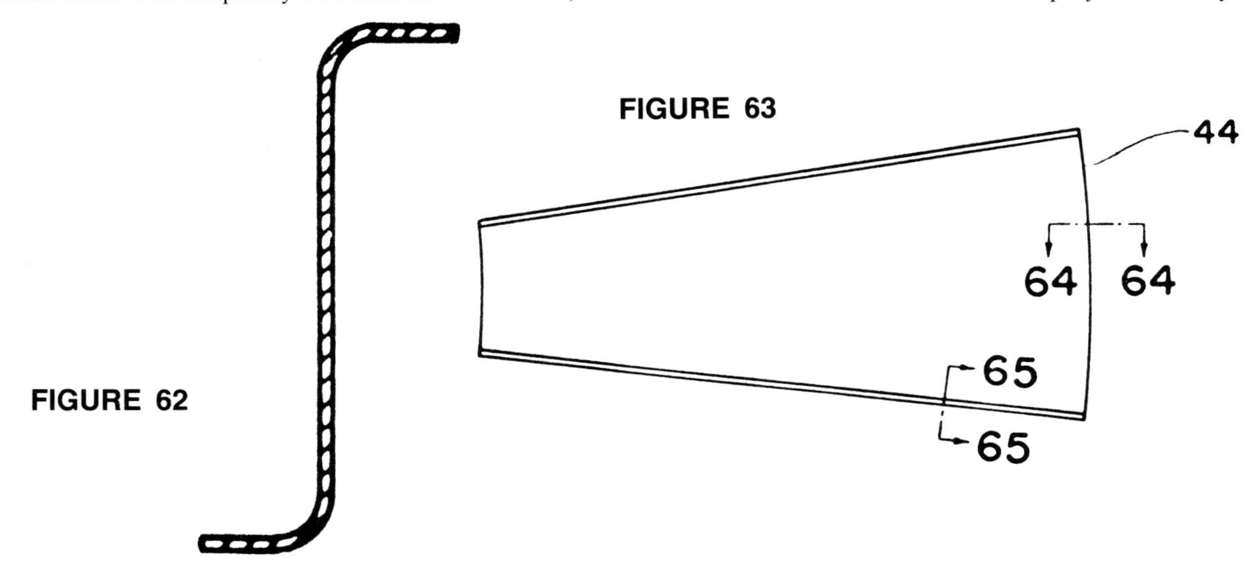

FIGURE 62

FIGURE 63

through the outer skin 162. This latch bar is provided on at least its exterior with a handle 173 so that by pulling on this handle the inner skin 166 and its frame 165 may be pulled outwardly so as to telescope inside the outer door closure part for permitting sliding of the door. The latch bar 172 is provided with detents 174 cooperating with a lock 175 so as to permit its being locked with the inner door skin and its frame either in or out.

The bottom of the door closure frame 161 mounts a grooved roller 176 which rides beneath a trackway 171A which extends up into its groove, this preventing the door from swinging outwardly. A further means for locking the door is provided by locking bars 177A that slide transversely in the door frame edge and into holes formed in the edge of the inner door frame 165, in the manner of a bank vault door lock. These bars are reciprocated by connecting arms 177 worked by a lever 178 which may be swung by a locking rotative knob 179, a reinforcement bar 180 providing extra stiffness for the skin section 160 adjacent the shaft of this knob 179. This door closure and door frame may be completely assembled in the factory if desired, or it may be an assembly of parts designed for assembly in the field at the time the house is assembled.

There is a depending bottom cowling 181 extending completely around the house, this cowling being an assembly of aluminum alloy strip sections that are fastened together during assembly of the house and arranged with its upper edge fastened to the flange 92 of the outer deck ring 29 by means of rivets or the like provided with spacers 182 so that the upper edge of the cowling is spaced away from the bottom periphery of the house side. This cowling 181 is made from sections formed to give the cowling a cross section wherein the cowling goes vertically downwardly a short distance and then smoothly curves inwardly to a level beneath the bottom of the outer deck ring 29, the cowling functioning aerodynamically to deflect wind blowing against the house side bottom downwardly under the house. The spaced upper edge of this cowling 181 allows water running down the house outside to flow into the inside of this cowling and here a gutter 182A, made of rubber-like material, catches it for disposal as desired. The bottom of the house deck is designed to be spaced a foot or so above the ground.

The house is in the form of a perfect cylinder having domed cowling on its top and bottom and with the ventilator on its top well streamlined. Therefore, the house is capable of resisting high wind stresses, its contour providing streamlining no matter which way the wind blows.

The house being described is approximately 36 feet in diameter and provides considerable living space in its interior because of its cylindrical shape. Its inside is equipped with two bathrooms 183 and 184 of the type disclosed in Patent No. 2,220,482, these bathrooms being arranged in tandem on one side of the mast 2 and in conjunction with wall units 185, 186 and 187 dividing half the house into two bedrooms. These wall units 185, 186 and 187 do not extend from the bathrooms completely to the inside of the house wall, but terminate so as to provide spaces of about the widths of usual bedroom doors, thus providing doorways. These doorways are closed, in each instance, by doors 188 which may be of the type consisting of vertically pleated fabricate door closures which slide transversely of the doorway. Door closures of this type are commercially available and have the advantage that they are light in weight and compact for shipping. The wall units separating the bedrooms are not high enough to provide complete wall closures, this being done above each of them by panels 189 that fan outwardly above the units from the mast 2 and which have top edges coinciding in shape with that of the domed ceiling of the house. These panels may be made from aluminum sheet sections or plastic and may take various forms. Preferably they are of a vertically corrugated nature so that they may be made of thin metal yet still be comparatively rigid. As described in the aforementioned patent, the bathrooms are assembly units that may be shipped as compact groups of parts, and the wall units 185, 186 and 187 should be of a similar nature. One example of a suitable wall unit will be presently described.

The other half of the house includes the doorways through the house side, of which there are preferably two, each doorway being located relatively close to the wall units 185 and 187 facing away from the sleeping accommodations. Other wall units 190 and 191 are respectively arranged to make comparatively small segregated spaces opposite each door, the unit 190 providing for an entrance hall and the unit 191 providing a kitchen space. The space between the units 190 and 191 provides for a large living room, the attractiveness of which is enhanced by eliminating the ventilating panel, previously described, through a short arc of the living room wall, and bending down the windowsill 91 to encompass a portion fairly close to the desk, this being done by the use of suitably stamped windowsill sections, whereby to provide a large bay window effect 192, using wider transparent sheets to close this bay window 192.

Short wall sections 193 and 194 extend inwardly from

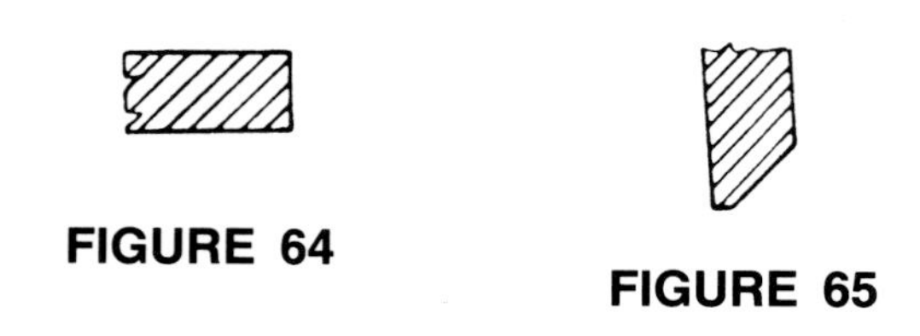

FIGURE 64

FIGURE 65

FIGURE 66

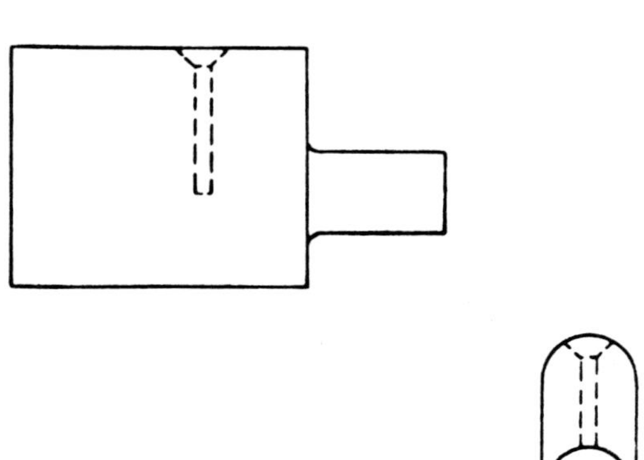

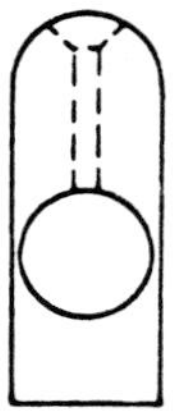

FIGURE 67

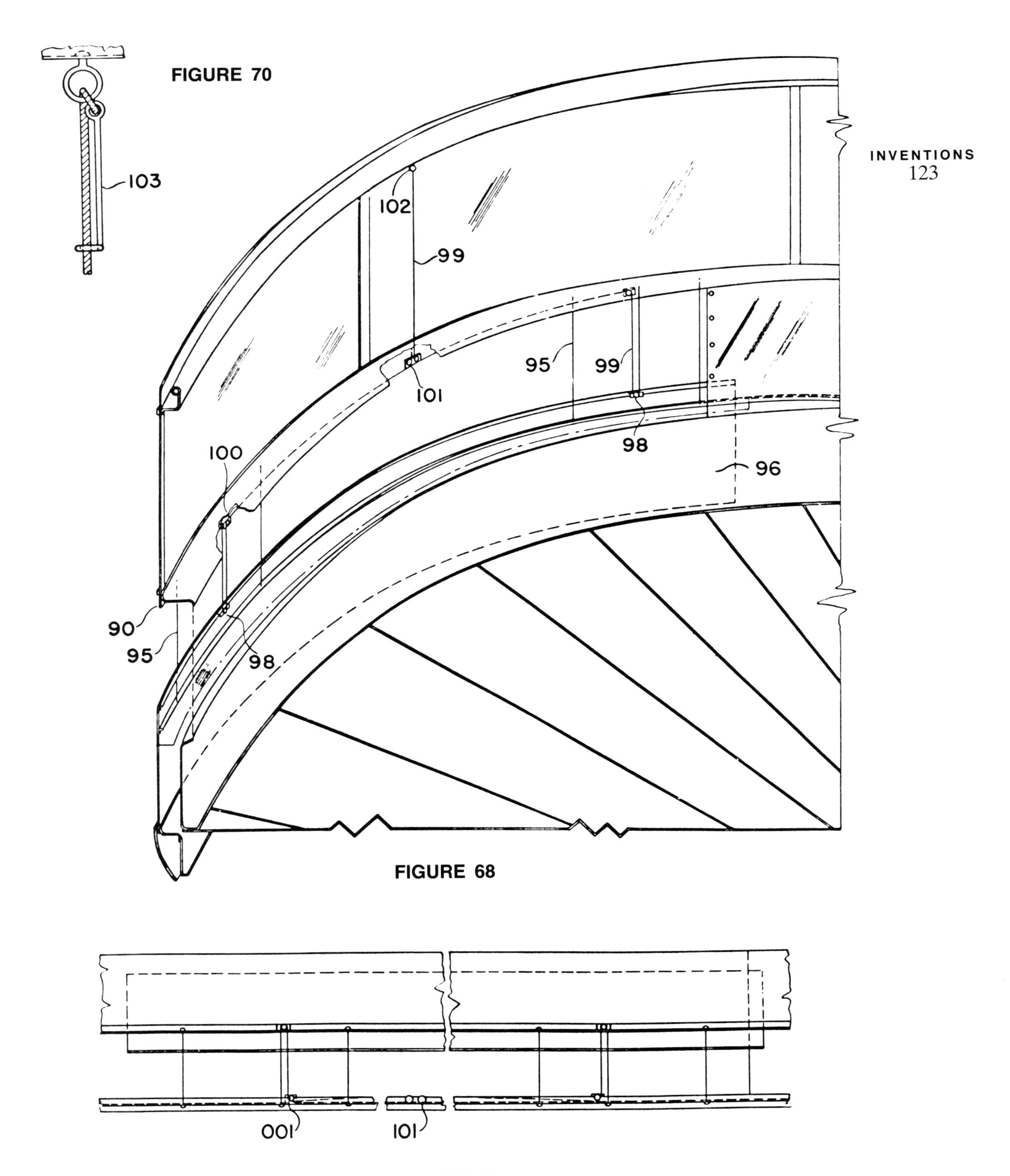

FIGURE 68

FIGURE 69

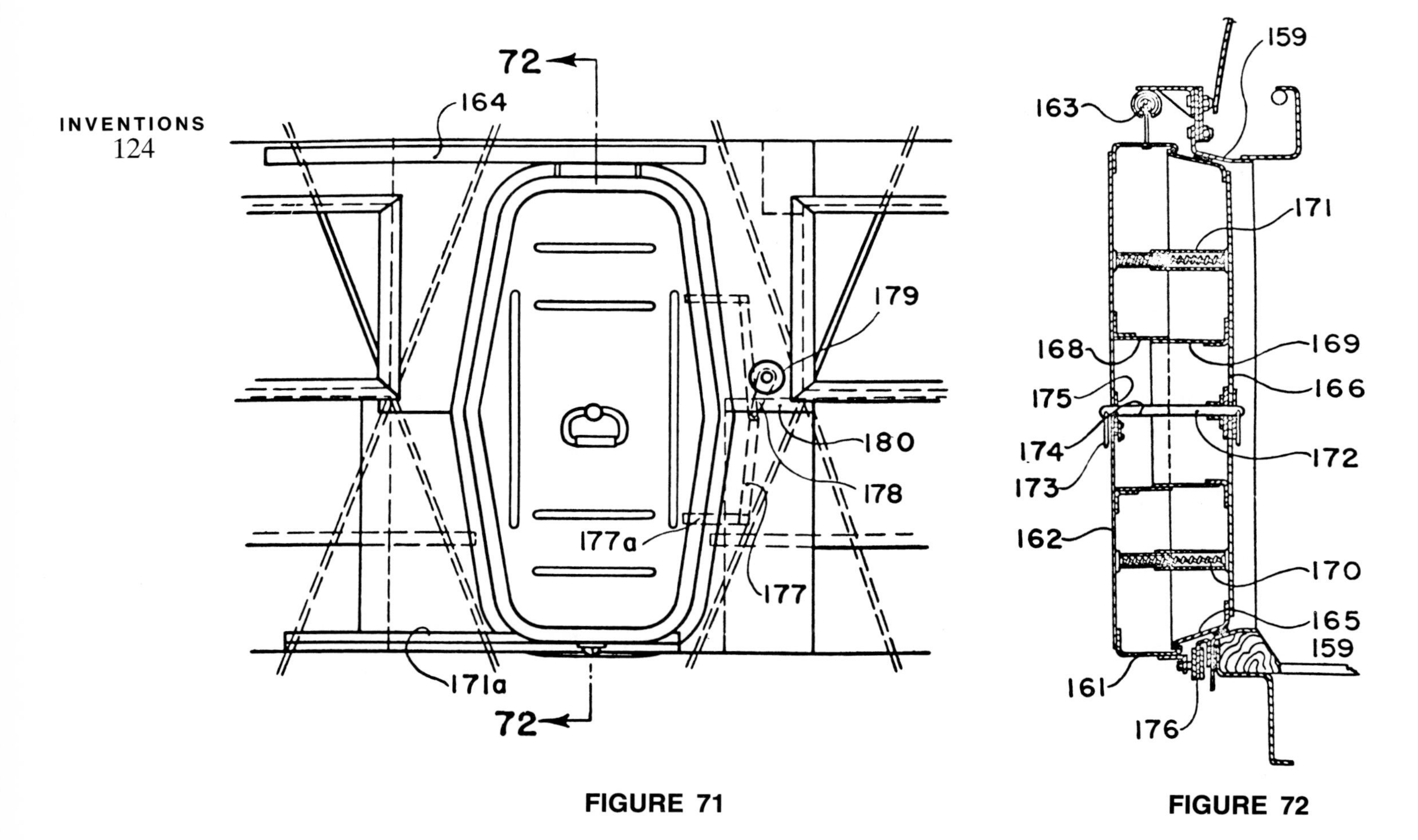

FIGURE 71

FIGURE 72

the inner edges of the units 190 and 191, respectively, to provide triangular spaces 195 and 196, and these spaces are provided with closures 197 and 198, respectively, so that they become vertical ducts extending from deck level to any height desired and which may be used as flues for aiding in the many possible heating and ventilating effects and for other purposes. For example, by having one of these flues opening from the ventilator bottom to the deck bottom, it is possible to prevent or reduce pressure differentials on the house outside even when the wind blows. This greatly reduces the heat loss from the house inside and makes the house more stable during windy conditions. It is unnecessary to provide closing walls above the units 190 and 191 or doors for the passageways between their outer ends and the inside wall of the house.

One or more of the wall units may comprise an assembly of structural shapes and sheet-metal skin sections providing a horizontally elongated enclosure internally provided with longitudinally extended upper and lower shafts 199 and 200 respectively positioned close to the top and bottom of the unit. These shafts are journalled to revolve and respectively mount axial spaced sprockets 201 and 202. Chains 203 ride these sprockets, the chains vertically extending from one sprocket to the other in each instance. The sprockets are keyed to at least one of the shafts so that the two radially spaced chains work in unison. Guideways 204 extend parallel and close to the straight sections of the chains 203 and carriages 205, fixed to the chains, mount spaced wheels 206 which ride these guideways 204. These carriages 205 are each in the form of a T with the rollers mounted to the ends of the head of the T and with the leg of the T pivotally fixed, at 207, centrally to a depending, side-opening basket 208. These baskets 208 are rectangular in cross section and their corners are provided with wheels 209 on their top corners at one end and 209A on their bottom corners at the other end, vertical guideways 210 being spaced outwardly from the guideways 204 and parallel to them at the basket end having the wheels 209 and other similar but lower guideways 210A being provided for the other basket end having the wheels 209A, and the wheels on the outsides of the basket, either going up or going down, riding these guideways so that the basket 208 is rigidly positioned against tilting at all times. The guideways 210 and 210A turn inwardly at their tops and bottoms toward the sprocket wheels to guide the wheels for which they are respectively provided as the baskets revolve around the sprocket wheels, it being understood that the curved upper and lower ends of the guideways are all curved properly to permit the baskets to revolve around the sprockets, the baskets being completely rigid while traveling. With this arrangement any of the wall units may be provided with a very great many shelves opening from the same side of the unit, the side of the latter being provided with a single opening coinciding in size and shape with that of the side opening of the various baskets and with which any of the baskets may be registered. The shelves provided by these baskets may be used to store clothing and the like, which will be much more easily found because it is only necessary to look into a single opening, while the baskets travel past the opening, until the article is found. Any convenient arrangement may be provided for powering one or both of the sprocket shafts to effect the necessary motion, and this may comprise an electric motor controlled from outside the wall unit. All of the parts described will, of course, be capable of assembly in the field or may be shipped as compact subassemblies.

These various wall units are made with a horizontal cross section that is in the form of a long oval so that even though the sides of the wall unit are made of very thin-gauged sheet metal they will not cause acoustical diaphragm noises. This principle applies to practically the entire house, all the major surfaces being curved, it following that the house provides for unusual quietness inside it as compared to the conventional house with flat walls. This use of curved surfaces has the further great advantage that the use of very thin-gauge sheet metal is completely practical, the curved surfaces giving it inherent rigidity. Very thin sections are made further practical by the fact that practically everything involved in the construction of the house, that is under any stresses whatsoever, is working in tension, excepting for the compression mast. All of the parts may be made from strong and highly tempered aluminum alloy, such as is used in aircraft construction, excepting for those parts subjected to unusual corrosion or requiring very great tensile strength. It is for this reason that the footing is made largely of stainless steel having an analysis designed particularly to resist corrosion, while the tension strands 3 are made of stainless steel having an analysis designed to provide maximum tensile strength. The use of these high-cost alloys in the construction of the house is economically tolerable only because the manufacture, packaging, and shipping of the house permits full use of mass production methods.

Due to the construction of the house there is a well-balanced thermal effect wherein there is practically no uncontrolled heat exchange between the inside of the house and the outside. This result is diminished if there is direct heat conduction between the inside and outside; there-

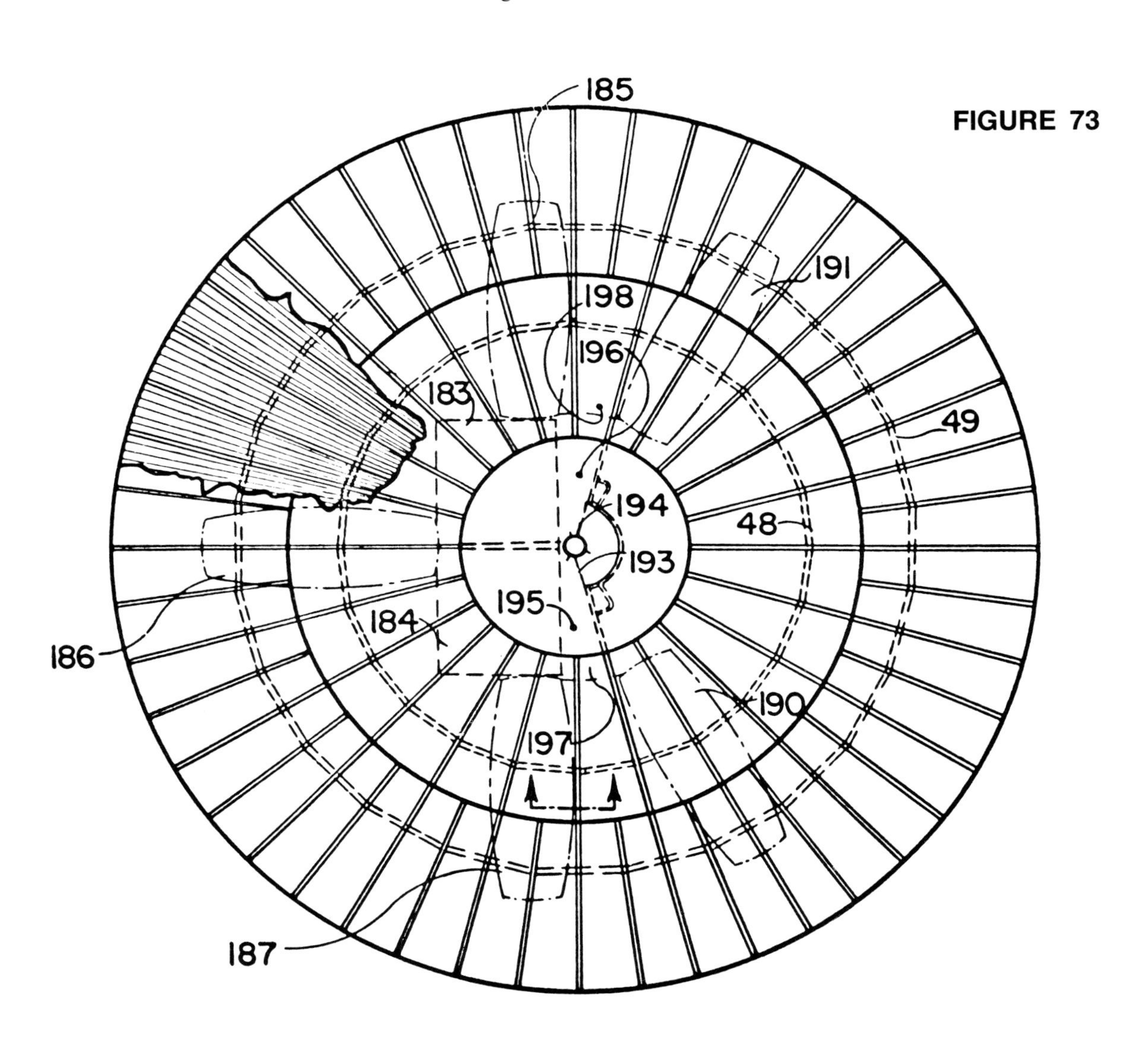

FIGURE 73

fore, wherever there are direct metallic heat-conducting paths it is preferred to break these paths by the use of thermal insulation, such as by the use of nonmetallic washers, inserts, sheets, and the like, depending on which is most effective.

It will be noted that there is a peripheral space around the house between the cowling lower edge and the side upper edge through which air may be drawn upwardly along the cowling bottom and out the ventilator, thus decreasing the tendency for condensation on the cowling bottom and reducing heating of the house inside in hot weather. Preferably aluminum foil is fitted over the tension strands, beneath the cowling, at least, by laying it thereon and bending its edges around the strands so as to form an intermediate wall between the cowling and the ceiling, the space between this intermediate foil wall communicating with the just-described peripheral space and the ventilator and confining the air sweeping under the cowling bottom to the space between this foil wall and the cowling. This arrangement provides a dead air space between the foil wall and the ceiling, and the foil acts as an infrared radiation reflector in both directions.

The present inventor believes that he is the inventor of this house disclosed herein in its entirety, of the various combinations and sub-combinations, and the various parts, and it is his intention to claim the house entirely, its various combinations and sub-combinations of parts and the parts themselves. The aerodynamic and thermal effects inside and outside of the house are his inventions, both as specifically described and as they are inherent, and the general conception of a house working dynamically like a machine in carrying the various loading stresses, and in coping with the elements and forces that must be handled by a completely satisfactory house, are all this inventor's conceptions, and this inventor intends to claim them, and all the rest, in every possible lawful manner. If the appended claims fail to cover every one of these phases and all possible other phases, and all equivalents of what is disclosed herein, specifically or inferentially, it is because of accident, inadvertence, or mistake, and not through intent.

FIGURE 74

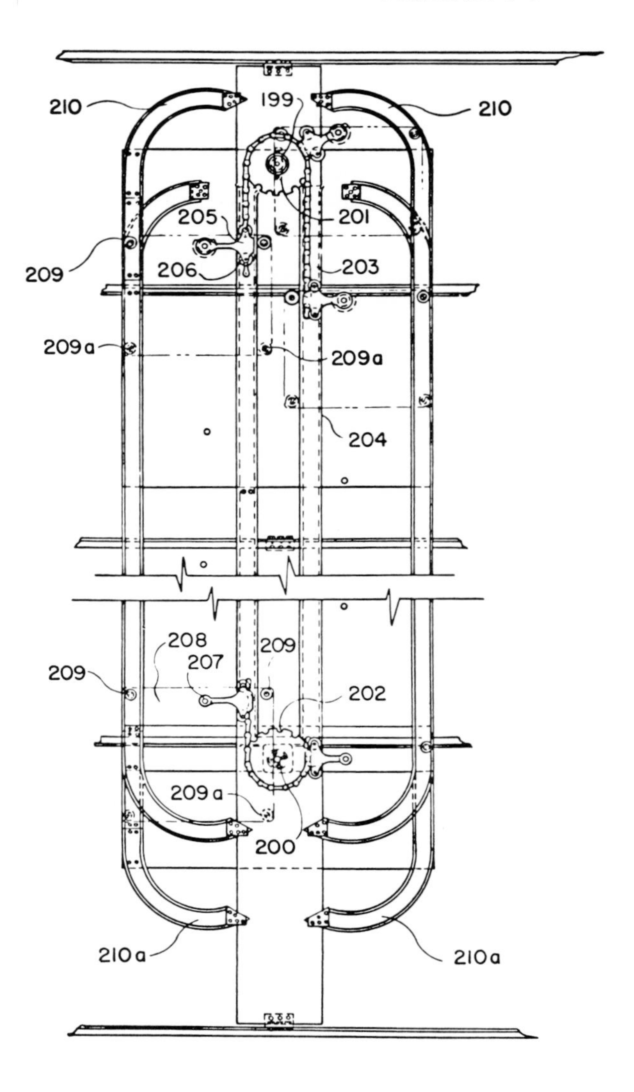

10▲GEODESIC DOME (1954)

U.S. PATENT—2,682,235

APPLICATION—DECEMBER 12, 1951

SERIAL NO.—261,168

PATENTED—JUNE 29, 1954

IN PRODUCING AND TESTING the Beech Aircraft prototype of the (Dymaxion) Fuller House, which was suspended around a mast, it was found necessary to stay the mast against wind-overturning in such a manner that the stays occupied a significant area of the house's interior. I saw that by developing the omni-triangulated tensegrity geodesic dome that it could serve as both domical enclosure and as its wind-stabilizing structure without any interior support. It became a spherical mast simultaneously omni-radiantly suspending its enclosed domain and omni-evenly distributing all local impact stresses to all its structural members.

At no time in my last 56 years have I paid any attention to conventional architecture's "orders" about the superficial appearance of my structures. I never try to anticipate what my structures are "going to look like." I am concerned only with providing comprehensive,

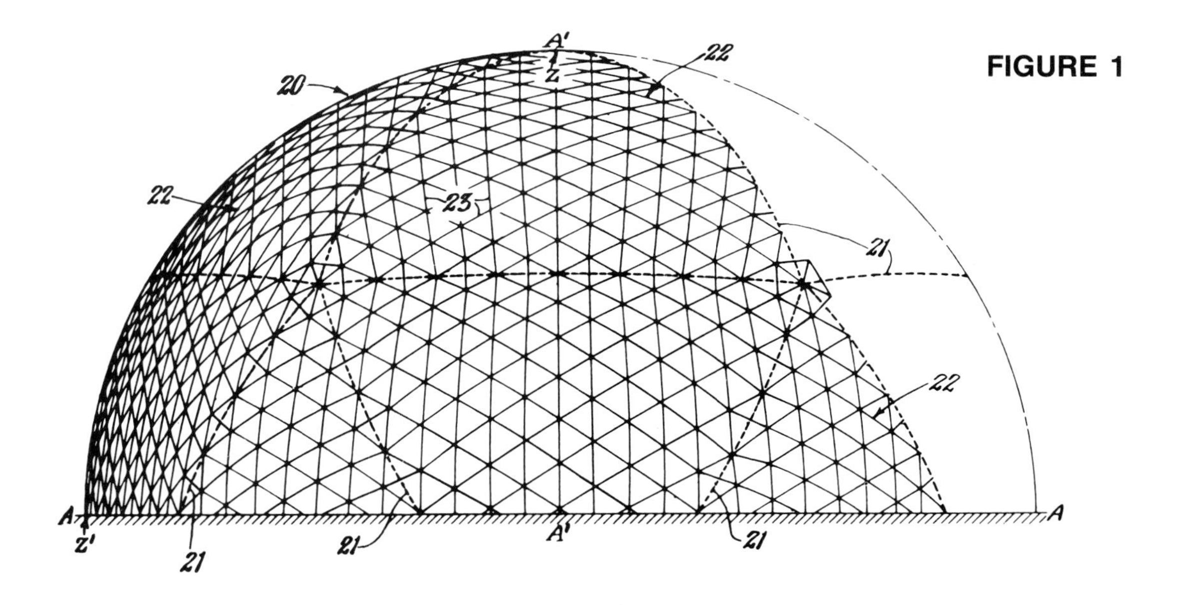

FIGURE 1

logical, pleasingly adequate, and most economical solutions to all design problems. I must be responsible for the method of production, assembly, installation, servicing, and transportability of all the parts and the behavior of the whole under all anticipatable conditions.

When the whole installation and assembly is complete and tested, and I can stand off and look at it as an operating reality, if it does not look beautiful to me, I know that I have failed to produce that which I undertook to do. Beauty to me must be a product result and not a purpose.

When I invented and developed my first clear-span, all-weather geodesic dome, the two largest domes in the world were both in Rome and were each 150 feet in diameter. They are St. Peter's, built around A.D. 1500, and the Pantheon, built around A.D 1. Each weighs approximately thirty thousand tons. In contrast, my first 150-foot-diameter geodesic all-weather dome installed in Hawaii weighs only thirty tons—one-thousandth the weight of its masonry counterpart. An earthquake would tumble both the Roman 150-footers, but would leave the geodesic unharmed.

FIGURE 2

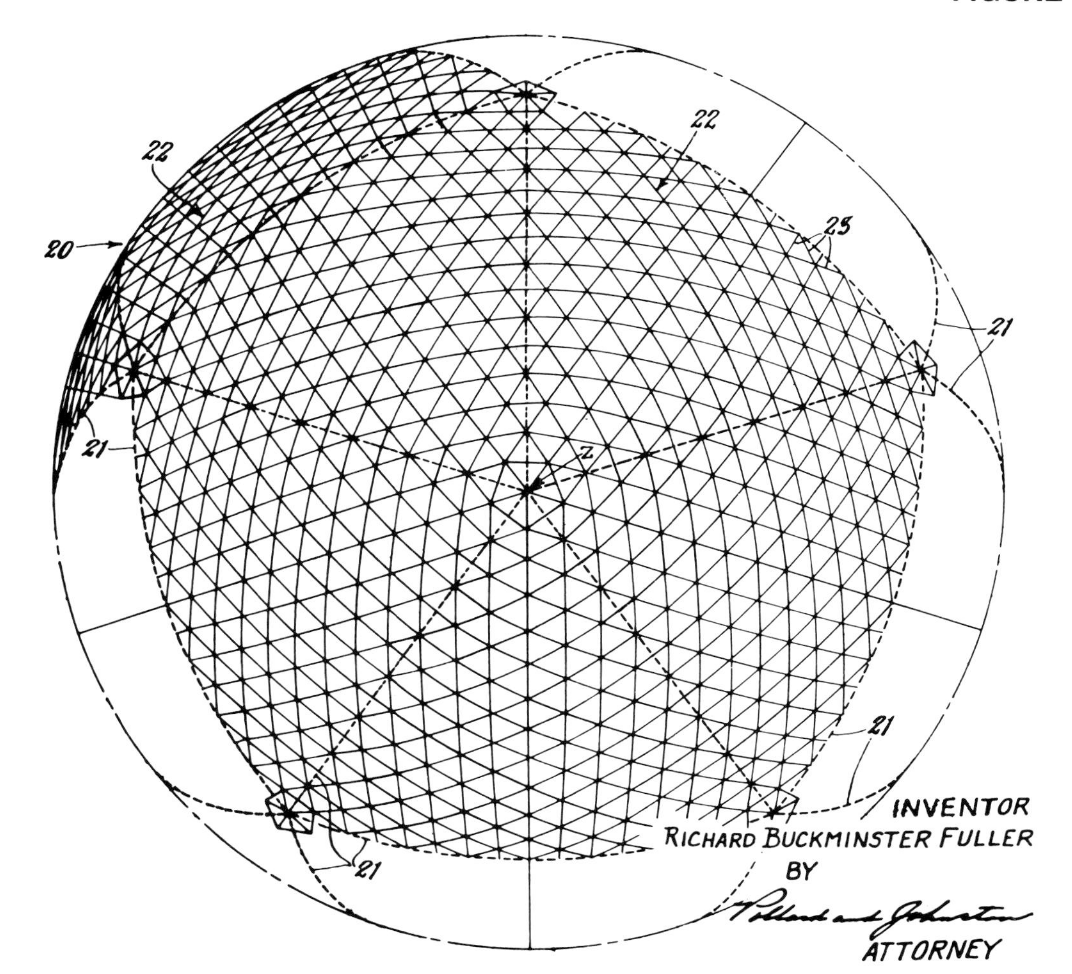

UNITED STATES PATENT OFFICE

Richard Buckminster Fuller, Forest Hills, N.Y.

BUILDING CONSTRUCTION

My invention relates to a framework for enclosing space.

SUMMARY

A good index to the performance of any building frame is the structural weight required to shelter a square foot of floor from the weather. In conventional wall and roof designs the figure is often 50 lbs. to the sq. ft. I have discovered how to do the job at around 0.78 lb. per sq. ft. by constructing a frame of generally spherical form in which the main structural elements are interconnected in a geodesic pattern of approximate great circle arcs intersecting to form a three-way grid, and covering or lining this frame with a skin of plastic material.

My "three-way grid" of structural members results in substantially uniform stressing of all members, and the framework itself acts almost as a membrane in absorbing and distributing loads. The resultant structure is a spidery framework of many light pieces, such as aluminum rods, tubes, sheets, or extruded sections, which so complement one another in the particular pattern of the finished assembly as to give an extremely favorable weight-strength ratio, and withstand high stresses. For example, the "8C270 Weatherbreak" constructed in accordance with my invention will support 7 lbs. with each ounce of structure and is able to withstand wind velocities up to 150 miles per hour. It is a dome 49 ft. in diameter, enclosing 20,815 cu. ft. of space, yet the frame is made of light short struts which pack into a bundle 2 ft. by 4 ft. by 5 ft., weighing only 1000 lbs. The plastic skin weighs 140 lbs., making the total weight of this "weatherbreak" a mere 1140 lbs.

Definition of terms

The basic and fundamental character of the inventive concept herein disclosed makes it desirable to define carefully certain terms some of which are used with special connotation, as follows:

Geodesic—Of or pertaining to great circles of a sphere, or of arcs of such circles; as a geodesic line, hence a line which is a great circle or arc thereof; and as a geodesic pattern, hence a pattern created by the intersections of great-circle lines or arcs, or their chords.

Spherical—Having the form of a sphere; includes bodies having the form of a portion of a sphere; also includes polygonal bodies whose sides are so numerous that they appear to be substantially spherical.

FIGURE 3

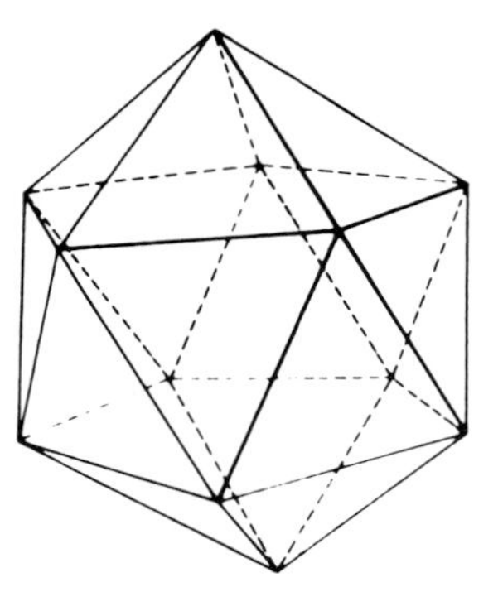

FIGURE 4

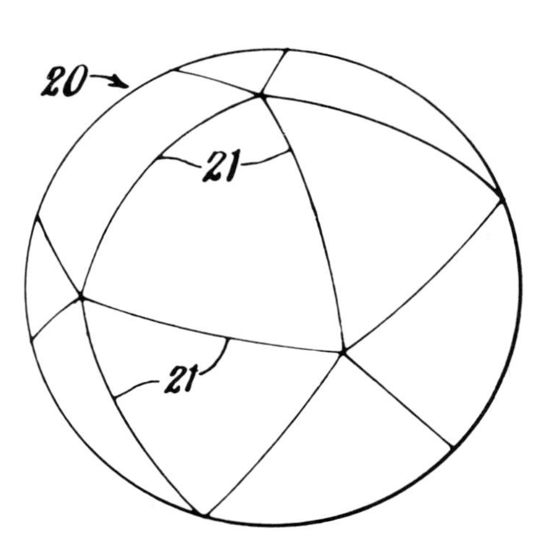

Icosahedron—A polyhedron of twenty faces.

Spherical icosahedron—An icosahedron "exploded" onto the surface of a sphere; bears the same relation to an icosahedron as a spherical triangle bears to a plane triangle; the sides of the faces of the spherical icosahedron are all geodesic lines.

Icosacap—Five spherical triangles of a spherical icosahedron, having a common vertex.

Grid—A pattern of intersecting members, lines or axes; usually intersecting great circles forming patterns made up of equilateral triangles, diamonds or hexagons.

Equilateral—Having all the sides approximately equal. The extent of variation in length of sides is determined trigonometrically or empirically by constructing three-way grids on the modularly-divided edges of the faces of a spherical icosahedron.

Modularly divided—Divided into modules, or units, of substantially equal length.

Framework—The frame of a structure for enclosing space; may be skeletal, as when made of interconnected struts; or continuous, as when made of interlocking or interconnected sheets or plates.

The meanings of these and other terms used in describing the invention will be more fully comprehended when considered with reference to the accompanying drawings and diagrams and the explanation thereof.

In its general arrangement, my building framework is one of generally spherical form in which the longitudinal centerlines of the main structural elements lie substantially in great circle planes whose intersections with a common sphere form grids comprising substantially equilateral spherical triangles. The visible pattern formed by the structural elements themselves does not necessarily show grids of equilateral triangles, for the visible grids may be equilateral triangles, equilateral diamonds or equilateral hexagons, the diamonds being made up of two equilateral triangles and the hexagons being made up of six equilateral triangles. The individual triangles, diamonds, or hexagons as the case may be, may be made of straight or flat elements, in which circumstance they define flat or plane figures; or they may be made of arcuate or spherical form to define spherical figures. Either way, the complete structure will be spherical, or substantially so. And either way, the individual structural elements are so arranged as to be aligned with great circles of a common sphere.

In my preferred construction, the grids are formed on the faces of a spherical icosahedron. Each of the twenty equal spherical equilateral triangles which form the "faces" of this construction is modularly divided along its edges. Lines connecting these modularly divided edges in a three-way great circle grid provide the outline for the plan of construction. I have found that if the structural members be aligned with the lines of the grids, the resulting framework will be characterized by more uniform stressing of the individual members than is possible with any construction heretofore known. The structural members may be aligned with all lines of the three-way grid, or just with selected ones of those lines. If the members are arcuate, or spherical, they will coincide with the grid lines; if they are straight, or flat, they will be chords of the great circles which are the grid lines.

A further general aspect of my preferred construction which may be noted here is that there is a "six-ness" throughout the pattern of structural elements on each face of the spherical icosahedron except that at each vertex, where five faces join at the center of an icosacap, there is a "five-ness." In the case of a skeletal framework made up of struts in a pattern of equilateral triangles, this "six-ness" is manifested by the fact that there are six such triangles around every vertex except at the vertexes of the icosacaps where the "five-ness" is manifested by the fact that there are only five such triangles around those vertexes. Similarly, in the case of a continuous framework made up of diamond-shaped sheets, we find a "five-ness" only at the vertexes of the icosacaps where five sheets toe in to the one point. This aspect of five-ness and six-ness will be described more in detail further on, and need be mentioned only in general terms here so as to lay a foundation for the description of various specific frameworks built according to the invention.

FIGURE 5

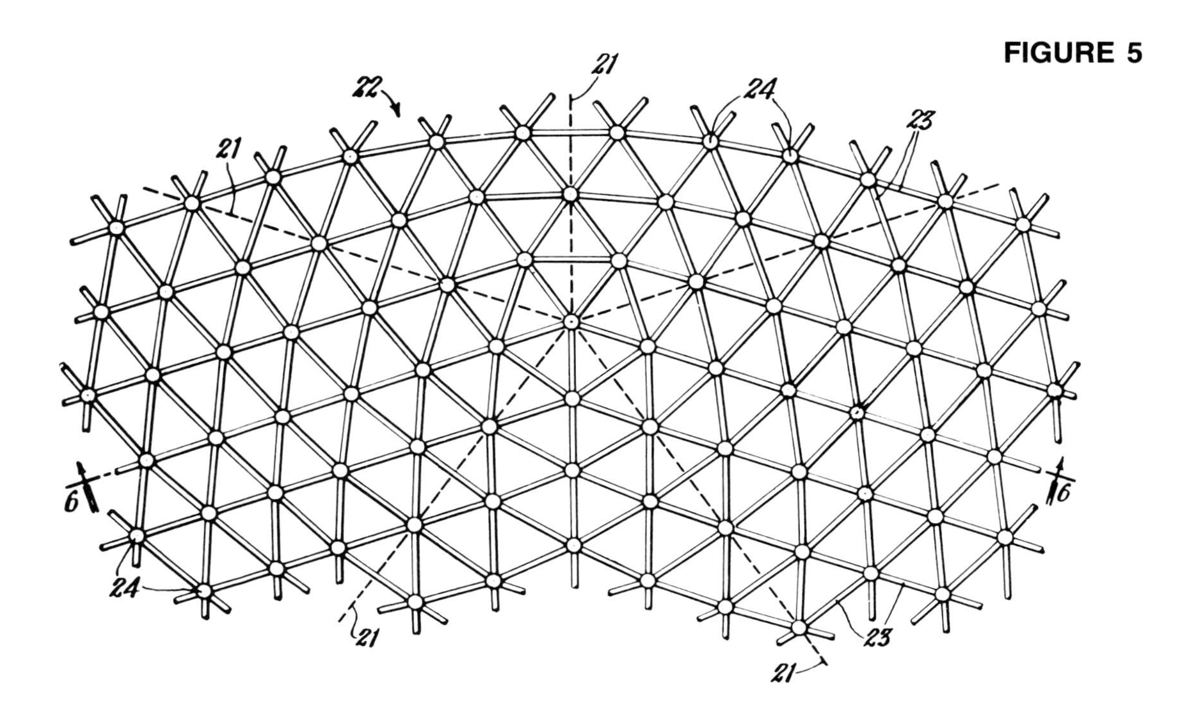

DESCRIPTION

In the drawings:

Fig. 1 is an elevational view of a building framework constructed in accordance with my invention and exemplifying a preferred form thereof; and Fig. 2 is a top

FIGURE 6

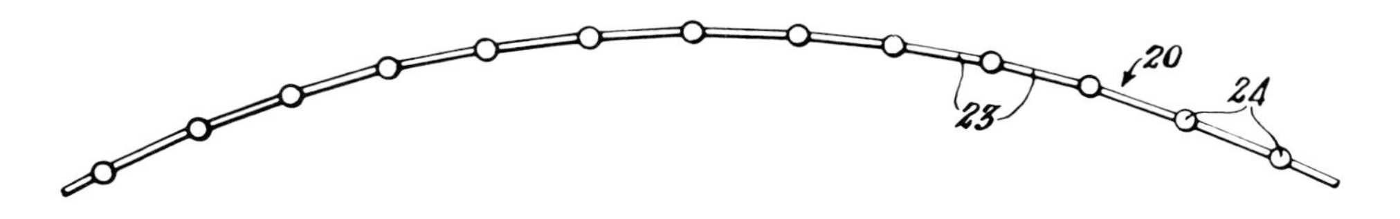

plan view of the same framework. These views are necessarily somewhat schematic because of the limitations imposed by the smallness of the scale to which they are drawn, making it impossible to show the detail of individual struts or of the fastenings which hold them together.

Fig. 3 is a diagrammatic perspective view of an icosahedron; and Fig. 4 a view of the same icosahedron after it has been "exploded" onto the surface of a sphere. These views are included to explain the structural basis of the main outlines of the framework of Figs. 1 and 2.

Fig. 5 is a detail plan view of a portion of the framework of Figs. 1 and 2, being that portion which immediately surrounds the top central vertex, i.e., the central part of the icosacap seen in Fig. 2.

Fig. 6 is a vertical section on the line 6—6 of Fig. 5.

Figs. 7 and 8 are detail views of my preferred form of strut fastening. Fig. 7 is a central vertical cross-sectional view through the fastening with two struts fixed therein, one of these being shown in the central section and the other in elevation. Fig. 8 is a horizontal sectional view taken as indicated at 8—8 in Fig. 7, one of the struts being shown in elevation.

Fig. 9 is a detail sectional view taken on the line 9—9 of Fig. 7.

Fig. 10 is a diagrammatic plan view of a modified construction in which the vertexes of the pentagons and adjoining hexagons are offset inwardly to form an "involuted" truss-like structure. This view represents a portion of the framework similar to that shown in Fig. 5. However, instead of showing the framework itself, the planes of the equilateral triangles formed by the struts of the framework are shown as though they were triangular panels so as to permit shading of the view in such a way as to pictorialize the resulting "dimpled" surface.

Fig. 11 is a diagrammatic cross-sectional view taken on the line 11—11 of Fig. 10.

Fig. 12 is a diagrammatic cross-sectional view similar to Fig. 11 showing a further modification in which the vertexes of the pentagons and adjoining hexagons are offset outwardly to form an "involuted" truss-like structure which is the inside-out of the structure of Fig. 11.

Fig. 13 is a fragmentary plan view of another embodiment of the invention, and Fig. 14 is a transverse sectional view of the same, taken as indicated at 14—14 in Fig. 13. Figs. 13a and 13b are detail perspective views of certain component parts of the truss illustrated in Figs. 13 and 14.

Fig. 15 is a diagrammatic plan view illustrating a portion of a framework in which the main structural elements consist of interconnected sheets.

Fig. 16 is a detail plan view of one of the sheets used in the framework of Fig. 15, and Fig. 17 is a detail longitudinal sectional view of such a sheet taken as indicated at 17—17 in Fig. 16.

Fig. 18 is a further detail view of the Fig. 15 framework showing the manner in which four adjacent sheets are interconnected or interlocked.

The framework construction illustrated in Figs. 1 to 9 inclusive is representative of the best mode devised by me of carrying out my invention particularly as utilized in

FIGURE 7

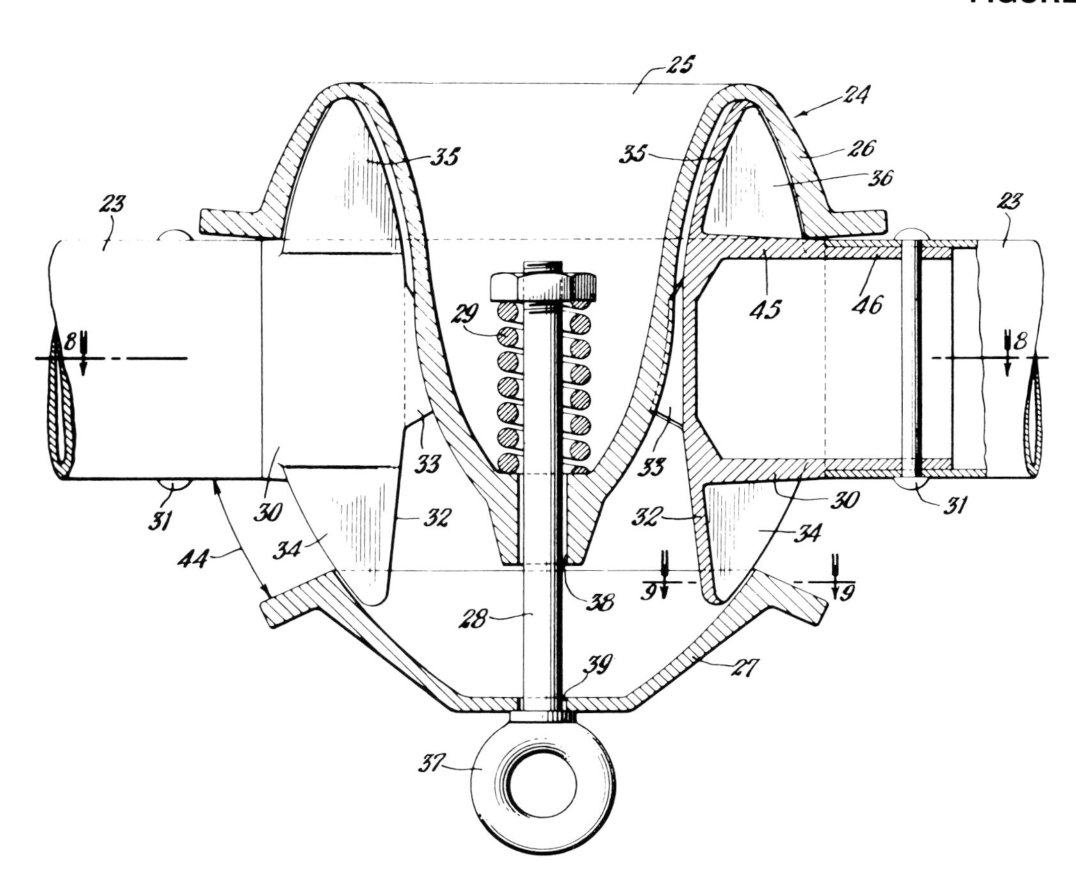

structures up to approximately 50 feet in diameter. The struts which comprise the structural elements of this framework form a portion of a spherical icosahedron 20 whose modularly divided edges 21 are interconnected by three-way great circle grids 22. A spherical icosahedron has been defined above as an icosahedron "exploded" onto the surface of a sphere. This definition will be further explained by reference to Figs. 3 and 4, Fig. 3 being a diagrammatic perspective view of an icosahedron, and Fig. 4 a view of the same icosahedron exploded or projected onto the surface of a sphere. The icosahedron has twenty equal equilateral triangular "faces." The spherical icosahedron has twenty equal equilateral spherical triangular "faces." As used here, the term "face" refers to an imaginary spherical surface bounded by the sides or edges of one of the twenty spherical triangles.

The edges of each spherical triangle are modularly divided and are interconnected by the three-way great circle grids 22 previously mentioned. These grids are formed of a series of struts each of which constitutes one side of one of the substantially equilateral triangles defined by the lines of the grid. Each strut 23 is aligned with a great circle of the spherical icosahedron. Otherwise stated, the longitudinal centerline of each strut, or main structural element 23, lies substantially in a great circle plane. In the complete framework, the longitudinal centerlines of the main structural elements 23 lie substantially in great circle planes whose intersections with a common sphere form grids 22 comprising substantially equilateral triangles.

The number of modules into which each edge of the spherical icosahedron is divided is largely a matter of choice. In the framework of Figs. 1, 2, 5 and 6, the number is 16. Therefore we say the frequency is 16. But it might be 8 or 4 or some other number. Generally speaking, the larger the structure the greater will be the frequency selected in order to keep the sizes of individual struts within practicable limits for ease of manufacture, storage, packing, shipment, handling, and erection. I prefer to use light metal pieces for the struts, e.g., aluminum tubes as shown in Figs. 7 and 8. One metal alloy presently considered most suitable is the aluminum alloy known generally under the designation 61ST. A tubular strut size found satisfactory for structures 40 ft. in diameter is approximately 4 ft. long by 1⅞″ outside diameter by 0.032″ wall thickness. In general, I prefer to use struts which have a ratio of 24 units in length to 1 unit in transverse dimension, i.e., the "slenderness" ratio is 24 to 1. The frequency of the pattern as above defined can be selected with view to maintaining the optimum slenderness ratio for each size of framework.

The struts 23 may be interconnected by sliding joints

locked by gravity compression acting throughout the great-circle pattern of the framework as a whole. In erecting the framework it is best to start by assembling the struts which are to form the very top of the dome, i.e., at the center of the icosacap seen in Fig. 2. This can be done on the ground. Working radially outward in all directions, the dome will begin to take form and will gradually be lifted as the work proceeds until in the end it rests with its lowermost struts against the ground or on a suitable foundation prepared to receive it. It may be locked to the foundation by great circle bands or cables preferably extending along the great circle lines which define the edges of the icosahedron. If a poured concrete foundation is used, the lowermost struts and fastenings, or the ends of such struts, may be embedded in the foundation, in which case the concrete, or portions thereof, is poured after erection of the framework has been completed. Alternatively, the lowermost struts and/or the fastenings may be anchored to individual concrete foundation posts or to eye-bolts or other fastenings in such posts. In this arrangement any suitable auxiliary fastenings may be used to lock the framework to the foundation fastenings, such as bolts, cables, turnbuckle rods, etc., this being largely a matter of choice depending upon the type of construction best suited for a particular purpose.

Referring again particularly to Figs. 1 and 2, the ground line or foundation line is indicated at A—A in Fig. 1, and the components of the framework are so oriented that the midpoint of the pentagon at the center of an icosacap coincides with the zenith Z of the framework. In some cases, however, it may be preferred to shift the orientation of the framework, as for example to an orientation which would result from using the line A′—A′ as the ground or foundation line, in which case the zenith Z′ would no longer coincide with the midpoint of an icosacap but instead would coincide with a point within one of the spherical triangles which form the faces of the icosahedron. If the sheet of the drawing on which Fig. 1 appears to be turned so that its right-hand longer edge becomes the bottom of the sheet, that part of Fig. 1 which is bounded by Z′A′A′, becomes the right-hand portion of the reoriented framework. Note that with this particular orientation, the base line A′A′ is a geodesic line completely defined by struts of the framework. This provides a convenient foundation line and one which lends itself well to anchoring of the framework to its foundation. In Figs. 1 and 2, however, I have chosen the zenith Z orientation in order to provide a clear illustration in Fig. 2 of a complete icosacap as defined hereinabove.

One of the characteristics of the completed framework is that it is virtually self-locking. Once properly as-

FIGURE 8

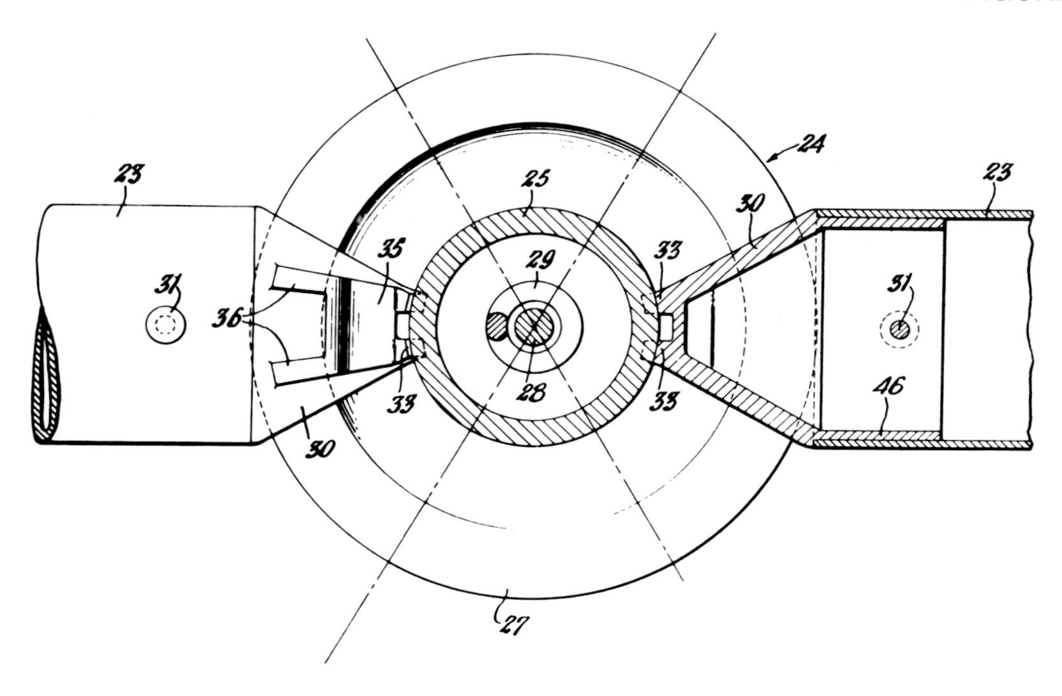

sembled in the manner described, it will not come apart except by more or less uniform expansion of all its parts. However, because the framework is somewhat resilient, localized forces acting outwardly against the inside of the structure may under certain conditions tend to expand one portion of the framework and produce what might be described as something akin to a blowout in a pneumatic tire. To resist such forces, and to assist in holding the struts together during erection, it is best that the means for fastening the ends of the struts be such as to lock them positively in place, in this respect supplementing the self-locking action described above, and giving added strength to the framework by reason of the fixed-end construction thus provided.

My preferred form of fastening is shown in Figs. 7–9. It is a ball-like "fist" configuration designated generally at 24, comprising complementary parts 25 and 27. In the specific construction illustrated, 25 is the outer part and 27 the inner part. These parts are clamped together by means of a bolt 28, a coil spring 29 being provided to afford a certain amount of resiliency in the fastening, which is particularly useful during erection of the structure. As seen in Fig. 7, outer part 25 is in the general form of an inverted bell, the edge of which is turned back on itself to provide a curved flange 26 complementary to the curved flange of inner part 27. Affixed to each end of each strut 23 is an attaching member 30 having a tubular body portion 45, the shouldered end 46 of which fits within the end of the strut. Attaching member 30 may be secured to the strut 23 by means of a rivet, pin, or bolt 31. Each fastening member has an inwardly extending lug 32 and an outwardly extending lug 35. Lug 32 has a pair of flanges 34 with arcuate edges conforming to the arc of the inner surface of inner part 27 of the fastener. Similarly lug 35 has a pair of flanges 36 whose arcuate edges conform to the inner surface of flange 26 of outer part 25. A pair of flanges 33 at the end of fastening 30 have arcuate edges conforming to the outer surface of the bell-like portion of outer fastening 25. The arrangement is such that the longitudinal centerline of struts meeting at any particular fastening 24, 27 can be adjusted to different angles so that the struts will form chords of great circles of the framework as a whole. As the framework is erected, it will tend to assume the general spherical form of the dome for which parts have been designed. Once it has assumed such form, the individual fastenings are tightened, compressing the coil springs 29 to the desired extent. If the fastenings are tightened to the extent which

FIGURE 9

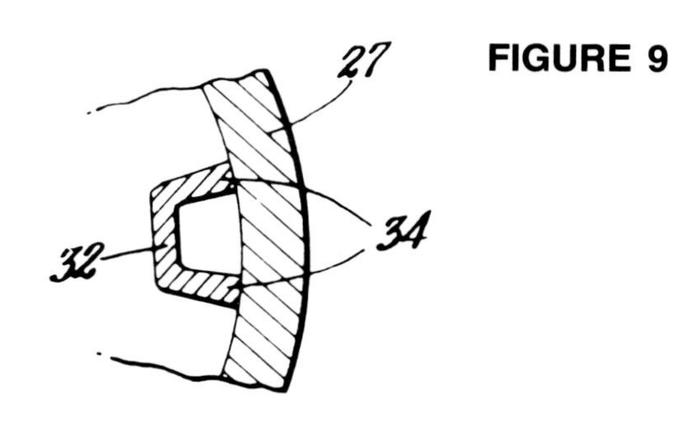

FIGURE 10

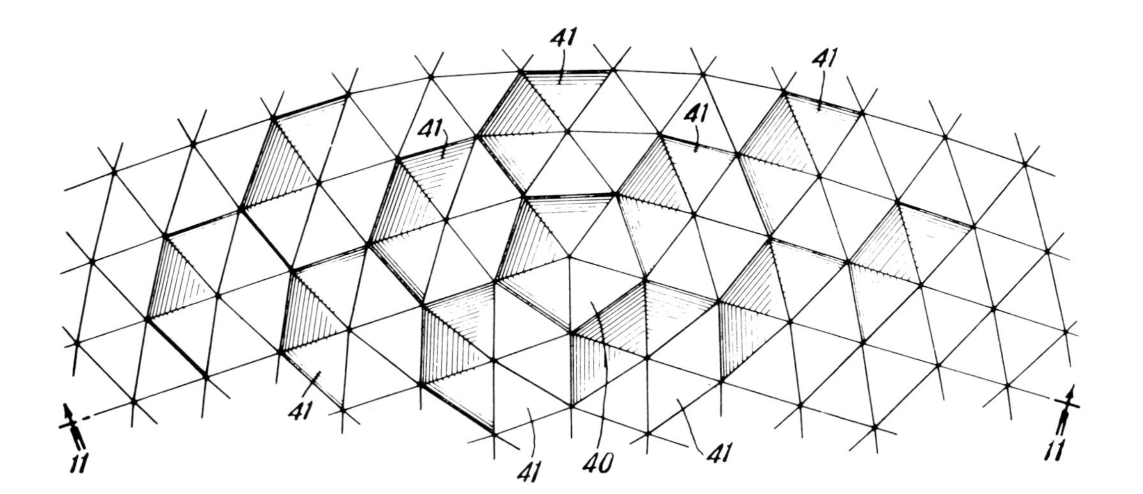

FIGURE 11

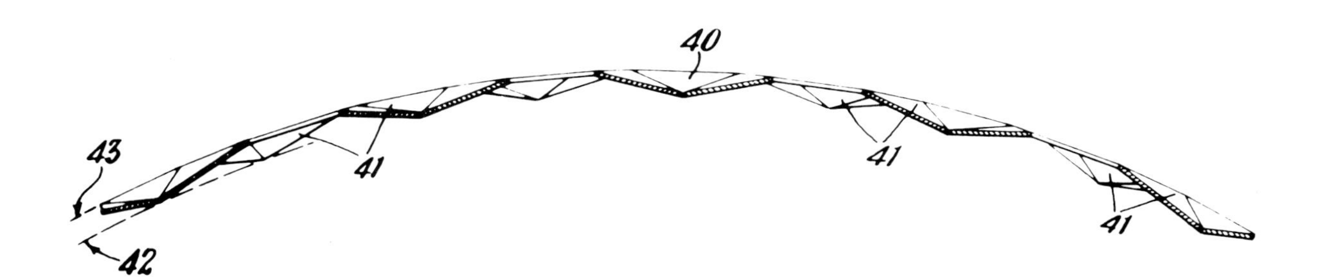

FIGURE 12

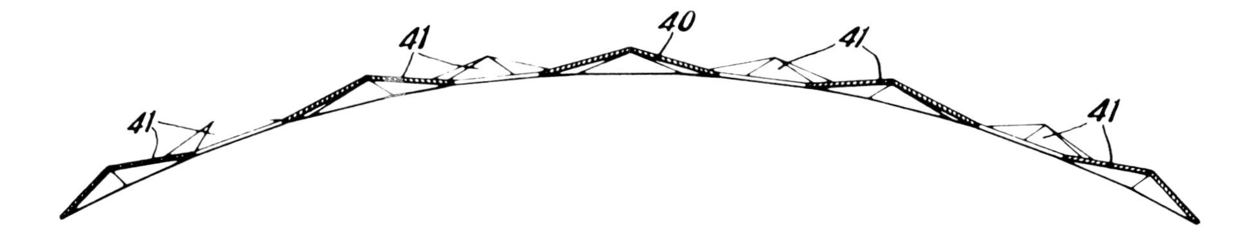

FIGURE 13

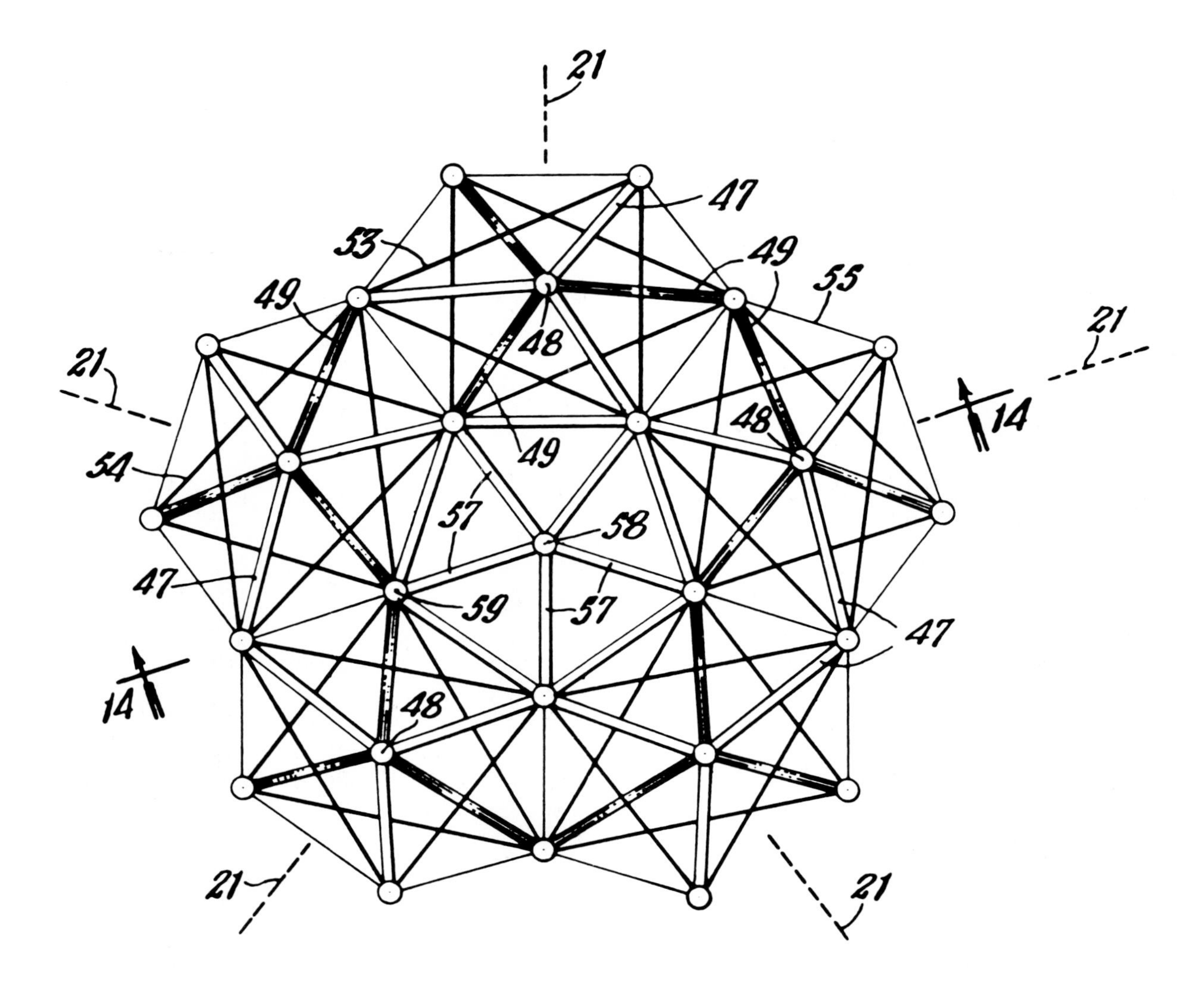

compresses springs 29 until they are driven solid, maximum rigidity is obtained. However, if greater flexibility is desired in the completed structure, bolts 28 will be tightened to a lesser extent, in which case the springs 28 will not be driven solid. Suitable lock nuts or lock washers may be used to hold the parts in the desired final adjustment.

In Fig. 7, bolt 28 is provided with an eye 37 at its inner end which is useful in attaching the plastic skin inside of the framework. Bolt 28 passes through openings 38, 39 in the outer and inner parts 25, 27 respectively of fastening 24.

Reference is now made to the modified construction illustrated in Figs. 10 and 11. Fig. 10 represents a portion of the framework similar to that shown in Fig. 5. However, instead of showing the framework itself, the planes of the equilateral triangles formed by the struts of the framework are shown as though they were triangular panels (instead of spaces outlined by the struts). This has been done to permit use of shading in such a way as to pictorialize the "dimpled" surfaces of this particular framework. The "dimples" are formed by inwardly offsetting the vertexes of the pentagons 40 and adjoining hexagons 41 to form what I term an "involuted" truss-like structure. This places all the inwardly offset vertexes substantially in a spherical surface 42 concentric with the main spherical surface 43. The main spherical surface is defined by the ends of the struts of the bases of the pentagons 40 and hexagons 41. The resulting structure is like a spherical truss defining inner and outer substantially spherical surfaces of concentric spheres. This framework based on two spheres is somewhat stiffer and less resilient than the framework of Figs. 1 and 2 based on a single sphere, and I consider the former best suited for the construction of domes in sizes ranging between approximately 50 and 140 feet in diameter. The struts which extend inwardly to the vertexes or points of the dimples are made somewhat longer than they would be in the single sphere construction so that, upon assembly, formation of the dimples is an inherent function of the lengths of the respective struts.

In the further modification illustrated by Fig. 12, the dimples are inverted. This framework, and the framework of Figs. 10 and 11, can be made of the same kind of struts described with reference to Figs. 5 and 6, and can be put together with the same type of fastening described

FIGURE 13a

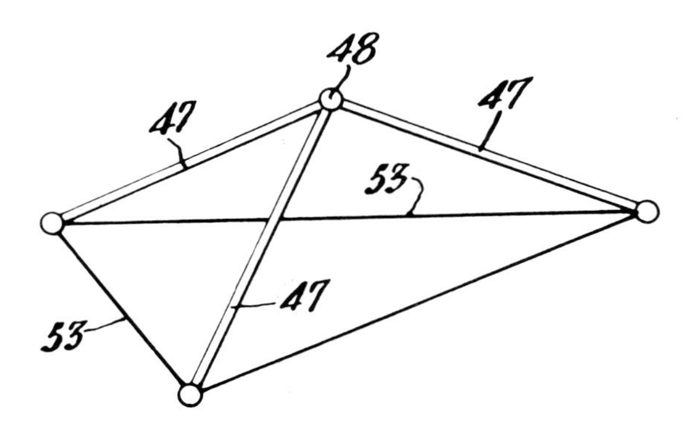

FIGURE 13b

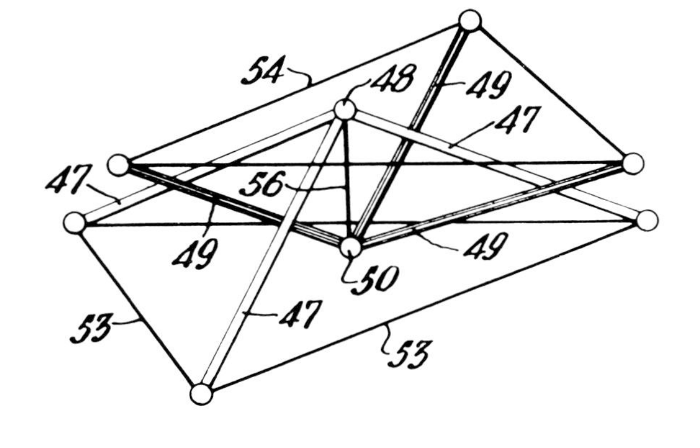

with reference to Figs. 7 and 8, although if desired other forms of struts and fastenings can be used within the limits of the appended claims. It will be observed that the fastenings of Figs. 7 and 8 allow for offsetting of the selected vertexes. (Note the clearance at 44.) Note also that in these modified constructions the struts which extend to the offset vertexes, while no longer lying substantially in a spherical surface, still are aligned with great circles of a common sphere; and such struts still lie substantially in great circle planes whose intersections with a common sphere form grids comprising equilateral triangles. These inwardly extending struts also are chords of great circles of the framework.

Reference is now made to another embodiment of the invention as illustrated in Figs. 13 and 14. This is a variant of the frameworks of Figs. 10–12; like them, it is based on two spheres. However, the framework of Figs. 13 and 14 is more complex, and comprises a truss formed of compression and tension members. I consider this type of framework best suited for the construction of domes in sizes from approximately 140 feet in diameter and upward. Fig. 13 covers a small area of the framework centering about the pentagon at the midpoint of an icosacap, i.e., an area corresponding to the central portion of Fig. 5 (except of course, that this is a different type of framework than that shown in Fig. 5). The framework is made up of struts similar to the struts 23 described with reference to the framework of Figs. 1, 2, 5, and 6, these struts being connected together by fastenings which may be similar to those described with reference to Figs. 7–9. The framework may be considered as made up of a series of tripods, one of which is shown in Fig. 13*a,* consisting of three struts 47 joined at the center of the tripod by fastening 48. This particular tripod may be described as an outwardly pointing tripod. Its central vertex as represented by fastening 48 lies in the main, or outer, spherical surface 51 and its base lies in spherical surface 52 concentric with surface 51. Arranged in complementary fashion to the outwardly pointing tripods are inwardly pointing tripods made up of three struts 49 joined together by fastening 50. Its central vertex as represented by fastening 50 lies in the inner spherical surface 52 and its base lies in the outer spherical surface 51. Two such complementary tripods are shown in Fig. 13*b*. The feet of the outwardly pointing tripods are joined together by tension members 53 which may be made of wires or cables. The feet of the inwardly pointing tripods are connected by similar tension members 54.

FIGURE 14

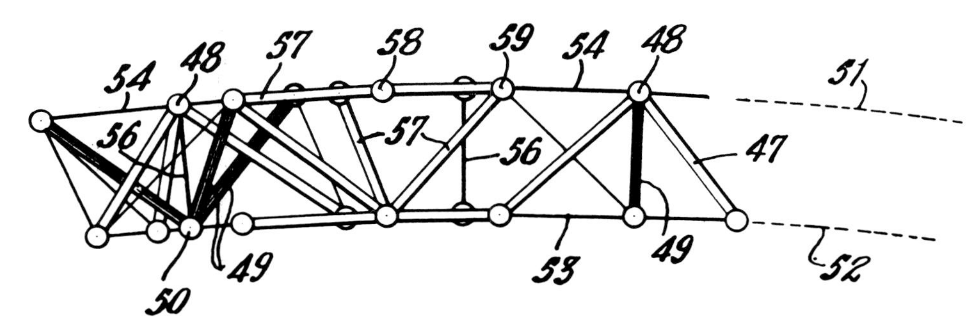

In Figs. 13 and 14, the struts 47 of all of the outwardly pointing tripods have been shown without any surface shading so that they appear light in the drawing, whereas the struts 49 of all the inwardly pointing tripods have been shown with surface shading so that they appear dark in the drawing. Thus, the "light" tripods are disposed with their vertexes in spherical surface 51 and the "dark" are disposed with their vertexes in spherical surface 52. Tension members 55 extend diagonally between the respective feet of the light and dark tripods. These tension members, as viewed in plan in Fig. 13, present a hexagonal outline, alternate corners of which are connected by the aforesaid tension members 53 and 54, tension members 53 forming a triangle made up of chords of spherical surface 52 and tension members 54 forming a triangle made up of chords of spherical surface 51. The resultant basic pattern of the outer spherical icosahedron in surface 51 is the same as that illustrated in Fig. 5. The same is true with respect to the resultant basic pattern of the inner spherical icosahedron in surface 52. In effect, therefore, we have here two concentric spherical icosahedrons joined by diagonal struts and tension members. The framework is tightened into a final rigid structure by means of tension members 56 extending radially with respect to spherical surfaces 51 and 52 between the fastenings 48 and 50 at the apexes of the light and dark tripods respectively. If desired, turnbuckles may be used in these tension members to secure the desired final tension to hold the structure with the proper degree of rigidity.

At the vertexes of the icosacap, the framework assumes a pentagonal form as clearly shown at the center of Fig. 13. At such points in the structure we have an inwardly (or outwardly) pointing pentagonal strut arrangement in place of the two complementary tripods which characterize the rest of the framework where the pattern is hexagonal. I prefer to bridge over the outer side, or

FIGURE 15

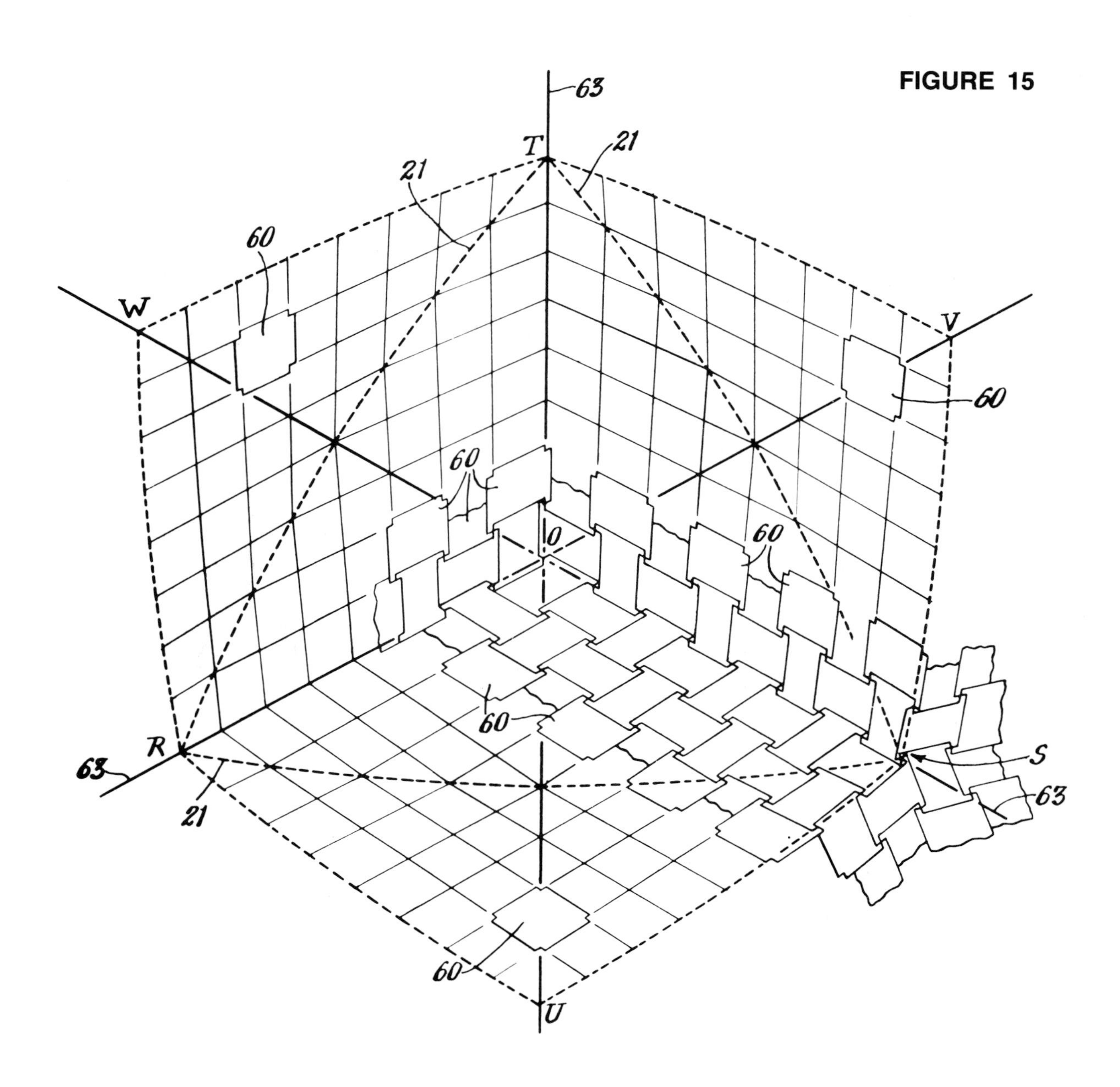

base, of the pentagonal strut arrangement at each vertex. In the specific framework shown, this bridging consists of five struts 57 joined together by fastenings 58 at the vertexes of the spherical icosahedron, and joined by fastenings 59 to the feet of the light and dark tripods immediately adjoining the respective pentagons.

In all of the forms of framework I have described, the lengths of the individual struts are substantially equal, but not precisely so. The slight differences in the lengths of different struts in a given framework determine the radius of the dome and whether it is based on one or two spheres. The number of different lengths of strut in any given framework based on a spherical icosahedron varies in accordance with the number of units, or modules, into which the edges of the spherical icosahedron are divided, i.e., in accordance with what I have previously termed the "frequency" of the three-way great circle grids. I have found that with a frequency of 16, as described in connection with the dome illustrated in Figs. 1, 2, 5 and 6, all conditions of the framework design are satisfied with 56 different lengths of strut. The same framework when built of grids having a frequency of 8 can be constructed from struts in 16 different lengths. With a frequency of 4, only 5 different lengths would be used. I have found, further, that there need never be any greater complication as to number of lengths of struts than that represented by a frequency of 16.

The slight differences between the lengths of the individual struts in turn create slight differences between the angles of the substantially equilateral triangles and this has the result of forming a spherical grid structure in which all the main structural members are in geodesic alignment or are chords of great circles of a common sphere. One way of determining the strut lengths is to construct a paperboard hemisphere to a scale of, say, 1 inch to 1 foot, and lay out the vertexes of one of the faces of a spherical icosahedron on its surface. These vertexes are next connected by drawing great circle lines (spherical straight lines) therebetween. The edges of the triangle defined by these lines are next divided equally into the number of units represented by the selected grid frequency. The division points are then connected by drawing great circle lines in the manner clearly shown in Figs. 1 and 2. (Note that the points along one edge are connected to every second point on another edge.) We now have a completed three-way grid pattern. Finally the length of the chordal struts is measured directly with the use of ordinary draftsman's dividers, allowance being made for the strut fastenings.

Figs. 15 to 18 inclusive illustrate another embodiment of my invention in which the main structural elements of the framework consist of interconnected sheets 60 of metal, plastic or other suitable material. The longitudinal centerlines (17—17, Fig. 16) of these sheets lie substantially in great-circle planes whose intersections with a common sphere form grids comprising substantially equilateral triangles. As shown the sheets are substantially in

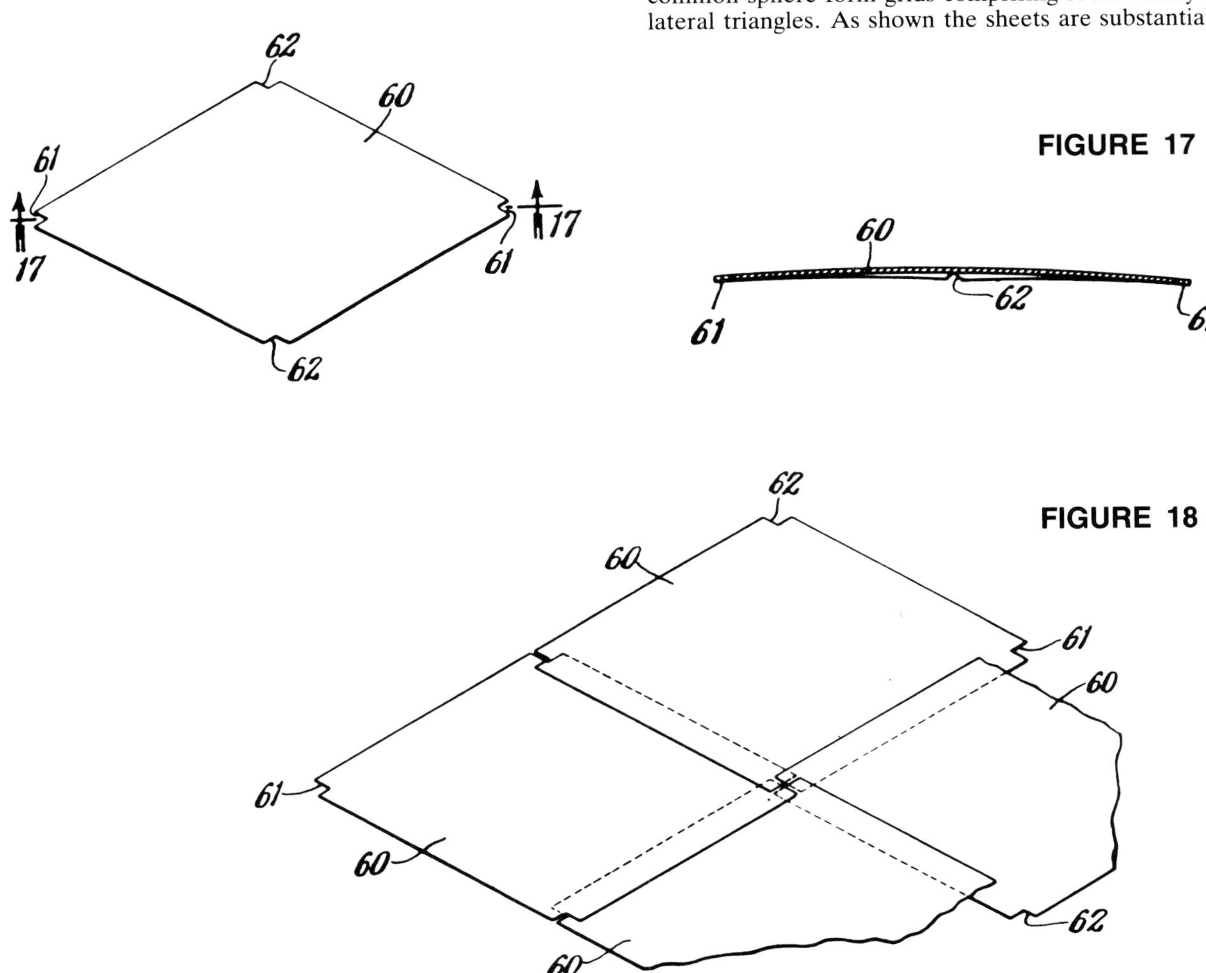

the form of equilateral diamonds whose minor axes are approximately equal in length to the sides. The corners of the sheets 60 are notched for interlocking engagement with the notches of adjacent sheets. The corners of the notches 61 and 62 lie substantially in great circle planes whose intersections with a common sphere form grids of substantially equilateral spherical triangles. Thus the sheets 60, like the struts 23 of the framework illustrated in Figs. 1 and 2, are in geodesic alignment. The alignment is such that the longitudinal centerlines of the sheets (and also their edges) are arranged in geodesic lines. Thus these sheets create the same sort of three-way grid pattern as I have described with reference to the several forms of framework in which struts are employed.

Here again, all the main structural elements are of almost the same size, the variation being determinable mathematically or by graphic solution as before. The frequency of the grids is a matter for selection in accordance with the special requirements of particular structures. As with the struts, the frequency will determine the number of different diamond sizes to be used in a given framework design. With a frequency of 16, for example, there will be 20 sizes or types of diamond per sphere. With hexagonal sheets on the same three-way grids, and with a frequency of 16, there will be 10 types per sphere, consisting of 9 types of approximately hexagonal sheets and 1 pentagonal sheet. Other forms and arrangements are possible.

Particular attention is directed to the manner in which the three-way grid pattern is built up in this form of my framework. Fig. 15 shows one complete face, or spherical triangle RST, of the spherical icosahedron, plus one-third of each adjoining face of the same, namely the additional areas RUS, SVT and TWR, or the total area RUSVTW. Geodesic lines 63, 63, 63 extend from each vertex of RST through the mid-point of the opposite side. SV, VT, TW, etc., are corresponding geodesic lines of the adjoining faces of the spherical icosahedron. Within area RUSO all of the sheets 60 are arranged with their longitudinal centerlines extending in one general direction. The same is true within areas SVTO and TWRO, except that in each case the general direction is different. Along lines RO, SO and TO, the sheets of the respective adjoining areas come together at an angle approximately equal to one of the spherical angles of spherical triangle RST. This can best be understood by noting the diamond patterns of the construction lines where they extend beyond the area covered by the sheets 60. Note that at the vertexes of the icosacaps (as at S for example), five sheets 61 toe in to a common point. Elsewhere throughout the framework as shown in Fig. 18 four sheets toe in to a common point, except at the center O of the triangle RST where only three sheets toe in at a common point. Thus there are five sheets around each of the vertexes R, S and T, three sheets around centers O, U, V and W and four sheets around all intermediate points.

With this general type of construction, I have discovered the possibility of making all the sheets exactly identical in overall size, the variation in type being secured by varying the sizes or depths of the notches 61 and 62. If the overlapping edges of adjacent sheets are riveted together, the holes for the rivets will be drilled on slightly different patterns to suit the different types and keep the fastenings in geodesic alignment. Thus all the sheets are sheared out to one size, and the manufacture of the different types for a particular sphere is completed by using adjustable jigs (or a series of different jigs) for the notching and/or drilling or punching tools. This greatly simplifies manufacture.

I prefer to form or press the sheets to a compound curvature conforming to the surface of the spherical icosahedron on which they are based.

Domes constructed in accordance with Figs. 15–18 may be erected by first assembling on the ground those sheets which are at the vertex of an icosacap, namely at that vertex which will be uppermost in the completed framework. Then, working around peripherally, additional sheets are interlocked and/or riveted together, raising the partially completed dome as the work progreses.

It is possible to begin interlocking the sheets in either a clockwise overlapping relationship, or in a counterclockwise overlapping relationship. By a clockwise overlapping relationship, I mean that at any given point where a group of sheets toe in to a common vertex, the edge of each successive sheet of the group is on top of the preceding sheet as we move around the vertex in a clockwise direction. By a counterclockwise overlapping relationship, I mean that at any given point where a group of sheets toe in to a common vertex, the edge of each successive sheet of the group is on top of the preceding sheet as we move around in a counterclockwise direction. This imposes what I term a "turbining" action in the framework, and the turbining action will be either clockwise or counterclockwise according as the overlapping relationship is either clockwise or counterclockwise. These turbining actions produce a highly effective locking action in the framework as a whole.

Geodesic locking bands or cables may be tensioned over the completed dome and anchored to a suitable foundation.

Geodesic frameworks constructed in accordance with my invention, if made of struts universally jointed at the vertexes of the triangles can be folded into a compact bundle without taking apart any but the final locking elements. This form of my invention is ideally suited for use as temporary shelters which are to be moved from place to place, such as huts, hangars, mess-halls, and headquarters units for army encampments.

The frameworks may be covered with plastic skins, inside or outside or both, or with other materials. Openings for access, light, sun and air are provided as desired.

The terms and expressions which I have employed are used in a descriptive and not a limiting sense, and I have no intention of excluding such equivalents of the invention described, or of portions thereof, as fall within the purview of the claims.

11▲PAPERBOARD DOME (1959)

U.S. PATENT—2,881,717

APPLICATION—JANUARY 24, 1955

SERIAL NO.—483,619

PATENTED—APRIL 14, 1959

BECAUSE OF THE TRIANGULATION of the geodesic dome, double-fluted corrugated paperboard proved to be particularly suitable for producing geodesic domes of thirty feet and under. Double-fluted corrugated paperboard is made passing through rollers, and because of this, the rollers can print the folding patterns for the paperboard dome. There's no way in which man can mass-produce so rapidly as with the printing press. The corrugated paperboard domes are produced at many universities, having first started with Yale University in 1951. They have to be covered with waterproof material, as the paperboard itself is hydroscopic. At the Massachusetts Institute of Technology, I worked with paperboard domes with the Forest Products laboratories and in Wisconsin with the Paper Industry Research Laboratories. We found it would be possible to introduce a compression-providing chemistry at the "beater" stage of the paper manufacturing. The corrugated paperboard used by the armed forces in World War II had an extremely high wet tensile strength, so that cartons could be dropped off on beaches, become very greatly wettened, and not lose their tensile coherence. What they did not have was high wet compression strength. Because they were bound together with steel straps and so forth, the lack of compression stiffness was irrelevant. It is, however, of complete relevance in producing dome structures.

The big paper-manufacturing factories had so much business that they could not afford to stop any of their projects and introduce the stiffening agent at the beater stage, so we've had to postpone that through the years. However, the high tensile strength of the paperboard made it extremely suitable for dome building, particularly when we have inwardly-turned great circle planes planted surrounding the triangles.

Many paperboard domes were built; one of the most notable was done by a New York City Lower East Side gang who covered a twenty-foot-diameter dome with polyethylene sheets of very low cost, and then with aluminum foil, and then with a complete enclosure of half-inch-mesh chicken wire. Then they started cementing the dome outside, troweling the cement onto the chicken wire mesh, and went progressively around the dome with fast-setting cement. However, they waited until the second day before they began to trowel in the higher ranges of the dome, and finally on the third day, they were able to cement in the top, which gave them a very extraordinary fireproof structure, beautifully insulated on the inside, with a complete moisture barrier, and the aluminum foil acted as a radiation barrier reflecting away the unwanted Sun heat in the summer, and retaining the generated heat inside in the winter.

UNITED STATES PATENT OFFICE

Richard Buckminster Fuller, Forest Hills, N.Y.

BUILDING CONSTRUCTION

My invention relates to a framework for enclosing space, and to building construction members for use in such a framework and for other uses.

SUMMARY

Frameworks used in buildings, and for roofs and shelters of all kinds, commonly are constructed of steel, aluminum, wood and the like, and the design of such frameworks is based upon long familiar principles in which relatively heavy compression members are employed in columns, beams and girders. Because of columnar, or compression, loading of such members, they must in general be heavy and stiff to resist buckling under compression. As a result, the ratio of weight to strength is high. In times past, efforts to gain a more favorable weight-strength ratio have brought about the development of the I-beam and the lattice girder; and in more recent times the continuance of such efforts has led to the development of light metals and their strong alloys for use in place of steel where cost does not preclude availing of such lighter weight but more expensive substitutes. Apparently those concerned with the development of frameworks for building construction have not perceived any practicable means to eliminate the use of materials and columnar forms which have high compressive strength. However, I have discovered how to make a building construction member which possesses design characteristics such that the compressive strength of component portions thereof is of relatively small, and perhaps almost negligible, concern. It has been a primary object of my invention to achieve such design characteristics in a general purpose building construction member.

I have found further that, by following the design features disclosed herein, it becomes thoroughly practicable to construct building frames of paperboard and the like. So, in popular phraseology, one might say that I have discovered how to make "cardboard" building members and frames—members possessing such surprising strength characteristics as to permit the use of cardboard in lieu of wood, aluminum, steel or other materials which, according to normal concepts of strength of materials, are not at all to be associated with such flimsy stuff as cardboard.

According to my invention there is provided a building construction member in the form of a polygonal frame, such as a triangle, having hollow sides perferably triangular in cross-section and centrally reinforced. This member is formed from a single flat piece of paperboard or the like comprising a series of elongated panels arranged end to end with transverse fold lines between the panels, and lateral extensions adjoining the panels along longitudinal fold lines and including fold lines spaced from and substantially parallel to the longitudinal fold lines. The lateral extensions fold about both the longitudinal and parallel fold lines to form a hollow box-like structure which in turn folds about the transverse fold lines between the panels to bring the outer ends of the structure together to form the triangular, or other polygonal, frame.

The ends of the lateral extensions are formed with flaps infolding along diagonal lines to form abutting mitered surfaces at the ends of the hollow sides of the polygonal frame. An important feature of my construction resides in gauging the angle of these diagonal fold lines to the number of sides of the polygon to "crowd" the folded lateral extensions within the perimeter formed by the interconnected panels, and set up the mitered joints of the polygon under a degree of pressure for increasing the unitary strength of the frame as set up for use. The elongated panels forming the perimeter of the frame may thus be placed under a degree of tension, tending toward a uniform distribution of stresses throughout the frame, with utilization to a marked extent of the tensile strength of the paperboard in what may be regarded as a skin-stressed frame member.

For further maximum realization of tensile properties as well as the compressive strength of the box-form frame portions, at least the outer surfaces of the polygonal frame are coated with a thin plastic film of high tensile strength, such as a vinyl resin which can be brushed, sprayed or otherwise applied to the completed frame members, or which can be applied to one or both sides of the paperboard from which the blanks for the frame members are cut. Extrusion coating can be used to apply the plastic reinforcement to the paperboard sheets, or the paperboard can be impregnated with the plastic. I have found, for example, that a triangular frame member formed in accordance with my invention, when brush painted with a thin film of vinyl plastic acquires astonishing strength characteristics. Thus, such a frame member having nearly equal sides averaging about 19.4 inches in length, the cross-section of the sides being of hollow triangular form with a base width of about 5 inches, and an altitude of about 1.5 inches, is of such strength that when placed with one side of the triangle on the floor it is possible for a 200 pound man to step up on the opposite point of the triangle and place his full weight on it without the slightest visible evidence of distortion or buckling of any of the sides of the frame member. The paperboard used may consist of the ordinary kraft board used in standard forms of shipping containers, and known as corrugated fibreboard. In the example just given, I employed a double-faced sheet in which the smooth outer sheets, known in the paperboard box business as "liners" consisted of a 38 lb. kraft board (38 lbs. per 1000 square feet) made from sulphate virgin pulp and glued to a 26 lb. kraft corrugated member having approximately 55 flutes per lineal foot (known in the trade as a type "B" flute), the board having a thickness of about 0.100″ (100 point).

The many desirable properties of my new building construction members are realized most fully when these members are assembled in such a manner as to form a building framework constructed in accordance with my prior Patent No. 2,682,235, issued June 29, 1954. For this purpose the building members described herein may be made as frames having three, four, five or six sides corresponding to the triangles, diamonds, pentagons and hexagons described in said patent. Alternatively, the diamonds may be made from two triangular frames secured together back to back, and the pentagons and hexagons can be made, respectively, from five and six triangular frames secured together in rosette fashion. In such constructions, each component member reinforces and lends strength to the structure as a whole, and the tensile strength of the extremely thin paperboard and plastic sections is utilized to tremendous advantage, and to an extent not realized in the conventional forms of structures heretofore known or used.

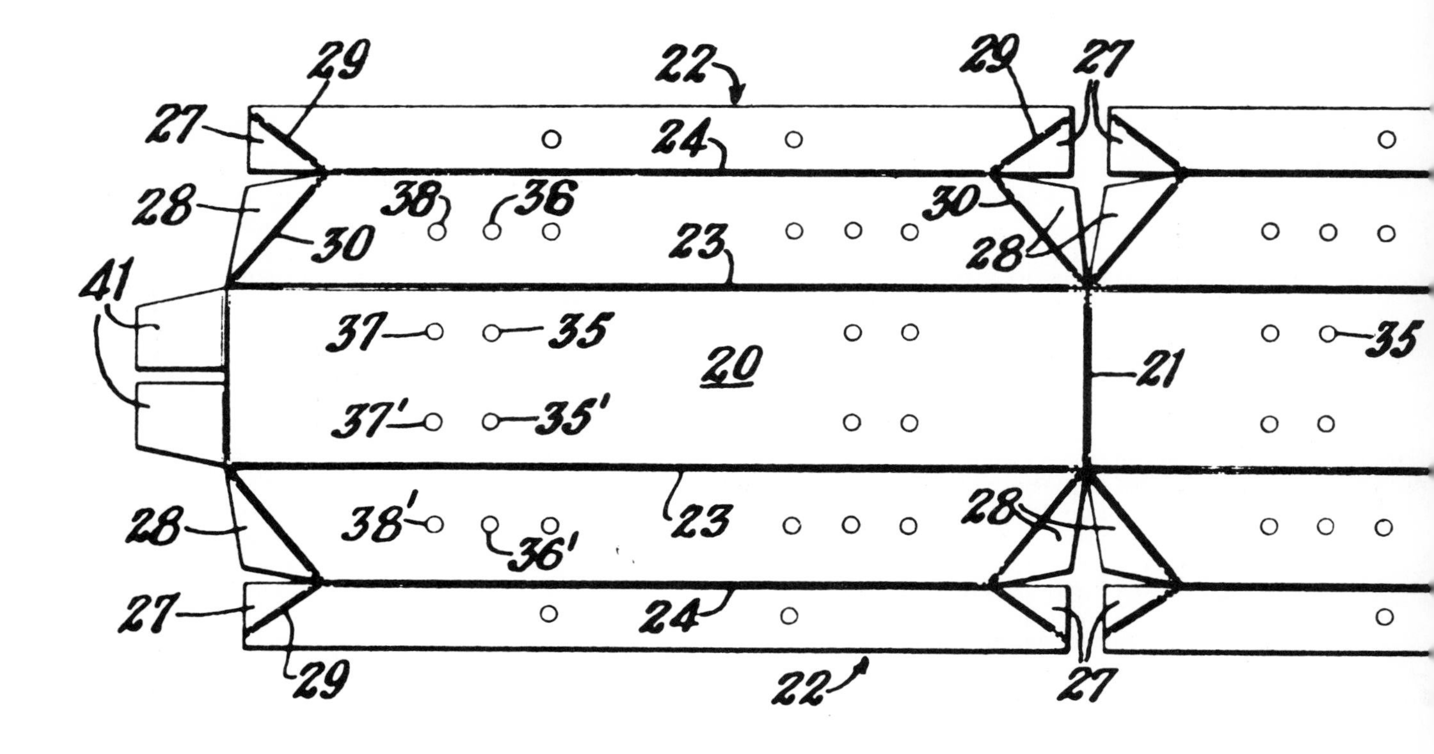

These and other features and advantages of my invention will appear more fully in the detailed description which follows.

Description

In the drawings:

Fig. 1 is a face view of a paperboard blank designed to be folded into a triangular frame with hollow sides, representative of the best mode contemplated by me of carrying out my invention.

Fig. 2 is an enlarged cross-sectional view along the line 2—2 of Fig. 1.

Fig. 3 is a view similar to Fig. 2, illustrating the first part of the operation of folding the blank.

Fig. 4 is a similar view illustrating a more advanced stage of the folding operation.

Fig. 5 is a similar view illustrating the form of the end of one of the hollow sides of the polygonal frame at the completion of the folding operation.

Fig. 6 is a face view of the completed triangular frame.

Fig. 7 is an enlarged cross-sectional view taken on the line 7—7 of Fig. 6.

Fig. 8 is a view illustrating one way in which triangular frames formed in the manner illustrated in the preceding views can be assembled as part of the frame of a building.

Fig. 9 is an enlarged cross-sectional view taken as indicated at 9—9 in Fig. 8.

Fig. 10 is a side elevational view of one of the component frames of Fig. 9, taken as indicated at 10—10 in Fig. 9, with the alignment pins removed.

Fig. 11 is a view similar to Fig. 9, illustrating another alignment of the component frames.

Figure 2

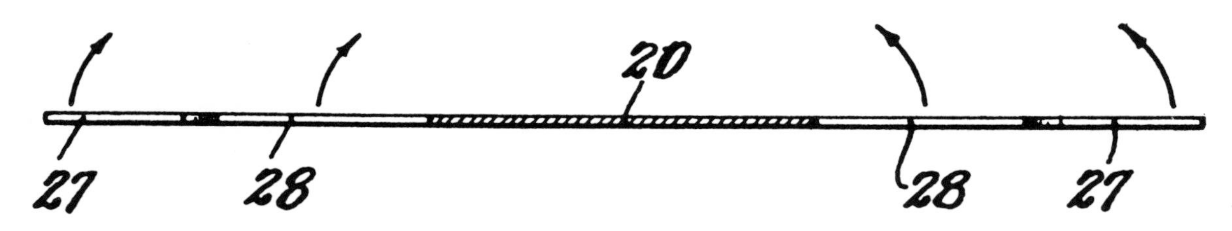

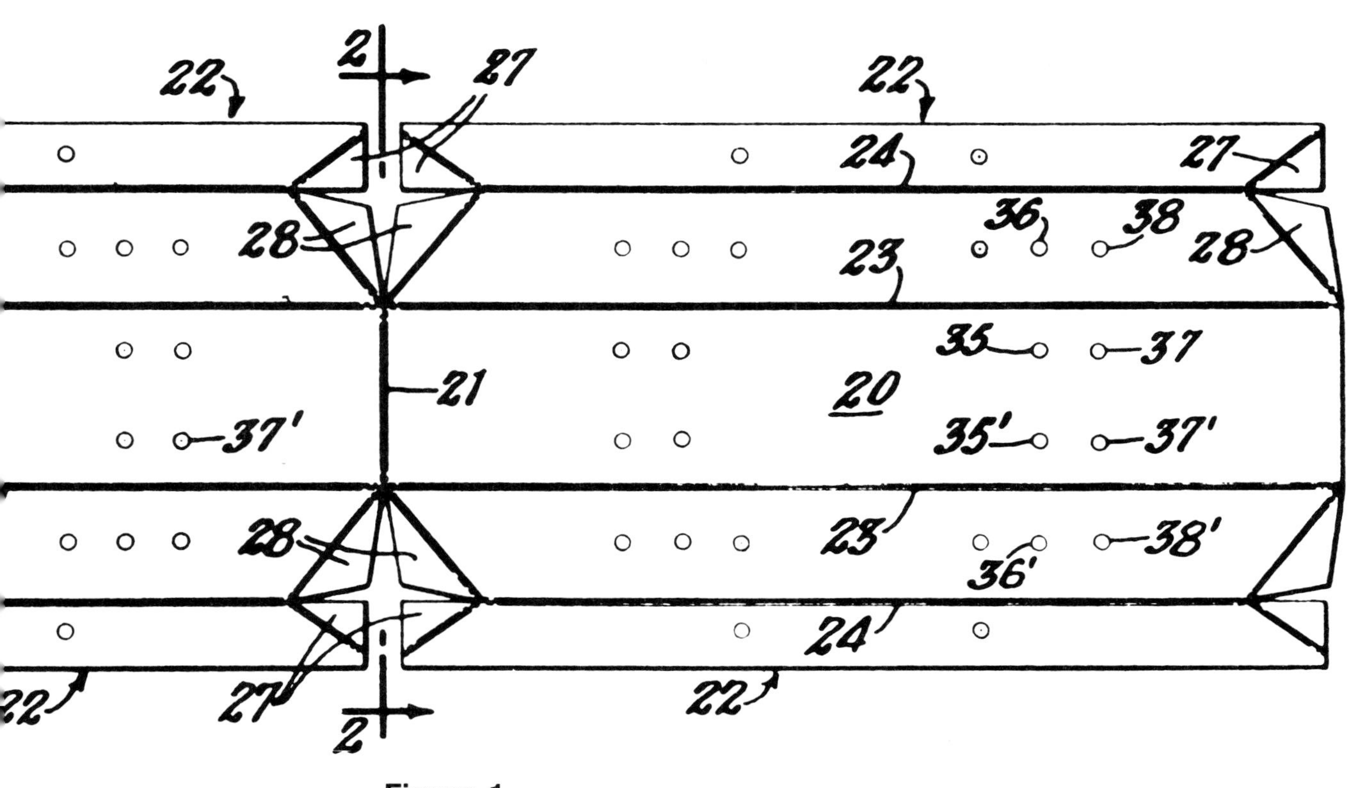

Figure 1

Fig. 12 is a side elevational view of one of the component frames of Fig. 11, taken as indicated at 12—12 in Fig. 11, with the alignment pins removed.

Fig. 13 is a view of a portion of a frame similar to that shown in Fig. 6, and having a panel closing the central opening thereof.

Fig. 14 is an enlarged cross-sectional view taken as indicated at 14—14 in Fig. 13.

Fig. 15 is a face view of a paperboard blank of a modified construction.

Fig. 16 is a face view of a building construction member formed from the blank of Fig. 15.

Fig. 17 is an enlarged cross-sectional view taken on the line 17—17 of Fig. 16.

Fig. 18 is a face view of a building construction member having four sides instead of three, formed from a blank similar to that shown in Fig. 1 or that shown in Fig. 15.

Referring to Fig. 1, my invention comprises, in its general arrangement, a building construction member

Figure 3

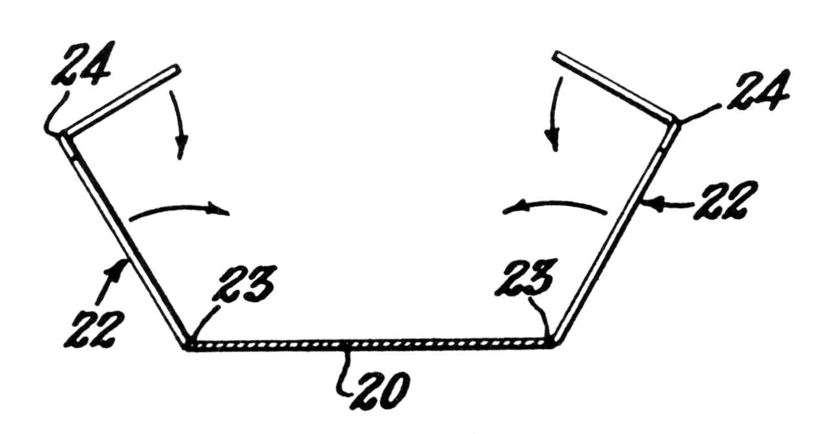

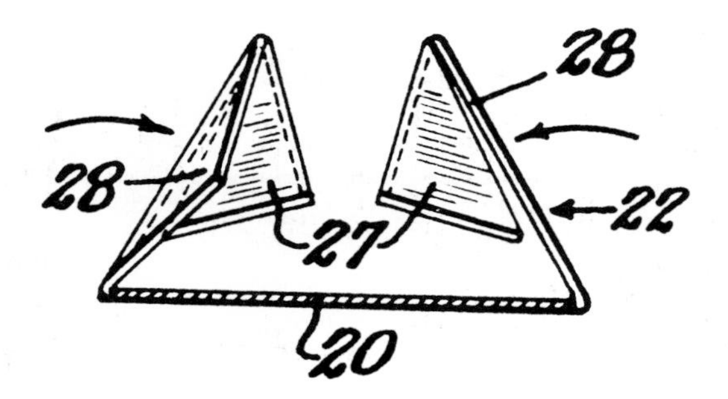

Figure 4

formed essentially from a single flat piece of paperboard or the like, comprising a series of elongated panels 20 arranged end to end, transverse fold lines 21 between the meeting ends of said panels, and lateral extensions 22 adjoining the elongated panels along longitudinal fold lines 23 and including fold lines 24 spaced from and substantially parallel to the fold lines 23, the lateral extensions folding about the longitudinal and parallel fold lines (Figs. 3, 4 and 5) to form a hollow box-like structure 25 (Fig. 7) which in turn folds about the transverse fold lines 21 to bring the outer ends thereof together to form a polygonal frame 26 (Fig. 6) with hollow sides. The ends of the lateral extensions preferably are formed with flaps 27 and 28 infolding along diagonal fold lines 29 and 30 to form abutting mitered surfaces 31 at the ends of the hollow sides of the triangular, or other polygonal, frame. As shown, these flaps are formed in pairs with the respective members of each pair located at opposite sides of the aforesaid parallel fold lines 24.

The angle of the diagonal fold lines 29 and 30 will depend upon the number of sides there are to be in the polygonal frame and, more specifically, upon the vertex angle, or angles, between each two sides, and the angle of these diagonal fold lines is additionally gauged (after taking account of the number of sides, i.e. the vertex angles) to "crowd" the folded lateral extensions 22 within the perimeter panels 20, 20, etc., of the polygon and set up the mitered joints of the polygon under a degree of pressure. This has the result of locking the several components of the frame together for maximum strength as a unit, and tends to distribute stresses encountered by any one part of the frame to the other parts thereof. This may be due in part at least to the fast that the perimeter panels 20, 20, etc., are placed under a degree of tension. Whatever the cause, I have found that building construction members made as I have described, possess astonishing strength, particularly when they are assembled in a generally spherical form in such a manner that the sides of the polygons lie substantially in planes whose intersections with the spherical surface define great circles of that surface. Means are provided for securing together the outer ends of the hollow box-like structure. In the preferred construction shown, a pair of end flaps (Fig. 1) 41 are formed as extensions of one of the perimeter panels 20 to be tucked into openings 42 (Fig. 5, and cf. also Fig. 6). Adhesive tapes or other securing means (not shown) may also be applied around the meeting ends of the structure.

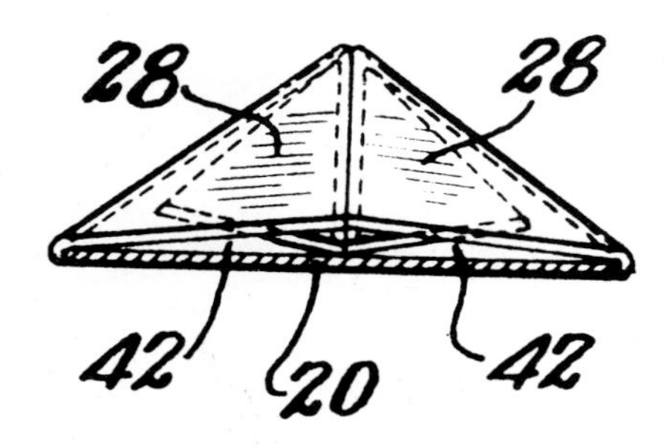

Figure 5

In my preferred construction, additional strength is imparted by providing the lateral extensions 22 at both sides of the elongated panels 20. These extensions are folded first diagonally inwardly as shown at 32 in Fig. 7, and

Figure 6

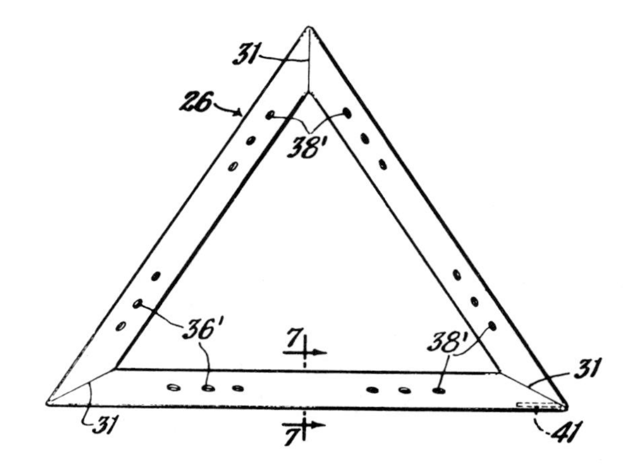

then outwardly to form a central reinforcement 33, 33 comprised by the outwardly folded portions of the lateral extensions. These outwardly folded portions bear at their ends against the inside of the elongated panels 20 and divide the box-like structure into two triangular compartments. If desired, a locking pin or pins 43 may be provided, extending through suitable openings formed in the portions 32 and 33 of the lateral extensions.

The manner in which the lateral extensions 22 and their infolding flaps 27, 28 are folded to form the hollow box-like structure is further revealed in Figs. 2, 3, 4 and 5. In the first three of these views, the arrows denote the directions of folding. Flaps 27 go inside of, and reinforce, flaps 28 to form the abutting mitered surfaces 31 to which reference has been made.

Figs. 13 and 14 show how a panel 34 may be used to close the center of the polygonal frame, the outer portions of this panel being received between the outwardly folded portions 33, 33 of the lateral extensions 22. By inserting panel 34 during the operation of folding the sides of the hollow box-like structure about the transverse fold lines 21, the panel is permanently locked in place. This panel further serves to reinforce the construction, and it may be made of any desired material such as paperboard, beaver board, plastic, aluminum foil, etc., and may be opaque, translucent, or transparent in accordance with the functions to be served, i.e. whether for letting in, or obstructing the passage of, light, conducting, or insulating against the passage of, heat, etc.

The modified construction illustrated in Figs. 15, 16 and 17 is substantially identical with the preferred form described above, with the exception that the lateral extensions 22 are present on only one side of the elongated panels 20, and are replaced by flaps 44 which are secured to extensions 22 in the overlapping relation shown in Fig. 17, as by means of a plastic, or other, adhesive, or by means of metal staples, or both, or otherwise, as desired,

In Fig. 8 I have shown a group of six triangular frames, which may be of either of the types disclosed in Figs. 6 and 16, assembled to form a hexagon according to one of the sub-components of a building construction framework made in accordance with my prior Patent No. 2,682,235, aforesaid. Other arrangements are possible so as to form, for example, diamonds made from two triangular frames secured together back to back, or pentagons made from five triangular frames secured together in rosette fashion similar to Fig. 8 (compare Figs. 2 and 5 of my patent aforesaid, with reference to the pentagon at the zenith Z and at each junction of the vertexes of the great circle triangles of the spherical icosahedron).

Fig. 18 shows another of the many possible forms of polyhedral building construction members which can be formed in either of the manners described, i.e. by utilizing blanks of the general form of either Fig. 1 or Fig. 15. In this case one additional perimeter panel 20 is added to the blank and the lengths of the respective panels are adjusted to the desired measurements. It may be observed here that my building construction members have been found suitable not only for use as components of building frames in which the members form a permanent part of the structure, but also may be used as temporary supporting members in constructing openings such as doors, windows, etc., in erecting buildings of conventional types.

In assembling my improved construction members, various arrangements are possible for holding the members together; I have employed wooden pins, both alone and in conjunction with metal clips. Also, I have found it efficacious to bind triangular frames together in diamond, pentagon, and hexagon assemblies, by passing glass fibre tapes around the complete periphery of the diamond, pentagon or hexagon. Such glass fibre tape is available as a "Scotch" tape and is exceedingly strong in tension, thus contributing importantly to the over-all strength of the completed framework, particularly when it is constructed in accordance with my prior Patent No. 2,682,235, aforesaid. In such constructions the several components so complement one another in the particular pattern of the finished assembly as to enable it to withstand high stresses, and the framework itself acts almost as a membrane in absorbing and distributing loads.

Figure 8

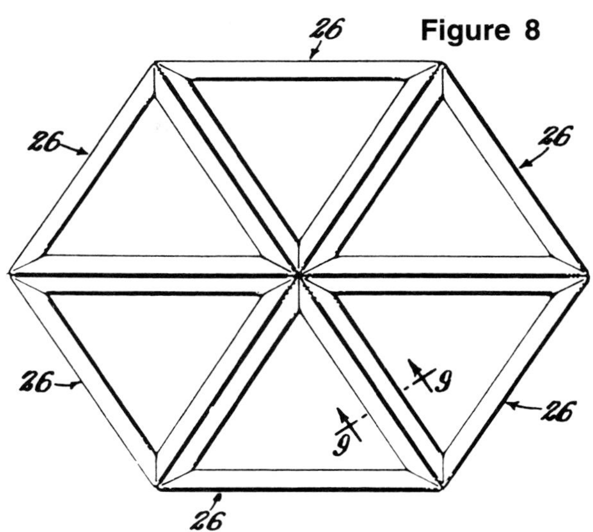

Figure 7

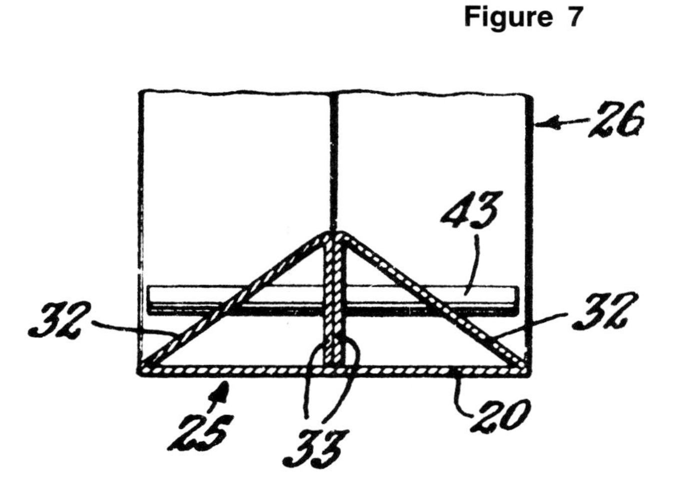

Another feature of my present invention resides in the provision of means for predetermining the correct angular relationship between each building construction member and its neighbors. This feature will be described with reference to Figs. 9 to 12 inclusive. Fig. 9 is an enlarged detail cross-sectional view taken as indicated at 9—9 in Fig. 8 and shows two triangular frame members 26 back to back, or substantially so. The construction of these frame members is the same as has been described with reference to Figs. 1 to 7 inclusive. The elongated panels 20 and lateral extensions 22 of one of the frames 26 are provided with perforations to receive alignment pins extending through similar perforations in the other polygonal frame 26 of like construction. These perforations are slightly out of alignment with the planes of the respective frames to predetermine the angular relationship of the two frames. More than one set of such perforations may be provided so that by selecting a particular set, one particular angular disposition of the adjacent frames is obtained whereas by selecting a different set of perforations, another angular disposition of the adjacent frames is obtained. Thus, by following a plan or set of specifications indicating the correct placement of the pins, structures of predetermined form are obtained without requiring knowledge or skill on the part of the workmen. I prefer to form the perforations by making cut lines in the paperboard without punching out the centers. Then, by punching out only those openings through which pins are to be passed during erection of the structure, the correct relation of the several components will be obtained without even the need to examine the specifications on the job.

Referring further to Figs. 9 to 12 inclusive, it will be observed that I have provided a building construction

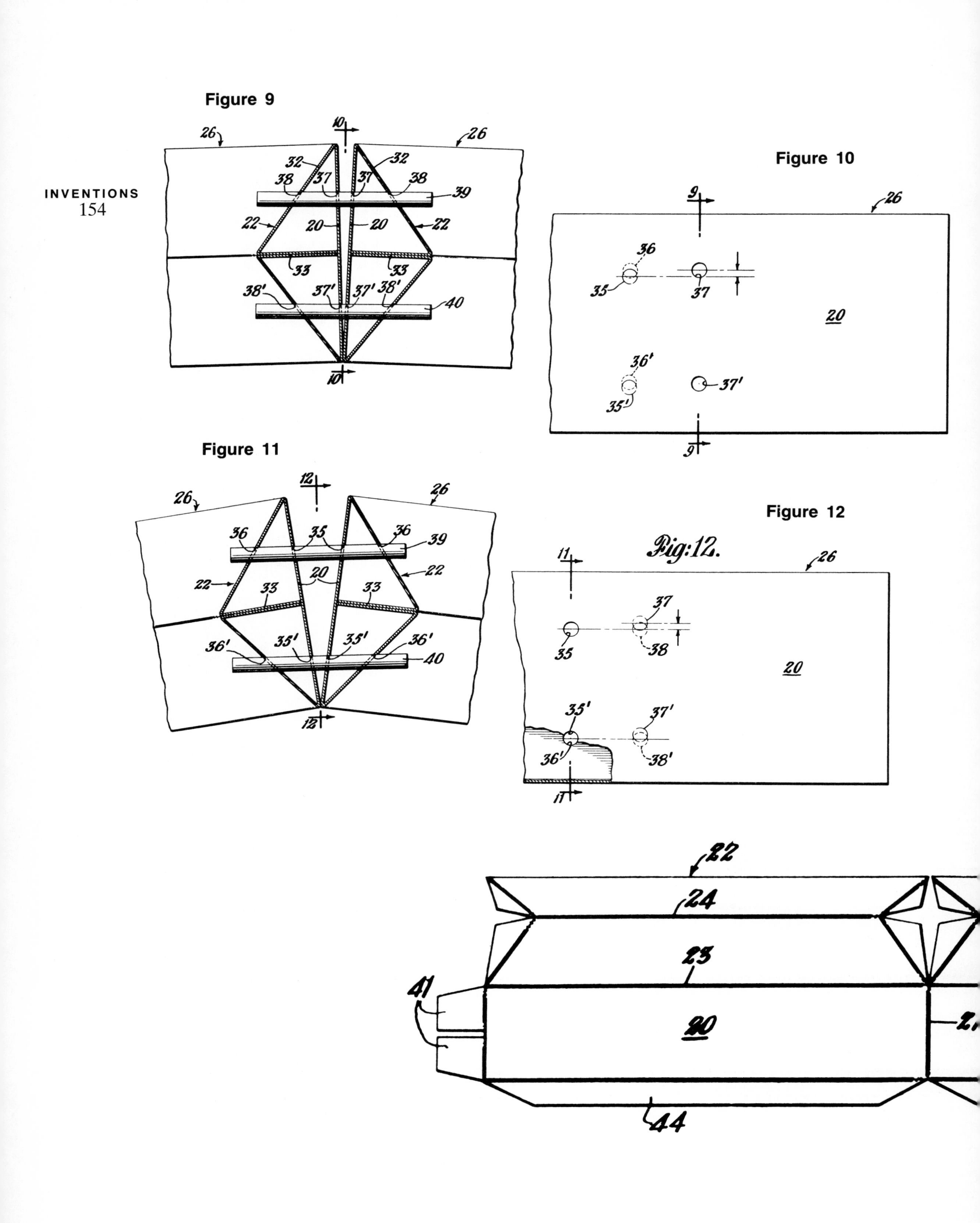
Figure 9
Figure 10
Figure 11
Figure 12
Fig:12.

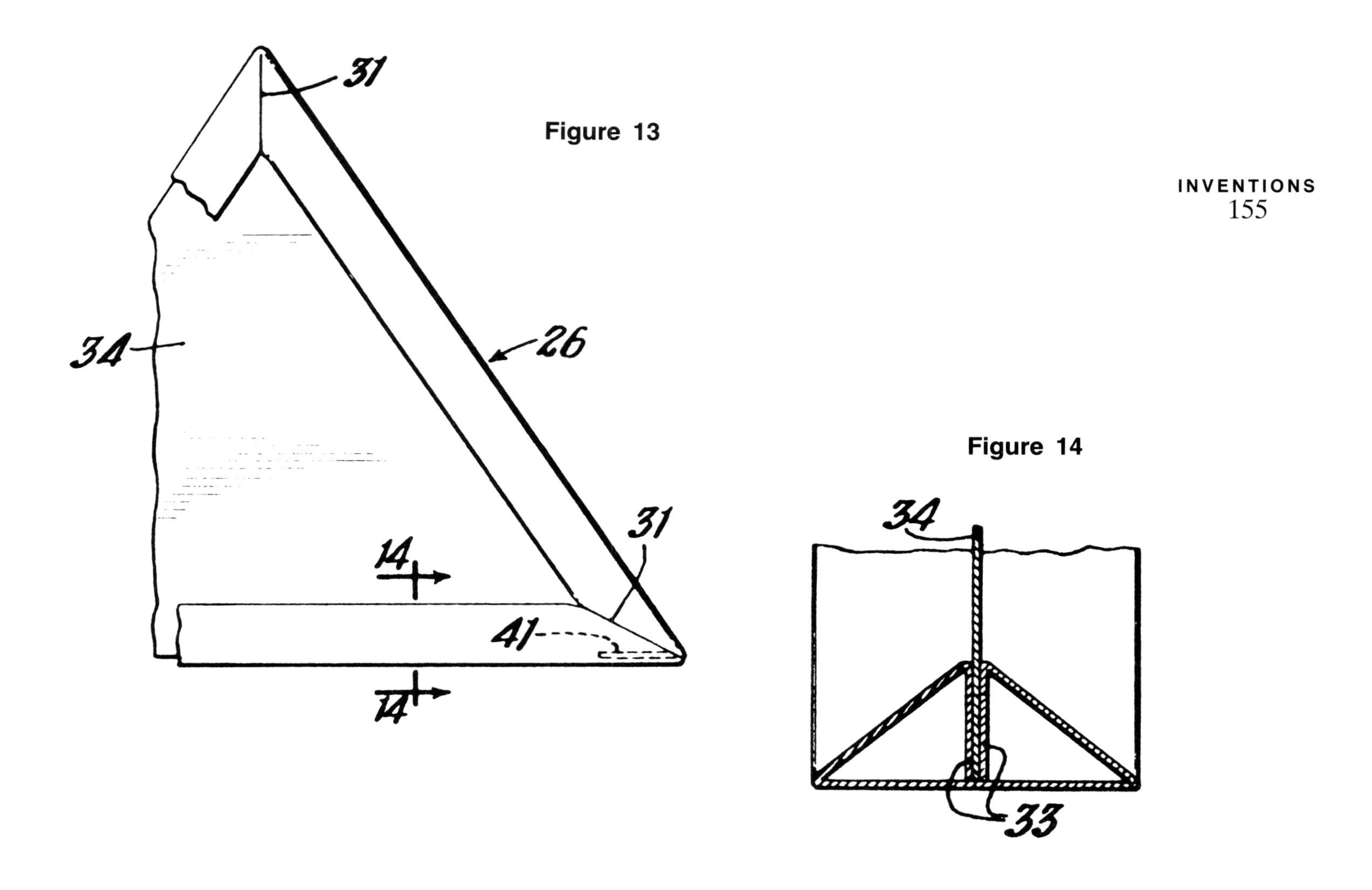
Figure 13
31
26
34
14
31
41
14
Figure 14
34
33

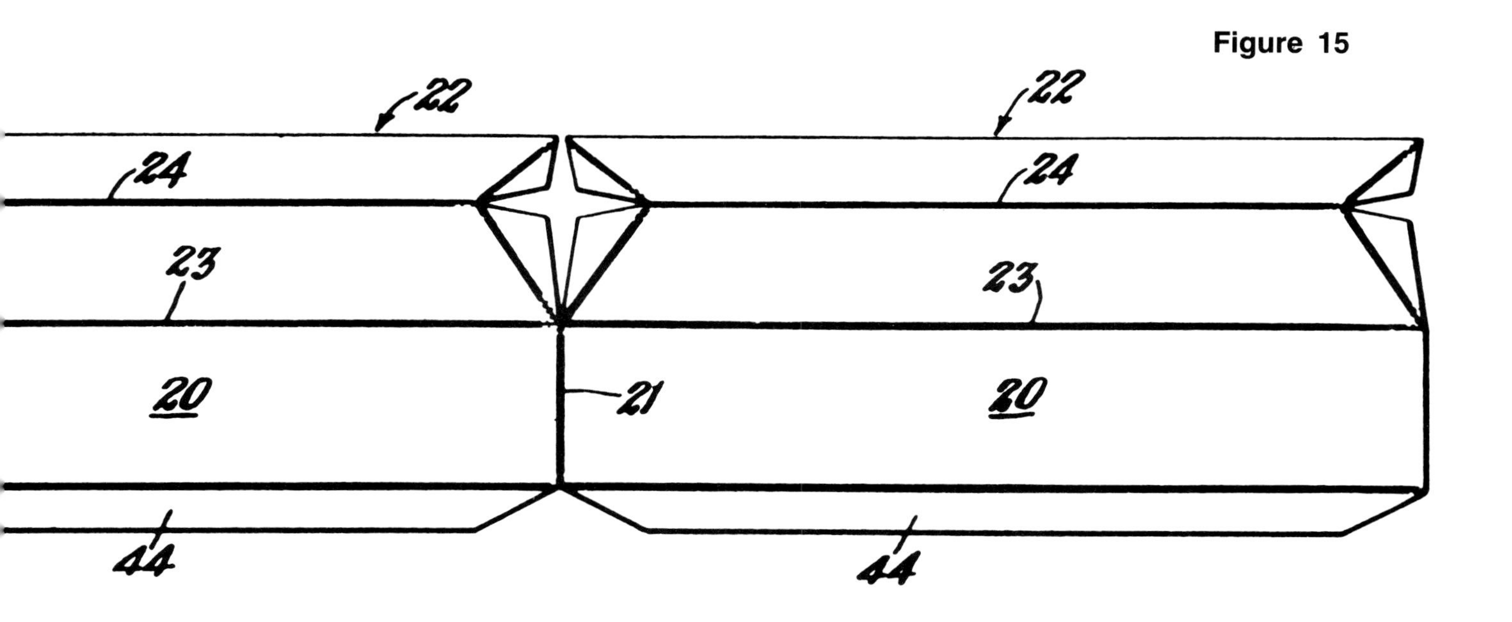
Figure 15
22
22
24
24
23
23
20
21
20
44
44

Figure 16

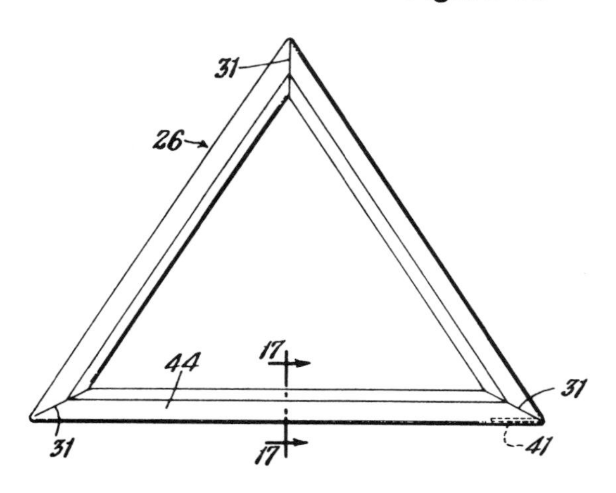

member of the character described, in which the elongated panels 20 and lateral extensions 22 are provided with a number of perforations, the alignment of which, when the member is set up for use, being as follows: one perforation 37 in a panel 20 opposite a perforation 38 in an extension of such panel comprising one pair of perforations to receive an alignment pin 39, and another perforation 35 in said panel 20 opposite another perforation 36 in said extension of such panel comprising a second pair of perforations alternatively to receive alignment pin 39, the pin axis of said one pair of perforations being at a slight angle to the pin axis of said second pair of perforations, and the pin axis of at least one of said two pairs being at a slight angle to the plane of the polygonal frame. In the example illustrated in Figs. 9 and 10, one of the two pairs of perforations, viz. perforations 35, 36 is blind, while the other pair, viz. perforations 37, 38 is completely punched out selectively to predetermine the angular relationship between said frame and another frame of like construction. In the example illustrated in Figs. 11 and 12, perforations 35 and 36 have been completely punched out while perforations 37 and 38 are blind, to predetermine another angular relationship between the two frames. Thus, by placing the pins 39 through the punched out perforations 37, 38 of Figs. 9 and 10, the condition illustrated in Fig. 9 is obtained, whereas by placing the pin 39 in the punched out perforations 35 and 36 of Figs. 11 and 12, the condition illustrated in Fig. 11 is obtained. In the first instance the angularity between the planes of the two frames is slight whereas in Fig. 11 the angularity is somewhat greater. The difference in the two angularities has been shown with considerable exaggeration in these views in order to reveal the principle of operation more clearly. It will be understood that additional apertures may be provided in order to afford a larger range of angular relationships to choose from. The gore between the backs of the two frames may, if desired, be filled with a triangular paperboard element or other filler piece. Also, a metal band or clip may be passed around the meeting edges or corners of the frames to be held in place by the alignment piece, or otherwise, as may be desired.

A second set of perforations 35′, 36′, and 37′, 38′ may be employed to receive, selectively, an additional pin 40,

Figure 18

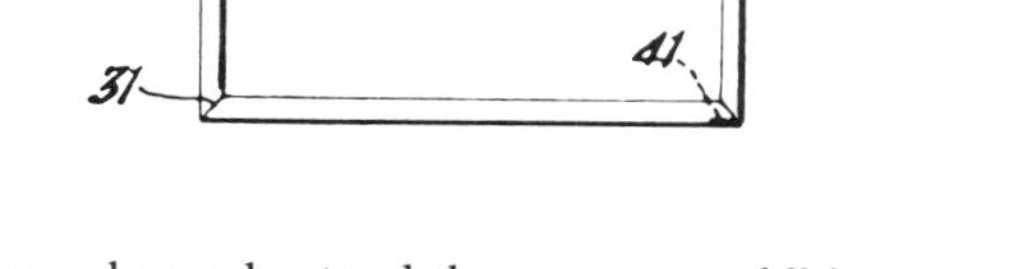

Figure 17

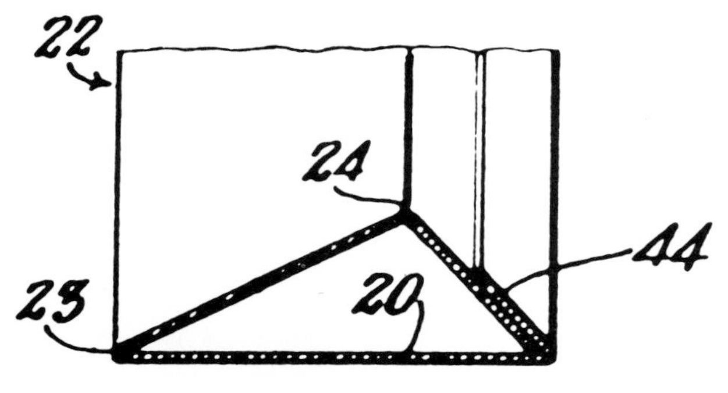

and it may be understood that as many additional sets of such perforations can be used as will be best suited for a particular utilization of my invention.

In summarizing my invention at the beginning of this specification, I have referred to the coating of the frames, or the paperboard from which they are formed, with a thin plastic film, such as a vinyl resin, which can be brushed, sprayed, or otherwise applied. I have found that a surprising degree of strength can be imparted to the paperboard structure by the use of vinyl plastic films which are on the order of 1⁄16 inch to 1⁄64 inch in thickness, although other thicknesses may be employed as desired, and the frames can be used without any coating whatsoever, particularly if an impregnated or waterproof paperboard is employed.

The terms and expressions which I have employed are used in a descriptive and not a limiting sense, and I have no intention of excluding such equivalents of the invention described, or of portions thereof, as fall within the purview of the claims.

12▲PLYDOME (1959)

U.S. PATENT—2,905,113

APPLICATION—APRIL 22, 1957

SERIAL NO.—654,166

PATENTED—SEPTEMBER 22, 1959

I FOUND THAT I COULD TAKE 4′ × 8′ plywood sheets and with the spherical trigonometry properly calculated put holes in the plywood sheets to match with the other sheets to produce a geodesic dome.

Take a common postcard and, with a ruler and a pencil, make marks at the midpoints of the long edges of the card. Then, with the ruler, connect the two midpoints, dividing the card into two rectilinear halves. Then measure the midpoints of both of the short edges of the postcard. Then, with the ruler, draw four connections between the midpoints of the short edges and the midpoints of the long edges; you'll find this makes a diamond, not only a diamond but also a line dividing the diamond's shortest axis, producing two triangles. Now bend the postcard on the four edges of the diamond and the midpoint of the diamond. You'll find that you can take these postcards, if you do the spherical trigonometry properly, and interconnect them in such a manner as to make three of them make a triangle, and you'll find they'll come together to make an icosahedral geodesic sphere. This being so, I saw that plywood could be bent enough in its own right so that with proper holes, located by spherical trigonometry, one could bolt plywood sheets together to form a sphere.

UNITED STATES PATENT OFFICE

Richard Buckminster Fuller, New York, N.Y.

SELF-STRUTTED GEODESIC PLYDOME

The invention relates to geodesic and synergetic construction of dome-shaped enclosures.

SUMMARY

Fundamental concepts of geodesic and synergetic construction are described in my prior patent, No. 2,682,235, granted June 29, 1954, and in my copending applications for patent, Serial No. 563,931, filed February 7, 1956, and Serial No. 643,403, filed March 1, 1957. As the invention of my prior patent has become known and used throughout the world, it can be assumed that the reader will be familiar with geodesic dome construction and the principal characteristics which distinguish it from the older architectural forms; so these characteristics will here be reviewed only briefly. For a comprehensive review, reference is made to Patent No. 2,682,235, aforesaid.

In geodesic construction, the building framework is one of generally spherical form in which the longitudinal centerlines of the main structural elements lie substantially in great circle planes whose intersections with a common sphere form grids comprising substantially equilateral spherical triangles. ["Great circle planes" are defined as planes whose intersections with a sphere are great circles. Such planes pass through the center of the sphere. The earth's equator and the meridians of the globe are representative of great circles in the ordinary accepted meaning of this term.] The grids can, for example, be formed on the faces of a spherical equilateral icosahedron. Each of the twenty equal spherical equilateral triangles which form the faces of the icosa is modularly divided along its edges. Lines connecting these modularly divided edges in a three-way great circle grid provide the outline for the plan of construction. Each of the smaller triangles formed by the three-way grid is approximately equilateral, i.e. its sides are approximately equal. The extent of variation in length is determined trigonometrically or by graphic solution of the grids as drawn upon the modularly divided edges of an icosahedron outlined upon the surface of a scale model sphere. It will be found that at each vertex of the icosa five of the grid triangles form a pentagon, whereas elsewhere throughout the pattern the grid triangles group themselves into hexagons, this being one of the distinguishing characteristics of three-way grid construction.

My present invention arises in the discovery that when perfectly flat rectangular sheets are shingled together in a three-way grid pattern and are fastened together where they overlap in the areas of the geodesic lines of the pattern, a new phenomenon occurs: there are induced in each flat rectangular sheet, elements of five cylindrical struts defining two triangles of the grid edge to edge in diamond pattern. The effect is to produce a three-way geodesic pattern of cylindrical struts by inductive action so that, when the sheets are fastened together in the particular manner described, the struts are created in situ. Thus the flat rectangular sheets are triangulated into an inherently strutted spherical form to produce what we may for simplicity term a self-strutted geodesic plydome. The flat sheets become inherently geodesic; they become both roof and beam, both wall and column, and in each case the braces as well. They become the weatherbreak and its supporting frame or truss all in one. The inherent three-way grid of cylindrical struts causes the structure as a whole to act almost as a membrane in absorbing and distributing loads, and results in a more uniform stressing of all of the sheets. The entire structure is skin stressed, taut and alive. Dead weight is virtually non-existent. Technically, we say that the structure possesses high tensile integrity in a discontinuous compression system.

FIGURE 1

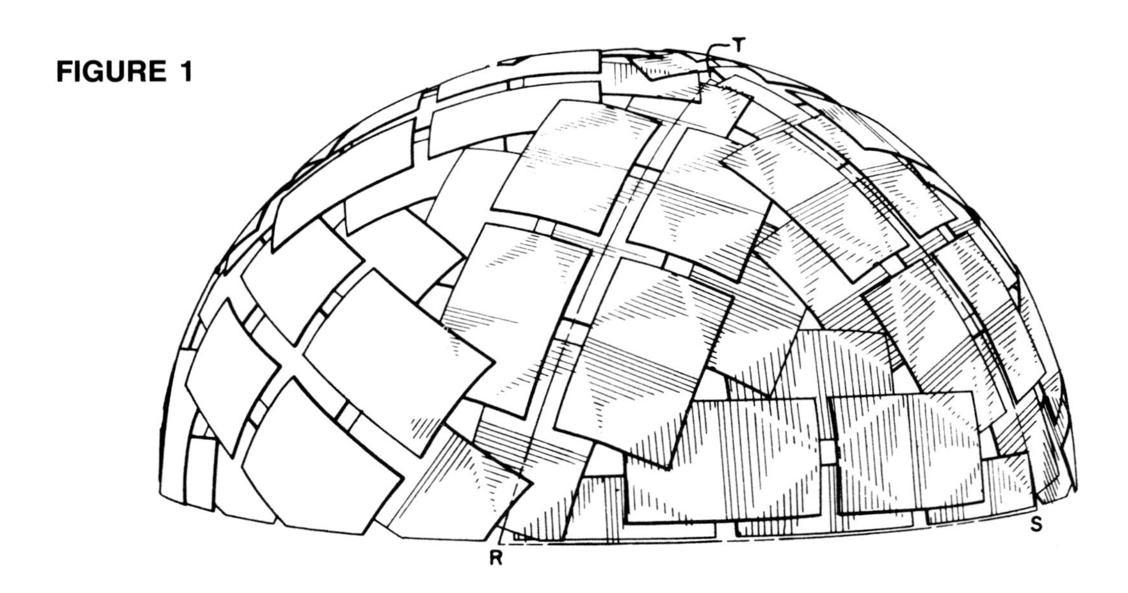

Description

With reference to the accompanying drawings, I shall now describe the best mode contemplated by me for carrying out my invention.

FIGURE 2

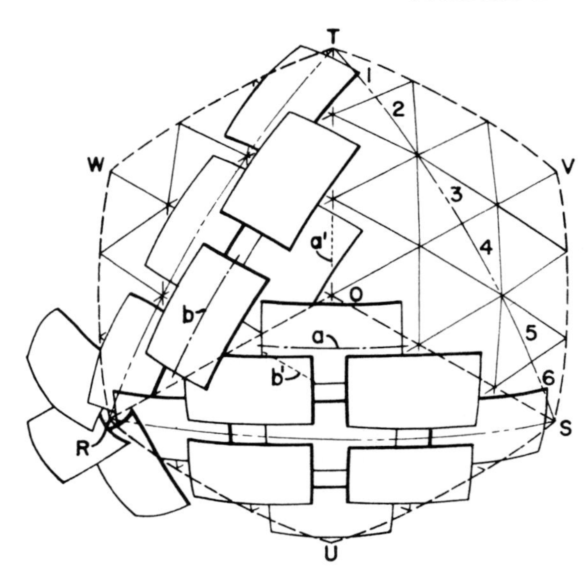

Fig. 1 is a perspective view of a geodesic plydome embodying my invention in a preferred form.

Fig. 2 is a detail perspective view of a portion of the Fig. 1 construction overlaid upon a diagrammatic representation of a three-way grid as an imaginary projection of the induced strutting of the dome. The area comprised is representative of one full face of the icosa with adjacent one-third sectors of adjacent faces. Combining the one-third sectors lying at each side of the respective meeting edges of the adjacent faces, we get three "large diamonds"; and

Fig. 3 is an enlarged detail view of the sheets which go to make up one of these large diamonds. Here the sheets are shown as they would appear when laid out flat and before they are fastened together.

Fig. 4 is a view similar to Fig. 3. Imagine that this big diamond is now a part of the completely assembled dome, and notice how the structure has inductively produced five struts in each of the sheets.

FIGURE 4

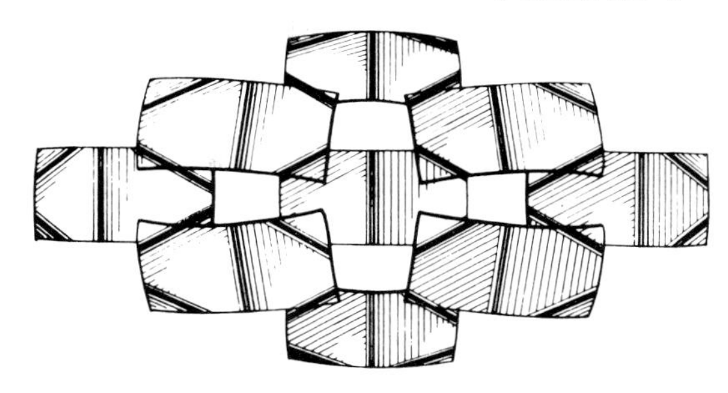

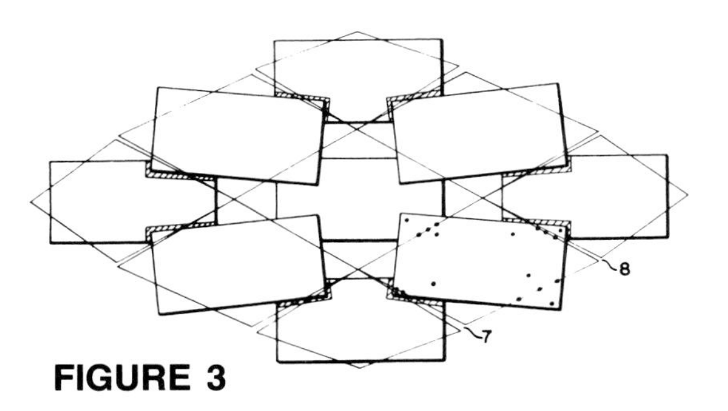

FIGURE 3

Figs. 5–8 inclusive, show icosa segments of several modified constructions in which pyramidal groupings of the triangular grid faces defined by the induced struts produce in and out convolutions of the spherical surface. In these several constructions the apexes of the pyramids define one sphere and the bases of the pyramids another. Which of the two spheres is the larger depends on whether the apexes of the pyramids project outwardly or extend inwardly. The sides of the pyramids may be regarded as struts connecting elements of the inner and outer spheres and thus creating a truss.

Fig. 5 follows the same sheet arangement as in Figs. 1–4. Because of the convoluted, or involute-evolute construction, we get hexagonal and pentagonal pyramids (pentagons at the vertexes of the icosa), which for simplicity I term a hexpent configuration. Here the apexes of the hexes and pents project outwardly (or upwardly from the plane of the drawing). Notice that a strut is induced along the short axis of each sheet.

Fig. 6 shows a modified hexpent pattern in which the sheets "toe in" to the apexes of the pyramids.

Fig. 7, like Fig. 5, has the same sheet arrangement as in Figs. 1–4. Here the induced geodesic triangles of the sheets form an inverted tetrahedron at the center of each face of the icosa, and one of the induced struts extends the long way of each sheet.

Fig. 8 has a sheet arrangement which may be compared to that of Fig. 6, but with one of the struts extending the long way of the sheet there is formed a pattern of inverted hexpent pyramids.

The construction shown in each of Figs. 5 to 8 inclusive may be turned inwardly or outwardly. For example if we think of Fig. 5 as representing the outer surface of a dome, we have pyramids projecting outwardly with their apexes in an outer sphere and their bases (or the corners of their bases) in an inner sphere. Or if we think of Fig. 5 as representing the inner surface of a dome, we have pyramids extending inwardly with their apexes in an inner sphere and their bases (or the corners of their bases) in an outer sphere.

In Figs. 1, 2, 5 and 6, the dot and dash lines RST represent one of the equilateral spherical triangles of a spherical icosahedron. In Fig. 2, the dotted lines OSVT, OTWR and ORUS each define an area combining one-third sectors lying at each side of one of the meeting edges of the adjacent faces. Thus area OSVT combines a one-third sector of icosa triangle RST, namely the sector OST with a one-third sector TSV of the adjacent icosa triangle. I call the combined sector areas "large diamonds." It is helpful to see the large diamonds when analyzing the structure as a whole, because, once the eye becomes practiced at picking them out, both the pattern of the icosa faces and of the induced three-way strutting is more easily discerned. This is especially so in the cases of Figs. 1–4, Fig. 5 and Fig. 7, in each of which all of the sheets are arranged approximately parallel to the major axis of the large diamond. This brings the major axes into focus, outlining the icosa faces. Then the eye finds the center of the icosa face, further identifiable by the small triangular opening at O, surrounded by a series of kite-shaped openings at the meeting edges of adjacent large diamonds and by square openings at the edges of the icosa triangle. It is suggested that a brief study of these characteristic formations with reference to Fig. 1 will be of much help in acquiring a general grasp of the geodesic alignment of the sheets themselves, and later of the induced geodesic three-way grid strutting across the corners and centers of each sheet.

Now, if we are proceeding by the graphic solution method, we first lay out the icosa faces on a scale model sphere, then divide the edges of one of the faces into the desired number of equal parts, or modules, which determines what I call the "frequency" of the three-way grid. For example, in Fig. 2 I have shown the dot and dash line ST divided into six modules numbered 1 to 6 for identification, providing a six-frequency grid. With the three edges of the icosa face so divided it is necessary only to join each point of one edge with every second point on another to produce the three-way grid shown, comprising three sets of great circle arcs intersecting to form a pattern of substantially equilateral triangles. Now we lay out the sheets on the grid pattern as shown in Fig. 2, centering the short axis of each sheet on alternate grid lines and working outwardly from the major axis of a large diamond toward its edges. With the frequency of six we get first a row of three sheets in spaced end-to-end arrangement, a row of two sheets at either side of this, and overlapping at the corners, and finally a single sheet coming

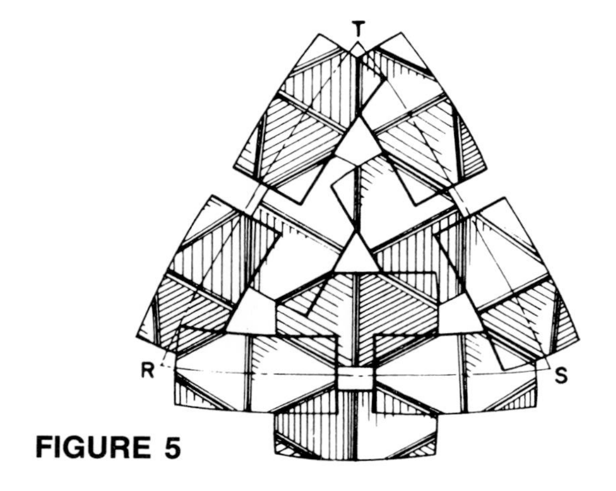

FIGURE 5

pend upon the design of the particular dome, extent of overlap of the sheets, thickness of the sheets and possibly other factors. In some cases the radius of the bend may be so large that the strutting is not clearly visible, or is perhaps only visible to a practiced eye. I have had the draftsman try to simulate the photographic appearance of the particular dome represented in Fig. 1, where the geodesic strutting shows up in the highlighted portions of the shaded areas. In Fig. 4 the effect has been considerably exaggerated in order to bring out the point. The self-strutting phenomenon takes place during assembly of the sheets according to their coding and fastening them together in the designated areas of the grid lines and at their corners as marked for factory-drilled for the fastenings. When Fig. 4 is imagined as a part of the completely assembled dome, a comparison of Figs. 3 and 4 will help to give an idea of the inductive strutting action. Fig. 3 is a static assembly of related parts which "know" the three-way geodesic grid pattern of the icosa; Fig. 4 a dynamic resolution of the pattern into *(a)* spherical form, *(b)* with

up to the center of each icosa face, as clearly shown in Fig. 2.

Notice that the longitudinal centerlines of the sheets (see the representative centerlines *a* and *b* in Fig. 2) lie substantially along great circles of the sphere, or lie substantially in great circle planes whose intersections with a common sphere form grids comprising substantially equilateral triangles. The sheets are now marked for interconnection along the lines of the three-way grid previously laid out. These lines of interconnection will be found to be substantially normal to the aforesaid intersections. Thus the line of interconnection marked on the three-way grid at *a′* is normal to centerline intersection *a,* and that marked at *b′* normal to *b,* etc. It will be found that the markings for interconnection of the sheets will vary from one sheet to another depending upon its position in the pattern. The number of different sheet markings depends upon the frequency of the grid. With the frequency of six shown in Figs. 1–4 there will be three different sheet markings, or types of sheet. It is desirable to label, or color-code the sheets to show how they are to be put together.

The sheets are now marked, or perforated, for the fastenings, following the designs laid out as above. Such perforations are shown by the black dots in a sheet at the lower right of Fig. 3. Notice that additional perforations are provided near the corners of the sheets so that the sheets will be fastened together both in the areas of the grid lines and also at points substantially removed from said lines. This not only "buttons down" the corners of the sheets, but assists importantly in creating the induced struts in the completed structure.

Turning now to Fig. 3, we see the sheets for a large diamond as they would appear when laid out flat and before they are fastened together. The shaded areas at the outer overlapping corners show the amount of increased overlap which occurs when the sheets are brought into position for fastening them together. Once they have been brought into position and fastened, the sinuses 7, 8, etc., between the grid line markings close up and the structure assumes its desired spherical form. Concurrently, there are induced in each flat rectangular sheet, elements of five cylindrical struts defining two triangles of the geodesic grid edge to edge in diamond pattern. As shown in Fig. 4, four of these struts cross the corners of the sheet and the fifth extends the short way of the sheet to form the base of the two geodesic triangles. These struts, as may be discerned from the shading, are simply bends in the sheet. The sharpness of the bend will de-

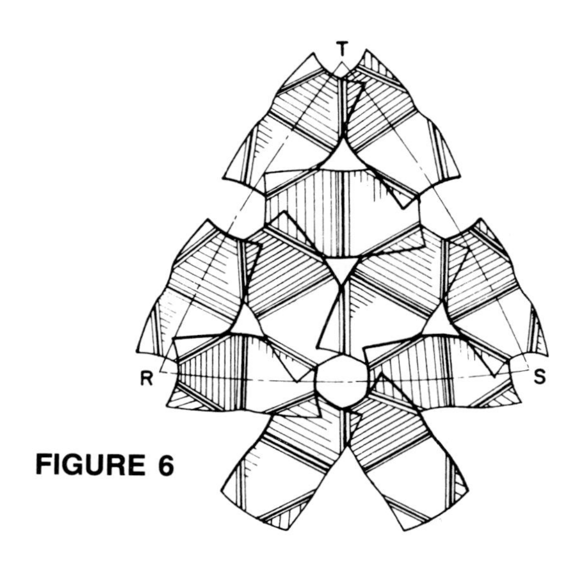

FIGURE 6

inherent struts expressing the pattern in terms of gentle bends in the sheets, each bend comprising elements of a cylindrical surface. It seems remarkable that the bends locate themselves, at least in part, even in the double thickness of the overlapping corners of the sheets where it might have been supposed that the stiffness of the double thick portions would suggest a greater resistance to bending. This result implies strongly that the inherent structuring of the geodesic grid pattern is so natural and strong in its tendency to produce a perfect self-supporting sphere that it departs from behavior patterns predicted from ordinary principles of mechanics and strength of materials. Since the behavior of the system as a whole is unpredicted from its parts, we say that the resulting structure is "synergetic." Such structures are vastly stronger, pound for pound, than any heretofore known.

The curve of the bends in the sheets, variable according to the factors named in the preceding paragraph, may comprise elements of a circular cylinder or elements of a cylinder of varying radius. This is to say, the radius of curvature of a particular strut need not be uniform. To some degree this factor may be influenced by the leverage imposed by the overlapping areas where the sheets are fastened together, and can vary according to the extent of such overlapping areas. Such leverage may throw the sharpest curvature of a bend a little to one side of the geodesic line, but the strut will in every case remain substantially a true geodesic line in the sense that its axis will lie in a plane whose intersection with a sphere is an element of a great circle. The strut itself becomes a chord of that sphere.

To keep the drawings clear and readable, the fastenings have been omitted, except as the holes for them have been depicted in Fig. 3 and as the geodesic grid lines used in locating them are shown in Figs. 2 and 3. The fastenings themselves may be of any conventional type, and in some constructions it would be feasible to use adhesive means for holding the sheets together in the same geodesic alignment.

The sheets may be of any desired material, such as plywood, aluminum, steel, plastics, plastic-coated wallboard, composites of plywood and aluminum, plywood and aluminum sheet or foil, etc. I have found that marine plywood in standard sheet sizes has excellent characteristics for induced strutting.

If desired, the openings between the sheets can be closed up, this being merely a function of the selected frequency of the grid in relation to sheet size. The proportions of the sheets also are subject to variation, but I recommend adherence to substantially a three to five ratio between width and length as giving best results for most building purposes. It is even possible to use sheets of other forms than rectangular, but an essential advantage of my construction is that is permits the use of plain rectangular sheets which are so readily available, stack so compactly for shipment and are least expensive. If the openings between the sheets are not closed up by the boards themselves, they may be used as skylights, and I have had good results with the use of thin skins of transparent mylar plastic for covering the openings. In some cases it may be desired to use an overall plastic inner or outer lining to weatherproof the dome; or weatherproofing may be secured by sealing the joints with plastic compounds or tape, and painting. Also, the overlapping of the sheets one upon another can be arranged so that the entire structure is weathershingled to shed water outwardly and downwardly over the surface of the dome. Such shingling of the sheets can also be arranged to cover the openings where they come together, or additional sheets can be slipped in to shingle over the openings.

By laying out the three-way grid pattern so that the radial lines of the polygons are longer than the lines forming the sides of the bases of the polygons, we obtain the hexpent and tetrahedral forms of the multiple-sphere trussed constructions explained in the outline description of Figs. 5 to 8, inclusive q.v. Thus in Fig. 5 we have, in a four-frequency grid design, a typical hexagonal pyramid with its apex at the center of the icosa face RST, this pyramid being formed by three sheets, and the two induced triangles of each sheet making two sides of the pyramid. Pentagonal pyramids occur at each vertex of the icosa. The pattern is one comprising hexagonal and pentagonal pyramids the apexes of which define an outer sphere, and the corners of the bases of which define an inner sphere.

In Fig. 6, we again have a pattern of hexagonal and pentagonal pyramids, but here six sheets toe in to the apex of a hexagonal pyramid and five sheets toe in to the apex of a pentagonal pyramid. In both this view and Fig. 5, one of the induced struts extends the short way of the sheet and in this respect there is a similarity to the neutral, or one-sphere, form of Figs. 1–4.

In Fig. 7, the induced geodesic triangles of the sheets form an inverted tetrahedron at the center of the icosa face, and one of the induced struts of each sheet extends the long way of the sheet to form the common base of the two induced triangles. Each triangle is two frequency modules wide, one frequency module high.

In Fig. 8, we again have a pattern of hexagonal and pentagonal pyramids, six sheets toeing in to the apex of a hexagonal pyramid and five sheets toeing in to the apex of a pentagonal pyramid, and one of the induced struts extending the long way of the sheet.

The terms and expressions which I have employed are used in a descriptive and not a limiting sense, and I have no intention of excluding such equivalents of the invention described, or of portions thereof, as fall within the scope of the claims.

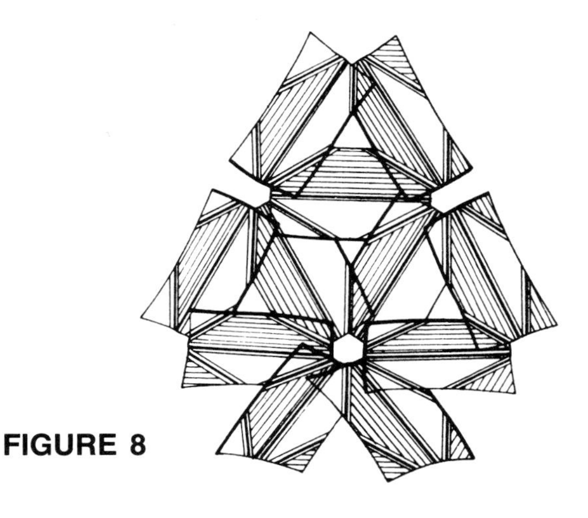
FIGURE 8

FIGURE 7

13▲CATENARY (GEODESIC TENT) (1959)

U.S. PATENT—2,914,074

APPLICATION—MARCH 1, 1957

SERIAL NO.—643,403

PATENTED—NOVEMBER 24, 1959

GEODESIC DOMES, 1144 feet in magnitude, made out of aluminum or steel tubular struts and hubs, were used for the United States international trade fair domes. We found it appropriate to produce fabric tents with the fabric cut out very economically in geodesic patterns, which patterning would produce a tent with vertexes corresponding to vertexes of the tubular dome structure but a little bit smaller than the structure. With the fabric so cut out to produce hyperbolic paraboloid surfaces with conic convergences corresponding to the vertexes of the geodesic tubular frame, we found it possible to mount powerful aluminum pads at the vertexes of the parabolic cones of the fabric to distribute the stress. We then had stainless steel chains emanating outwardly from the tent (from the center of each of the circular reinforcings at the outer end of the parabolic cones). We then found it appropriate to make keyholes in the hubs of the outer tubular geodesic dome to which the stainless steel chain pattern could be pulled outwardly by individuals climbing on the outside of the tubular dome. In this way we were able to stretch the interior tent outwardly toward the tubular geodesic frame. Due to the hyperbolic paraboloid surfaces the dome would

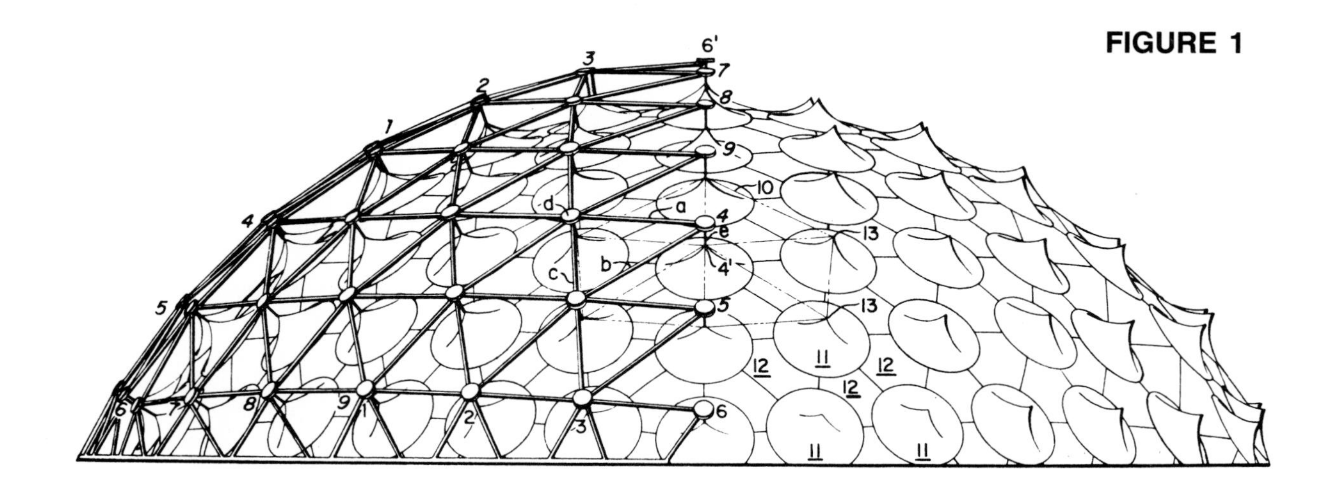

FIGURE 1

never flutter in any wind. This proved to be extremely satisfactory for a number of the international trade fair domes.

One of these trade fair domes ended up in Alaska and went undamaged through several winters with heavy snow-loadings until it was ultimately dismantled.

UNITED STATES PATENT OFFICE

Richard Buckminster Fuller, Forest Hills, N.Y.

GEODESIC TENT

This invention relates to geodesic building structures in the form of spherical tents.

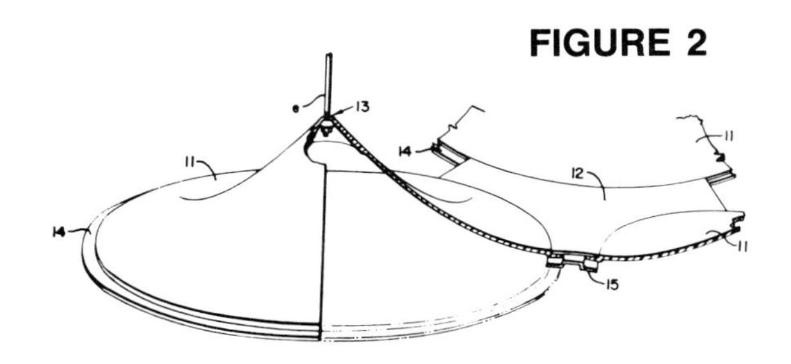

FIGURE 2

SUMMARY

Heretofore, as described in my prior Patent No. 2,682,235, granted June 29, 1954, I have discovered how to create building structures in which the main structural elements are interconnected in a geodesic pattern of approximately great circle arcs intersecting to form a three-way grid of substantially equilateral triangles. Such building structures may consist of skeletal frameworks made of interconnected struts, or of interlocking or interconnected sheets or plates, or of molded plastic sections fastened together along flanged edges, or of flexible fabrics or plastic skins conforming in pattern or behavior to the three-way grid geodesic construction. Also I had found that a very special relationship exists between a geodesic building structure made of interconnected struts and a complementary geodesic building structure made of flexible fabrics or plastic skins where these two structural components are made to conform in structure, pattern, or behavior to a mutual three-way great circle synergy. My present invention is concerned with an improved geodesic skin or tent construction which gives a new and synergetic stress distribution—synergetic in the sense that the behavior of the skin under stress is unpredicted by its several parts, and that there is imparted to the structure a strength beyond that which would be calculated using accepted values of strengths of materials and usual methods of stress analysis and computation. Fundamentally, I accomplish improved results by "tailoring" the several pieces which go to make up the tent in such a fashion as to yield an omni-triangulated suspension pattern. The pieces themselves may or may not be of generally triangular form, but the suspension pattern should be so in any case. The tailoring is such as to include an omni-triangulated pattern of suspension points extending over substantially the entire area of the tent, with a predetermined dip in the fabric between one suspension point and another. This dip produces a catenary curve, or an approximation thereof, between each pair of adjacent suspension points. Around each point of suspension the structural form is essentially that generated by revolution of a catenary segment about the catenary suspension point. This form approximates a cone and for simplicity is sometimes referred to herein as "conical." In some instances a truly conical form can be used, so I employ the term "conical" as including both a true cone and such pyramidal or catenary forms as will be described with reference to the several exemplary embodiments shown in my drawings.

Description

The accompanying drawings illustrate the best mode contemplated by me for carrying out my invention according to several preferred embodiments thereof.

Fig. 1 is a side elevational view of a domical structure embodying the invention. It shows an exterior geodesic framework with a geodesic tent supported within it. Part of the frame has been removed to show more clearly the catenary form of the tent.

Fig. 2 is a detail perspective view of one of the conical elements of the structure together with one of the connecting pieces of generally triangular, or diaper, form, and adjoining portions of other conical elements.

Fig. 3 is a detail view of a triangular piece for a tent of modified construction.

Fig. 4 is a detail view of an assembly of flat triangular pieces for a tent of another modified construction.

Fig. 5 is a schematic diagram illustrating the fundamental stress pattern of the several catenary constructions of Figs. 1 to 4 inclusive.

Fig. 6 is a comparative sketch to show (by correlation with Fig. 5) the relationship between the cone-diaper

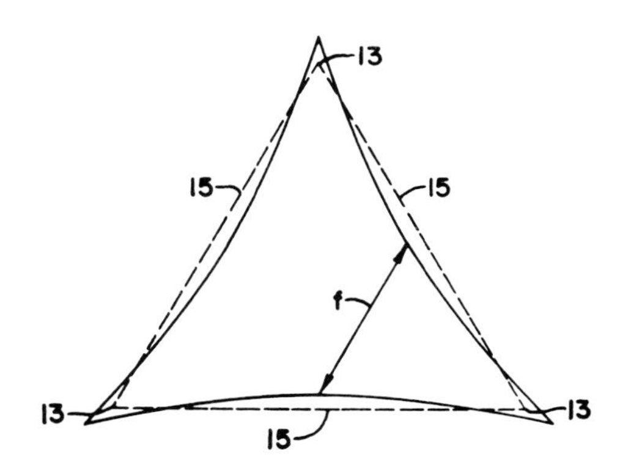

FIGURE 3

form of Figs. 1–2 and the approximation of that form with the use of flat pieces of tent material according to Fig. 4.

Fig. 7 is a photographic reproduction of a completed dome in which the tent structure is made of tailored triangles of the general form typified in Fig. 3.

Fig. 8 is a photographic reproduction of a portion of the interior of a tent made of tailored triangles of the general form typified in Fig. 4.

Reference is made to Fig. 1 which shows a geodesic building structure made of interconnected struts and a complementary geodesic building structure in the form of a tent supported within the first named structure. A portion of the outer building structure has been removed to show more clearly the catenary form of the tent. The framework of the outer supporting structure is constructed on the pattern which I have described as comprising approximately great circle arcs intersecting to

FIGURE 4

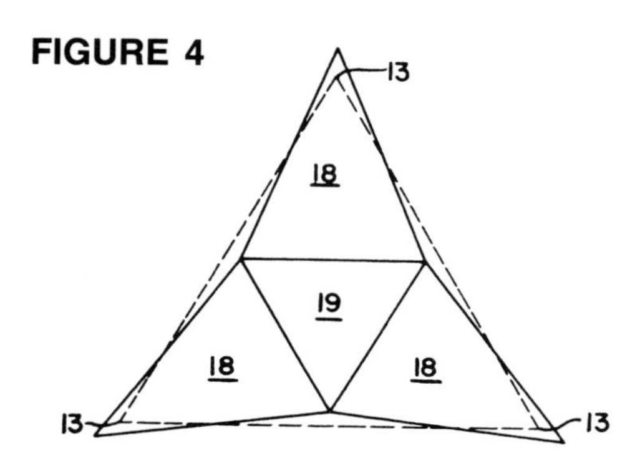

form a three-way grid of substantially equilateral triangles. In the particular embodiment selected for illustration, the three-way grids are formed on the faces of a spherical icosahedron. Each of the equal spherical equilateral triangles of this construction is modularly divided along its edges. Great circle arcs connecting these modularly divided edges in a three-way great circle grid provide the outline for the plan of construction. Thus on the spherical equilateral triangle shown at 6, 6, 6′ in Fig. 1, we have a series of great circle arcs 1—1, 2—2, 3—3 etc., a second series of great circle arcs 4—4, 5—5, 6—6 etc., and a third series of great circle arcs 7—7, 8—8, 9—9 etc., each series paralleling one of the sides of the spherical triangle 6, 6, 6′, and the three series of arcs intersecting to form an omni-triangulated pattern in which the triangles form pentagons at each of the vertexes, 6, 6, and 6′, and hexagons throughout the rest of the pattern. The structural members *a, b* and *c* of the framework are aligned with the lines of the grids. In the particular construction shown, these structural members are considered as being in the form of tubular struts connected at points of intersection by hub-like members *d*. The inner building structure or tent is suspended within this framework from the hubs *d* by suspension cords or rods *e* so that the tent will have a pattern of suspension points which is complementary to the three-way grid of the supporting structure. In the omni-triangulated pattern of the tent structure, the broken line 10 represents a hexagon centered on suspension point 4′ located radially inward from suspension point 4 of the outer supporting framework (radially with reference to the center of the spherical icosahedron).

In the embodiment of Figs. 1 and 2, the tent is made up of conical pieces 11 and connecting pieces 12 of generally triangular or diaper form. The conical pieces are essentially of the form generated by revolution of a catenary segment about the catenary suspension point, so the apexes of the generated forms constitute the susspension points and these are arranged in accordance with the described geodesic three-way grid pattern. In this way I have provided a tent of generally spherical form tailored to an omni-triangulated pattern of projecting points of suspension formed by the apexes of forms generated by revolution of catenary segments about the respective catenary suspension points. The spaces between the circular edges of the conical pieces are filled in by the diapers 12 which may be fastened adhesively or otherwise to flanges 14 at the bases of the conical pieces. Also, if desired, and as shown in Fig. 2, there may be connecting members 15 between the edges of adjacent diapers 12 and cones 11.

I have discovered that this construction results in the creation of what may be described as an inner sphere: regarding the several suspension points 13 as defining a sphere which I shall here refer to as the outer sphere, I find that the interconnected diapers 12 become stressed to the form of an inner sphere. Thus the tent structure as a whole uniquely combines the inwardly dipping catenary suspension lines between the omni-triangulated pattern of suspension points with the outwardly curved fabric of the "inner sphere." I have found that this combination of inwardly and outwardly curving lines of stress produces a tent of surprising strength and ridigity. Even when formed of the thinnest nylon skins, the tent is characterized by high strength and freedom from fluttering in the wind. A hypothesis for the behavior of fabric tents supported in geodesic frames may be made with reference to the schematic diagram of Fig. 5. The lines of stress may be said to flow in natural radial catenary lines 15 outwardly and downwardly from the points of suspension 13. These radial catenary lines immediately and precessively induce circular lines, or rings, 16 at 90° to their respective axial cones to which the radial lines distribute their loading. These rings then precessively beget in turn further outwardly radial lines, and the radial lines again precessively beget circumferential rings. I believe that this fundamental precessive regeneration may be compared with the behavior of circular wave propagation, so that my tent is capable of distributing loads in the most nearly even energy distribution outwardly to the largest rings surrounding each vertex or point of suspension. When the outermost rings of the series of concentric rings 16 formed in response to the vertex stressing finally become tangent to one another on the lines 17, they form a tangential hexagon and pentagon network throughout the whole geodesic tent whereby all loads are shared three ways by the synergetic three-way grid of omni-triangulated geodesic great circle system lines. While I have here suggested what presently seems to me to be the best possible explanation of observed surprise characteristics of my tent construction, I do not wish to be limited to this or any particular hypothesis of theory of stress behavior.

Another thing I have observed is that if a tent is constructed along geodesic lines, but without tailoring-in the catenary construction and without recognition of the inner sphere, there is created a natural tendency for the fabric of the tent to stretch into shapes somewhat approaching catenary curves. Such stretching thins out the fabric and weakens it, further demonstrating the value of providing a predetermined cone-catenary construction.

In Fig. 5 it will be noticed that the radial lines 15 between three adjacent suspension points 13 form a triangular figure comprised of three catenary curves. This figure is shown in Fig. 3. The dotted lines represent straight lines adjoining the three points of suspension 13, and may also be understood as representing each of the catenary curves 15 as seen from above in spherical plan view. When the fabric within the three catenary borders 15 is flattened out, we get a shape approximately as shown in the full lines in Fig. 3. Notice that the distance along a line *f* between two of the concavely curved edges appears shorter than that between the sides of the dotted triangle. This is because line *f* is approximately in the plane of the smaller inner sphere. However the fabric when laid out flat to the full line position of Fig. 3 extends beyond the corners of the suspension points 13. According to another embodiment of my invention, a tent of the basic cone-catenary construction is made up of triangular pieces of the form shown in Fig. 3, seamed together along their concavely curved edges to produce the structure illustrated in Fig. 7. In this view the shadows cast by the outer supporting framework on the nylon fabric of the tent reveal approximately the cone-catenary form of the tent. Fig. 7 represents a practical example of the utility and value of my tent construction as it has been applied to an 8,000 square foot geodesic dome which was erected for the United States Government pavilion at the International Trade Fair in Kabul, Afghanistan. The tent was seamed together from the triangular pieces of nylon of the form shown in Fig. 3. The fundamental characteristics of this modified construction are essentially the same as have been described with reference to the embodiment of Figs. 1 and 2, and the schematic diagram of Fig. 5.

According to another embodiment of my invention illustrated in Fig. 4, a similar result is attained by tailoring

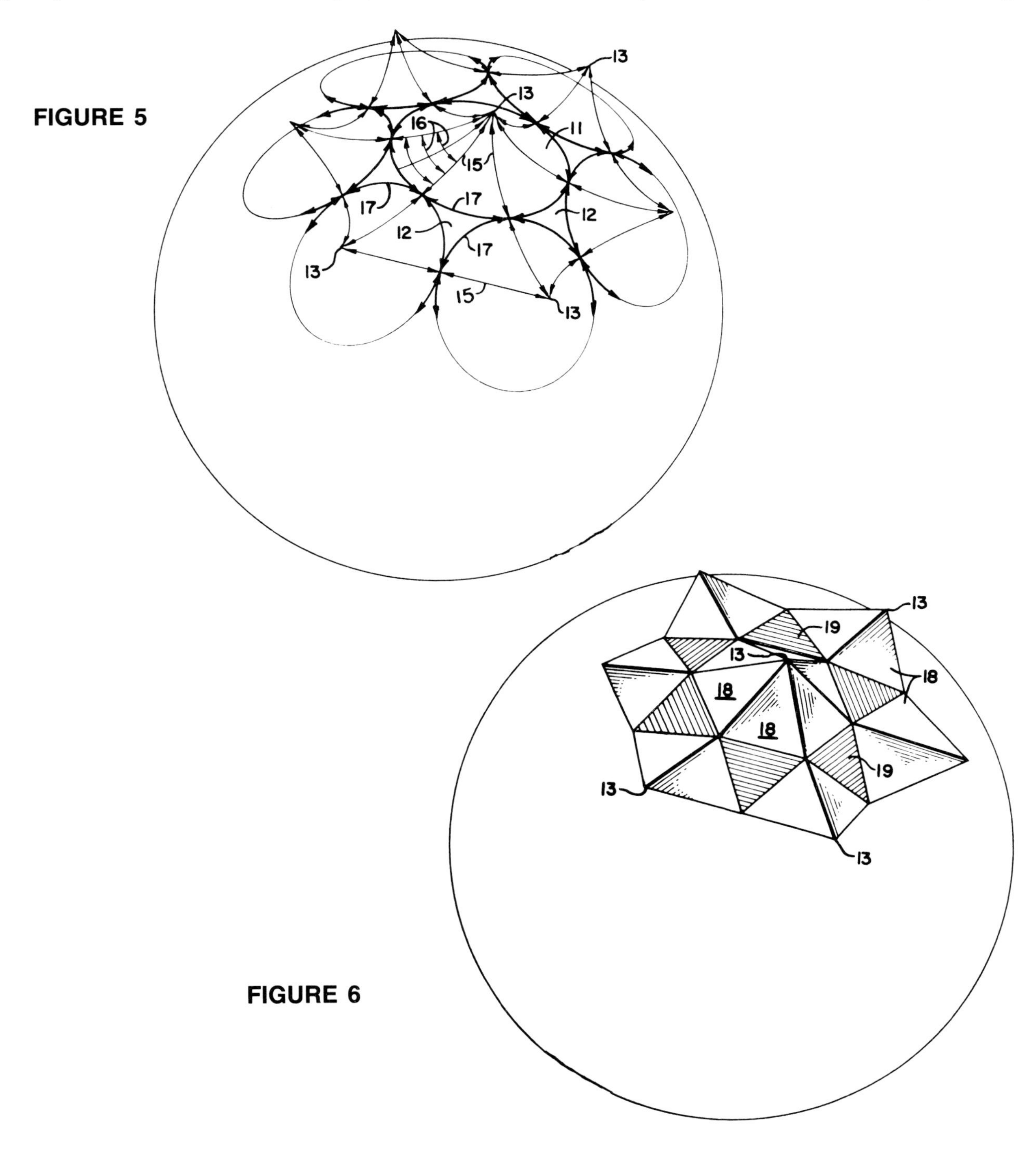

FIGURE 5

FIGURE 6

FIGURE 7

a catenary triangle from four flat triangles seamed together in the pattern shown here. The three outer triangles 18 constitute one-sixth or one-fifth of a pyramidal "cone" (depending on whether they are centered on one of the hexagons or one of the pentagons of the three-way grid system), and the center triangle 19 forms in effect the connecting piece or diaper of the Figs. 1–2 construction. In the diagram of Fig. 6, the "diapers" 19 have been shaded for easy recognition. By comparing Figs. 5 and 6, the basic equivalence between the cone-diaper construction and the Fig. 4 construction can be readily discerned. How closely the flat triangular pieces 18 and 19 of the Fig. 4 construction approximate the cone-diaper construction in terms of its effectiveness in creating the cone-catenary pattern is revealed in Fig. 8. This is a photographic view showing a portion of the interior of a tent made of tailored triangles of the general form typified in Fig. 4.

In each of the three specific embodiments I have described, it will be seen that the tent has a pattern of suspension points extending uniformly over substantially its entire area; in each the tent is tailored to dip inwardly between adjacent suspension points; each conforms to the omni-triangulated three-way grid pattern; and in each the inward dip of the fabric approximates a catenary curve between one suspension point and another. Other points or correspondence between the several embodiments may be discovered from the description which has preceded. The pieces which go to make up the tent may be fabricated in a variety of ways from thin flexible fabrics such as nylon skins, or from less flexible materials, or in some cases from rigid materials, as may be desired. The pieces may be stitched, glued or otherwise fastened together. They may be cut from fabric according to the patterns of Figs. 3 or 4, or they may be made in molded fiber glass or other molded materials in the form of Fig. 2. Thus in general the pieces which go to make up the tent may be either molded or made from flat sheet material, and in either case may be of any desired thickness and flexibility.

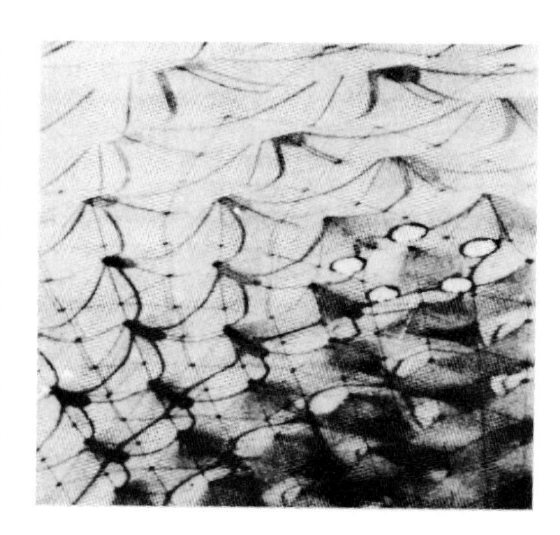

FIGURE 8

The terms and expressions which I have employed are used in a descriptive and not a limiting sense, and I have no intention of excluding such equivalents of the invention described, or portions thereof, as fall within the scope of the claims.

14▲OCTET TRUSS (1961)

U.S. PATENT—2,986,241

APPLICATION—FEBRUARY 7, 1956

SERIAL NO.—563,931

PATENTED—MAY 30, 1961

NATURE'S SIMPLEST STRUCTURAL SYSTEM in the Universe is the tetrahedron. The regular tetrahedron does not fill all-space by itself. The octahedron and tetrahedron complement one another to fill all-space. Together they produce the simplest, most powerful structural system in Universe.

I first used the name octahedron-tetrahedron truss. I then shortened it to the "octet truss." Having shipped it across state lines, I applied for a trademark for the octet truss.

It is interesting that at the present time, twenty-two years later, the U.S. government is specifying in its bids for space structures the octet truss (my copyrighted trademark name) for all the main structuring for space stations.

UNITED STATES PATENT OFFICE

Richard Buckminster Fuller, Forest Hills, N.Y. (407 S. Forest St., Carbondale, Ill.)

SYNERGETIC BUILDING CONSTRUCTION

My invention relates to a truss construction for building purposes; particularly to roof, wall and floor framework and to a combined roof and wall framework or wall and floor framework, etc.

SUMMARY

In my prior patent, No. 2,682,235, issued June 29, 1954, I have disclosed how to gain a surprisingly favorable weight-strength ratio in structures of generally spherical form, now widely known both here and abroad as "geodesic domes." The present invention is a discovery of how to gain an extremely favorable weight-strength ratio in structures of other forms, including those which are functionally conformed in shape for special purpose applications as well as more conventional forms based on the rectangular prism rather than sphere. In the sphere, the tremendous gain in the "ratio" accrued primarily from a unique arrangement of the main structural elements in which they are all aligned with great circles of a common sphere. In this sense, geodesic construction could be considered inapplicable as such to building frameworks of other than spherical form. However, I have found that if a flat roof, wall or floor framework is built up of struts (or sheets) of equal length (size) in such a fashion that such elements are comprised within a common octahedron-tetrahedron system, the strength of the framework is far greater than would be predictable using any conventional formulae based on resolution of forces and known values of strength of materials. In fact, my practical tests have shown that the actual strength of these "flat" one system octahedron-tetrahedron structures so far exceeds calculated values as to suggest a hypothesis that such structures are "synergetic" in the sense

FIGURE 1

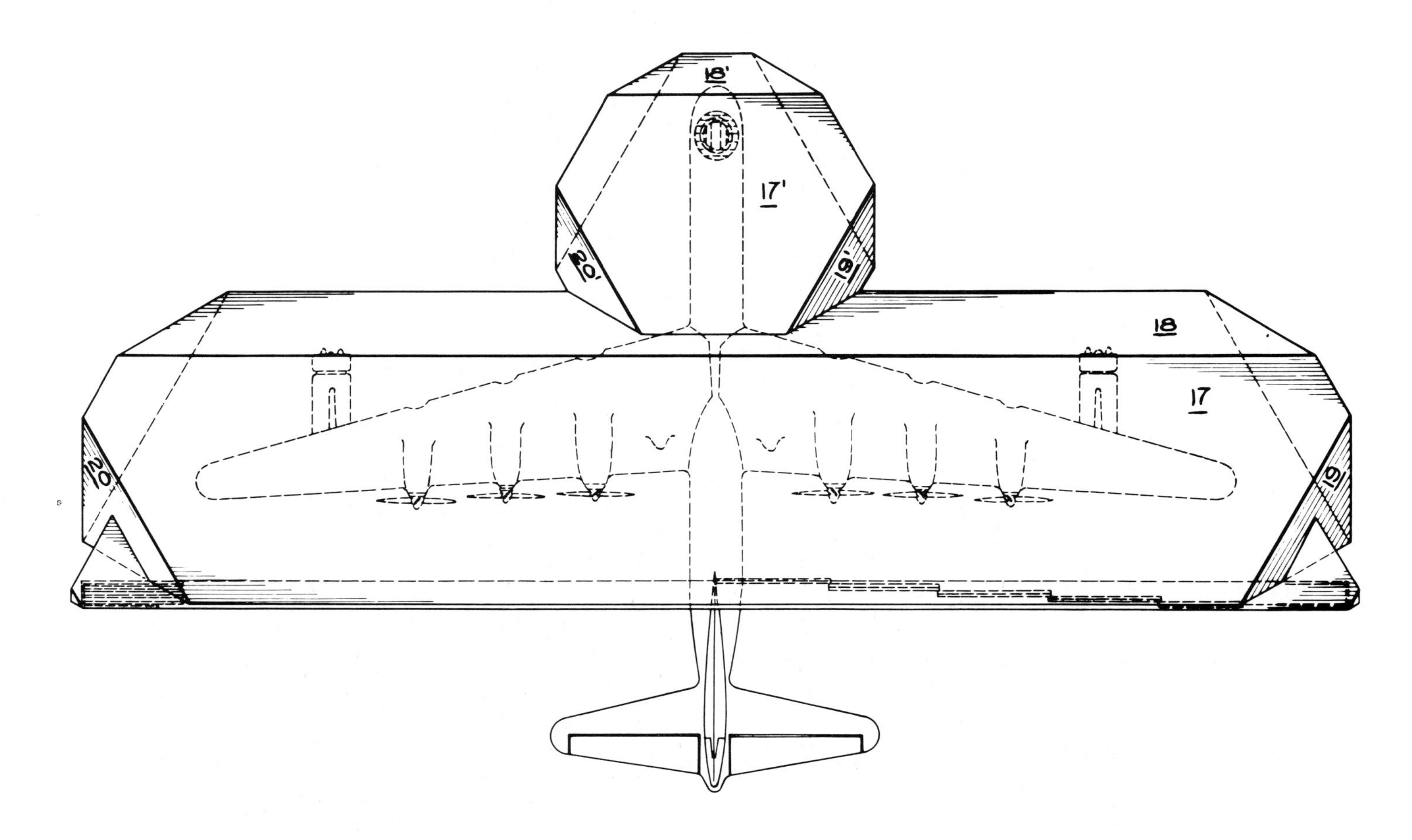

that we have a stress behavior in the system which is unpredicted by its parts.

In general, my invention consists of a roof, wall and/or floor framework consisting of a truss in which the main structural elements (e.g. struts or triangular sheet members) form equilateral triangles interconnected in a pattern consisting of octahedrons and tetrahedrons with the major axes of all octahedrons in parallelism throughout the framework. Thus all such structural elements are comprised within a single octahedron-tetrahedron system, and this apparently yields a new optimum of tensile-compressive integrity throughout the framework. Note that the singleness of the octa-tetra system or "octetruss," carries throughout the roof, wall and floor intersections. This is made possible by a novel alignment of the intersecting truss "surfaces" which holds to the integrity of the strength-creating octa-tetra system. The advantages of my construction are thus obtainable in combined roof-wall, wall-floor, and roof-wall-floor combination frameworks, as well as in individual floor, wall or roof frameworks. Consequently my invention will be found to provide a comprehensive solution to all building truss construction problems, yielding in each application a synergetic and essentially surprising result in terms of the fundamental weight-strength ratio.

Definitions of terms

Octahedron—A polyhedron having eight equal equilateral triangular plane faces or sides; may be skeletal, as when made of interconnected struts; or continuous, as when made of interlocking or interconnected sheets or plates; or partly skeletal and partly continuous.

Tetrahedron—A polyhedron having four equal equilateral triangular plane faces or sides. Like the octahedron, it may be skeletal, continuous, or a combination of the skeletal and continuous forms.

FIGURE 2

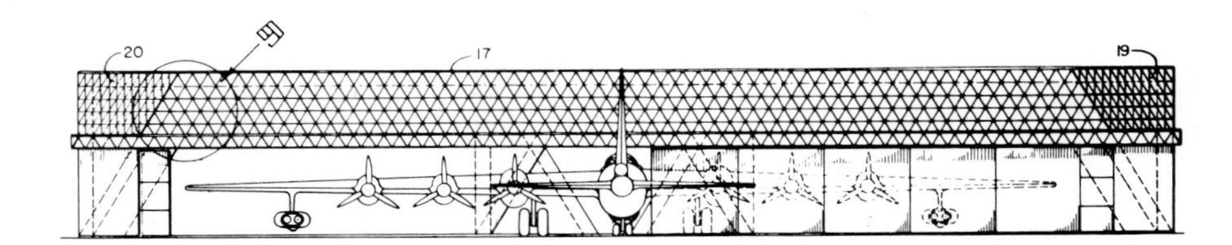

Octahedron-tetrahedron system—An assemblage of octahedrons and tetrahedrons in face to face relationship. Thus when four tetrahedrons are grouped to define a larger tetrahedron, the resulting central space is an octahedron; together, these figures are comprised in a single, or "common" octahedron-tetrahedron system.

Framework—The frame of a structure for enclosing space, or the frame of a roof, wall or floor; used to distinguish from individual frame components of a roof, wall or floor, so as to denote the whole as distinguished from its parts.

FIGURE 3

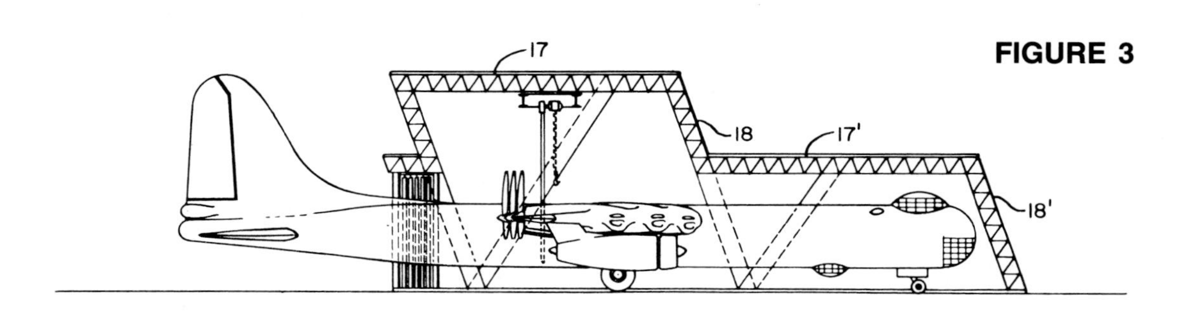

Synergy—The behavior of a system as a whole unpredicted by its parts.

DESCRIPTION

Fig. 1 is a plan view of a servicing dock for a B–36 bomber, the roof and walls of the dock being constructed in accordance with my invention.

Fig. 2 is a front elevational view of the same dock.

Fig. 3 is a vertical sectional view through the center of the dock.

Fig. 4 is a perspective view of a representative truss section of the roof and wall framework of the dock or other structural framework.

Fig. 5 is a perspective view of one of the octahedrons and a conjoined tetrahedron comprised in the truss of Fig. 4.

Fig. 6 is a schematic view of the octahedron and tetrahedron of Fig. 5 separated for clear illustration.

Fig. 7 is a detail perspective view of one of the struts of the truss of Figs. 4 and 5.

Fig. 8 is an enlarged cross-sectional view on the line 8—8 of Fig. 7.

Fig. 9 is an enlarged detail of a representative wall and roof intersection corresponding to the portion shown within circle 9 in Fig. 2. This view illustrates how all the plane surfaces of the truss conform to a common octahedron-tetrahedron system so that vector equilibrium is obtained throughout both the walls and roof of the dock framework.

Fig. 10 is a side elevational view of a modified form of strut.

Fig. 11 is an end view of the strut of Fig. 10.

Fig. 12 is a plan view of a representative connection.

Fig. 13 is a side elevational view of the same connection.

Fig. 14 is a top perspective view of a modified form of truss.

Fig. 15 is a perspective view of one of the sheets or plates which go to make up the truss of Fig. 14.

Fig. 16 is a similar perspective view of four such plates assembled to form one of the octahedrons of the truss of Fig. 14.

Reference is made first to Figs. 5 and 6 to illustrate definitions given in the Summary. Fig. 5 shows an octahedron O and a conjoined tetrahedron T which may be imagined as being formed of a number of struts of equal length joined together at their ends in any suitable manner, as by fittings *f*. In Fig. 6 the octahedron and tetrahedron of Fig. 5 are separated for clear illustration of the forms of these two kinds of polyhedrons. Tetrahedron T has six struts and four equal equilateral triangular plane faces or sides. Octahedron O has twelve struts and eight equal equilateral triangular plane faces or sides.

In Fig. 6 three of the struts have been shown by dot-dash lines because, when the tetra T and octa O are conjoined as in Fig. 5, these three struts are common to T and O.

As the truss is assembled so as to extend or "grow" in other directions, we of course have common struts and common faces or sides between all of the conjoined octas and tetras in the complete framework, and if we adhere to the integrity of this "Octetruss" system, the structure will be characterized by complete vector equilibrium. Also, the major axes of all octahedrons will be in parallelism throughout the framework, whereby all of the structural elements will be comprised in a single octahedron-tetrahedron system of optimum tensile-compression integrity throughout. Further, the sides of the octas and tetras will lie in common planes forming plane surfaces of the truss. The arrangement can additionally be defined as a roof, wall and floor framework consisting of a truss in which the main structural elements form triangles interconnected in a pattern defining four unique planes intersecting one another at acute angles, all such planes conforming to a common system of polyhedrons. Each "unique" plane is considered as including planes parallel to it, the point being that there are planes extending in four distinct directions and symmetrically oriented with respect to one another. (In a cube or rectangular prism we would have only three unique planes.) The polyhedrons (octahedrons and tetrahedrons) may be skeletal, as when made of interconnected struts as shown in Figs. 5 and 6; or continuous, as when made of interlocking or interconnected sheets or plates as shown in Figs. 14, 15 and 16 to be described; or partly skeletal and partly continuous, as when made partly of interlocking sheets and partly of struts. Fig. 14 may be imagined as illustrative of a combination of the skeletal and continuous forms in that some of the sides of the polyhedrons are "open."

Again referring to Fig. 5, we may now proceed to the definition of the octahedron-tetrahedron system as given in the Summary: an assemblage of octahedrons and tetrahedrons in face-to-face relationship. When four tetrahedrons are grouped to define a larger tetrahedron, the resulting central space is an octahedron; together, these figures are comprised in a single, or "common" octahedron-tetrahedron system.

The four unique planes of the system will also be comprehended from Fig. 9 in which they are represented by AAA, BBB, CCC and DDD. Thus the points A,A and A together define a first plane, the points, B,B and B a second plane, and so on. A′A′A′ is a plane parallel to plane AAA and is therefore not a plane "unique" from plane AAA. So we have in effect a system made up of four unique sets of parallel symmetrically oriented, omni-triangulated, planes, and for simplicity say, merely, four unique planes.

Fig. 9 also illustrates the singleness of my octa-tetra system which carries throughout the roof, wall and floor intersections. This is made possible by a novel alignment of the intersecting truss "surfaces" AAA, BBB, etc., which alignment holds to the integrity of the strength-creating octa-tetra system. The advantages of my construction are thus obtainable in a roof-wall combination framework as shown in Fig. 9, and in wall-floor and roof-wall-floor combination frameworks, as well as in individual floor, wall or roof frameworks built of trusses such as shown in Fig. 4.

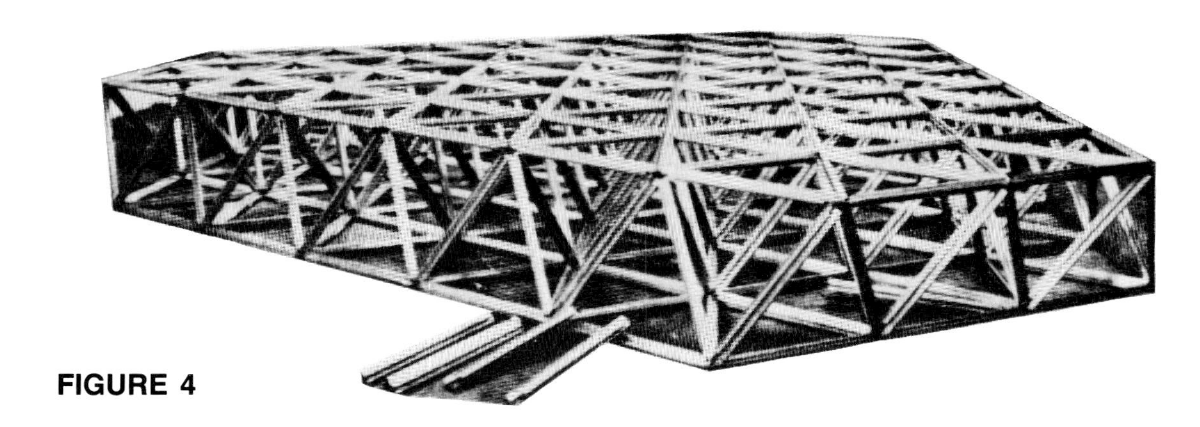

FIGURE 4

With these fundamental concepts in mind, it is now possible to comprehend the description of a particular embodiment of my invention made according to the best mode contemplated by me for carrying it out. This embodiment is illustrated in Figs. 1–9 inclusive which disclose how to construct a roof and wall framework for an airplane hangar or servicing dock. Represented is a "nose dock" for a B–36 bomber. The four unique planes of this dock framework are indicated at 17, 18, 19 and 20 (and again at 17′, 18′, 19′ and 20′). In Fig. 1 the dock is considered as roofed and sided, or "skinned"—i.e. roofed with corrugated aluminum sheet and having a polyester resin skin over the side walls of the framework. While my invention is applicable to the construction of frameworks, and truss elements therefor, of any desired material, or materials, the particular servicing dock here shown is considered to have a framework built of extruded aluminum struts, roofed with corrugated aluminum and sided with a plastic membrane.

In Fig. 2 it may be considered either that the plastic membrane is removed to show the "surface" elements of the framework, or that these surface elements are discernible through the plastic membrane cover. In any case it has been my purpose in Fig. 2 and in the Fig. 3 cross-sectional view, to reveal how it is that the single octahedron-tetrahedron system is carried throughout the roof and the various other plane surfaces of the frame-

FIGURE 5

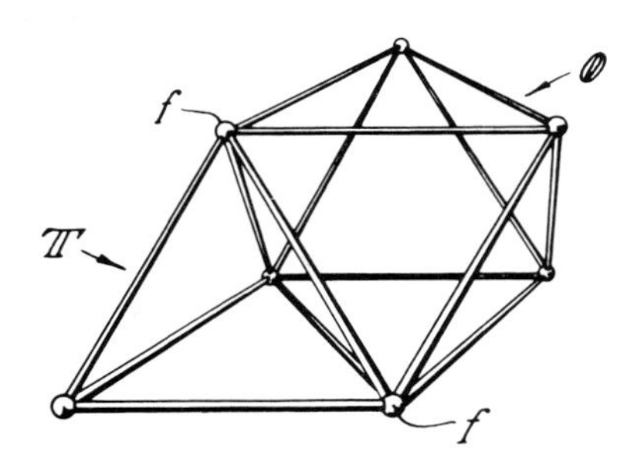

work. I wish to emphasize that this arrangement is not arbitrary, but rather is based upon my discovery that a system possessing this kind of design integrity yields strikingly improved results in terms of strength and lightness; also in terms of its low packaging cube; i.e., when disassembled, the modular parts pack for shipment into far less space as compared with its ultimate cubic enclosure. Other advantages flow from these, and as applied for example to airplane servicing docks and other military structures, introduce an entirely new concept in logistics. The extreme lightness in turn produces important reduction in foundation loading. The fact that all of the modular elements (struts or sheets as the case may be) are the same, simplifies erection by eliminating selectivity of parts. Any member is the right member and very little skill is required for assembly. The structure is adaptable

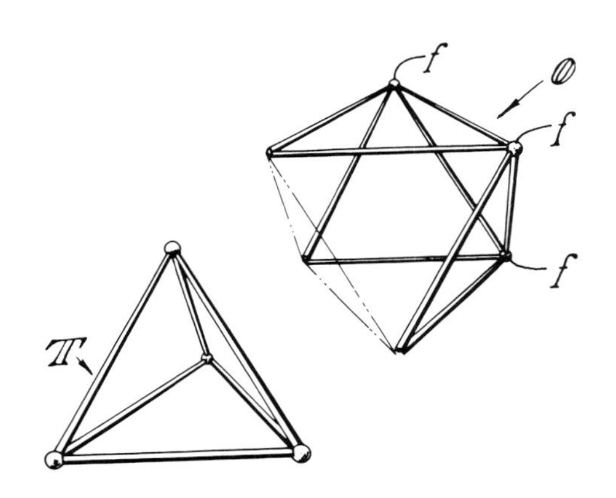

FIGURE 6

for many uses. It can form flat slabs for roof or floor construction. It can be made as a pitched roof, and can be adapted for use as a bridge or trestle for vehicle or pedestrian. Working floors and platforms for hangars and other buildings can be made from the same units as comprise the building structure itself, and can be comprised within the same octa-tetra system, further contributing to the realization of the advantages flowing from the unitary character of my system of construction.

In order to give some notion as to the practical advantages of my invention in the particular adaptation selected for illustration, I may cite that the servicing dock for the B–36 bomber is 296′ in length by 68′ in width, the nose section comprising an area approximately 54′ × 56′, providing an over-all covered area of 19,692 ft.^{2}. The construction is entirely of aluminum extrusions with a total strut weight of 74,595 lbs. The corrugated aluminum roof cover weighs 19,892 lbs. and the wall membrane 5,910 lbs. The total weight of the entire structure is 115,887 lbs., but if we subtract the weight of the door and track, the weight of the covered framework itself is just over 100,000 lbs. (100,397).

The significance of these figures will be understood when it is realized that the weight of the structure is a mere 0.115 lbs./ft.^{3}; that the foundation load is only 254 lbs. per lineal foot; and that when disassembled the modular parts pack for shipment into approximately $\frac{1}{350}$ of its ultimate cubic enclosure. Actually this entire hangar, including the struts, corrugated aluminum roof, membrane,

FIGURE 7

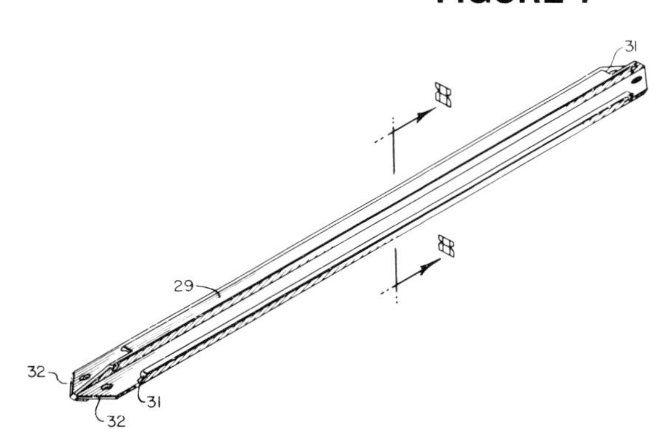

doors and tracks when packaged for shipment, can be carried by two trailer trucks with space to spare in the second truck. This is on the basis of a trailer truck of a capacity of 1536 cubic feet. The strength of the framework exceeds manyfold the results which would be calculated using any conventional engineering formulae based on revolution of forces and known values of strength of materials. This suggests that there is some kind of a stress behavior in the system as a whole which cannot be predicted, and which perhaps can only be described as "synergetic."

Note in Fig. 3 that the struts of the roof are in alignment with the struts in the walls so that we adhere to the common octahedron-tetrahedron system, producing a framework of novel integrality. In the specification and in the claims I employ the term "framework" in the sense defined in the Summary, as the frame of a structure for enclosing space, or the frame of a roof, wall or floor; distinguishing from individual frame components of a roof, wall or floor so as to denote the whole as distinguished from its parts. Fig. 9 further illustrates the detail of the system in which the major axes of all of the tetrahedrons are parallel throughout the framework, or in which all of the main structural elements (struts) form triangles interconnected in a pattern defining the four unique planes AAA, BBB, CCC and DDD intersecting one another at acute angles, all such planes conforming to the common system of polyhedrons.

In Fig. 4 we see in perspective a representative truss section of the roof and wall framework of the dock of Figs. 1–3. And in Fig. 5 we see one of the octahedrons O and one of the tetrahedrons T comprised in the truss of Fig. 4 (compare the exploded view, Fig. 6).

The struts may be of any desired form, but in Figs. 7 and 8 I have illustrated a feature of a preferred construction in which the struts are generally X shape in

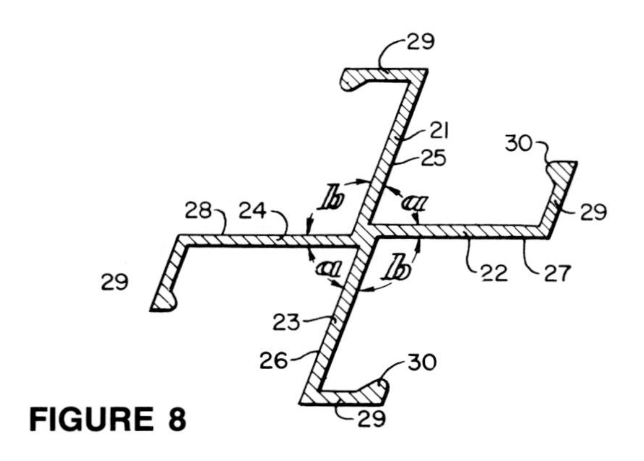

FIGURE 8

cross section, with the sides 21—24 of the X section of the struts disposed at such an angle to one another as to lie substantially in the planes of the sides of the octahedrons and tetrahedrons constituted by the respective struts. (Angles *a* approximately 70° 32′, angles *b* approximately 109° 28′.) Further, as seen best in Fig. 8, the sides of the X section of the struts are offset sufficiently to bring one surface of each into the plane of the center of the X. Otherwise stated, surface 25 of side 21 lies in the same plane as surface 26 of side 23. Similarly, surface 27 of side 22 lies in the same plane as surface 28 of side 24. The respective sides 21—24 may be provided with strengthening or stiffening flanges 29 swastika fashion, and the flanges may have inwardly extending projections 30 of bead-like conformation. I have had struts of this form fabricated successfully as aluminum alloy extrusions. They may, if desired, be extruded of magnesium or other alloys, or other materials. Flanges 29 should be cut away as shown at 31 in Fig. 7 at each end of the strut so as to avoid interference with the sides of interconnected struts where they are joined together. Also, the ends of the struts are cut back at an angle, as at 32, for the same purpose, and holes are drilled for bolts or rivets.

Another form of strut according to my invention, comprises aluminum tubing or the like, into the ends of which are inserted fittings so designed that the ends of the struts will be generally X shape in cross section. This modification is illustrated in Figs. 10 and 11, which may for example comprise an aluminum tube 33 to the ends of which are fastened the fittings which have tubular portions 34 to match the inside tube end, and flanges 35 disposed in the same manner as the sides 21—24 of the X section described with reference to Figs. 7 and 8. These fittings are drilled to receive the fastenings used to secure the various struts together at the intersections. (These are the fittings *f* previously referred to in describing Figs. 5 and 6.) They go together in the manner illustrated in Figs. 12 and 13, Fig. 12 being a plan view of a representative connection, and Fig. 13 a side elevational view of the same connection. This is a "9-point" connection providing six struts (six axii) radiating outwardly from the center of one of the hexagons that can be seen in the plane A′A′A′ of Fig. 9, and three struts extending downwardly from that

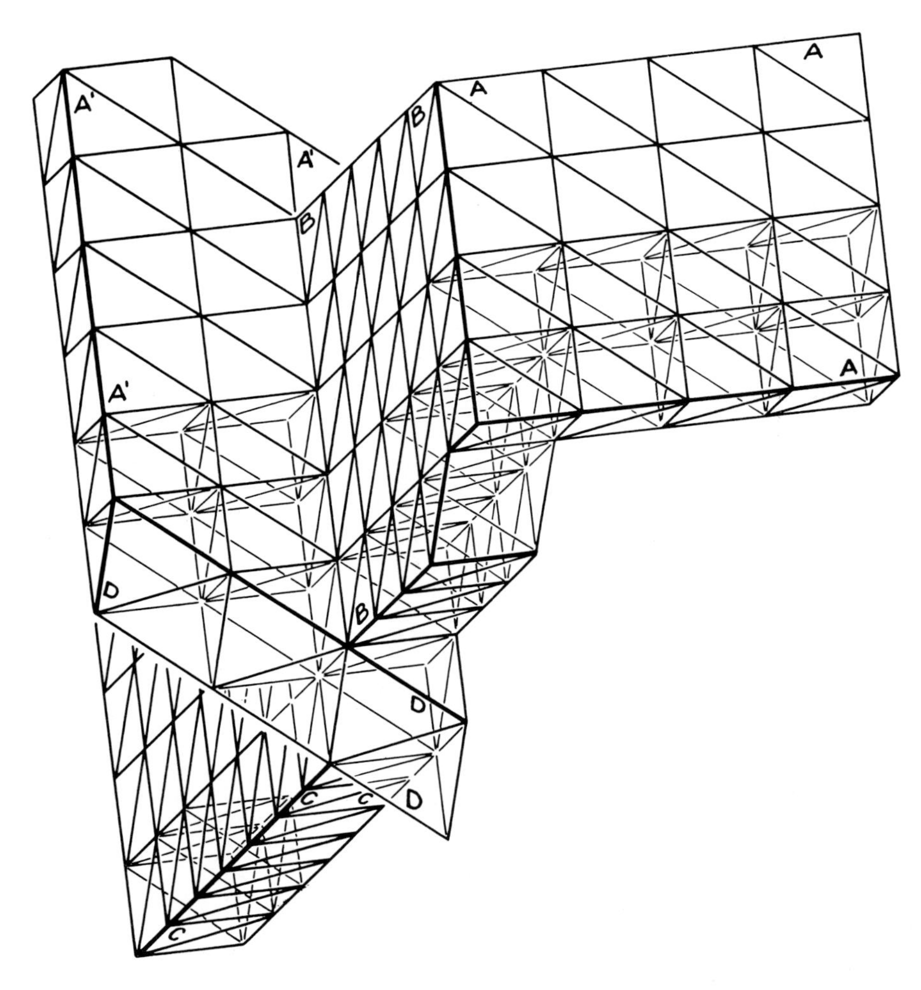

FIGURE 9

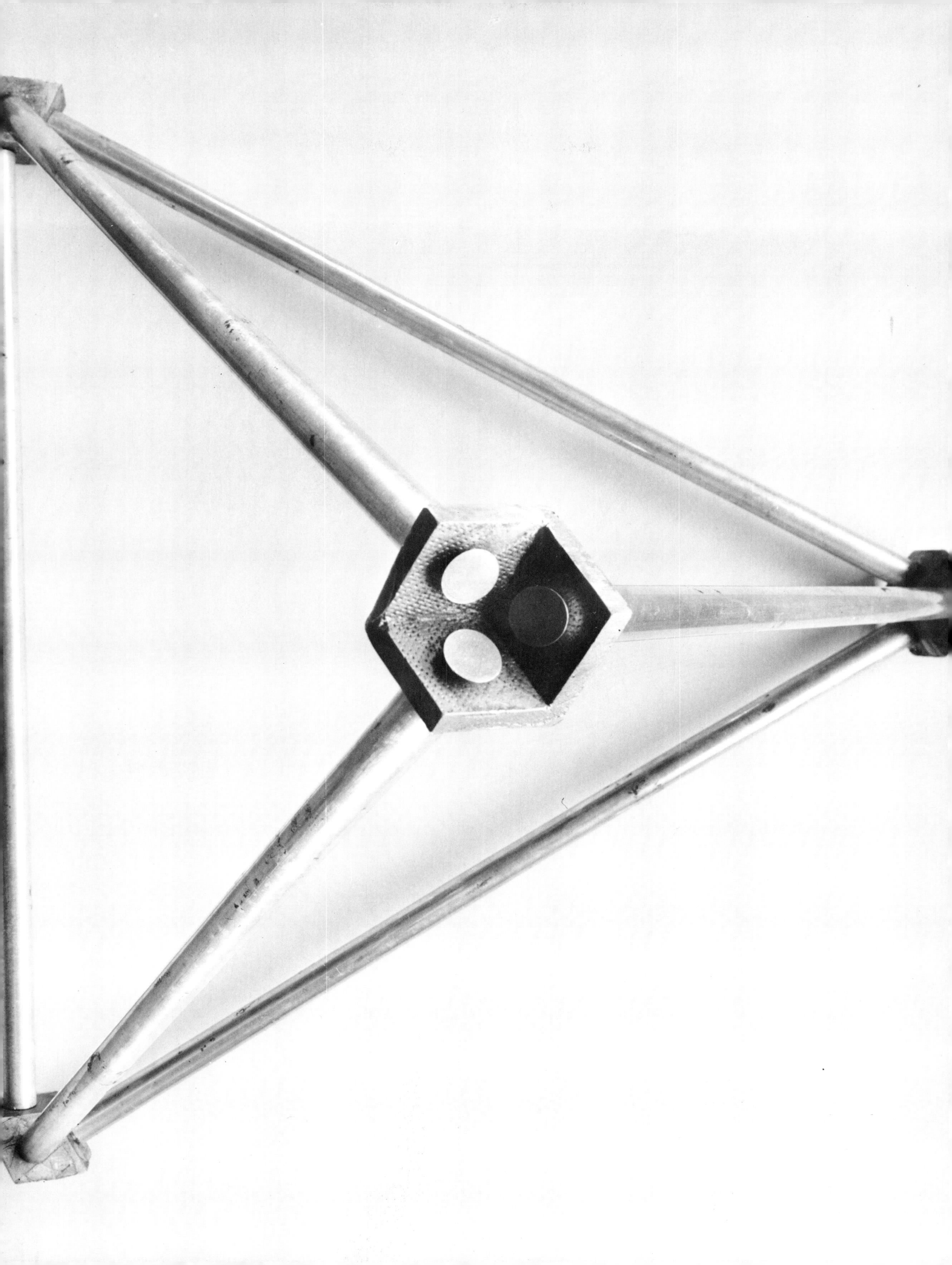

FIGURE 10

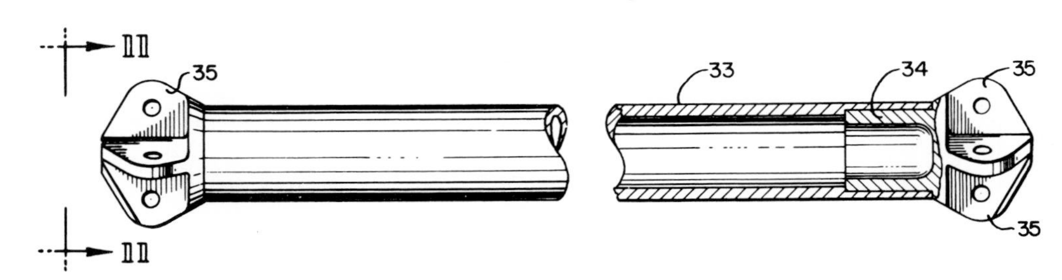

FIGURE 11

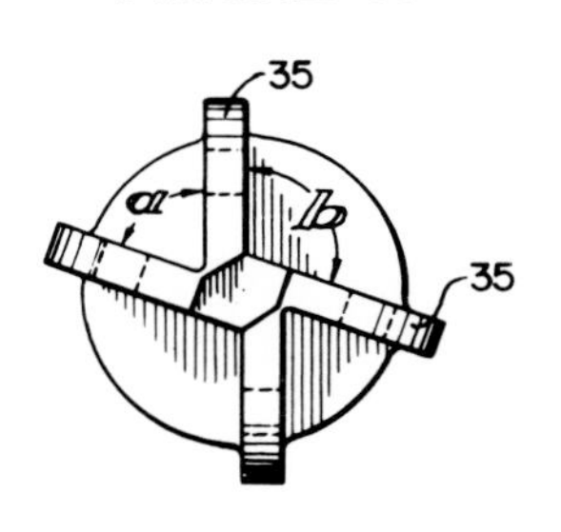

FIGURE 12

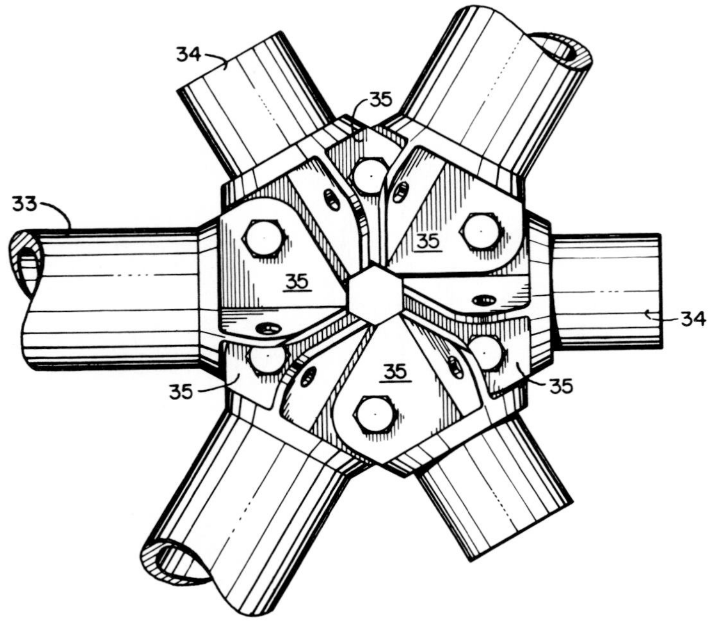

FIGURE 13

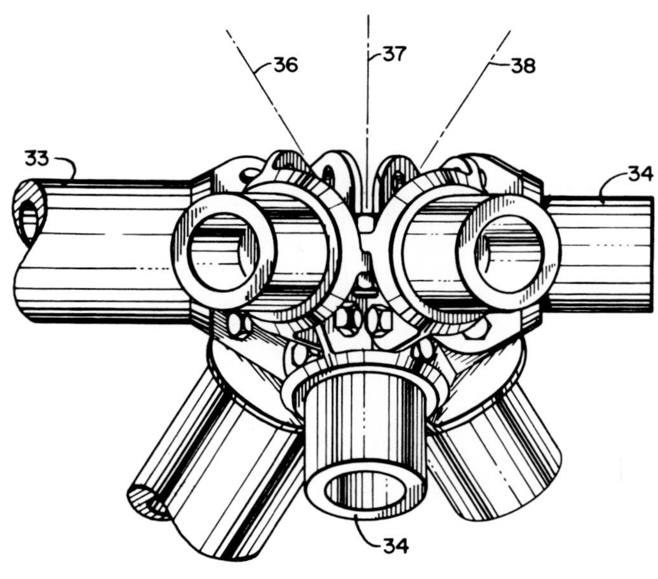

center as the apex of the tetrahedron directly below it.

In sections of the framework where we get in effect a double truss, as occurs whenever we come to a roof-wall or wall-floor intersection, there is required a full "12-point" (six axii) intersection, and the construction illustrated in Figs. 12 and 13 is adapted for such an intersection, the three additional struts coming into the intersection on the axes 36, 37 and 38 shown in Fig. 13. Note in Fig. 12 that flanges 35 of the fittings overlap in a uniform clockwise pattern. However, these fittings may if desired be designed for counter-clockwise overlapping. In either case the connection is characterized by what I term "plus and minus turbining." Inasmuch as two members can never go through the same point, they must always turbine right or left—as in the poles of a tepee. I believe the the particular construction which I have described

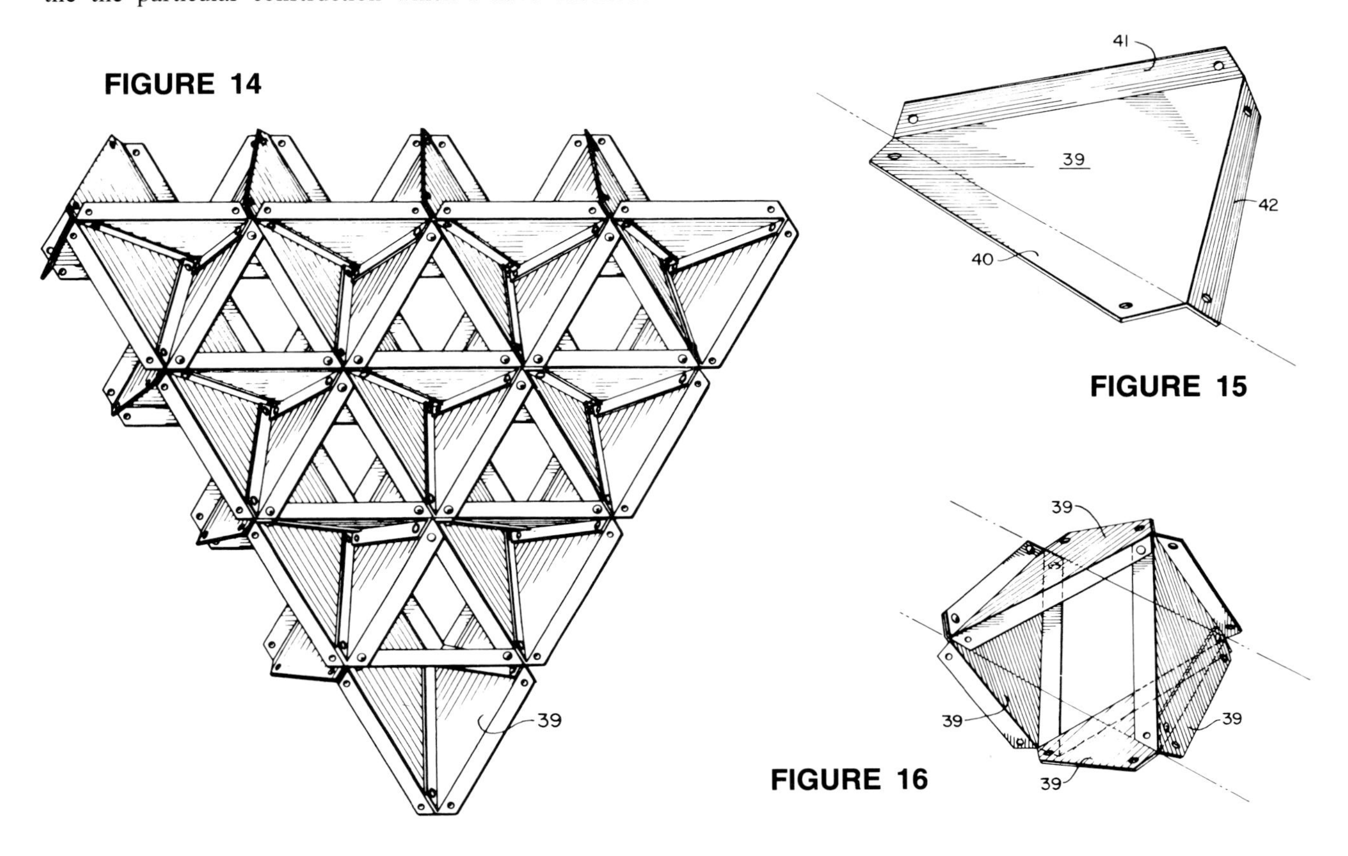

FIGURE 14

FIGURE 15

FIGURE 16

with reference to Figs. 12 and 13, as well as the connections made with the struts in Figs. 7 and 8, possess peculiar advantages when employed as the connecting points of the four unique planes of my framework, and that this so-called "turbining" construction contributes importantly to the surprising results which I have been able to attain with such frameworks. However, I have found it possible to obtain exceptional results also with other types of connections, so that the octa-tetra construction may be considered to have utility apart from the particular construction described, while in another aspect of my invention the connection and the octa-tetraconstruction are considered to possess a special coaction when both are employed together.

In some cases the surface aspect of my framework may present the plane which represents the middle of the octahedrons. In such a case we have an example of a one-half octahedron, or pentahedron. Such a system retains the feature of providing four unique planes intersecting one another at acute angles, in which all such planes conform to a common system of polyhedrons.

Figs. 14 through 16 illustrate a typical application of my invention to frameworks as built up of sheet modules instead of strut modules. Fig. 15 shows the module. It may, for example, be a thin sheet 39 of aluminum with a flange 40 extending from one edge thereof coplanar with the body of the sheet and flanges 41 and 42 extending from its other two edges at the proper angle to lie in the planes of the faces of the octahedrons and tetrahedrons of the system of the framework. Flange 41 extends upwardly and outwardly of the sheet, flange 42 downwardly and outwardly. The flanges may be apertured for fastenings as shown, but in some cases I prefer that the flanges of the aluminum sheets be held together with epoxy cement.

In Fig. 16 we see four of the aluminum sheets 39 assembled to form one of the octahedrons of the truss of Fig. 14. This truss is made up entirely of identical modules, just as the strut form of truss previously described is made up of identical modules. That is, one type and size of strut, or one type and size of sheet, does the job for the entire structure, floors, walls, and roof. Thus, in each case, I build a roof, wall or floor framework consisting of a truss in which the main structural elements form triangles interconnected in the pattern which has been fully described hereinabove. Sheet 39 is in the form of an equilateral triangle. Flanges 41 and 42 extend at an angle thereto of approximately 109° 28′.

The terms and descriptions which I have employed are used in a descriptive and not a limiting sense, and I have no intention of excluding such equivalents of the invention described, or of portions thereof, as fall within the scope of the claims.

15▲TENSEGRITY (1962)

U.S. PATENT—3,063,521

APPLICATION—AUGUST 31, 1959.

SERIAL NO.—837,073

PATENTED—NOVEMBER 13, 1962

NOTHING IN UNIVERSE TOUCHES anything else. The Greeks misassumed that there was something called a solid. Democritus thought it could be that there were some smallest things in those solids, to which he gave the name *atom.* Today we know that the electron is as remote from its nucleus as is the Earth from the Moon in respect to their diameters. We know that macroscosmically none of the celestial bodies touch each other. So, both microscosmically and macrocosmically nothing touches. It was Kepler who discovered the tensional coherence of the solar system despite the millions of miles that intervene between the celestial bodies. Isaac Newton hypothesized the formula for the rate of variance of the interattractiveness of celestial bodies, which proved on comprehensive employment to be correct.

From 1927 on I sought to discover how to produce what I call tensional integrity structures. I have said many times, at some of the top engineering schools in the world, structural analysis is predicated on compressional continuity and nature doesn't use it ever. There is no way for structural analysis to analyze a geodesic dome. This still continues to be true. I think it is reprehensible. The only way to analyze it is with pneumatics and hydraulics. At the molecular level, this is the method of quantum mechanics; it could never be done with crystalline continuity, for such continuity does not exist in Universe.

This brought about my tensional integrity structures, which name I contracted to *tensegrity* structures. Tensegrity structures are the essence of all geodesic domes. When we increase the frequency of modular subdivisions of geodesic domes the edges of the triangles, representing the chords of central angles, get shorter and shorter and the interval between the mid-chord and the mid-arc of the central angles also decrease with the increasing frequency of the modular subdivisions. Because the materials used in the construction of the dome have some substantial dimension, we get to the point where the high-frequency production of the arc-altitude is such that the materials (the individual tensegrity components) touch one another. Every one of these elements is where it wants to be within the structure—there are no tensions anywhere, no slacks, all of the stresses are absolutely even—so we then fasten the two structural components together where they touch and want to be. This takes out the springiness of the geodesic domes and makes them rigidified. Because tension and tensegrity have no limit of clear spanning, tensegrity structures open up completely clear-spanned dome structures of any size.

UNITED STATES PATENT OFFICE

Richard Buckminster Fuller, 104—01 Metropolitan Ave., Forest Hills 75, N.Y.

TENSILE-INTEGRITY STRUCTURES

The invention relates to a system of construction which utilizes the tensile properties of structural materials to the fullest advantage. It has special application to structures of vast proportions such as free-span domes capable of roofing a stadium or housing an entire village or city, and to mammoth air-flotable spheres as well as collapsible light weight structures adapted to be transported by rocket. In general, my invention is useful wherever it is advantageous to make the largest and strongest structure per pound of structural material employed. It is applicable also to geodesic structures such as described and claimed in my prior Patent No. 2,682, 235.

SUMMARY

The essence of my invention consists in the discovery of how to progressively reduce the aspect of compression in a structure so that, to a greater extent than has been found possible before, the structure will have the aspect of continuous tension throughout and the compression will be subjugated so that the compression elements become small islands in a sea of tension. This is to bring the slenderness, lightness and strength of the suspension bridge cable into the realm previously dominated by the compression column concept of building. The suspension bridge is fundamentally a tensioned structure through its use of the catenary curve of the cables between compression column towers. My invention is akin to taking some of the compression out of the "compression towers," i.e. the columns, walls and roofs, of a building, or even taking compression out of a single column or mast through the creation of a structure having discontinuous compression (as hereinafter defined) and continuous tension and wherein the islands of compression in the mast are progressively reduced in individual size and total mass.

As applied to a geodesic dome structure, I might describe my invention as a structure of generally spherical form comprising discontinuous compression columns arranged in an overall pattern of three-column tepees each column of each tepee being joined in "apparent" continuity to one column of one of three adjacent tepees to form what appears as a single column-like member, and the outer parts of the columns of each tepee being connected to one another only by tension elements. Tension elements also connect the outer parts of the columns to points on the columns in the region where they are joined together in apparent continuity.

As applied to structures generally, my invention con-

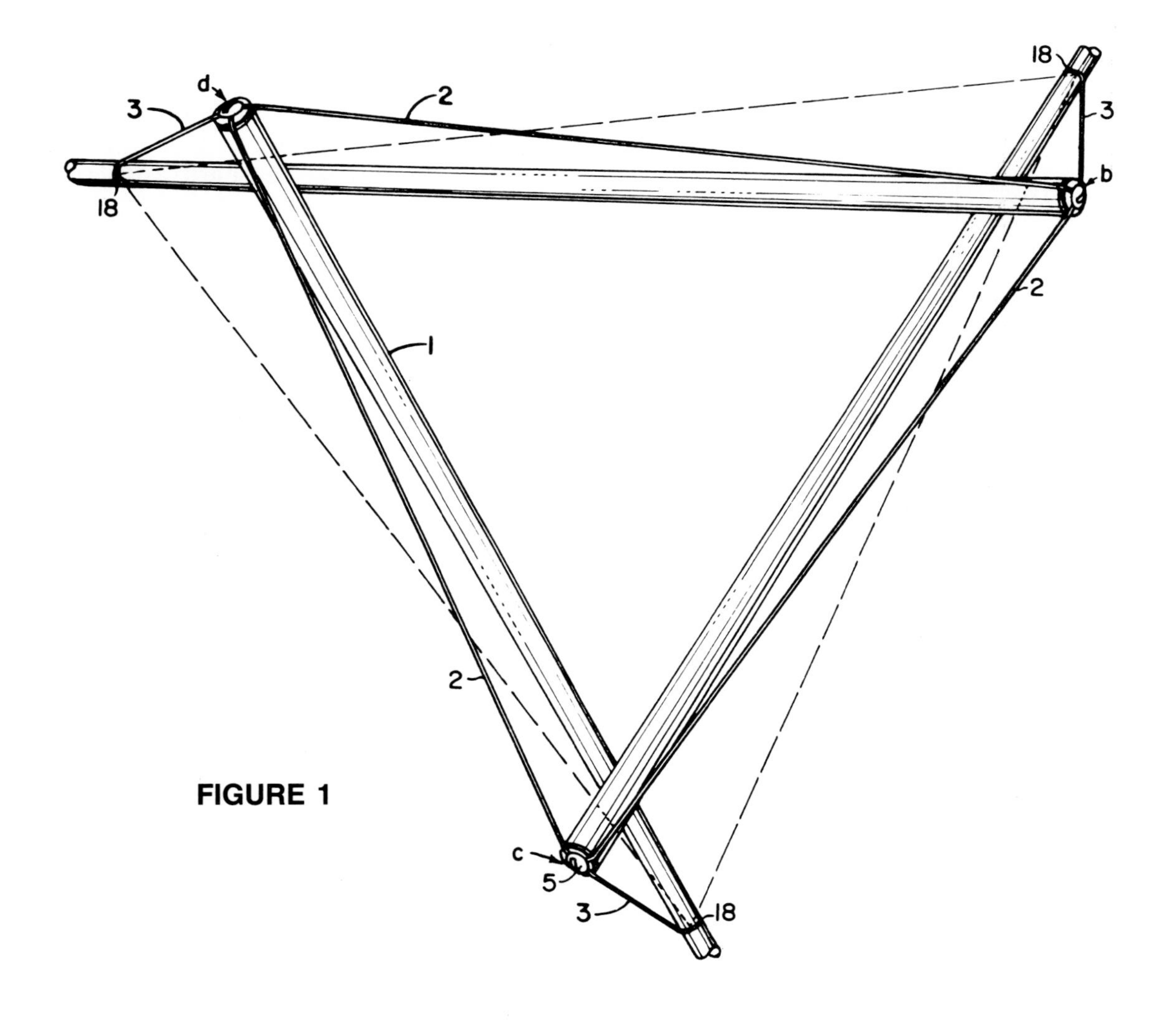

FIGURE 1

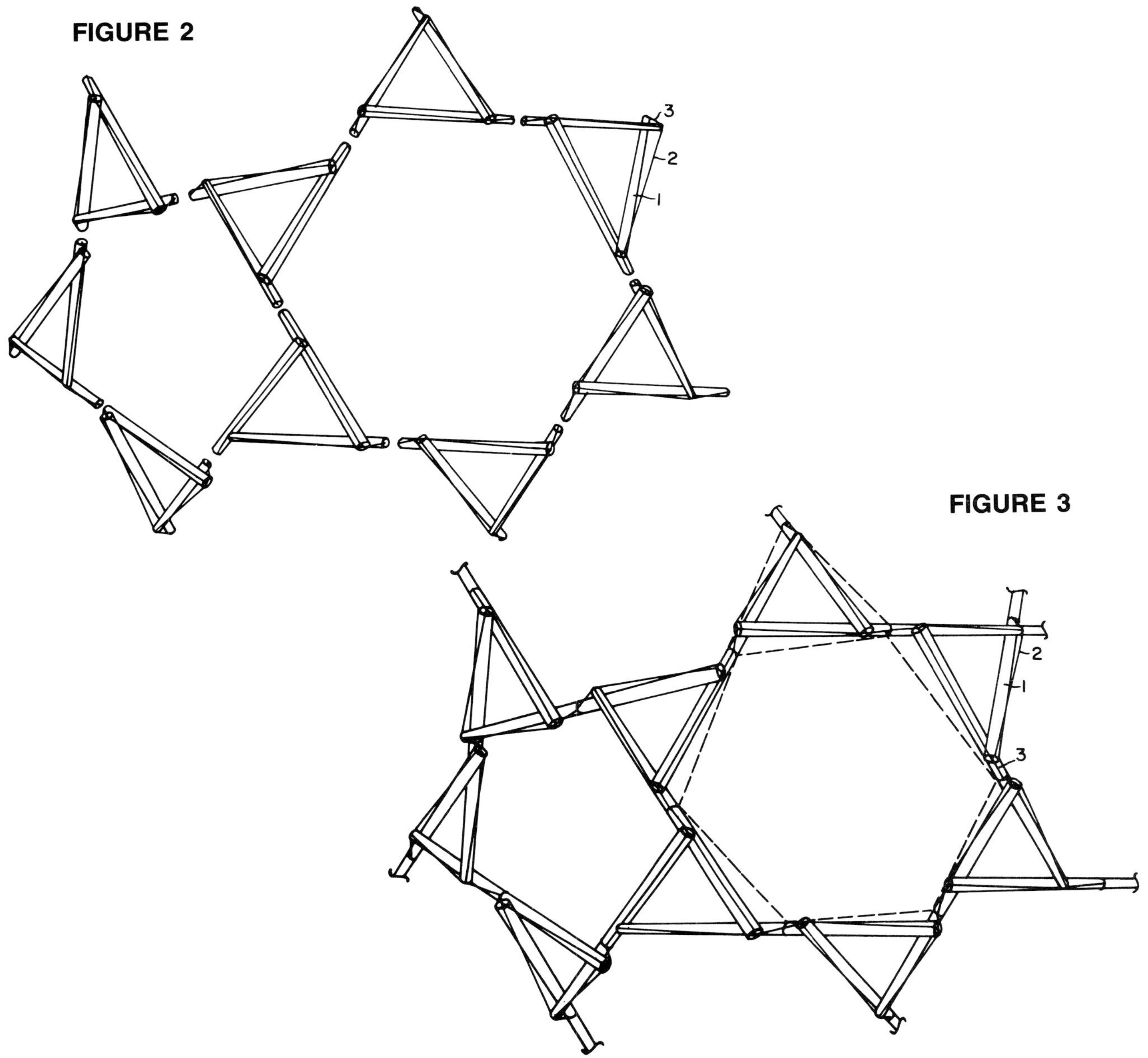

sists in a structure comprising a plurality of discontinuous compression columns arranged in groups of three nonconjunctive columns connected by tension elements forming tension triangles, columns of adjacent groups being joined together in apparent continuity as above described.

According to another aspect of my invention the structure comprises an assemblage of tension and compression components arranged in a discontinuous compression system in which the compression components comprise an assemblage of compression and tension components arranged in a discontinuous compression system whereby the islands of compression in the initial discontinuous compression system are progressively diminished in size and relative mass.

Description

Fig. 1 shows the plan of a three-column tepee with connecting tension elements, called a three-strut octahedral tensile integrity unit, or "tensegrity."

Fig. 2 shows an assemblage of the three-strut tensegrities of Fig. 1. This view is analytical, for in the actual structure struts of adjacent tensegrities are integrally joined in "apparent" compressional continuity. This actual structure is shown in

Fig. 3, which otherwise corresponds to Fig. 2, and shows the discontinuous compression structural complex.

Fig. 4 is a side elevational view of the strut and tension sling component of the discontinuous compression structural complex of Fig. 3, called a "boom."

Fig. 5 is a plan view of the boom of Fig. 4.

Fig. 6 is a sectional view taken on line 6—6 of Fig. 4.

Fig. 7 is a boom dimension schedule for a 270-boom tensegrity sphere.

Fig. 8 is a color code for assembling the booms of Fig. 7 according to one embodiment of my invention.

FIGURE 4

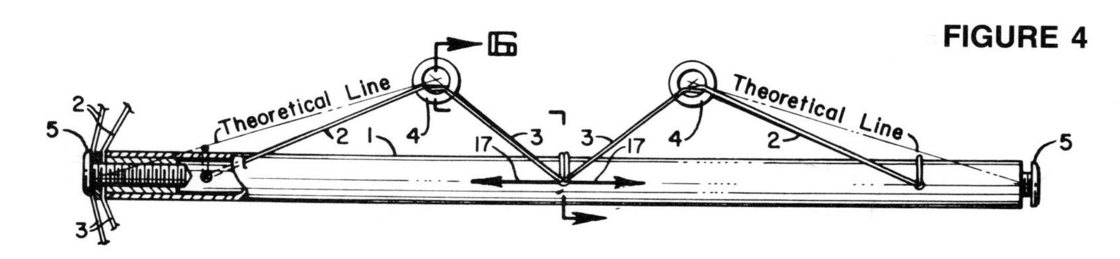

Fig. 9 is a further color code for assembling the booms.

Fig. 10 is a plan view of a 270-strut tensegrity obtained by following the color code of Figs. 8 and 9.

Fig. 11 is a plan view of a 270-strut isotropic tensegrity obtained by assembling booms which are all of exactly the same design.

Fig. 12 is a diagram to explain the turbining tendencies of the thrusts of the boom ends in the tensegrity of Fig. 11, known as a "single bonded" tensegrity.

Fig. 13 is a diagram to explain the turbining tendencies of the thrusts of the boom ends in the tensegrity of Fig. 10, known as a "double bonded" tensegrity.

Fig. 14 shows a modified form of strut, or boom.

Fig. 15 is a perspective view of another modified form of boom.

Fig. 16 is a perspective view showing how the booms of Fig. 15 are interconnected in a tensegrity complex.

Fig. 17 is a plan view of a 270-strut geodesic sphere tensegrity utilizing the complex of Fig. 16.

Fig. 18 is an isometric view of a discontinuous compression column which can be used to replace the compression strut of the component shown in Fig. 4.

Fig. 19 is an enlargement of a portion of Fig. 18 showing a single strut made as a discontinuous compression column according to what may be imagined as a reduced scale version of Fig. 18. The area of Fig. 18 covered by Fig. 19 is that indicated at 19 on Fig. 18.

Fig. 20 is a top perspective view of a further modified form of boom which includes surfacing elements for a goedesic dome.

Fig. 21 shows an assemblage of a number of identical booms of the form shown in Fig. 20.

Fig. 22 is an enlarged cross-sectional view taken as indicated at 22—22 in Fig. 21.

FIGURE 5

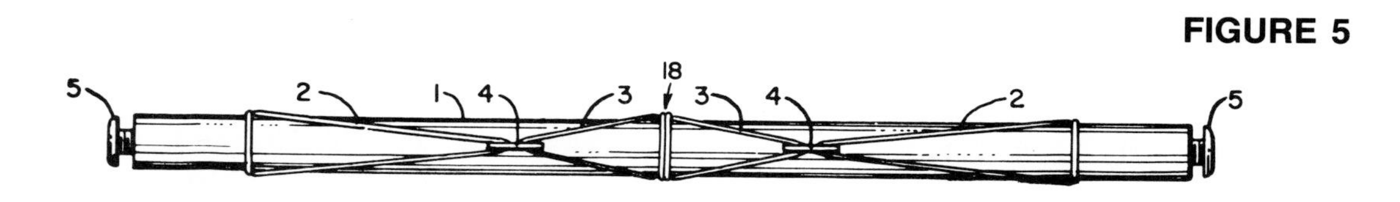

FIGURE 6

My tensegrity structure comprises a plurality of compression columns arranged in groups of three non-intersecting, or spaced, columns 1, crossed or overlapped to make a tripod as in an Indian tepee, Fig. 1, the columns of adjacent groups, Fig. 2, being integrally joined together, Fig. 3. Figs. 1 and 2 are analytical, Fig. 3 showing the actual structure. Fig. 1 illustrates diagrammatically the cohering principle of the primary system as one component of my unique tensile integrity complex. This primary system, a three-strut tensile integrity unit, is termed a "tensegrity." It will be observed from Fig. 1 that this primary system displays the six vertexes which are characteristic of the octahedron, a polyhedron having three axii, six vertexes, eight faces and twelve edges. In a spherical system made up of such primary octahedral tensegrities it becomes possible to omit tension wires which if present would lie along six of the twelve edges of each octahedron. The omission of such wires tends to obscure the visual appearance of the eight triangular faces of the octahedron, but does not destroy the octahedral aspect of the primary system that is necessarily fixed and predetermined by the presence in the system of the aforesaid six vertexes which characterize the octahedron. Columns 1 are connected by tension elements such as wires or cables 2 forming a tension triangle, a—b—c—, and tension elements 3 connect the vertexes a, b, c of the tension triangle to points 18 on the columns 1 in the region where they are joined to the columns of adjacent tensegrities. The dotted lines outlining the base of the tripod in Fig. 1 are theoretical and do not necessarily exist in the actual structure. If these dotted lines are considered to represent tension wires we have a complete primary octahedral tensegrity system in which the compression columns 1 are separated from one another by the tension elements 2 and 3. Tension elements 2 and 3 extend throughout the structure in a continuous network, whereas columns 1, being separated from one another, do not make a continuous network and may be described as compression columns arranged discontinuously, hence "discontinuous compression columns." Another aspect of discontinuous compression is found within a single column and will be understood from Fig. 3: the tension forces in wires 3, when resolved, have components 17, 17 acting in opposite directions. Thus the right-hand end of column 1 as viewed in Fig. 3 acts functionally as one compression column and the left-hand end acts as another compression column. Thus, if we disregard bending forces, the compression forces can approach theoretical zero at the point 18 in the column where the tension elements 3 are attached to it. This results in a separation of the compression forces in the two ends of the column. Because of this separation of the compression forces, we have again what may be described as "discontinuous compression," this time within a single column. Yet, because the two ends of the column are parts of one integral member, it appears as though there is just one compression member. Inasmuch as the foregoing analysis of the compression forces has revealed that there are really two separate compression columns within the one integral member, it seems helpful to think of the columns of the adjacent groups, Fig. 2, as being joined together, Fig. 3, in "apparent" continuity, for although the continuity in a structural sense is real, in respect of functions in compression it is not real. Hence the compression columns are said to be joined together in "apparent continuity." And because the tension network is continuous throughout the structure, whereas the columns are separated from one another and are supported so as to float in the network of tension wires, it has seemed appropriate to characterize the structure further as comprising compression elements which are like "islands" of compression in a "sea" of tension elements. If desired, the structure may contain tension wires as represented by the dotted lines in Figs. 1 and 3. Such wires when used will relieve the columns of bending stresses and complete the interior tensional integrity of the octahedral system. The compression members are said to be discontinuous, because no force of compression is transmitted from one to the other as they float in a sea of tension elements. Their positions are fixed by the octa system of the unit, although they have a turbining tendency as we shall see later. In the discontinuous compression assembly of Fig. 3, termed a com-

FIGURE 7

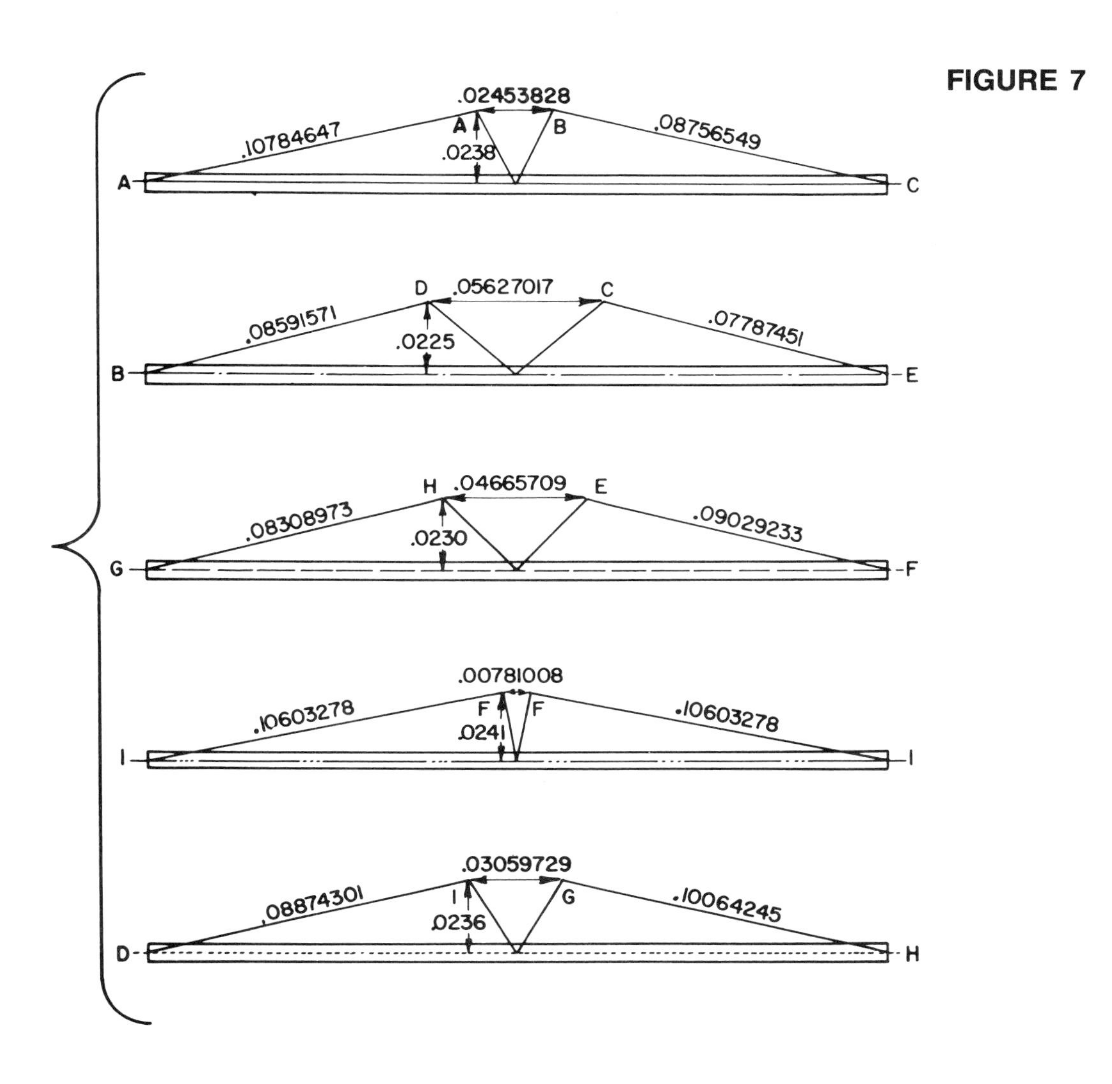

tension is created from end to end of each compression element by the tension elements 2 and 3.

In the integrated struts or columns of adjacent tensegrity units we have the prefabricated component of my structure shown in Fig. 4, termed a boom, which in combination with the tension elements 2 and 3, emerges as a characteristic form which I call a "B-boom" owing to its configural resemblance to the latter B. This boom is functionally an element of two adjacent primary tensegrities by reason of the separation of the compression forces in the manner already described. In the particular embodiment of the B-boom shown in Figs. 4—6, the compression column takes the form of a tubular strut such as a tube of aluminum or steel, drilled to receive the wire lacings of the tension slings 2, 3, and having a Roule plug 16 in each end to receive a threaded fastening 5. Washers 4 at predetermined apexes of the sling triangles are adapted to receive the fastening at the ends of mating booms. If there is slack in the system it can be taken up by additional washers at the ends of the tube. The wire slings can be double as shown, or single. The mating booms are secured together by inserting one of the fastenings 5 of one boom through one of the washers 4 of the other as shown in Figs. 1 and 4.

FIGURE 8

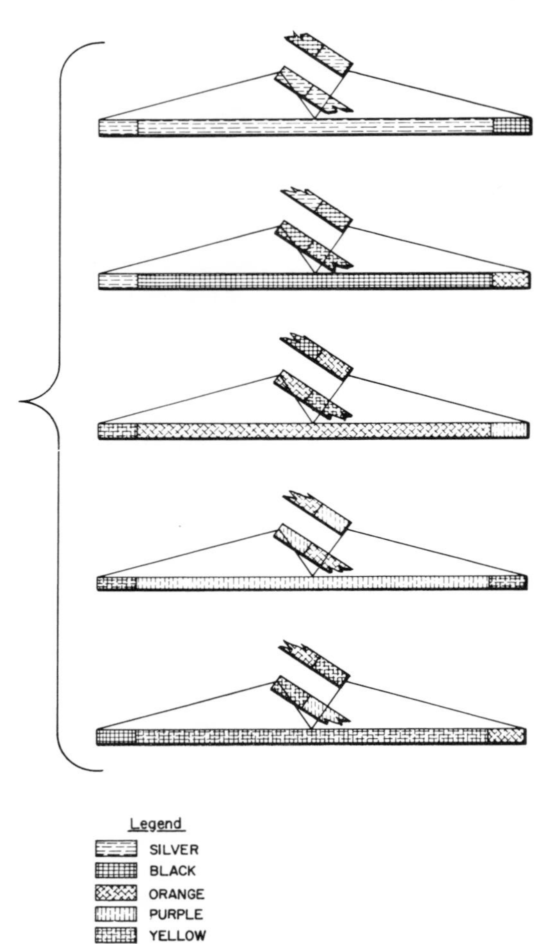

FIGURE 9

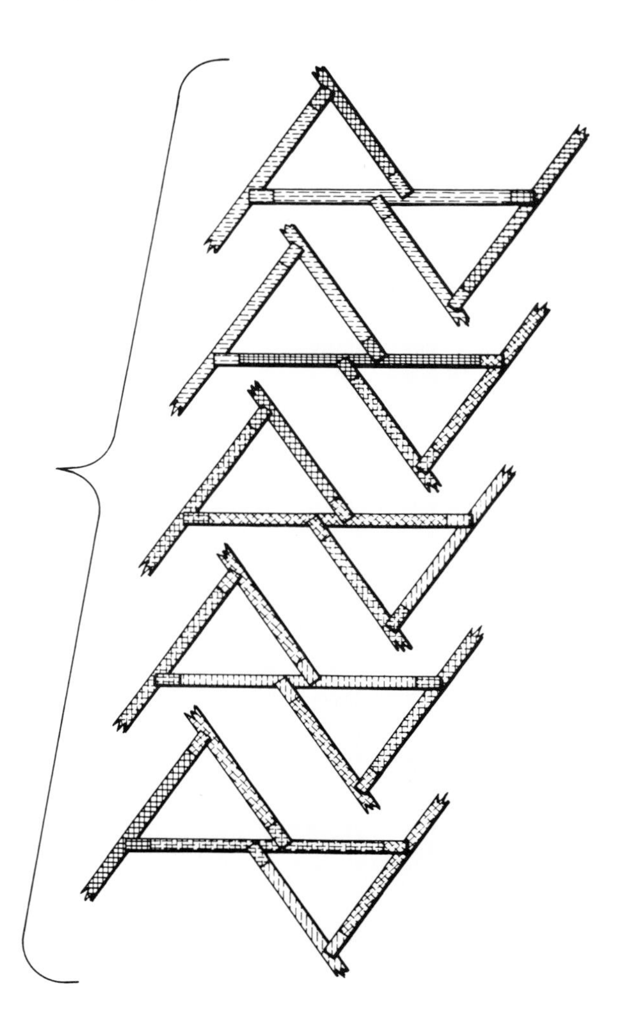

plex of the primary three-strut tensegreties, or a "complex tensegrity," the imaginary wires represented by the dotted lines in Fig. 1 are made unnecessary because of the arrangement of the primary tensegrities in a spherical system.

In our consideration of the complex tensegrity of Fig. 3, it will be observed that the terminal junctures of the several primary tensegrities are all in alignment, i.e. 180° junctures, and are apparently compressionally continuous by reason of the fact that each pair of columns 1 integrally joined together, appear as a single column-like member. Because of this seeming continuity of compression from one primary tensegrity to another and because the central coherence of the primary tensegrities is visibly discontinuous by reason of omission of the tension elements shown by the dotted lines in Figs. 1 and 3, the complex tensegrity presents a visibly deceptive appearance to the unwary observer in which the joined tripod legs of adjacent units appear as single units and, as such, appear to be the primary "elements" of the complex tensegrity, whereas, we have learned from our analysis of Fig. 1 and its relation to Figs. 2 and 3 that our elements are the three-strut octahedrons and that the cohering principle of the simplest elements is tensegrity. Notice that because the tension elements 3 connect the vertexes of the tension triangle to points on the struts 1 in the region where they are joined together, a continuity of

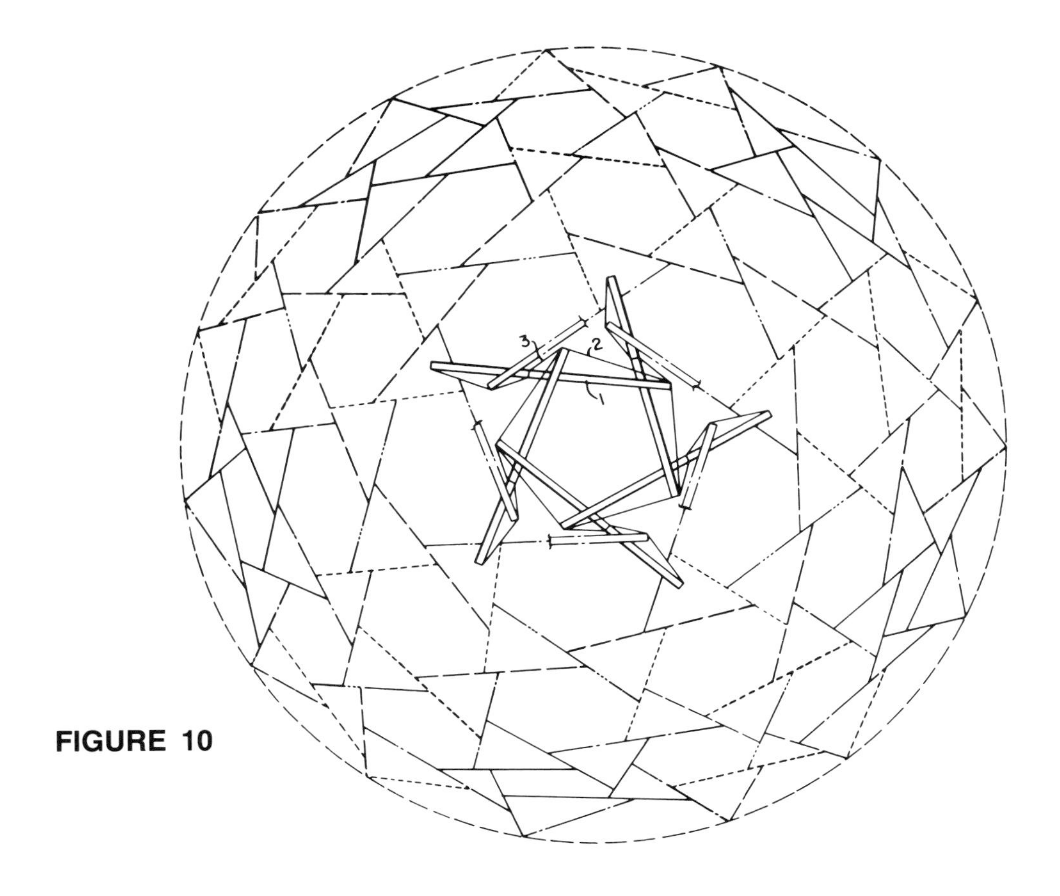

FIGURE 10

As an example to illustrate the best mode contemplated by me for carrying out my invention I shall describe the construction of a 270-boom tensegrity geodesic sphere based upon a six-frequency subdivision of an icosahedron. This sphere will be constructed of five different boom components. The design factors for the booms are given in Fig. 7. Dimensions A—A, B—C, etc., are for the theoretical lines designated in Fig. 4. The length of all booms will be such as to subtend an angle of 25° 14′ 30″ of the sphere and will of course vary according to the size of the sphere to be constructed. This length will be determined by simple trigonometric calculation in each case. Given the size of the sphere to be constructed, the factors shown in the diagrams are used as multipliers for direct calculation of the dimensions of the wire slings of the five boom components. Altitude factors are multiplied by the radius of the desired sphere, chord factors by the diameter.

The several boom components may then be colored according to the code given in Fig. 8, following which the manner of assembling them becomes merely a matter of matching colors. Fig. 10 also shows the manner of assembling the booms, in this case through identification of the several designs of boom components by a code consisting of solid and broken lines. The goedesic sphere tensegrity of Fig. 10 is known as a double bonded turbo triangles complex tensegrity (six frequency). "Double bonded" denotes the overlapping triangles of Fig. 10. Fig. 11 shows another embodiment of my invention also comprised of a 270-boom sphere known as a single bonded turbo triangles complex tensegrity. "Single bonded" denotes the non-overlapping triangles of this view. Comparing Figs. 10 and 11 it will be noticed that in the single bonded construction the triangles are spaced apart whereas in the double bonded construction they overlap. Where the triangles are spaced apart (Fig. 11) it will be seen that a point of attachment 18 (Fig. 1) of the tension elements 3 to the central portion of the column-like member 1 lies between the spaced triangles; and where the triangles overlap (Fig. 10) said point of attachment lies within the overlap. The significance of the difference between the single and double bonded constructions will be explained with reference to Figs. 12 and 13 which are diagrams of the respective constructions. In both constructions the axes of the struts of each group of three are in spaced overlapping relation such that the axial thrusts of the struts at the points where they are connected by the tension triangle elements will be additive to produce turbining forces tending to rotate the tension triangle. In the specific embodiments shown in Figs. 12 and 13 this turbining tendency is counterclockwise as indicated by the arrows A. In the single bonded tensegrity, Fig. 12, the corners of adjacent tension triangles are spaced apart whereby the thrusts of adjacent discontinuous struts of adjacent tension triangles will be additive to produce turbining forces B additive to those tending to rotate the triangle as indicated at C. In the double bonded tensegrity, Fig. 13, the corners of adjacent tension triangles are overlapped whereby the thrusts of adjacent discontinuous struts of adjacent tension triangles will be additive to produce turbining forces B opposed to those tending to rotate the tension triangle.

Fig. 14 illustrates a modified form of boom having a relatively wide girth at its center and tapered ends. Adjustable tie down sockets or dead eyes 5′ provide means at one side and centrally of the boom for tensionally securing it to the ends of the other booms of like construction. The boom may be made in two parts threaded into the connecting sleeve 6, furnishing means for adjusting the length of the boom. Holes 7 at the boom ends provide means for tying in to the dead eyes 5′ of adjacent booms.

Fig. 15 shows another form of boom consisting of a

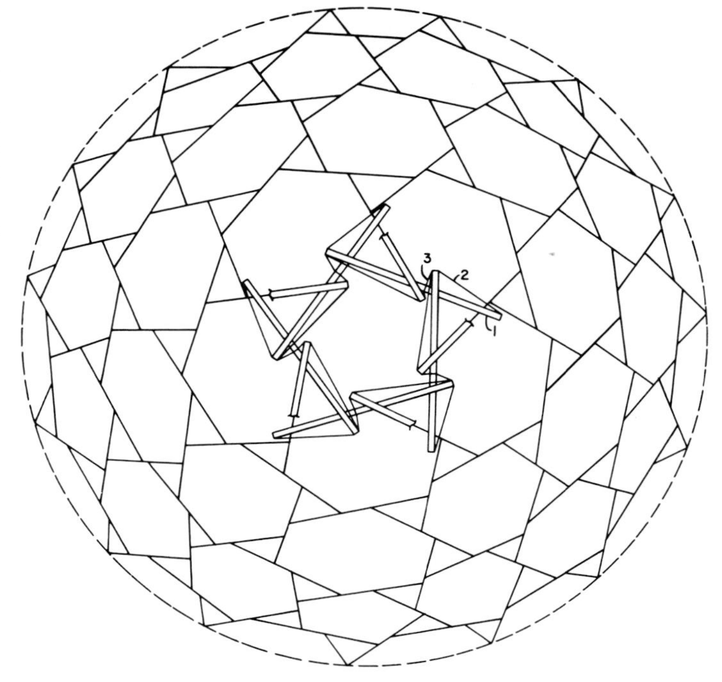

FIGURE 11

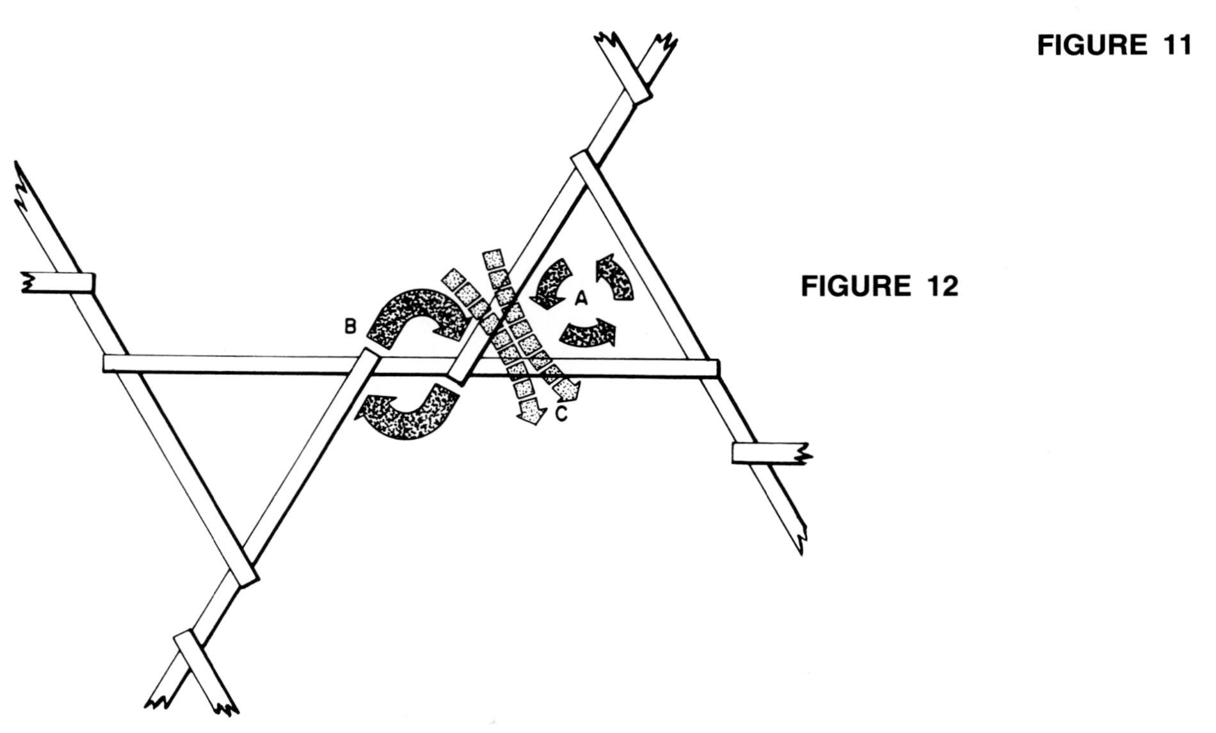

FIGURE 12

trough-shaped member whose base 8 forms a compression column. Elevated central portions of the edges of the trough-shaped members are formed with recesses 9 to receive the ends of the troughs of booms of like construction as in Fig. 16, and privide means for tensionally connecting the several booms in creating the primary and complex tensegrities I have described. Here the base of

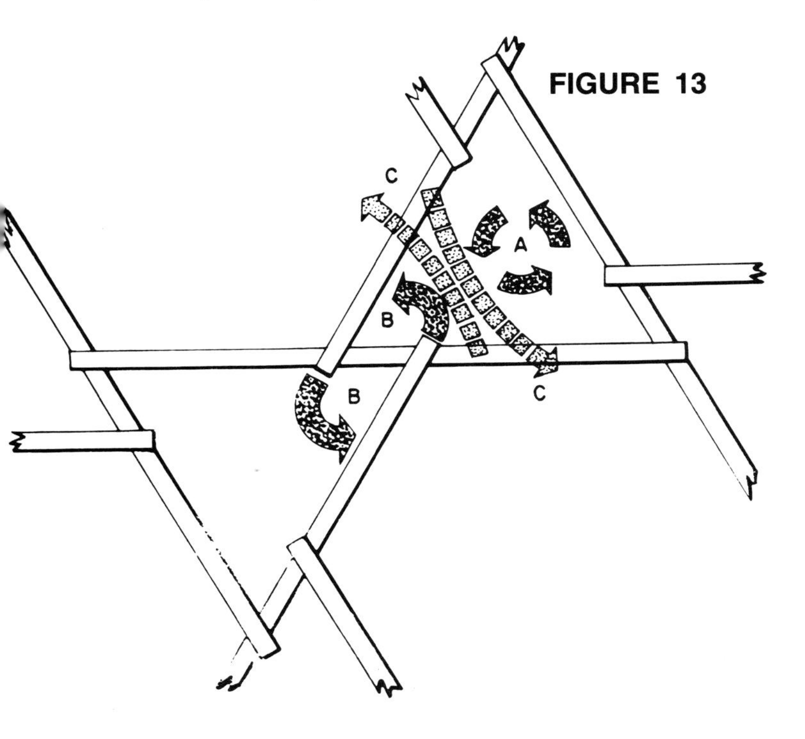

FIGURE 13

the boom forms the compression column and its upper edges form the tension triangles and the tension elements connecting the vertexes of the tension triangles to points on the columns in the region where the columns of adjacent groups are joined together in apparent continuity, i.e. centrally of the booms. The functional equivalence of

with single bonded positively turbining triangles constructed from the trough to gutter booms of Fig. 15. The triangles, hexagons and pentagons of the tensegrity complex may be covered with any suitable plastic or metal skin such as the flanged metal pans shown at 11, 12. The triangular pans may be flat. The hexagonal and pentagonal pans be made of flat sheets bent to suit the form of the hex and pent surfaces as shown. A complete watershed is afforded by this construction, each gutter boom draining into the center of an adjacent boom, the booms forming a spidery pattern of eaves troughs for the surfacing skins.

Fig. 18 shows a discontinuous compression structure in which the double lines represent compression struts and the single lines represent tension elements. If desired, the compression columns or struts 1 of the primary and complex tensegrities of Figs. 1—3 may be replaced by discontinuous compression struts made according to Fig. 18. There results a structure comprising an assemblage of tension and compression components 1, 2 and 3 arranged in a discontinuous compression system in which the compression components 1 themselves comprise an assemblage of compression and tension components arranged in the discontinuous compression system of Fig. 18 whereby the islands of compression (1) in the initial discontinuous compression system are progressively diminished. The diminished islands of compression represented by the struts shown in double lines in Fig. 18 may in turn be replaced by discontinuous compression struts made according to what may be imagined as a reduced scale version of Fig. 18, with the result shown in Fig. 19 which is an enlargement of a portion of Fig. 18.

Because of discontinuous compressions in my tensegrity complex, local tensions in the system can be tautened or released so as to permit a tensegrity sphere to be folded up for transport by plane, helicopter or rocket. Upon automatic release of tautened tensions and taking up of slack in released tensions for restoration of the original tension forces, the sphere (or dome) remembers its original shape and will resume it, thus providing a self-erecting structure.

With reference to Figs. 20–22, I shall now describe an-

FIGURE 14

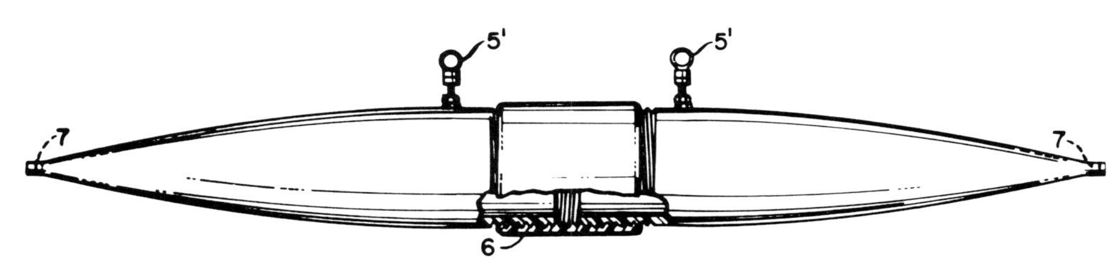

the trough boom to the strut and sling boom of Figs. 4—6 becomes apparent when the tensional stresses are resolved as indicated by the dotted lines T in Fig. 16 and the base of the trough is regarded as an apparently continuous compression strut C forming parts of two adjacent primary tensegrity tripods. Again the tension and compression aspects emerge in the characteristic form I have called a "B-boom." The booms may be trussed interiorly by perforated metal strips 10 welded or otherwise secured to the sides of the trough.

Fig. 17 shows a 270-strut geodesic sphere tensegrity

other embodiment of my invention in which the booms also include surfacing elements for a geodesic dome. This boom may comprise a trough-shaped member similar to the boom shown in Fig. 15 whose base 8′ forms a compression column and whose upper edges are designed to be stressed in tension. Weirs or recesses 9′ are adapted to receive the ends of the troughs of booms of like construction. A pair of generally triangular sheets 13, 14 extend laterally from each edge of the trough, forming a large triangle 13 and a small triangle 14 at each side of the boom. The large triangles are at opposite ends of the

FIGURE 15

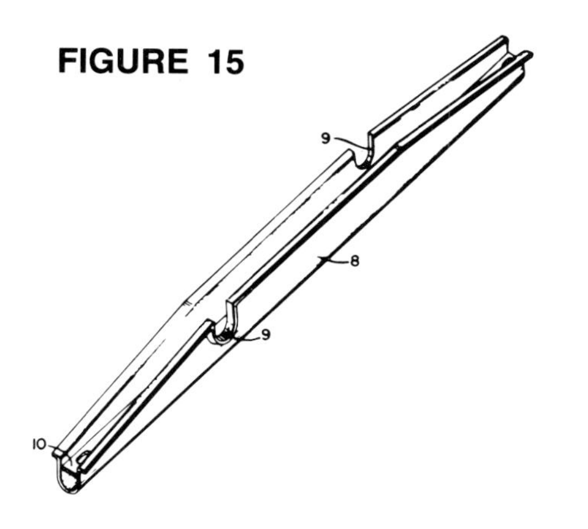

FIGURE 16

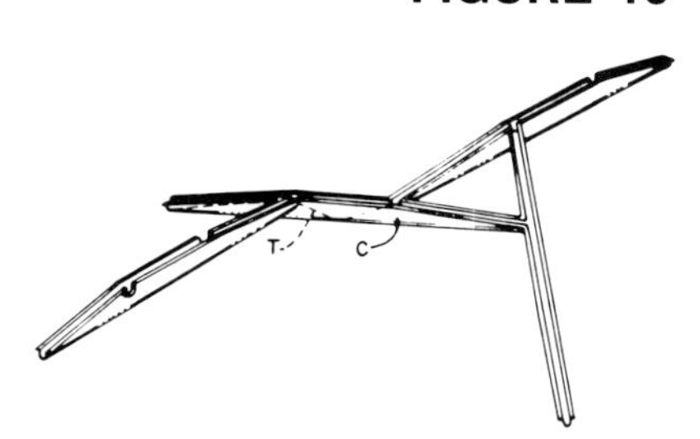

boom. The same is true of the small triangles. By reason of this peculiar arrangement, an assemblage of like components to form a geodesic tensile tensegrity, Fig. 21, finds five or six large triangles 13 mating to form the surface of a pentagon or hexagon as the case may be (see the pentagons and hexagons of the tensegrity of Fig. 11), and three small triangles 14 mating to form the surface of the triangles of the tensegrity. The edges of the triangles are preferably curled into a trough-like form as at 15 to assist in forming a watershed. Also, the triangles are preferably curved into petal-like form as shown to gain an overlapping iris pattern and to accommodate the surfacing elements to the form of the tensile part of the boom with its sloping ends.

It will be remembered that in describing the design of a 270-boom tensegrity geodesic sphere with reference to the design factors given in Fig. 7, it was disclosed that the length of all booms will be such as to subtend an angle of 25° 14′30″ of the sphere. Thus the five different boom components of Fig. 7 all utilize one common length of strut, so the 270-strut tensegrity of Fig. 10 is entirely constructed from boom components utilizing this one standard-length strut. I have found that this simplification can be carried one step further by constructing the 270-strut tensegrities of either Fig. 10 or Fig. 11 from one unique B-boom consisting of a standard-length strut and a standard form of tension sling. For example, I have con-

FIGURE 17

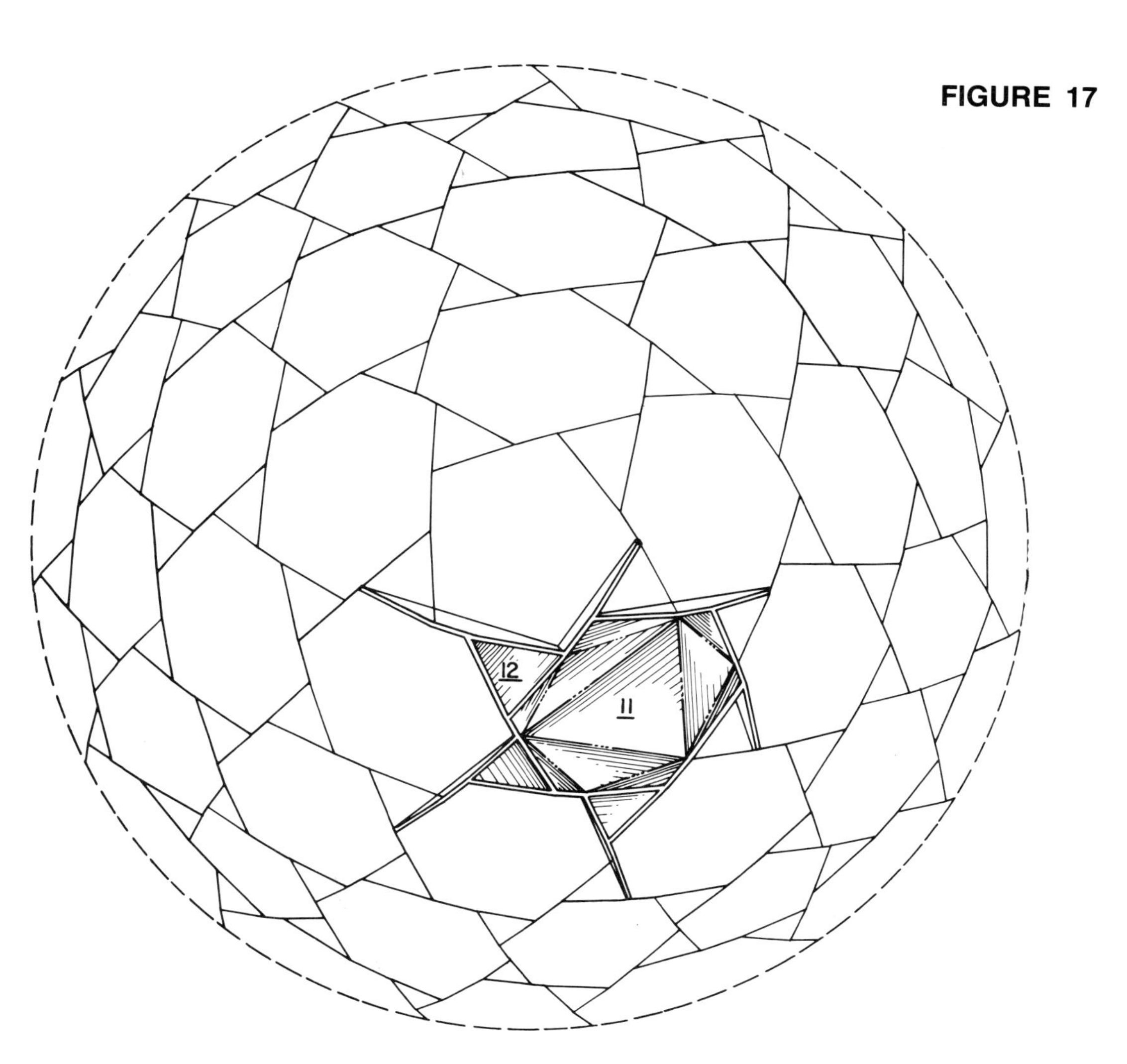

structed the double-bonded tensegrity of Fig. 10 entirely from one such unique B-boom design which will now be described in terms of its principal dimensions. The length of the boom, as before, will be such as to subtend an angle of 25° 14′30″ of the desired sphere. Referring to the boom at the top of Fig. 7, the factors for designing the unique boom become:

A—A .... .187
A—B .... .066
B—C .... .187
Altitude .... .024

Given the size of the sphere to be constructed, these factors are used as multipliers for direct calculation of the dimensions of the wire sling of the one unique boom component. The factors are simply multiplied by the radius of the desired sphere.

According to this further simplification, one unique B-boom alone implements any given size of spherical structure of visibly negligible radial variations in surface sphericity and of any frequency modular multiplication of self-divisioning to beyond billionfolds. Here emerges a surprising and significant attribute of my tensegrity sphere, for past concepts of identical subdivisioning of a sphere have suggested that the upper limit of such subdivisioning is sixty identical components or one hundred and twenty components comprising sixty positive and sixty negative ("mirror image") components. Tensegrity according to my present invention now permits spherical structuring in any size and in any frequency of subdivisioning in both single and multiple layer trussing with the use of only one unique boom component. Heretofore such uniformity could be enjoyed only in rectilinear structures for it had seemed that spherical structuring was excluded from such simple treatment. Yet I have found in tensegrity the key to a new uniformity in the erection of spherical structures which opens the door to the practical construction of spheres and spherical domes of virtually unlimited size without prohibitive complexity of componentation.

FIGURE 19

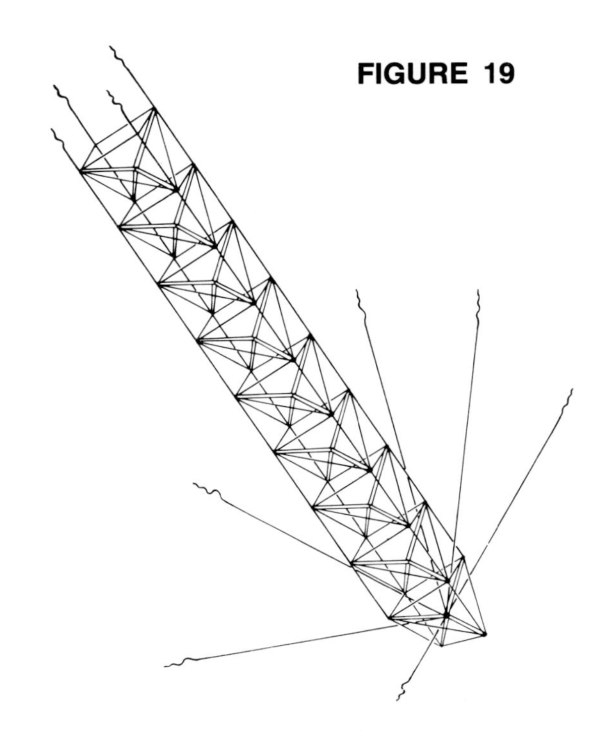

FIGURE 18

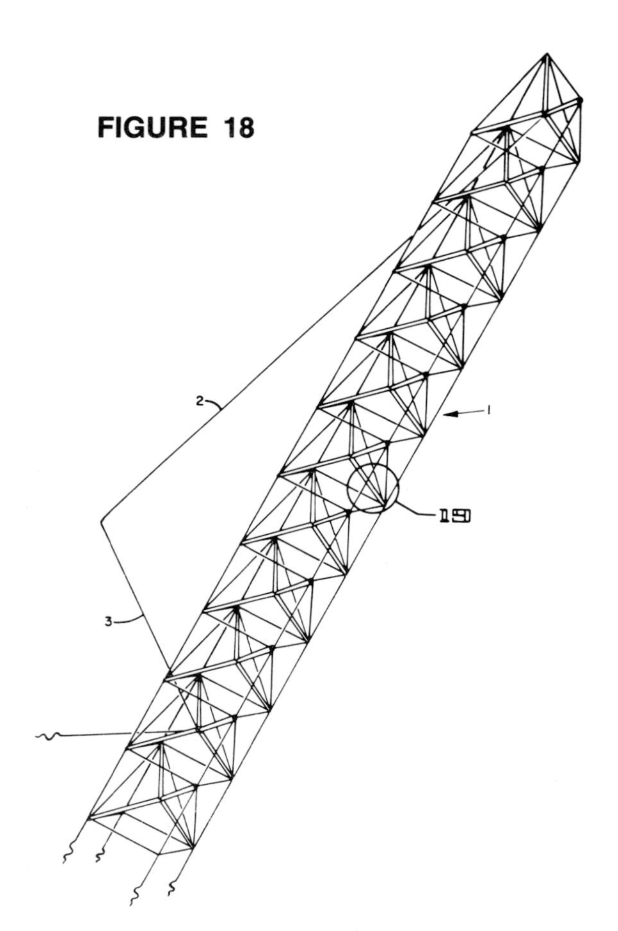

FIGURE 20

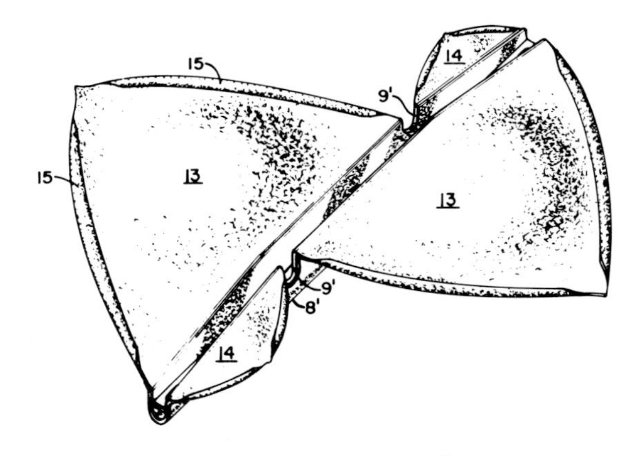

As the size of the structure and the number of identical components increase, central angles of the sphere subtended by individual components and the lengths of the chords represented by each component are so reduced in relation to the size of the sphere that the arc altitudes of the tension slings become negligible. The tension filled

FIGURE 21

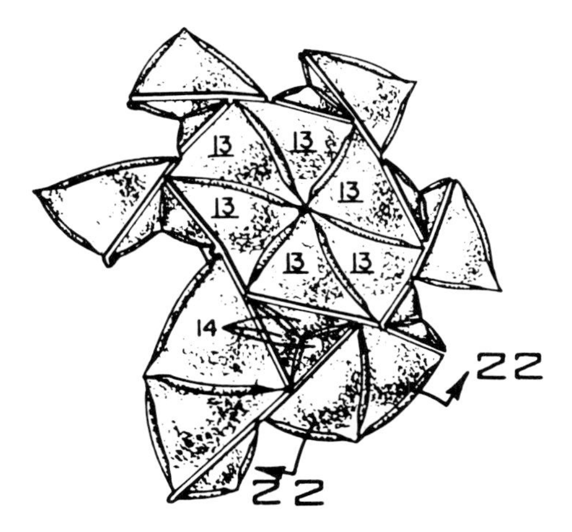

gap between adjacent booms then becomes virtually invisible and the booms seem deceptively to be arrayed in continuous compression contact. Yet they are not in such continuous compression contact and therefore are not subject to the circumferential shear stresses characteristic of more conventional compression systems. Instead, through the interaction of the tension slings, each boom is pulled radially inwardly in tensional coherence. In such high frequency arrays the tension slings of the booms can advantageously be constituted by integral flanges or fins on the booms as in Fig. 16, wherefore the discontinuous tensional nature of the structure becomes invisible and may only be apprehended upon analysis of the progressive stages of tensegrity from the tepee through the

FIGURE 22

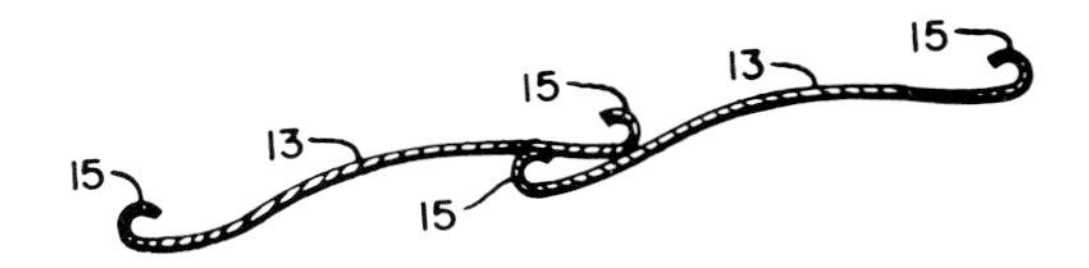

three-strut octahedral system as described with reference to Figs. 1–3.

The terms and expressions which I have employed are used in a descriptive and not a limiting sense, and I have no intention of excluding such equivalents of the invention described as fall within the scope of the claims.

16▲SUBMARISLE (UNDERSEA ISLAND) (1963)

U.S. PATENT—3,080,583

APPLICATION—JUNE 8, 1959

SERIAL NO.—818, 935

PATENTED—MARCH 12, 1963

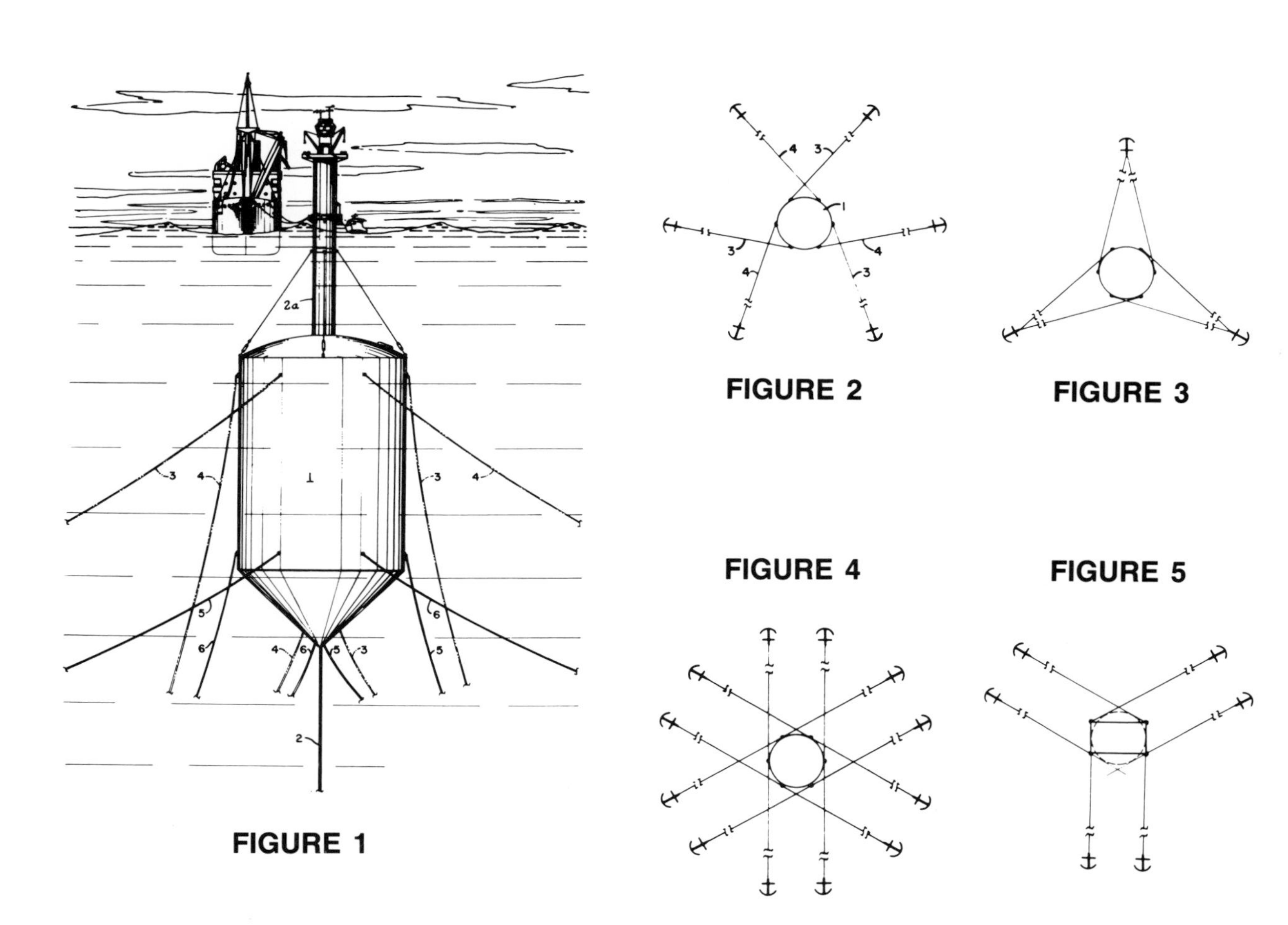

OCEAN WAVE TURBULENCE does not penetrate deeply below the surface. Fourteen or twenty feet below the surface there is no turbulence. Because humanity is developing more and more submarines for more and more military purposes, I could see that the law of shipbuilding advantage applied: every time you double the length of a ship you increase its volume and therefore its payload eight times while increasing its surface only four times. Every time you double the length of a ship, your payload-to-surface advantage doubles, therefore you have less surface to build and drive through the water. So giant submarines will become very effective around the world as transports operating entirely below any turbulence. I found humanity tending to venture on the sea more and more and felt it would be a very great advantage to build submarisles with lighthouses sticking up above the water and landing docks when there is not too much turbulence but making it possible for the submarines to dock below the surface. I could see that we needed to have islands such as the submarisle all around the world to make possible the changing of cargoes at various points.

UNITED STATES PATENT OFFICE

Richard Buckminster Fuller, 104—01 Metropolitan Ave., Forest Hills 75, N.Y.

UNDERSEA ISLAND

The invention relates to an undersea island and an improved anchoring system therefor. My undersea island has special application to offshore oil drilling rigs and I shall describe it with particular reference to this field of use. However, the invention will be valuable for other purposes such as its use as a manned operating base in explorations of the ocean bottom. Thus the invention can be applied to the purposes of the broad oceanographic program currently being advocated by the National Academy of Sciences.

The extension of offshore oil well drilling into deeper and less sheltered waters has created enormous engineering problems in the design, construction and maintenance of drilling platforms. The most acceptable solution to date has been the use of barges provided with equipment for erecting fixed platforms on pilings driven into the ocean floor. Sometimes a lower barge hull is sunk to the bottom. According to another system the barge hull is jacked up on the driven pilings to make an elevating deck on which the drilling installations are carried. See *Drillings,* Special American Petroleum Institute Section, February 26–28, 1958, Associated Publishers, Inc., Dallas, pp. 30–34. The cost of building and erecting such rigs is enormous, but this is only part of the story for it still leaves the problem of maintaining an installation which is exposed to the buffeting of wind and sea, and the losses sustained when the gales and hurricanes blow can be prodigious. One of the principal objects of my invention as applied to offshore oil well rigs has been to provide an installation which is less at the mercy of the elements.

An offshore oil well rig according to my invention comprises in essence an undersea island having an anchoring system which effectively restrains it against motion other than a moderate movement up or down as the tide comes in and runs out. The "island" is a submerged caisson within which is installed the derrick or its equivalent, and other installations common to oil drilling operations. A hollow communications shaft extends from within the caisson to a boarding platform above the surface of the sea. I am aware that it has been proposed heretofore to support well drilling apparatus above a submerged bouyant tank arrangement, but my invention is concerned more particularly with providing a more practical solution including an improved anchoring system and other improvements which make it feasible to move the heart of the well drilling operations into a caisson under the sea.

My invention in its general arrangement comprises an undersea island including a buoyant caisson submerged under the pull of anchor rodes extending from the sides of the caisson to anchors distributed around it, several of the rodes extending tangentially clockwise and several tangentially counterclockwise, the former constituting a group of rodes tending to produce clockwise rotation of the caisson about a vertical axis and the latter constituting a group of rodes tending to produce counterclockwise rotation. The one group of rodes creates a torque which is equal and opposite to that created by the other. This has the result that the caisson is fixed in the grip of opposed torques while yielding to slow vertical movement with the tides against the resilient pull of the catenary sags in the rodes. The anchor rodes preferably comprise three pairs of counter-torquing rodes secured to the caisson at a common horizontal level and three other

FIGURE 6

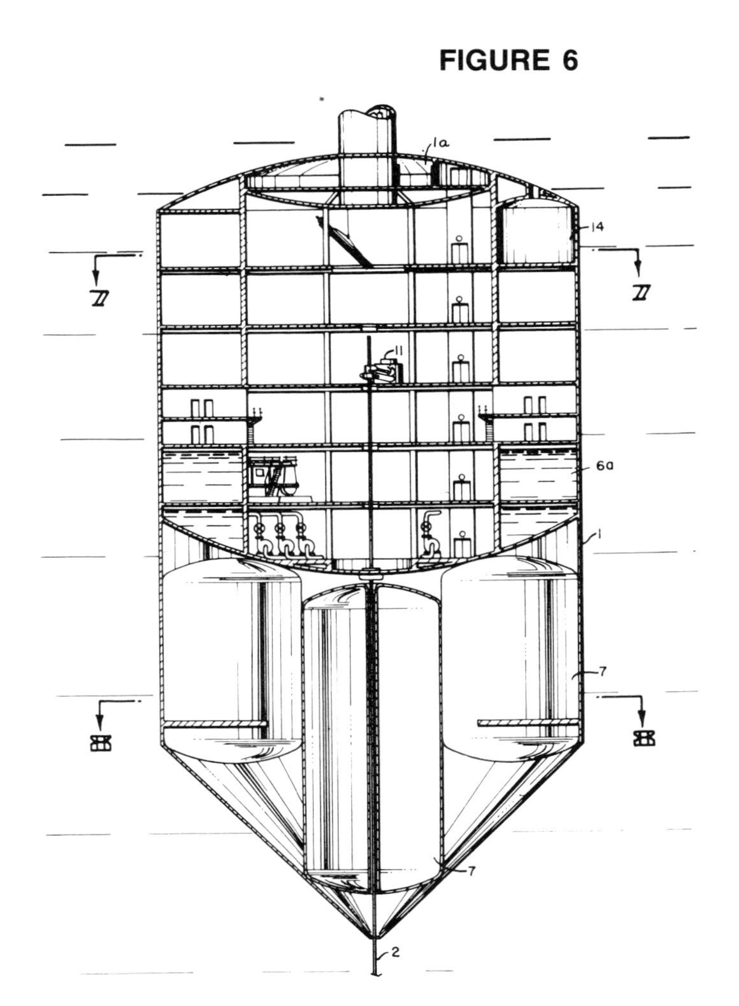

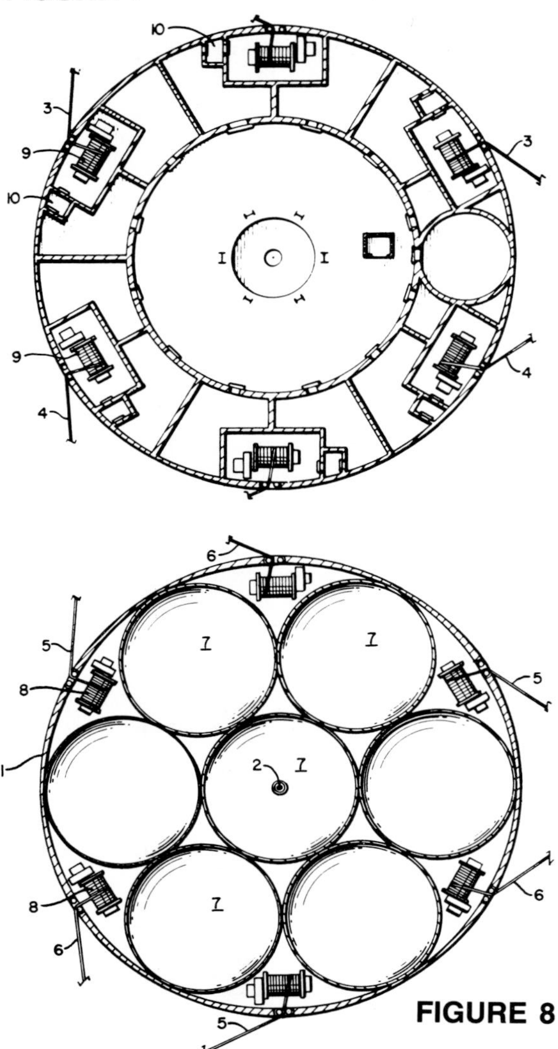

FIGURE 7

FIGURE 8

pairs of counter-torquing rodes secured to the caisson at another horizontal level spaced from the first. Or there may be just three criss-crossing pairs of rodes distributed so as to extend tangentially away from the sides of the island in several directions. The anchors for one rode of each pair define the base corners of a tetrahedron whose apex is the undersea island and the anchors of the other rode of each pair define a similar tetrahedron. Thus we have what may be described as two counter-torquing tetrahedra.

In its special application to an offshore oil well rig, my invention comprises a buoyant caisson submerged under the pull of anchor rodes arranged in the manner described above, marine oil well apparatus including derrick structure within the submerged caisson, and a hollow communications shaft extending from the caisson to a boarding platform above the surface of the sea. The caisson is fixed in the grip of opposed torques created by the oppositely extending tangential rodes while the shock of surface seas on the surface and near-surface parts of the rig is absorbed by the resilient pull of the catenary sags in the rodes.

In the drawings, wherein I have illustrated the best mode contemplated by me for carrying out my invention:

Fig. 1 is a side elevational view of an offshore oil well rig embodying my invention in one of its preferred forms.

Fig. 2 is a schematic plan view showing three criss-crossing pairs of counter-torquing anchor rodes. Also this view may be understood as representing a plan view of the anchor rode arrangement for the caisson of the offshore oil well rig of Fig. 1 in which there are six such criss-crossing pairs of rodes, three pairs at each of two different horizontal levels.

Fig. 3 illustrates a special case in which the three (or six) pairs of criss-crossing rodes extend to three points of anchorage instead of six (or twelve). The doubling up at the anchors does not detract from the efficacy of the arrangement as there are still two counter-torquing tetrahedra in the system.

Fig. 4, another schematic plan view, shows a twelve anchor system.

Fig. 5 is included to show how my counter-torquing system is applicable regardless of the form of the caisson.

Fig. 6 is a central vertical cross-sectional view of the well rig caisson of Fig. 1.

Fig. 7 is a horizontal cross-sectional view taken on line 7—7 of Fig. 6.

Fig. 8 is a similar view on line 8—8 of Fig. 6 (bouyancy control tanks shown in top plan).

Fig. 9 is an enlarged view of the upper part of the rig of

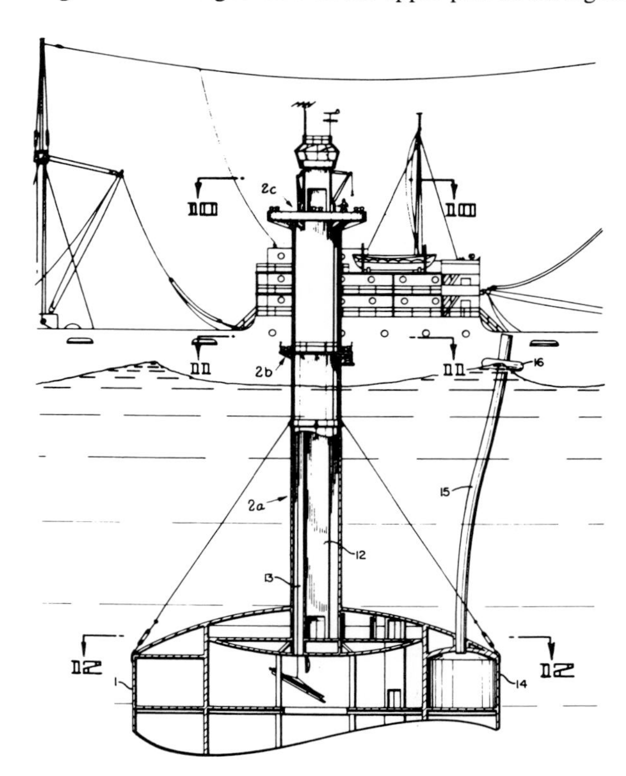

FIGURE 9

Fig. 1, the upper part of the caisson and part of the communications tower broken away in vertical cross section.

Fig. 10 is a horizontal cross-sectional view at enlarged scale taken on line 10—10 of Fig. 9.

Fig. 11 is a similar view on line 11—11 of Fig. 9.

Fig. 12 is a similar view on line 12—12 of Fig. 9.

Fig. 13 is a side elevational view of the oil well rig floating as a surface ship in a horizontal position favorable for towing between drilling sites.

Fig. 14 is a side elevational view illustrating the adaptation of my specially anchored caisson used as a submerged operating base with provision for harboring submarines.

Fig. 15 is a detail vertical cross-sectional view taken on line 15—15 of Fig. 14.

Fig. 16 is a detail horizontal cross-sectional view taken on line 16—16 of Fig. 15.

Fig. 17 is a detail horizontal cross-sectional view showing the slip for receiving the access hatchway of the submarine.

Fig. 18 is a detail vertical cross-sectional view taken on line 18—18 of Fig. 17.

In Fig. 1 we see the submerged caisson 1 of an offshore oil well rig below which at 2 extends the drill pipe casing. The mast-like structure 2*a* above the casing is a hollow communications shaft extending from within the caisson to boarding and cargo loading decks 2*b* and 2*c* (Fig.9) above the surface of the sea. Boarding from a small boat and transfer of operational cargo, such as drill pipe, from a mother ship are indicated pictorially. This view also shows how the buoyant caisson 1 is submerged under the pull of anchor rodes distributed around it.

The arrangement of the rodes is of the greatest impor-

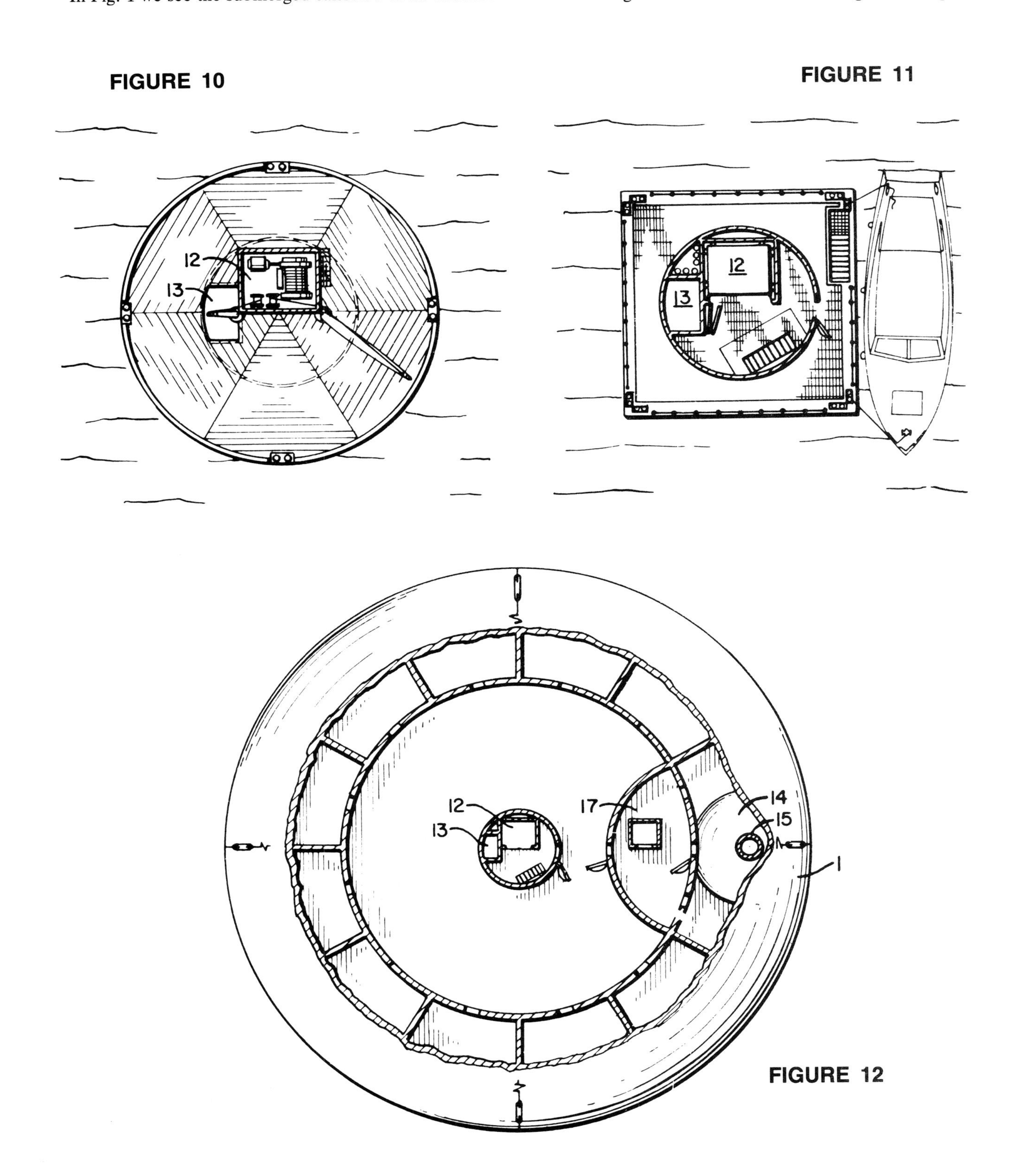

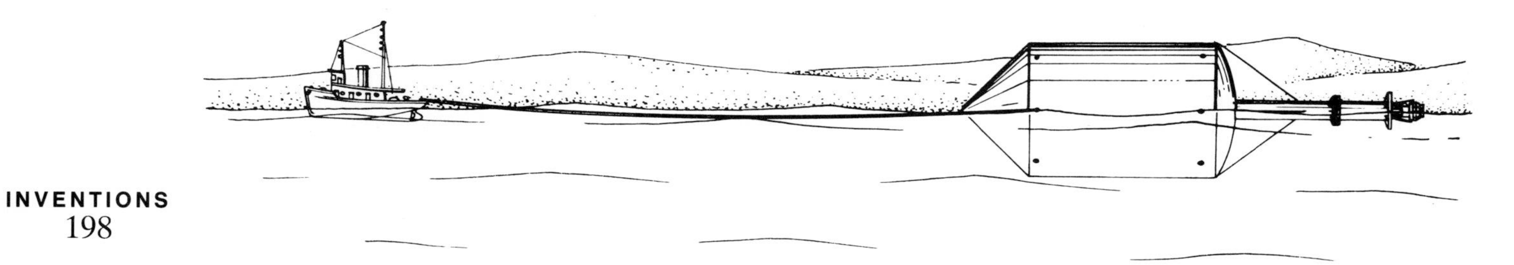

FIGURE 13

tance in achieving primary benefits of my invention. Referring to Figs. 1 and 2, we see the rodes extending from the sides of the caisson to anchors distributed around it. Because of the considerable distance between the caisson and the anchors, the full length of the rodes cannot be shown within the scale of the drawing; so these and succeeding views are to be read with the understanding that there will be long intermediate sections of the cables between the break lines as shown particularly in Figs. 2–5. The several rodes 3 extend tangentially clockwise and the several rodes 4 tangentially counterclockwise, Fig. 2, the former constituting a group of rodes tending to produce clockwise rotation of the caisson about a vertical axis and the latter group of rodes tending to produce counterclockwise rotation. The one group of rodes creates a torque which is equal and opposite to that created by the other. The result of this is to fix the caisson in the grip of opposed torques while it yields to slow vertical movement with the tides against the resilient pull of the catenary sags in the rodes. Notice that there are three pairs of counter-torquing rodes 3—4 as shown in Fig. 2. Notice also that the rodes of each pair 3—4 are in criss-crossing relationship. In the preferred arrangement shown in Fig. 1, there are three pairs, 3—4, of counter-torquing rodes secured to the caisson at a common horizontal level, and three other pairs, 5—6, secured to the caisson at another horizontal level spaced from the first. This arrangement produces vertical stability of the caisson as it is held within the frame of reference of the counter-torquing rodes. A peculiar effect of the particular arrangement of the rodes is that the caisson is resiliently but closely restrained in what amounts practically to a position of immobility, yet allowing for extremely slow vertical movement with the tides. If desired the vertical movement can be controlled during drilling operations by pumping sea water in or out of the flotation tanks associated with the caisson. But regardless of this the caisson will always be fixed in the grip of the opposed torques of the several pairs, preferably three or six, of criss-crossing rodes.

It is also to be observed that the anchors for one rode of each pair, i.e. the rodes 3 for example, define the base corners of a tetrahedron whose apex is the caisson or undersea island, while the anchors of the other rode 4 of each pair, define a similar tetrahedron. Thus we have a system which may be described and understood as consisting of two counter-torquing tetrahedra. As such the system yields improved, and I might say unique, results in terms of stability. It will be appreciated that the characteristic of stability and the degree in perfection of this characteristic is of extreme importance when my anchoring system is used as applied to offshore well drilling rigs. If, as in accordance with my invention, the drilling rigs are to be freed from such operational limitations as the driving of pilings into the ocean floor, the buoyant caisson which takes the place of the fixed platforms supported on such pilings must have unusual stability of position as compared with ships at anchor. Perfection of

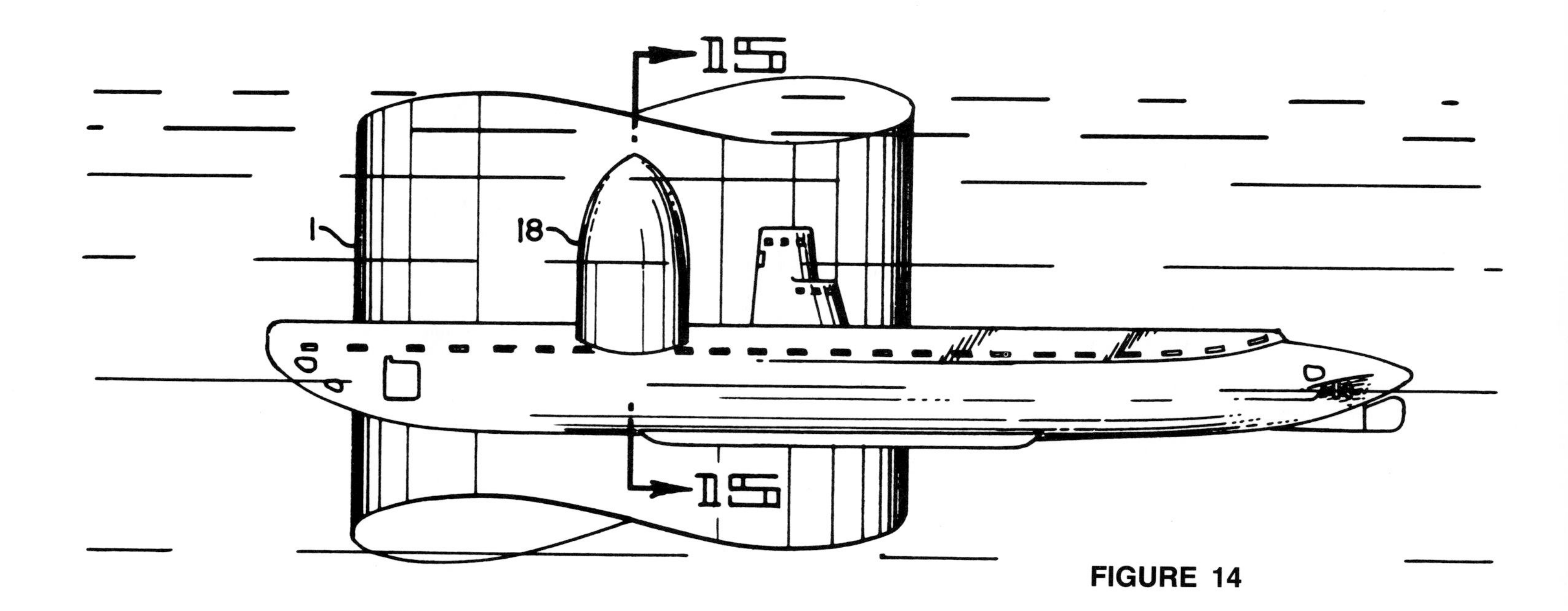

FIGURE 14

stability also becomes a matter of vital concern when my undersea island is used as a manned operating base in explorations of the ocean bottom or as an undersea naval base including provision for undersea docking of submarines.

Fig. 3 illustrates a special case in which the three pairs of criss-crossing rodes extend to three points of anchorage instead of six; when there are three pairs of rodes at seperated levels as in Fig. 1, we would have according to the Fig. 3 case, six pairs of criss-crossing rodes extending to either three or six points of anchorage instead of six or twelve. The doubling up at the anchors does not detract from the efficacy of the arrangement as there are still at least two counter-torquing tetrahedra in the system.

Fig. 4 shows a twelve anchor system resulting in two sets of counter-torquing tetrahedra. . Essentially this is a system which simply multiplies the system of Fig. 2 by the factor 2.

Fig. 5 shows how my counter-torquing system is applicable regardless of the form of the caisson, the caisson in this instance being retangular as viewed in plan.

I have described the rodes as extending "tangentially clockwise" and "tangentially counterclockwise." In this context my use of the descriptive "tangentially" is not restricted to the exact "tangent" of the mathematician. And, of course, it will be understood that when we have a rectangular caisson as in Fig. 5, there can be no mathematical tangent; nevertheless there is in practical effect and in substance the same "tangential" arrangement as obtains in Figs. 2,3 and 4. I direct attention also to the fact that in Fig. 2, the angle between the rodes and the radius intersecting their points of attachment is greater than 90°; in Fig. 3, less than 90°; and in Fig. 4, exactly 90°. And in Fig. 5, with reference to the imaginary superimposed circle, there is an exact 90° tangency to a theoretical point of attachment of the rodes. Thus, according to my definition, a disposition of the anchor rodes which is substantially tangent as distinguished from radial is regarded as being "tangential" and will confer the primary benefits of my invention.

Inside the well rig caisson 1 buoyancy regulating tanks 7, Fig. 6 and 8, provide a controlled buoyancy. Sea water is pumped in and out of selected tanks by means of suitable pumping machinery such as used in controlling water ballast in ships. Fig. 8. shows the buoyancy tanks disposed in closest packing arrangement. In this arrangement lateral stability can be effectively controlled in all directions. This arrangement also is well adapted for the accommodation of anchor winches uniformly spaced around the inside periphery of the caisson. Thus, in Fig. 8, we see the winches 8 for anchor rodes 5 and 6 mounted in flooded spaces within the lower compartment of caisson 1. These winches may be conveniently operated by hyraulic drives controlled from within the water-tight operating space in the upper part of the caisson. The tangential arrangement of the anchor rodes creates large stresses in the shell of the caisson. To accept such loading the thickness of the shell plating is built up adjacent the hawse openings for the rodes. Suitable sheaves or rollers may be provided at the sides of the hawse openings to avoid undue wear on the anchor cables as the rodes are tightened or slackened to balance the torques in the two tetrahedral systems and regulate the catenary sags in the rodes.

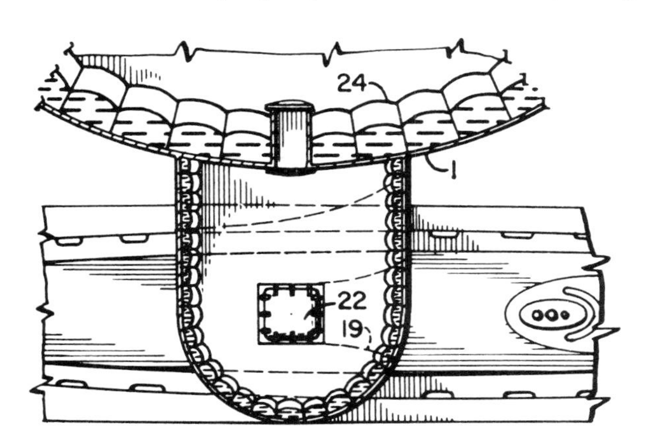

FIGURE 16

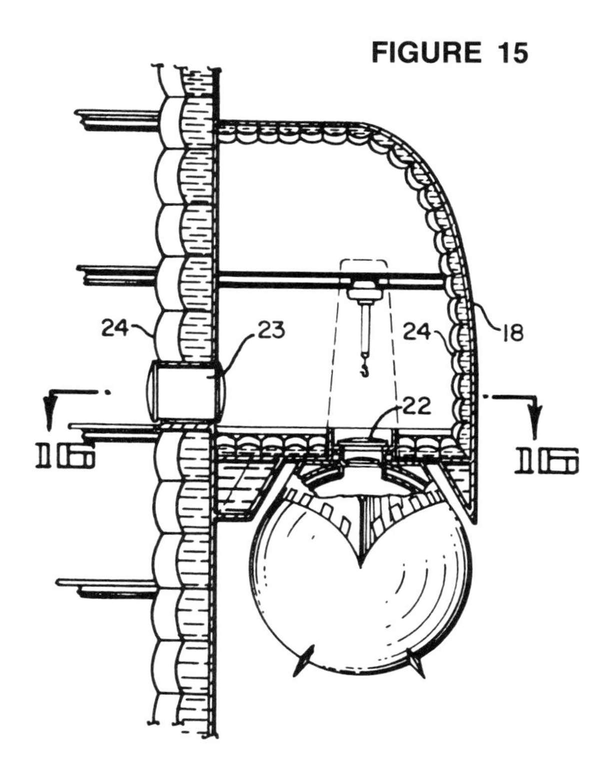

FIGURE 15

In Fig. 7 we see another set of anchor winches 9, for the rodes 3 and 4, arranged in flooded spaces sealed off from the interior working spaces of the caisson. Air lock chambers may be provided for access to the winch compartments, such chambers being shown at 10 in Fig. 7. The interior working space of the caisson preferably is decked and bulkheaded as shown in Fig. 6, and provided with intercommunicating elevators and stairways. Openings in the floor below the rotary table and draw-works 11 of the well drilling rig accommodate the drill pipe and its casing. This arrangement eliminates the need for separate derrick structure, the decked and bulkheaded caisson itself serving in effect as the derrick of the rig.

The hollow communications shaft which extends from within the caisson to the decks above the surface of the sea will be provided with an elevator 12 for personnel and freight, and may also have a shaft 13 (Figs. 9 through 12) through which long sections of drill pipe and casing may be lowered from the cargo receiving deck, Fig. 10. Water-tight compartments 1*a* at the top of the caisson provide a means of temporarily sealing off the working spaces from the communications shaft in the event that this shaft should be damaged and flooded as a result of collision or from other causes. The main cargo deck of the communications shaft will be located some 60 feet above mean high tide so as to be beyond the reach of damage by high seas. The lower deck, Fig. 11, used principally for purposes of boarding from small craft, will be closer to the surface of the sea but it is small and has an open work deck so as to offer little resistance to the pounding of the waves.

The oil well rig will include the various kinds of machinery and auxiliary equipment customary to offshore oil well operations including the aforementioned rotary table and draw-works 11, mud pumps, mud tank, blowout preventers, tanks, etc. Suitable means will be provided for conveying the oil to tankers brought alongside. During periods when weather conditions do not permit tankers to be in a position to take on oil, the oil may be flowed into storage tanks *6a* within the caisson, buoyancy

FIGURE 17

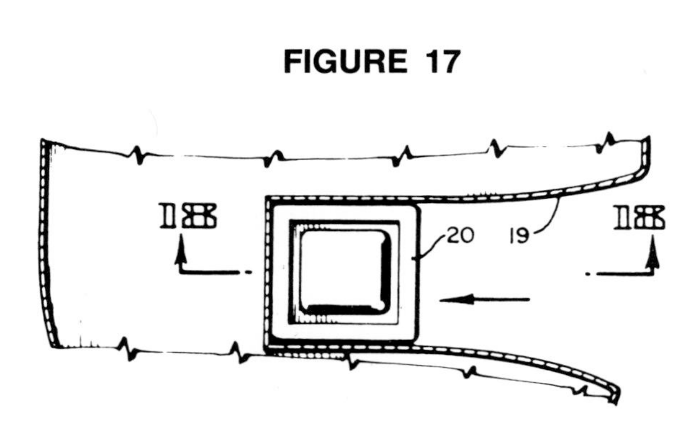

meanwhile being maintained by pumping sea water from within the buoyancy tanks 7.

In the preferred construction shown I have provided a chamber 14 and safety bulkhead 17 (see Fig. 12, the upper part of Fig. 6, and the lower part of Fig. 9) from which extends a flexible air supply and escape shaft 15 floated on the surface of the sea by a suitable raft 16 through which crew members may be hoisted on bosun's chairs lowered from a rescue vessel.

By regulating the ballasting and buoyancy of the rig, it can be floated in the position shown in Fig. 13 for towing to and from anchorages. A pair of anchor rodes at op-

FIGURE 18

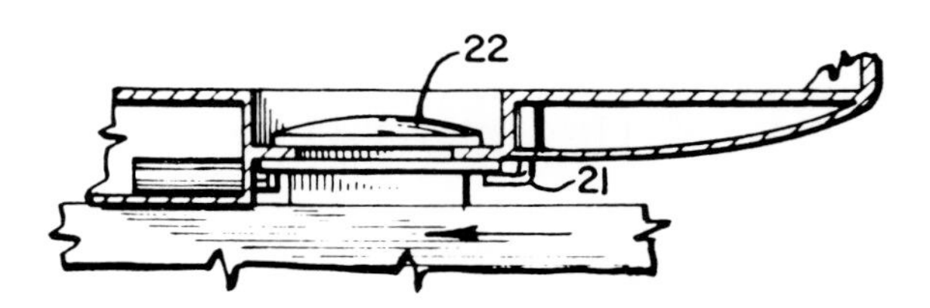

posite sides of the lower end of the caisson can be tied together as a towline and the coniform shape of the lower end of the caisson furnishes fair water entry as the bow of a barge.

With reference to Figs. 14–18, I shall now describe the application of my undersea island as a manned operating base provided with submarine berthing and communication facilities. The caisson 1, Fig. 14, may be of the general form described with reference to Fig. 1, and it will be anchored as before with the counter-torquing rode system. At a point preferably well above the points of attachment or entry of the rodes is a berth for a submarine such as may be comprised within the blister 13. The bottom of the blister has a slip 19, Fig. 17, having a flared entrance functioning like a ferry slip to guide the submarine into its berth. The sides of a hatchway 20 fit within the slip 19. The hatchway 20 being located on the deck of the submarine ahead of its conning tower, the submarine can be visually conned to its berth. Suitable dogs 21, Fig. 18, operated from within the blister secure the hatchway against the bottom of the blister adjacent an entrance hatch 22 thereof. The blister compartment 18 gives access to the caisson proper through a bulkhead compartment 23. The shell of the caisson and blister preferably is triple and bulkheaded or quilted as at 24. The outer set of quilted spaces is filled with liquid, and the inner set with air, providing a double shell, liquid and air, to make it torpedo proof.

The terms and expressions which I have employed are used in a descriptive and not a limiting sense, and I have no intention of excluding such equivalents of the invention described as fall within the scope of the claims.

17▲ASPENSION (GEODESIC STRUCTURES)(1964)

U.S. PATENT—3,139,957

APPLICATION—JANUARY 24, 1961

SERIAL NO.—88,245

PATENTED—JULY 7, 1964

I FOUND IN THE WORLD of tensegrity it was also feasible to produce what I call the aspension dome, a dome that could be progressively assembled on the ground from its maximum-diameter base and continually hoisted aloft. This is really an alternative to the tensegrity geodesics. It has an accordian-opening effect such as the foldable Japanese lantern, which can be progressively pulled open to provide space.

FIGURE 1

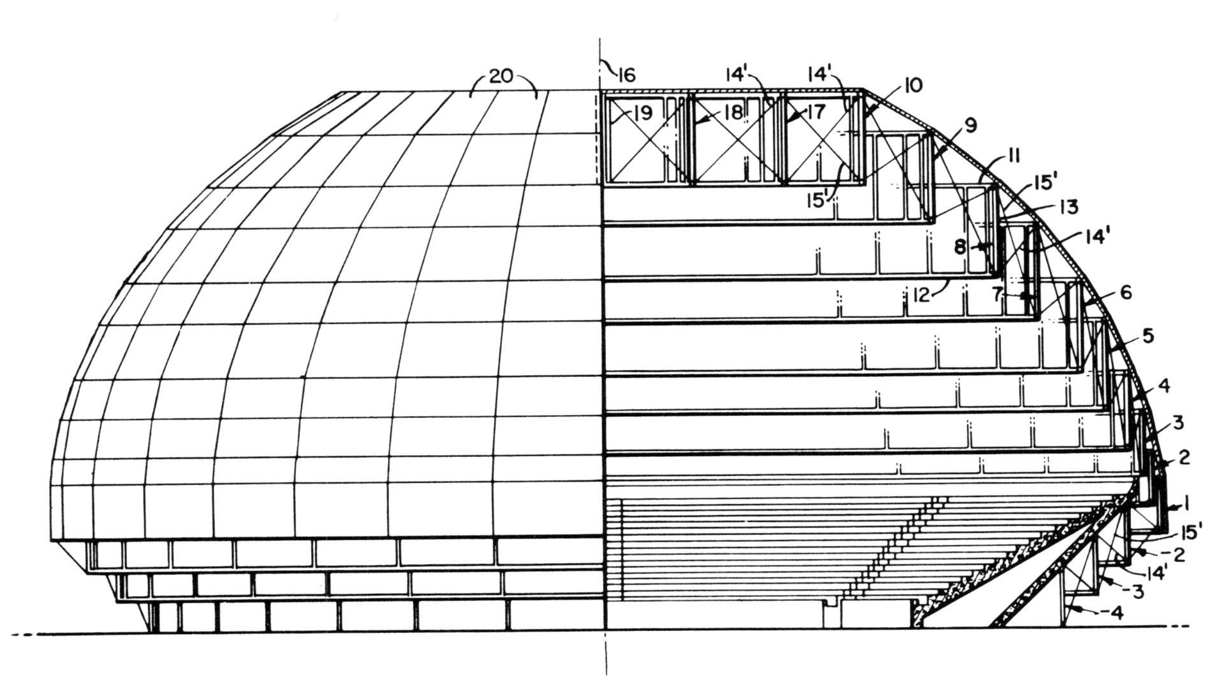

UNITED STATES PATENT OFFICE

Richard Buckminster Fuller, 407 S. Forest St., Carbondale, Ill.

SUSPENSION BUILDING

My invention relates to building construction.

SUMMARY

I have discovered how to make building structures and components possessing in substantial measure the advantages of catenary suspension heretofore confined principally to the suspension bridge. The catenary cables of the suspension bridge sag downwardly to the mid-point of the bridge, and would seem to possess no utility in any structure which arches upwardly. So it has been a surprise to me to find that there is a way by which a catenary suspension system can be converted into an arched structure of domical or polygonal form. By breaking up the suspension cables into increments suspending an ascending series of polygonal or circular frames stepped upwardly one within another, altitude is gained, replacing the catenary sag of the bridge cables with a rising, arched, suspension system. My new structure comprises a series of box frames of polygonal, cylindrical or other form, these frames being of progressively varying sizes arranged in a concentric array of sequentially different heights above a common plane of reference and in vertically overlapping spaced relation to one another. Tension elements such as flexible cables or wires extend between and are secured to adjacent pairs of the box frames in the series. These tension elements include tension members extending downwardly from their points of securement to one frame of a pair to their respective points of securement to the other frame of the pair. In this manner, successive frames of the series are suspended one from another in either a rising, descending or

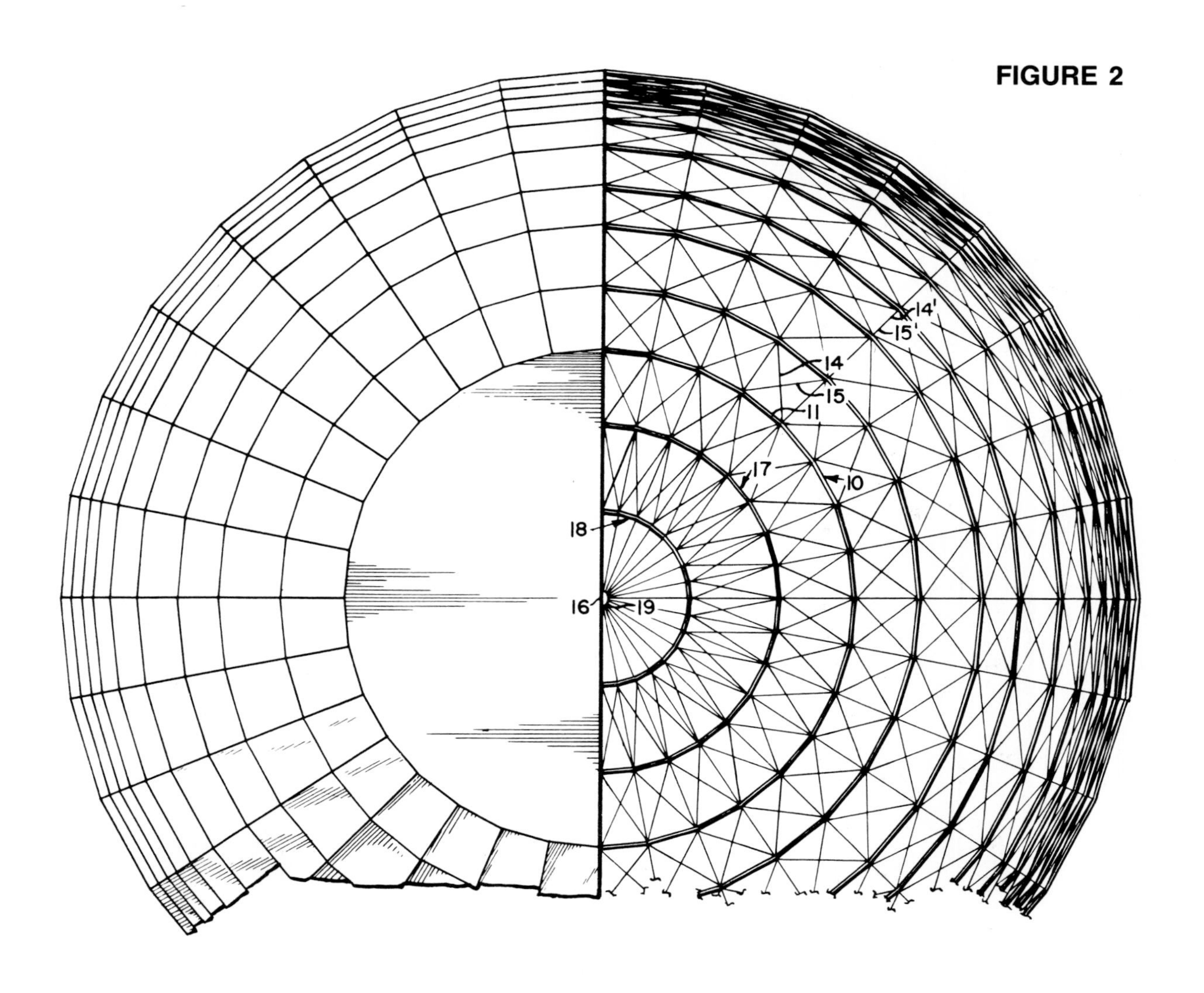

FIGURE 2

FIGURE 3

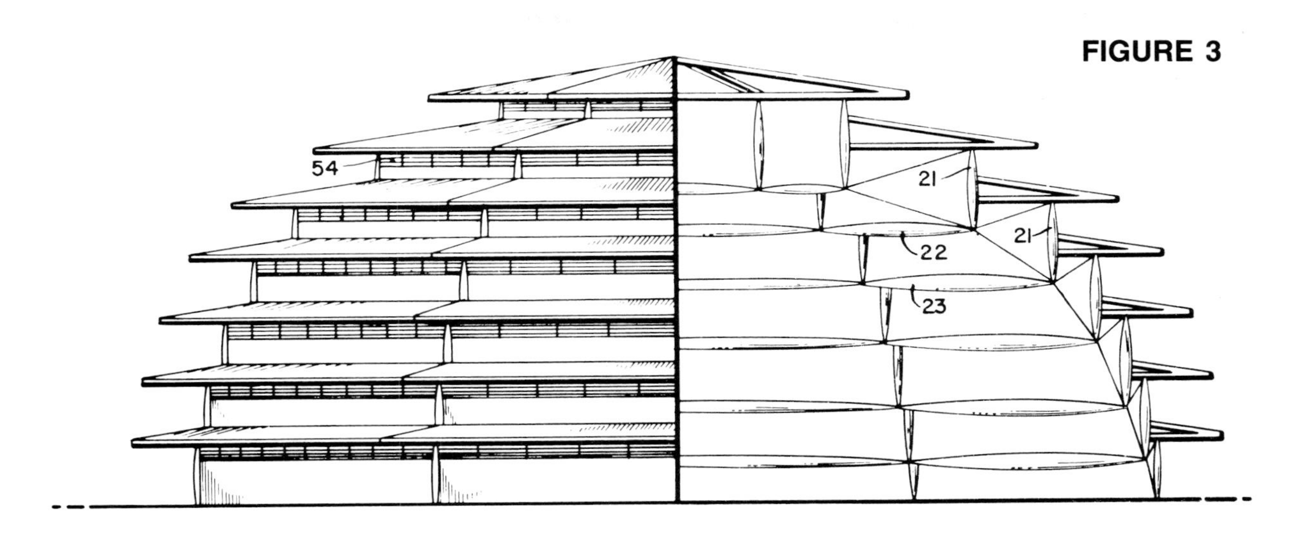

level series as may be desired. Other tension members extend upwardly from their points of securement to one frame of a pair of their respective points of securement to the other frame of a pair whereby successive frames in the series are anchored down one to another. The suspension members and anchor members complement one another in performing a third function, the provision of a tensioned buttress against tilting of the compression members comprised in the frames. Thus are satisfied the three functions of (1) suspension, (2) anchoring and (3) buttressing, all by the simplest arrangement of tension wires in combination with the series of frames as disclosed. Additionally, these tension wires act to stabilize the frames themselves, and furnish a tension system efficiently disposed for resisting torquing and counter-torquing of the frames about the central axis of the structure. These can be catalogued as functions (4) and (5). Still other tension members may be provided that will be disposed in radial planes containing the central axis of the structure. In the case of a structure utilizing a series of frames of polygonal form, the several tension members are secured to the vertexes of the polygons. Those tension members lying in radial planes will apply tension forces to the upper peripheries of the frames and thereby reduce compressive forces imposed on such peripheries by others of the tension members.

Description

With reference to the drawings I shall now describe the best mode contemplated by me for carrying out my invention.

Fig. 1 is an elevational view partly in vertical section illustrating the application of my invention to a roofed stadium or concert hall.

Fig. 2 is a plan view of the Fig. 1 structure with the right-hand portion of the roof covering removed to reveal the suspension frame construction, taken as indicated at 2—2 in Fig. 1.

Fig. 3 is an elevational view illustrating another embodiment of the invention, the right-hand portion being shown in diagrammatic vertical section.

Fig. 4 is a similar view of another embodiment.

Fig. 5 is a plan view of the suspension frame system of the Fig. 4 structure, taken as indicated at 5—5 in Fig. 4.

Fig. 6 is a diagrammatic vertical section of a still further embodiment of the invention.

Fig. 7 is a plan view of the Fig. 6 structure, roof covering omitted, taken as indicated at 7—7 in Fig. 6.

Figs. 8 to 17 inclusive are diagrammatic representations illustrative of the general scope of applicability of my suspension building system to various building forms with the use of different kinds of box structures and with several alternative dispositions of the suspension and anchoring cables,

Fig. 8 being a vertical sectional view illustrative of flanged cylindrical box sections,

Fig. 8a a plan view illustrative of an asymmetrical form of my construction,

Fig. 9 a vertical sectional view illustrative of flanged box sections of frusto-conical form,

Fig. 10 illustrative of the use of similar box sections of inverted frusto-conical form,

Fig. 11 a plan view illustrative of a special elongated form of box section,

Fig. 12 an isometric perspective view of another form of framing and one alternative arrangement of the suspension and anchoring cables,

FIGURE 4

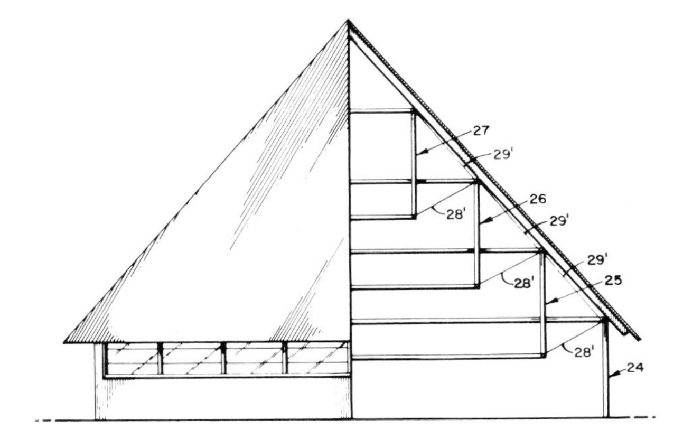

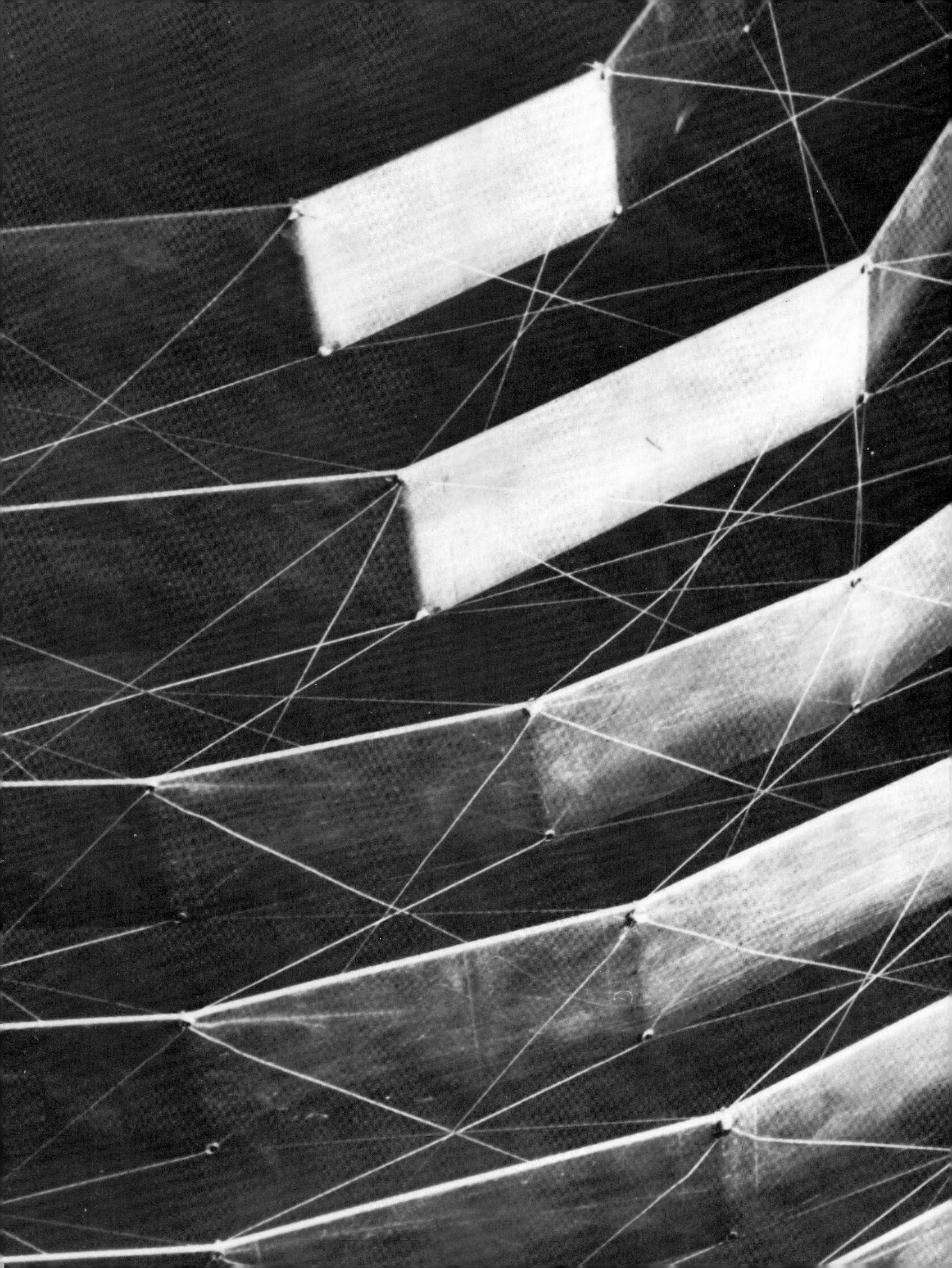

Fig. 13 a diagram analyzing the vertical components of the reach of the cables of Fig. 12,

Fig. 14 a partial view similar to Fig. 12 with an alternative disposition of the cables,

Fig. 15 a diagram analyzing the vertical components of the reach of the cables of Fig. 14,

Fig. 16 a view similar to Fig. 14 showing a still further disposition of the cables, and

Fig. 17 a diagram analyzing the vertical components of the reach of the cables of Fig. 16.

Fig. 18 is a perspective view illustrating the application of my invention to a spherical or hemispherical structure based on the spherical octahedron.

Fig. 19 is a detail cross-sectional view taken on the line 19—19 of Fig. 18.

In Figs. 1 and 2 we see a structure embodying my invention in one of its preferred forms as applied, for example, to a roofed sports arena or stadium. The structure comprises a series of polygonol box frames 1 to 10 inclusive, –2, –3 and –4. In this particular construction each frame is comprised of upper chord members 11 and lower chord members 12 forming the upper and lower peripheries of the respective frames, and vertical compression columns or struts 13 extending between the vertexes of the polygon formed by the upper chord members 11 and the corresponding vertexes of the polygon formed by the lower chord members 12. Considering an individual frame as a whole, such frame is an annular form faceted into rectangles around its rim or periphery by the members 11, 12 and 13. The series of box frames 1 to 10 inclusive, –2, –3 and –4, are of progressively varying sizes and they are arranged in a concentric array at sequentially different heights above a common plane of reference (for instance, the horizontal plane of the foundation) and in vertically overlapping spaced relation one to another. Tension elements extend between and are secured to adjacent pairs of box frames in the series, these tension elements including tension members 14, 14′ which extend downwardly from their points of securement to the lower frame of a pair to their respective points of securement to the upper frame of the pair whereby successive frames in the series are suspended one from another [function (1) of the Summary given at the beginning of this specification].

The tension elements also include other tension members 15, 15′ extending downwardly from their points of securement to the lower frame of a pair to their respective points of securement to the upper frame of the pair whereby successive frames in the series are anchored down one to another [function (2) of the Summary]. Tension members 14 and 15 may be arranged in criss-crossing pairs as shown in Fig. 2 In this particular embodiment the same tension members are crisscrossed when viewed in plan, Fig. 2, as well as when viewed in elevation.

The downwardly extending tension members may include members disposed in radial planes containing the central axis 16 of the structure, and the upwardly extending tension members 15 also may include members disposed in such radial planes. In Fig. 1 the particular tension members 14′ and 15′ which come into view are those which lie in one of such radial planes. In the preferred construction shown, the several tension members 14, 15, 14′, 15′ are secured to the respective polygonal box frames at the vertexes thereof. Also in this particular construction the domical structure formed by the series of suspended box frames is flattened on top by the provision of a series of box frames 10, 17 and 18 of progressively varying sizes arranged in a concentric array above the common horizontal plane in reference and in horizontally spaced relation one to another. As in the rest of the structure, we have the tension members 14, 14′ function-

FIGURE 5

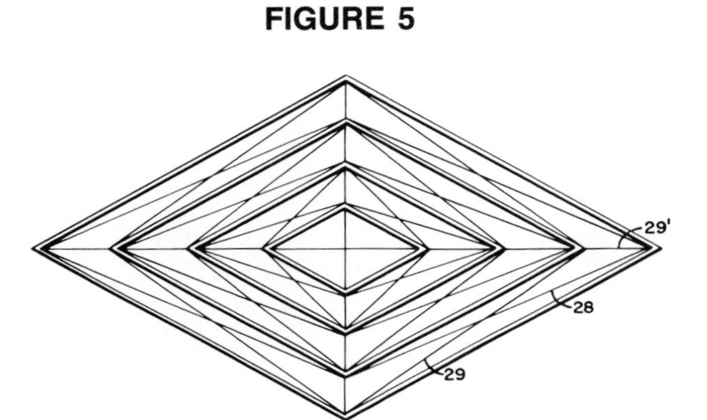

FIGURE 6

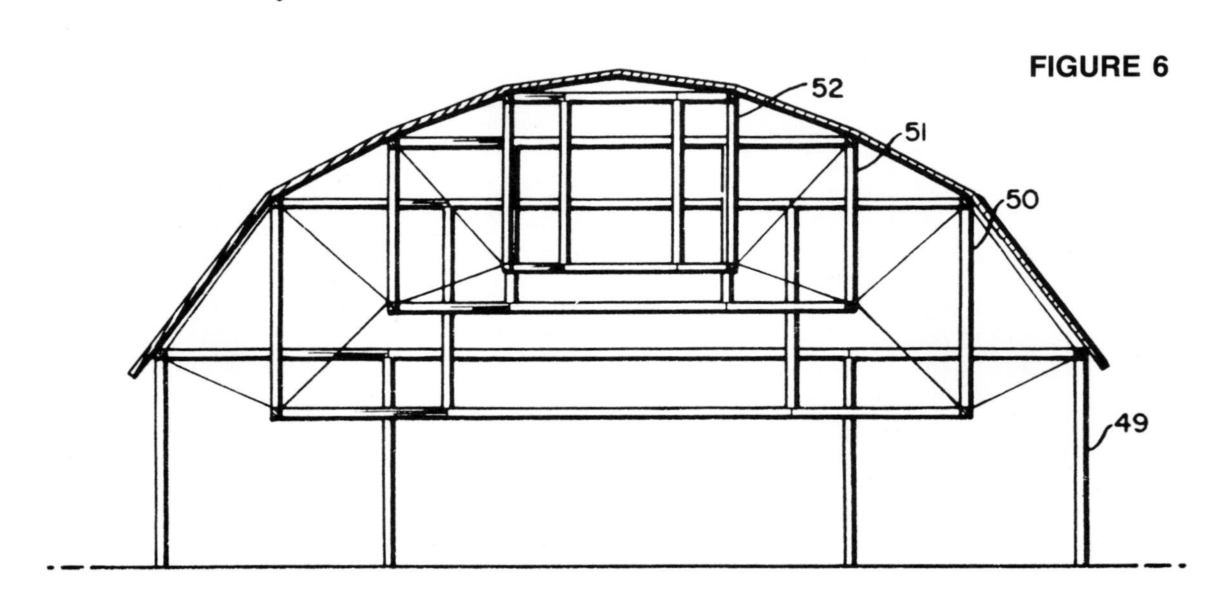

FIGURE 7

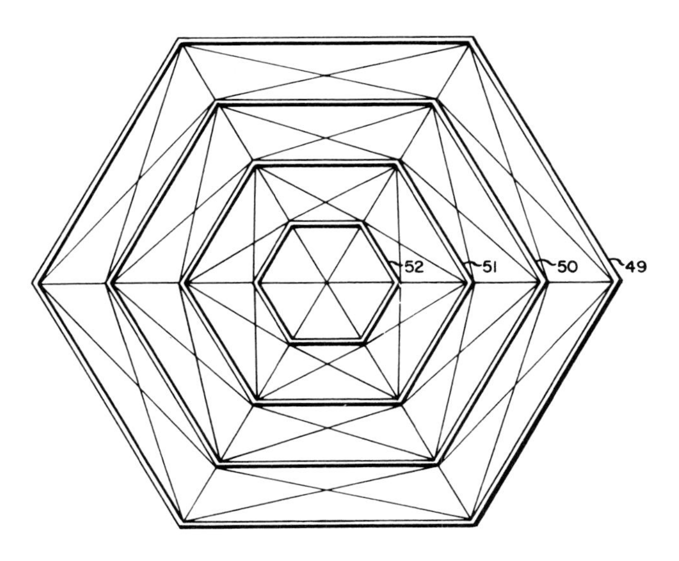

ing as suspension members, and 15, 15′ functioning as anchoring members. All such tension members 14, 14′, 15, 15′ complement one another in performing a third function, the provision of a tensioned buttress against tilting of the compression members 13 comprised in the frames [function (3) of the preceding Summary]. Additionally, these tension members act to stabilize the frames themselves and furnish a tension system efficiently disposed for resisting torquing and counter-torquing of the frames about the central axis 16 of the structure [function (4) and (5) of the Summary]. Further, the tension members 14′, 15′ which extend outwardly from the upper and lower peripheries of the frames will apply tension forces which will reduce compressive forces imposed on such peripheries by others of the tension members.

Annular frames 1 to 10 inclusive are illustrative of a rising series of frames, annular frames 1, –2, –3 and –4 illustrative of a descending series of frames, and annular frames 10, 17 and 18 illustrative of a level series. The

FIGURE 9

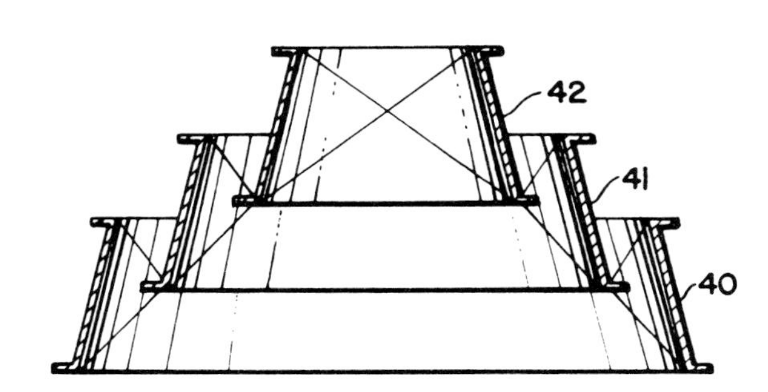

FIGURE 8

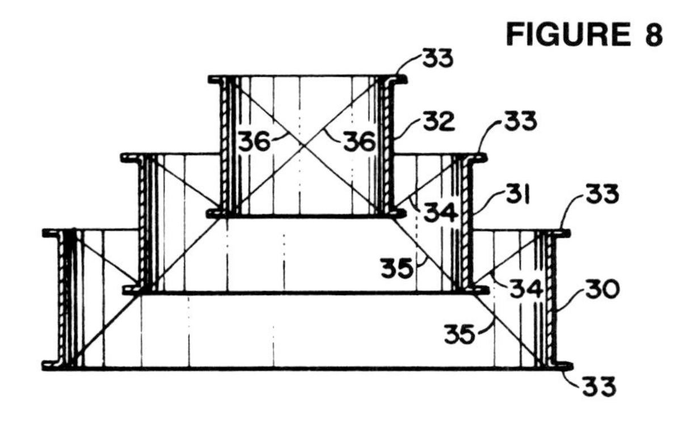

FIGURE 10

FIGURE 8a

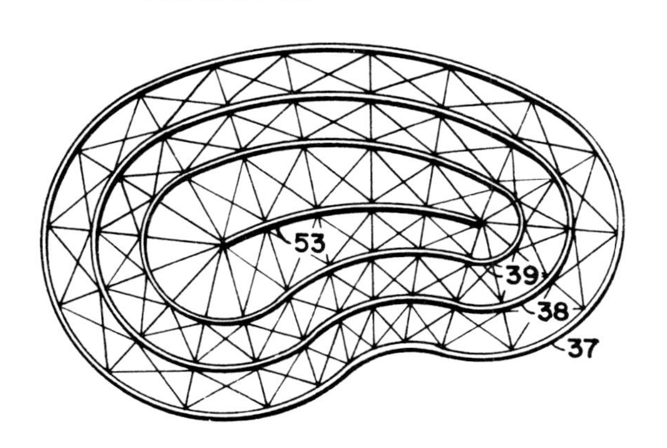

level series may terminate in a single central strut or annulus 19. Beginning at the base of the structure we have the annular frame –4 from which is suspended at a higher elevation the frame –3 from which in turn is suspended the frame –2 and then in succession we have suspended each from the other in the order named the frames 1, 2, 3 and so on, altitude being gained meanwhile in the progressively stepped suspension system. Similarly, frame 17 is suspended from frame 10, frame 18 from 17, and center post 19 from frame 18. As in the suspension bridge the tension aspect of the structure is highly significant through utilization of the more favorable tensile strengths of materials as contrasted with their vastly inferior compressive strengths. It would be difficult to imagine an upside down suspension bridge and yet that is in effect what I have achieved when my structure is considered in any

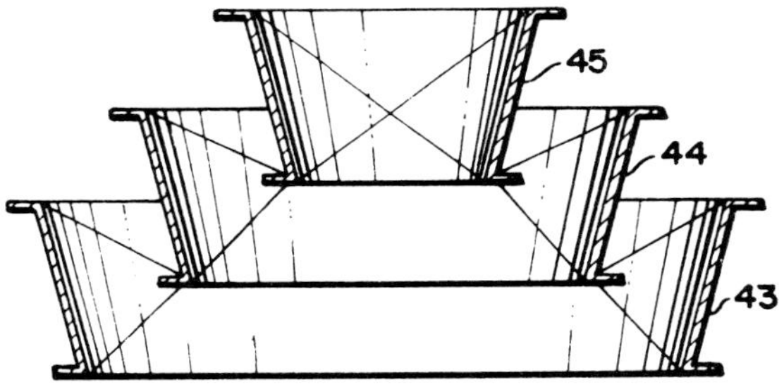

FIGURE 11

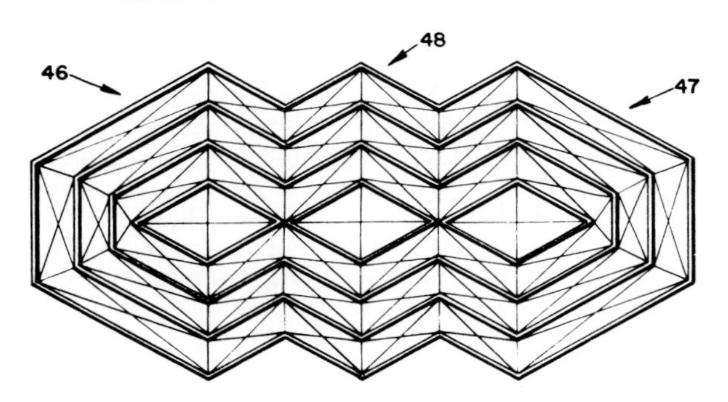

given vertical profile. Altitude is gained by breaking up the suspension system into steps and stability of the stepping is attained by the three dimensional aspect of the structure as contrasted with the essentially two dimensional system of the suspension bridge. The end result is a tremendous saving through the realization of a larger building per unit weight of the material employed in its construction, this in turn being the benefit conferred by my discovery of how to get more use from the most favorable property of materials, namely tensile strength. Essentially I have devised a building construction which in large measure replaces heavy columns and beams with

FIGURE 13

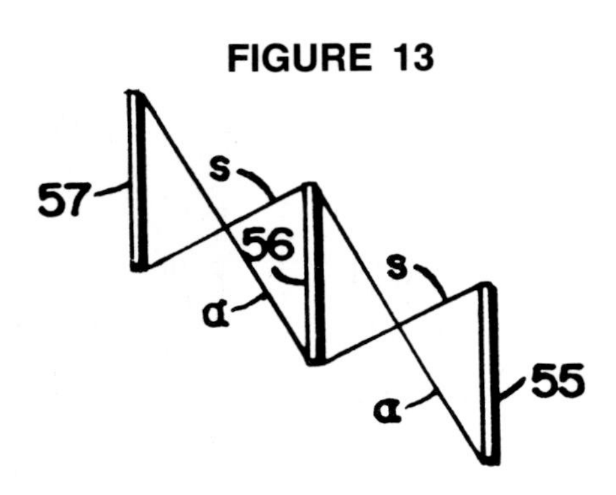

a light spidery array of tension wires and cables.

The structure can be roofed in many ways. In the embodiment of Figs. 1 and 2, roofing sheets 20 of trapezoidal form are shown. The method by which these are attached and the detail of the construction are not an essential part of my present invention.

In Fig. 3 I have shown a modified frame construction in which the compression members 21, 22, etc. are comprised of elliptical struts designed for increased efficiency in the use of materials in compression. Pagoda roofing is here employed to advantage as an efficient watershed with provision for good ventilation by means of windows or louvres 54 set well under the overhangs in the roof.

In Figs. 4 and 5 the polygonal frames comprise a series of diamond shaped frames 24 to 27 inclusive. Suspension wires 28, 28′ suspend each successively higher frame from the lower-numbered frame of the series. Anchoring wires 29, 29′ disposed in radial planes containing the central axis of the structure are provided as before. Both the wires 28, 28′ and the wires, 29, 29′ are secured to the upper edge of the lower frame of each pair of frames, in this respect representing a variation from the crisscrossing arrangement of Figs. 1 and 2. The roof structure here simplifies into a generally pyramidal form.

It is essential to understand that the box or annular frames of my invention may be either skeletal as when made of struts or tubes as I have described, or continuous as when made in the form of sheets, plates or panels.

In Fig. 8 we see an example of the latter construction in which the frames or annuli are comprised of simple cylinders 30, 31 and 32. These may have marginal flanges 33 as shown. In the embodiment typified here the suspension wires 34 extend downwardly from the upper edge of

FIGURE 12

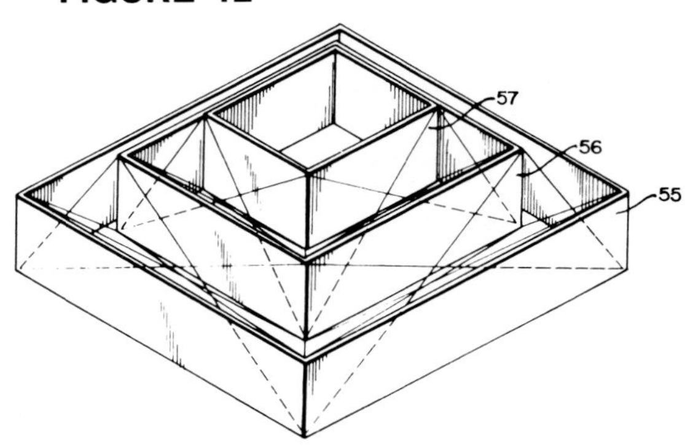

the lower frame of a pair to the lower edge of the upper frame of the pair, and the anchor cables 35 extend from the lower edge of the lower frame of a pair to the lower edge of the upper frame of the pair (instead of to the upper edge of the pair as in the embodiment of Figs. 1 and 2). The system of wires 35 may be projected as at 36 if desired, providing a continuous line of tension from the top to the base of the structure.

Fig. 8a shows the application of my invention to an assymmetrical structure comprising the frames 37, 38, 39, 53, which may be of any desired peripheral contour.

Fig. 9 represents another kind of continuous annular frame 40, 41, 42, being one of frusto-conical form.

In Fig. 10 the frusto-conical frames 43, 44, 45 are inverted from the position of those in Fig. 9. If the sheet

FIGURE 14

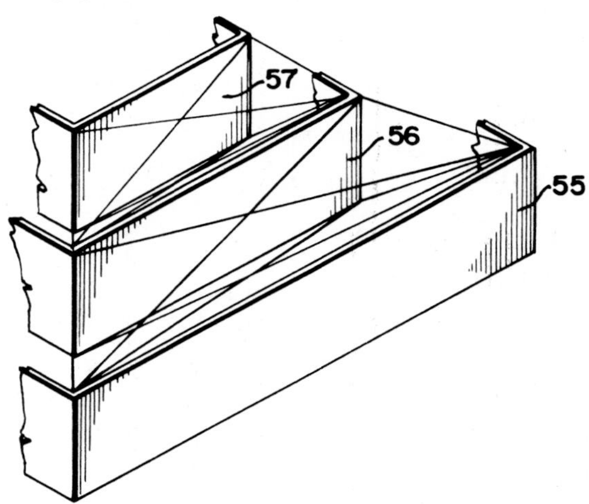

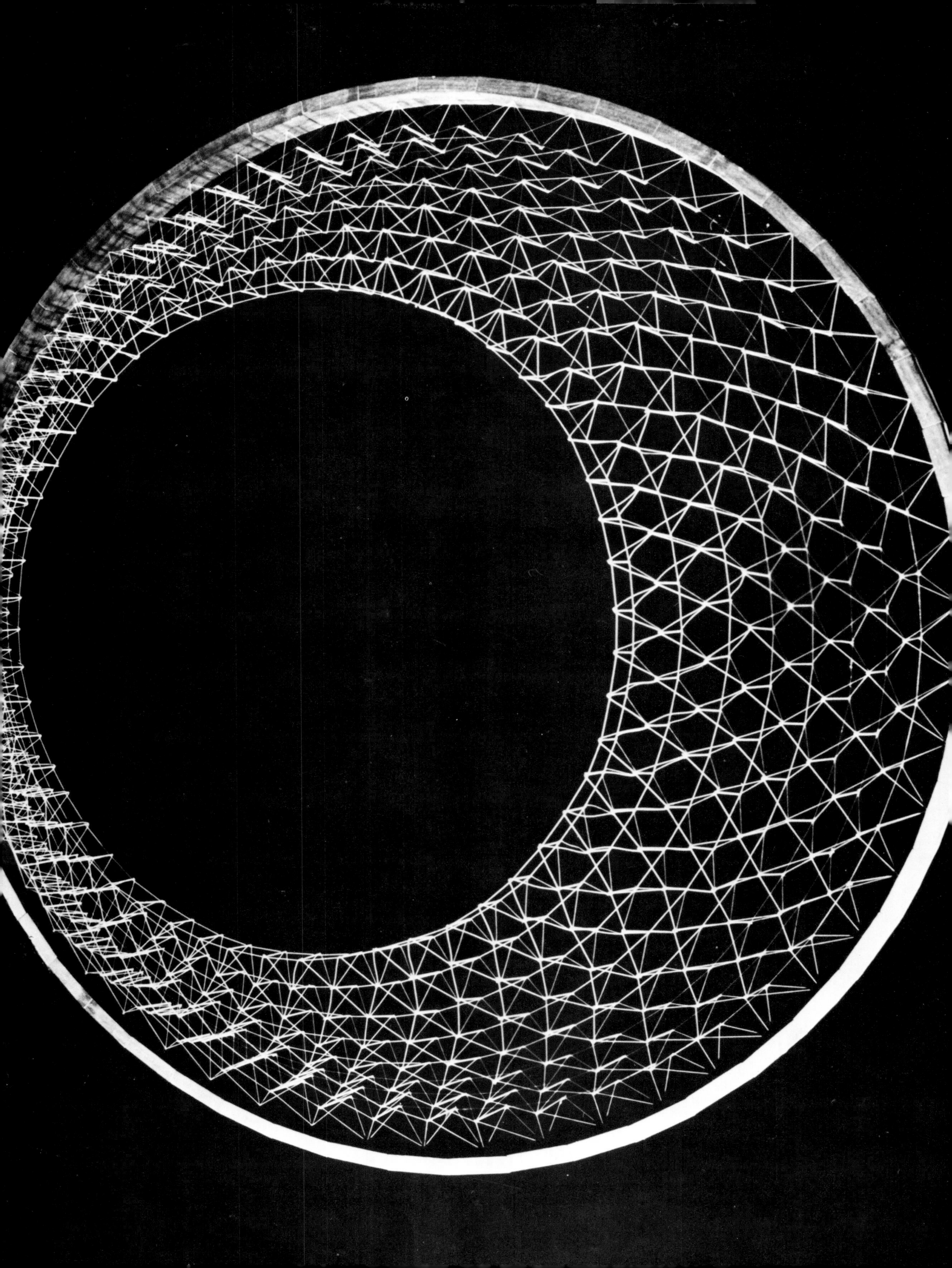

FIGURE 15

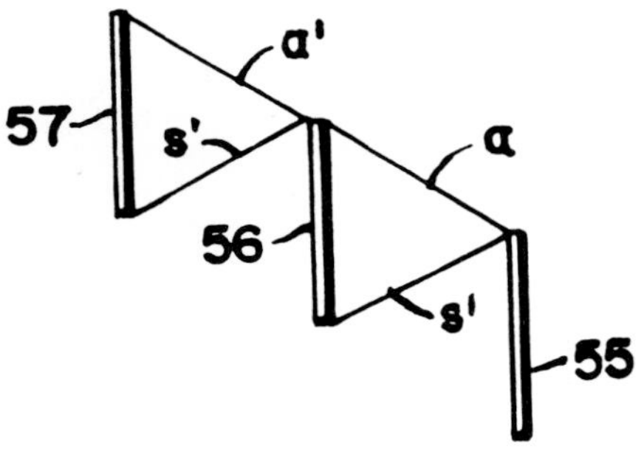

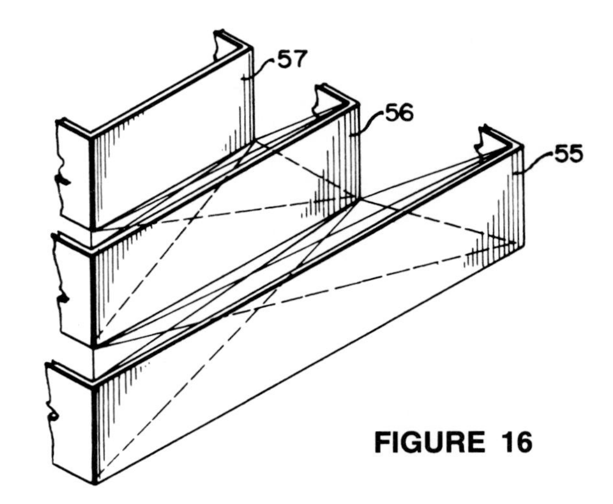

FIGURE 16

containing Figs. 8, 9, and 10 is inverted, we see three additional variants in which either cylindrical or frustoconical frames are arranged in what I have previously referred to herein as a descending series. This would equate to the series −2, −3, −4 in Fig. 1.

Fig. 11 shows the application of my invention to an elongated structure of involute-evolute form comprising end sections 46, 47, of generally hexagonal pyramidal form and a connecting section of involute-evolute form, affording added strength to such an elongated structure. The disposition of the suspending-anchoring-buttressing cables may be generally similar to that in the structures already described. The construction of the frames themselves as seen in elevation will be generally similar to that described with reference to Fig.1.

Figs. 6 and 7 shows the application of my invention to a structure of hexagonal form comprising the series of hexagonal frames 49, 50, 51, 52, each higher-numbered frame being suspended from the lower-numbered frame of the series. Here the extent of vertical overlapping between successive frames varies whereas the horizontal spacing remains constant. This is in contradistinction to the arrangement of Fig. 1 in which the horizontal spacing between successive frames varies. The feasibility of varying both vertical overlap and horizontal spacing lends great flexibility to the design of the structure.

Fig. 12 shows the application of my invention to a rectangular form of structure comprising the series of rectangular frames 55, 56, 57. As before, these frames can be considered to be made up of struts in an open or skeletal framework or to be continuous as when made in the form of sheets or panels. The frames themselves may if desired be triangulated through the use of diagonal wires extending between the diagonally opposite corners of each side or facet of the frame. Fig. 13 may be considered as a diagrammatic section taken either on the central diagonal plane or a central plane bisecting the sides of the frames, s being the suspension members and a the anchoring members. Regarding Fig. 13 as the diagonal section, s and a are both disposed in a vertical plane containing the axis of the structure. Regarding Fig. 13 as the section bisecting the sides of the frames, s and a are crisscrossed both horizontally and vertically as will be understood from Fig. 12. In either case Fig. 13 gives a picture of the vertical components of the reach of the cables.

Figs. 12–13, Figs. 14–15 and Figs. 16–17 are comparative views illustrative of the general scope of applicability of my suspension building system to structures utilizing several alternative dispositions of the suspension and anchoring cables. Figs. 14 and 16 represent portions of the same frame arrangement shown in Fig. 12 but the arrangement of the tension members is different. The difference in this arrangement may best be discerned in Fig. 15 as compared with Fig. 13. In Fig. 15 suspension cables s' extend in the same manner as the cables s of Fig. 13 but the anchoring cables a' extend between the upper corners of the successive frames in the series. Figs. 16 and 17 show another alternative arrangement of the tension members in which the anchoring cables a'' extend between the lower corners of the successive frames in the series. Thus in the construction of Figs. 12–13 we have vertically crisscross cables whereas in Figs. 14–15 and Figs. 16–17, these cables do not cross one another although the tensioning pattern in all cases is triangulated, resulting in an omni-triangulated building construction. The differing patterns of triangulation may be combined in various ways, using one system at one side or facet of the structure and an alternative system in the adjacent side as may be desired. Another alternative arrangement would be to use only suspension elements s, s' or s'' at one side of the series of frames and only anchoring elements a, a' or a'' in the side adjacent and continuing this alternating suspension and anchoring in succeeding sides.

FIGURE 17

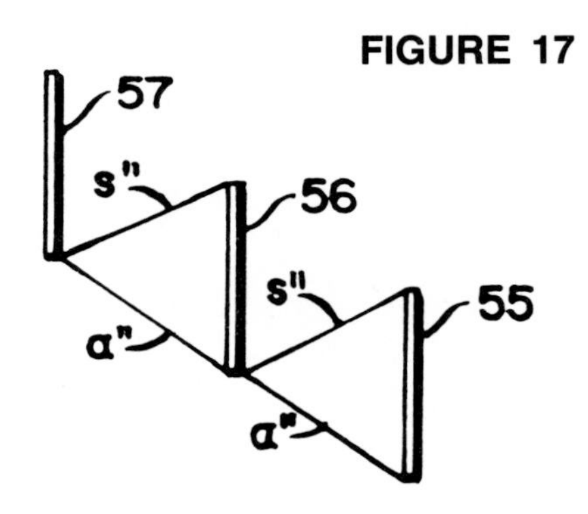

Figs. 18 and 19 show the application of my invention to a spherical structure or to a segment of spherical structure based on the spherical octahedron. Here we have a hemispherical dome comprising four spherical triangles of the octahedron defined by the intersections of three great circles, 58, 59, 60. Relating these to the globe we have the great circle 58, at the equator, 59 at 0° and 180° longitude, and 60 at 90° and 270° of longitude. Two of the spherical triangles of the octahedron appear at 61 and 62. Each is subdivided by means of any selected number of spherical triangles of progressively smaller sizes arranged in a concentric array. These several concentric triangles form the inner or outer edges of frames of either skeletal or continuous form. Considering the frame array for spherical face 61 of the octahedron, we have the frames 63, 64, 65, 66. Considering the several frames of the series to be of the same radial depth (67 = 68 = 69 = 70), and reorienting the structure so that a line containing the center of spherical face 61 and the center of the sphere is vertical, we have, as in other forms of my invention, a series of box frames of progressively varying sizes arranged in a concentric array at sequentially different heights above a common plane of reference and in vertically overlapping spaced relation one to another. Also, as before, tension elements 71, 72, Fig. 19, extend between and are secured to adjacent pairs of the frames in the series. Tension elements 71 extend downwardly, i.e., inwardly of the sphere, from their points of securement to one frame of a pair to their respective points of securement to the other frame of the pair whereby successive frames in the series are suspended one from another. Tension members 72 extend upwardly from their points of securement to the lower frame of a pair to their respective points of securement to the upper frame of the pair whereby successive frames in the series are anchored down, i.e. inwardly of the sphere, one to another. It will be understood that the several tension elements serve all of the five functions described hereinabove, namely those of suspension, anchoring, buttressing, stabilizing of the frames themselves and resisting torquing and countertorquing of the frames about a line passing through the centers of the sphere and of the octahedral face. It may be mentioned that in Fig. 18 I have for simplicity of illustration omitted tension elements 71 and 72, the general arrangement of which will be understood from the description of the other forms of my invention. Also, it will be understood that any of the described alternate arrangements of the tension elements may be employed as desired.

FIGURE 18

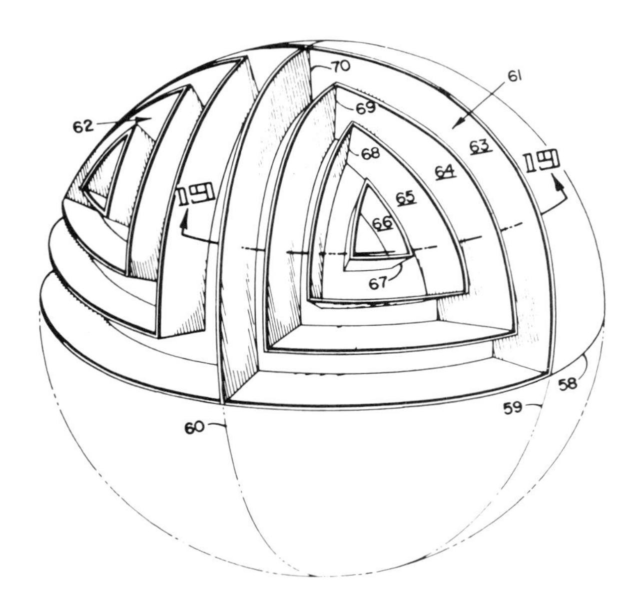

FIGURE 19

The terms and expressions which I have employed are used in a descriptive and not a limiting sense, and I have no intention of excluding such equivalents of the invention described as fall within the scope of the claims.

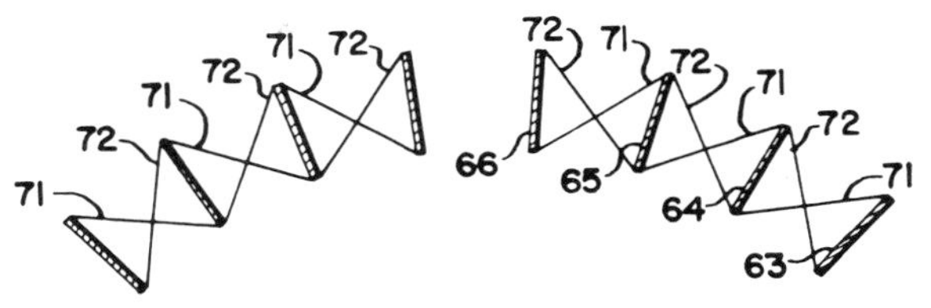

18▲MONOHEX (GEODESIC STRUCTURES) (1965)

U.S. PATENT—3,197,927

APPLICATION—DECEMBER 19, 1961

SERIAL NO.—160,450

PATENTED—AUGUST 3, 1965

MONOHEX CAME ABOUT BECAUSE an omni-triangulated pattern of geodesics can have its triangles grouped together in diamond forms, but they can also be grouped together to form hexagons and pentagons. There will always be twelve pentagons in a sphere; all the rest will be hexagons. This dome can also be composed of pentagons, hexagons, and triangles. I found that the hexagons and pentagons made it possible to have circular openings inside the pentagons and hexagons with triangular structuring between them. This seemed to be very desirable because the holes could be made by cylinders, i.e., the surface of the sphere turning into a cylinder at the hole and protruding outwardly, providing compound curvature with very great strength. I found that we could then make triangular units intervening between the hexagons and pentagons that join together to form the cylindrical openings, which components could nest comfortably when disassembled; we sometimes call this the Fly's Eye Dome.

UNITED STATES PATENT OFFICE

Richard Buckminster Fuller, 407 S. Forest St., Carbondale, Ill.

GEODESIC STRUCTURES

The invention relates to geodesic structures as comprised in the fabrication of domes, spheres, and spherical segments or truncations, as used in buildings and other architectural forms, or for other purposes.

The fundamentals of geodesic structures are described in my prior Patent No. 2,682,235, granted June 29, 1954, in which are defined some of the more common terms in the relatively new art of geodesics. The present specification is addressed primarily to those who have a practical working knowledge of geodesic construction although it should be readily understood as well by architects and engineers following a background reading of the prior patent aforesaid.

All omni-triangulated geodesic grids may have their triangles collected into hexagons plus twelve pentagons or into omni-diamond patterns as well as in simple omni-triangulated patterns. Thus when we look at a typical geodesic "dome," we may see an over-all pattern which emerges as a honeycomb of hexagons (with twelve pentagons as aforesaid), or as diamonds, or as simple triangles. Sometimes the hexagons and pentagons will be of pyramidal form and thus have triangular facets. My pres-

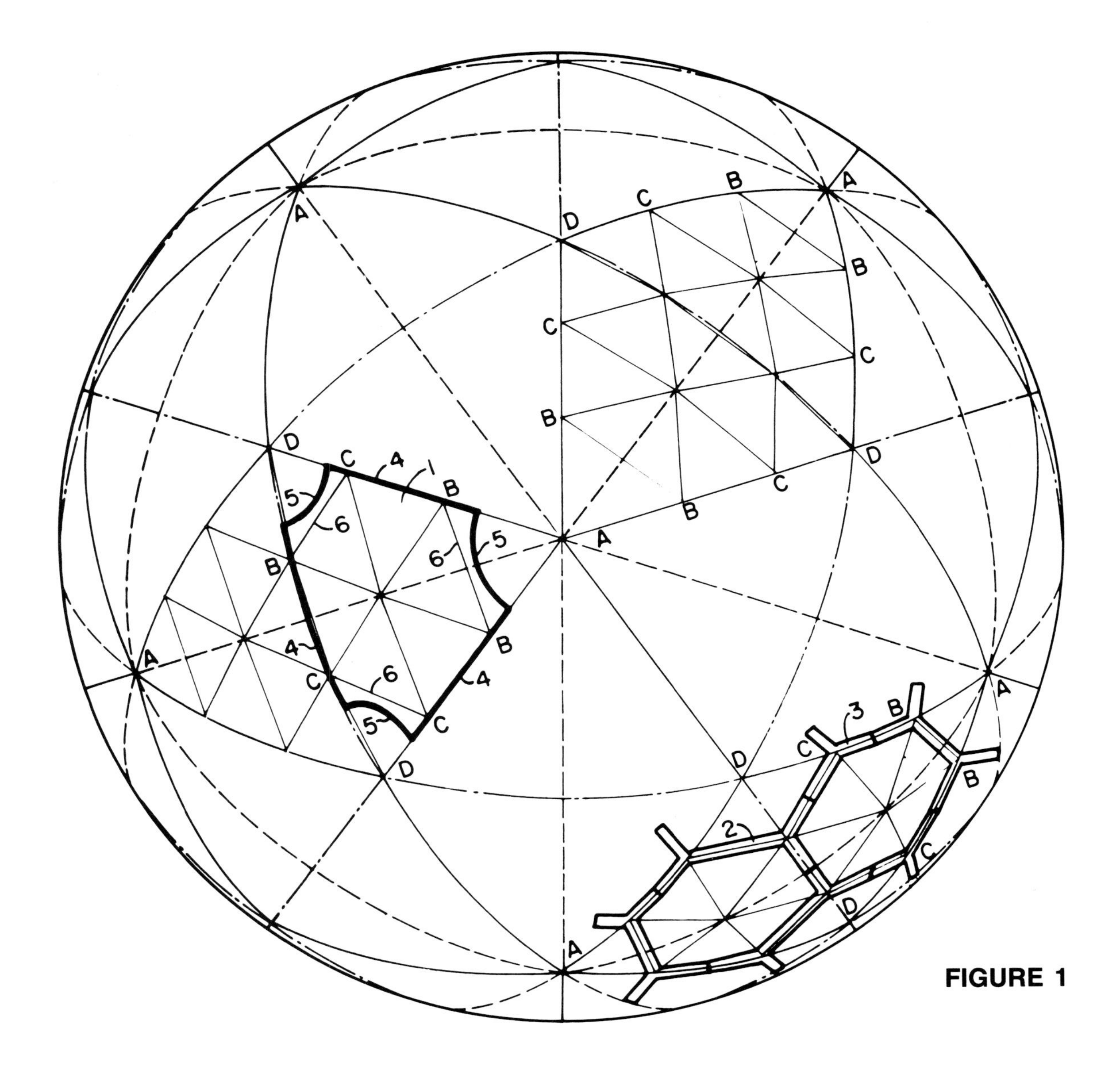

FIGURE 1

ent invention relates to a particularly advantageous development of the hexagonal-pentagonal manifestation of the geodesic pattern, having regard however to the possibility that the hexagons in the pattern may be faceted so as to reveal a typical pyramidal sub-form.

The "building blocks" of my present invention may consist of hexagonal sheets or plates, or of members which go together to form a framework defining openings of hexagonal and pentagonal form. In either case, openings in the spherical shell will appear at the vertexes and centers of the icosahedron triangles. These openings may be either circular or polygonal in form as desired. In all cases it is contemplated that the openings may be closed to form a continuous roof or wall as by the means which will be described herein.

Whether assembled initially from sheets or plates on the one hand or as an open framework on the other, the construction comprises tension rings of either circular or polygonal form, and in certain of my preferred constructions these tension rings will be interconnected with one another to form a comprehensive geodesic network of tension elements conforming to the geodesic pattern of the "building blocks."

In general my invention consists of a structure of generally spherical form comprising a plurality of members joined together in a geodesic pattern of hexagons and pentagons and which includes tension elements disposed around the perimeters of areas within the geodesic pattern, these being areas whose borders are defined by edges of adjoining members. The tension elements are arranged to transmit forces from member to member throughout the geodesic pattern of hexagons and pentagons. This general statement will be easier to understand after study of the accompanying drawings and detailed description illustrative of the best mode contemplated by me for carrying out my invention.

In the drawings:

Fig. 1 is a diagram showing my preferred geodesic layout as applied to a 6-frequency breakdown known in the art as the "6v triacon" ("v" standing for frequency).

Fig. 2 is a diagram to explain the spherical geometry of the great circle geodesic frame of reference.

FIGURE 2

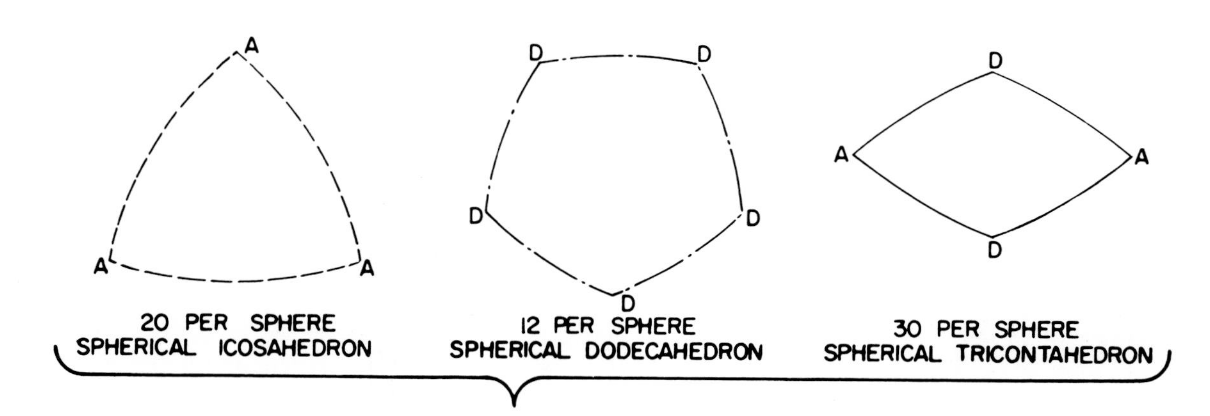

FIGURE 3

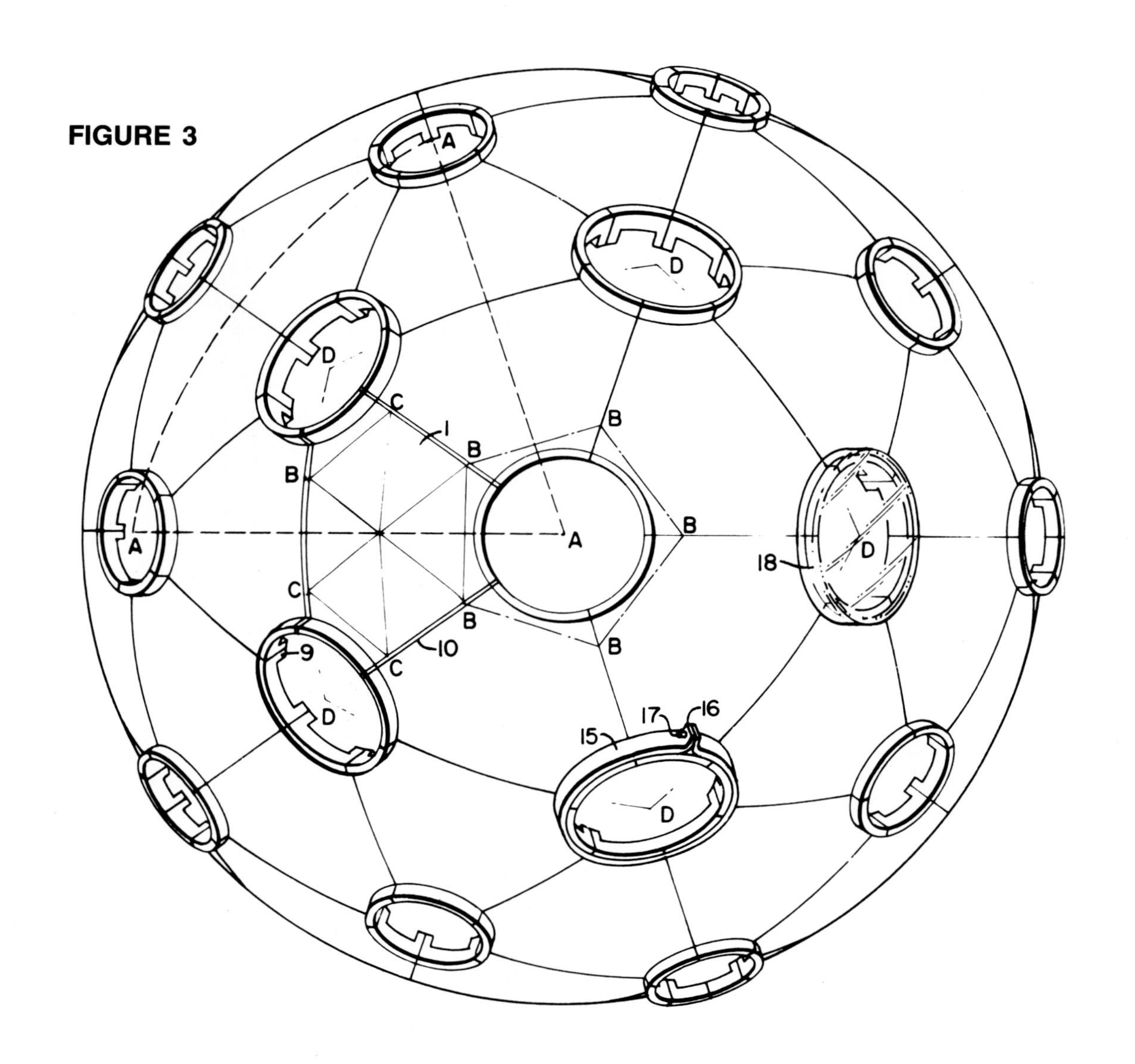

Fig. 3 is a view of one form of structure embodying my invention. It may be considered as showing either the top plan view of a hemispheric dome or simply as a side view of a spherical structure.

Fig. 4 is a perspective view of a hemispheric dome constructed in accordance with another embodiment of my invention.

Figs. 5 to 10 inclusive are detail views of various types of members used as the principal components of the structures; Figs. 5 to 8 relating to components for the type of structure shown in Fig. 3, and Figs. 9 and 10 to components for the type of structure shown in Fig. 4.

Fig. 5 shows a component of generally spherical form,

Fig. 6 a component of hexagonal pyramidal form,

Fig. 7 a component of flat hexagonal form, and

Fig. 8 a component similar to Fig. 7 such as may be formed of sheet metal or the like.

Fig. 9 shows a component of a framework constructed in accordance with my invention in its general relationship to two adjacent components of like construction.

Fig. 10 is a detail plan view of the primary component of Fig. 9 showing a system of pre- or post-tensioning rods which may be used therein.

Fig. 11 is a view of a sub-component assembly for producing members similar to those of Figs. 5, 6 and 7.

Fig. 12 is a view taken as indicated at 12—12 in Fig. 11 following assembly and grouting of sub-components and main components.

Fig. 13 is a detail cross-sectional view taken on the line 13—13 of Fig. 12.

Fig. 14 is a view of a sub-component for producing members similar to those of Fig. 8 or 10, but of modified construction.

Fig. 15 shows one form of component produced from an assembly of Fig. 14 sub-components.

Fig. 16 shows another form of component produced from an assembly of Fig. 14 sub-components.

Referring to Figs. 1 and 2, the dotted lines represent the sides of the triangles AAA of a spherical icosahedron, of which there are twenty triangles per sphere. The dot and dash lines represent the sides of the pentagons DDD, etc., of the spherical dodecahedron, of which there are twelve pentagons per sphere. The full lines represent the sides of the diamonds ADAD of the spherical tricontahedron, of which there are thirty diamonds per sphere.

Upon one of the diamond faces of a tricontahedron is shown a typical layout for a 3-way grid of the type known in the art as a "triacon" breakdown ABCD. This 3-way grid may equally be considered as being based upon any of the three geodesic forms shown in Figs. 1 and 2, namely the icosahedron, the dodecahedron or the tricon-

FIGURE 4

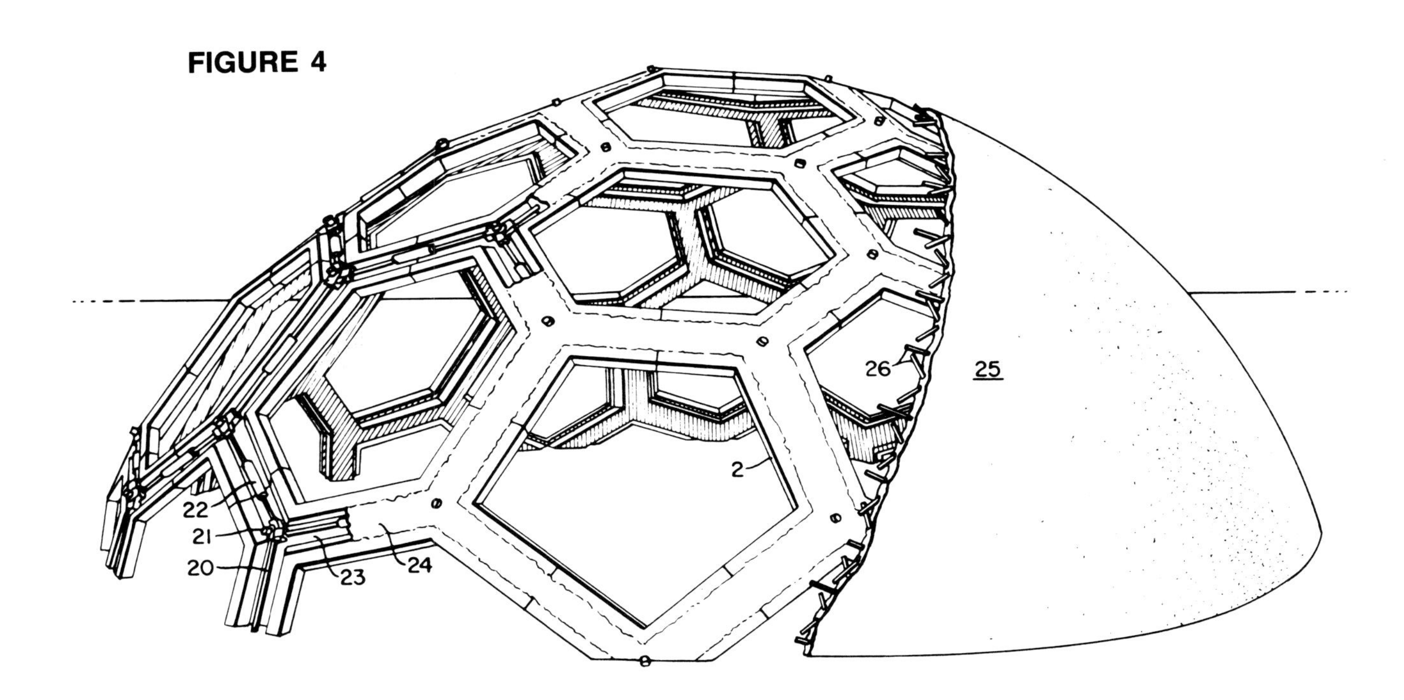

FIGURE 5

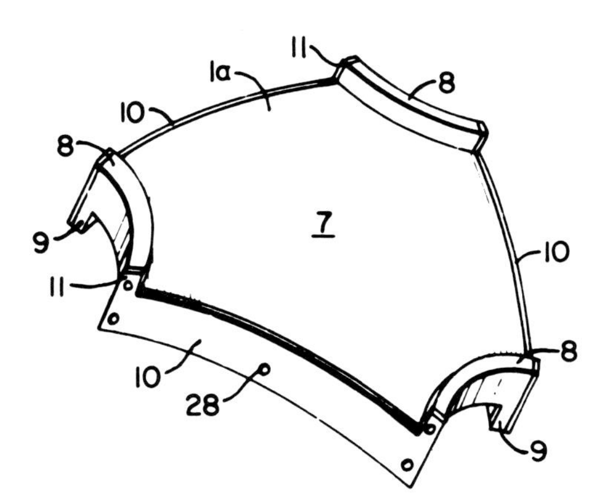

tahedron, the relationship between the three being shown as follows:

For simplicity the suffix "hedron" will be omitted from the descriptive names of the polyhedrons.

FIGURE 6

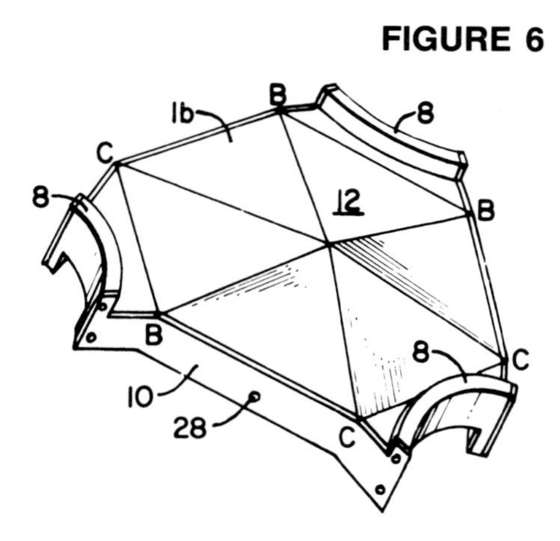

FIGURE 7

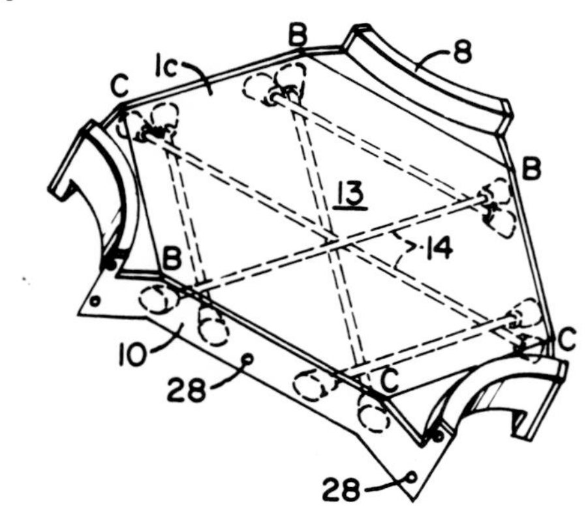

The vertexes of the triconta comprise:
Either—

2 icosa vertexes icosa
2 icosa centers relation

Or

2 dodeca vertexes dodeca
2 dodeca centers relation

Each icosa vertex is a dodeca center.
Each dodeca vertex is an icosa center.

From the foregoing analysis it will be understood that the icosahedron is the inversion of the dodecahedron or vice versa, and that the same division of the sphere results regardless of whether one considers that the breakdown has been based upon the icosahedron, the dodecahedron or the tricontahedron. Cf. Felix Klein, Elementary Mathematics from an Advanced Standpoint: Arithmetic, Algebra Analysis, translated from 3rd (1925) edition; New York, Dover Publications; p. 123.

The 3-way grid pattern ABCD has been shown in Fig. 1 on three of the faces of the tricontahedron. In the one which appears at the left center, there has been superimposed a heavy black line illustrating a sheet or plate-like member 1 of generally hexagonal form which constitutes

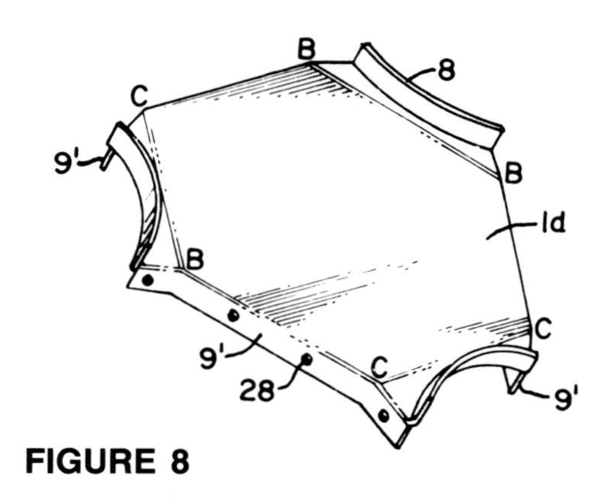

FIGURE 8

one of the components, or members, of the construction illustrated in Fig. 3. Upon the 3-way grid shown at the bottom of Fig. 1 is superimposed a framework comprising a plurality of members 2, 3. In the case of members 1, three of the edges, those designated 4, are aligned with the construction lines of the 3-way grid pattern, i.e. with the sides of the tricontahedron or the grid lines within. The circular sides 5 of member 1 may be considered as being related to grid lines 6 and, if desired, could be

FIGURE 9

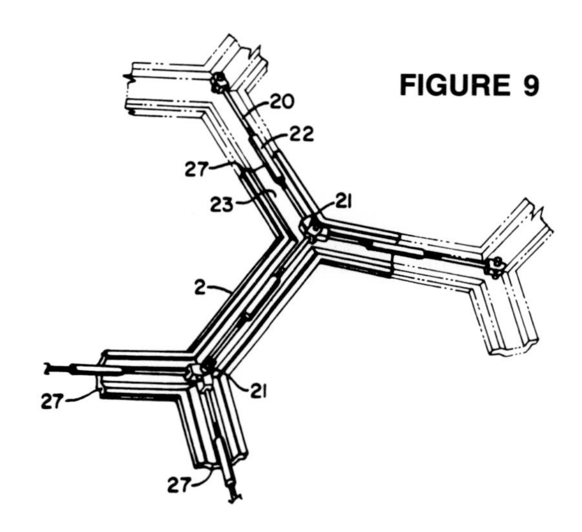

made to coincide with such grid lines, in which case the sides 5 would be straight instead of curved.

In the case of members 2 and 3, it is the centerlines of the members which coincide approximately with the grid lines. Thus, as in the case of the differing embodiments described in my prior Patent 2,682,235 aforesaid, the alignment of the structural members may be such that the longitudinal center lines thereof are arranged along geodesic lines or such that their edges or selected ones of their edges are arranged along geodesic lines.

In Fig. 3 we see a plurality of the members 1 assembled to form a dome or sphere as based upon what may be considered as the icosahedron pattern AAA or the dodecahedron pattern DDDDD or the tricontahedron ADAD. The members 1 may be of any of the several forms shown in Figs. 5 to 8 designated respectively 1*a*, 1 *b* 1 *c*, 1*d*. The Fig. 5 form has a spherical surface 7, upwardly extending flanges 8 at three sides, these being in the form of circular segments extending along the lines 5 of Fig. 1, and downwardly extending flanges 9 along the remaining edges of member 1*a*, namely the edges corresponding to 4 in Fig. 1. Member 1*a* may be made of any suitable construction material such as concrete, plastic resin, light-weight insulating concrete, etc., and may be reinforced along three of its sides as by means of metal strips, 10. As shown, these metal strips are formed with upwarding extending lugs 11 for a purpose which will be explained later.

The structural component in 1*b* shown in Fig. 6 may be used in conjunction with, or in place of, the component 1*a* of Fig. 5. In the 1*b* form, the construction is essentially the same except for the fact that the surface 12 is of pyramidal form instead of being spherical.

The further embodiment 1*c* shown in Fig. 7 has a flat surface 13 and is reinforced by a series of pre- or post-tensioned rods or elements 14 forming a 3-way grid harmonizing with the stress pattern of the 3-way grid ABCD previously described. Similar reinforcement may be used in the case of members 1*a* and 1*b*. Thus any of the members 1*a*, 1*b*, 1*c* may consist of pre- or post-tensioned bodies comprising tensioned elements arranged in a 3-way grid pattern complementing the geodesic pattern formed by such members. By this means each of the basic components of the structure may be locally stressed as distinguished from comprehensive stressing of the structure as a whole in a manner which remains to be described.

The form of component 1*d* shown in Fig. 8 may con-

FIGURE 10

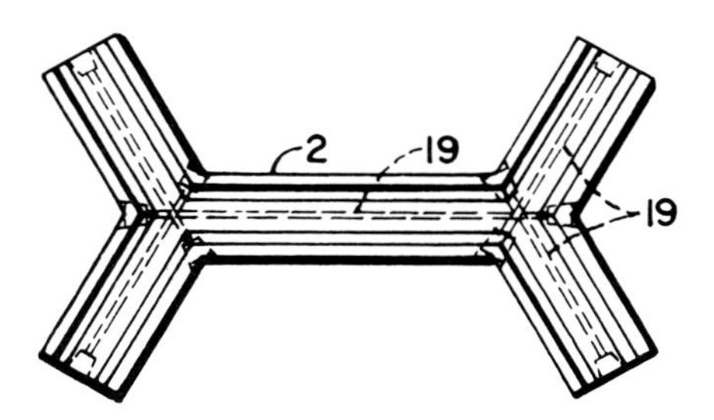

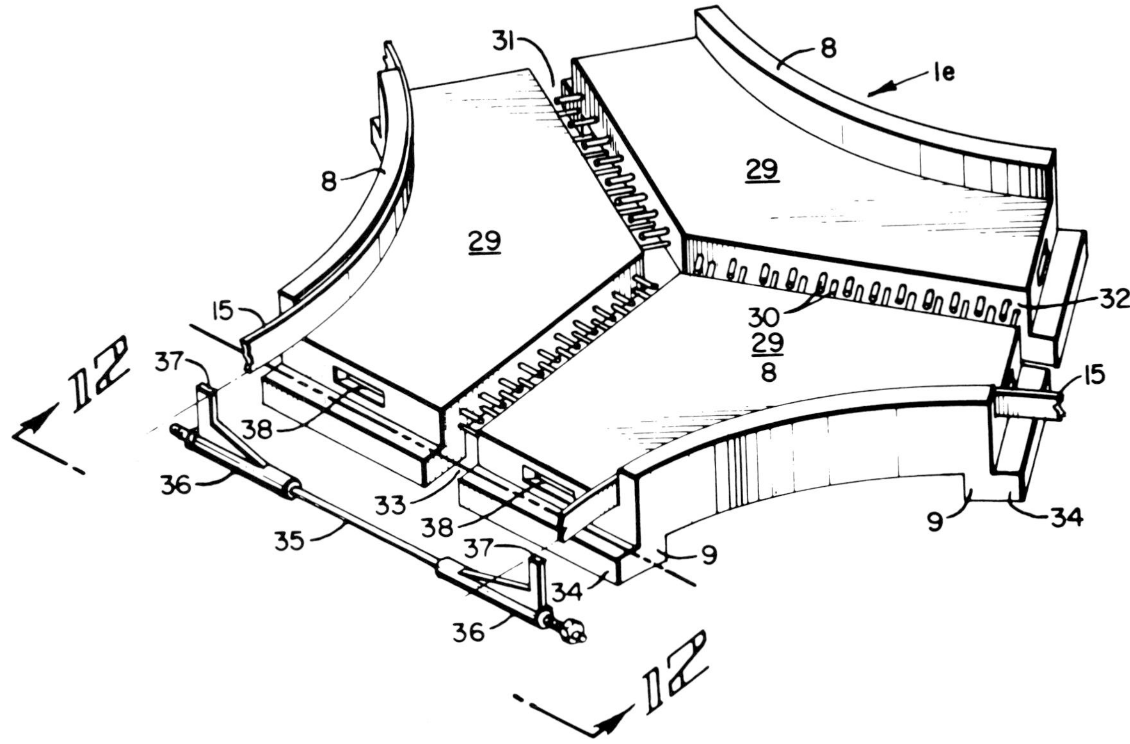

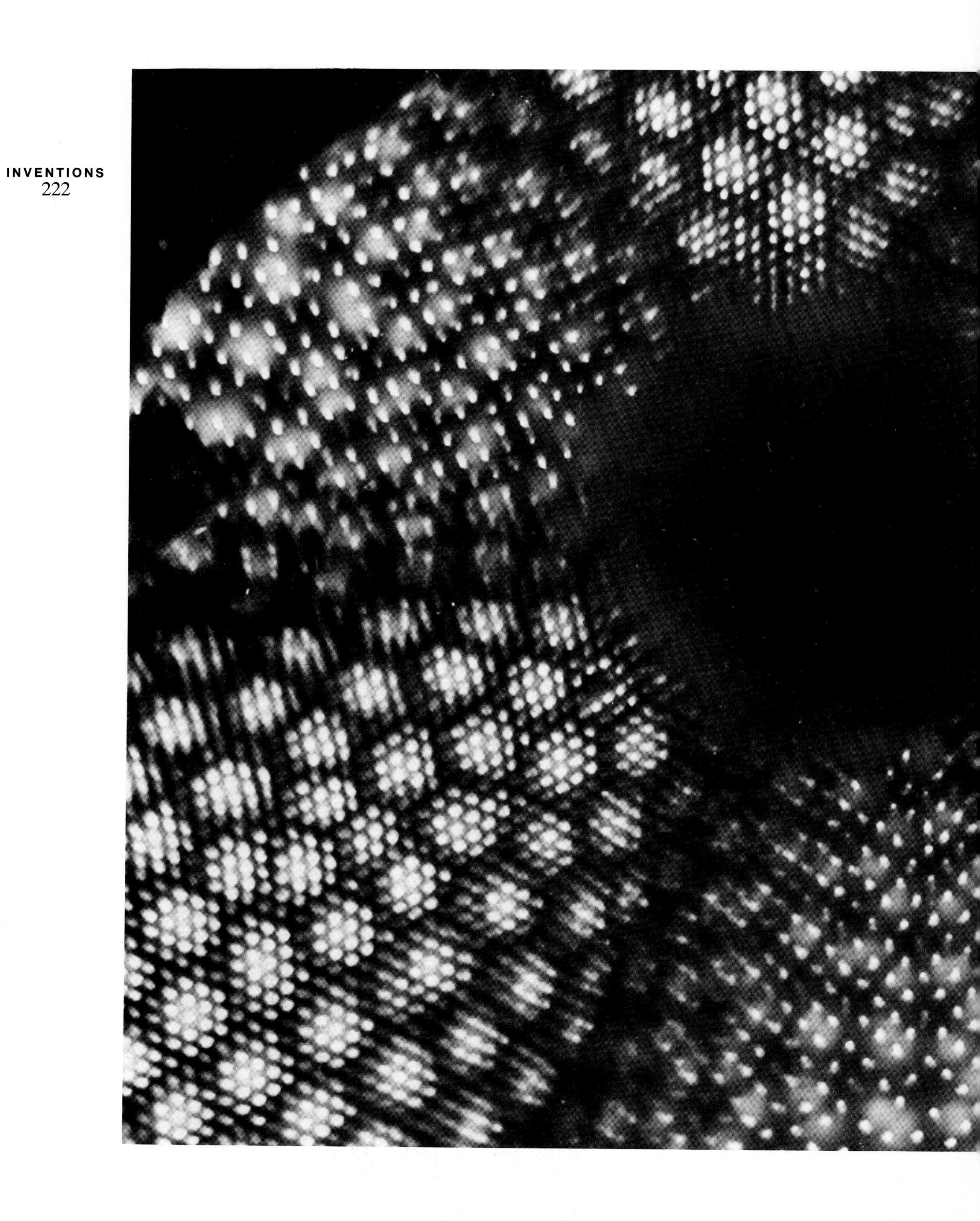

veniently be made from sheet metal such as a strong aluminum alloy. In this member the upwardly extending arcuate flanges 8 and downwardly extending flanges 9′ can be formed in a drawing press. In this construction the flanges 9′ take the place of flanges 9 of the components of Figs. 5–7 and also take the place of the separate metal strips 10 thereof.

The relationship of the several forms of basic components shown in Figs. 6–8 to the geodesic pattern of Fig. 1 and the corresponding geodesic structuring of Fig. 3 is shown by the corresponding grid line lettering BB, CC, BC, etc. The theoretical relationship is the same also in the case of the component of Fig. 5 although certain of the grid lines become imaginary by reason of the fact that the surface is spherical and therefore cannot reveal the grid lines which show up in the other forms. Nevertheless the 3-way stress pattern remains both as to the basic component itself and as to the pre- or post-tensioned rods 14 in the case where such rods are used.

It will be understood that each of the faces of the spherical tricontahedron of the structure may be laid out on the same 3-way grid shown on three of such faces in Fig. 1. The basic components are then assembled according to the pattern thus produced. In my preferred construction according to this embodiment of the invention, the basic components are held together by tension rings 15 made of a material of good tensile strength. These rings also preferably are constructed with suitable tightening or clamping means such as provided by the flanged ends 16 and a bolt 17.

Another way of holding the basic components together is by means of flanged "manhole covers" 18 which can be made of any suitable material and may be either transparent, translucent or opaque according to the intended use to which the structure is to be put. The flanges of the covers 18 may be arranged to engage the outer surfaces of the collar formed by the several arcuate segments 8 of the basic components or the flanges of the covers may be engaged over the tension ring 15. In the former case the covers may themselves constitute means for tensioning the basic components whereas in the latter case they may be used simply as a means for closing the openings in the structure.

From the foregoing description it will be discerned that I have provided a structure of generally spherical form comprising a plurality of members 1 which may be of one or more of the several forms 1*a*, 1*b*, 1*c* and 1*d*, such members being joined together in a geodesic pattern comprising hexagons and pentagons (the pentagons always coinciding with one of the openings in the structure), and tension elements such as 15 disposed around the perimeters of areas within the geodesic pattern whose borders are defined by edges 8, 8 etc., of adjoining members comprised in groups of such members surrounding the respective areas. The tension elements transmit forces from member to member throughout the geodesic pattern. The aforesaid tension elements 15 preferably are interconnected one to another as by means of the strip 10, the upwardly extending lugs 11 of which are engaged by the tension rings 15 to form a comprehensive geodesic network of tension elements conforming to the geodesic pattern of said members. Where this is done in combination with the use of the pre- or post-tensioning rods 14 previously described, the resulting structure is characterized by locally stressed basic components plus comprehensive stressing of the structure as a whole. In this construction the "holes" in the geodesic 3-way grids are always surrounded and "healed" through diffusion of the stress patterns in the members themselves. The true result of this can only be accurately described as an omniwave diffusion of the stress pattern as obtained by reason of the geodesic structuring described.

In the alternate embodiment illustrated diagramatically in the lower part of Fig. 1 and structurally in Figs. 4, 9 and 10, we have a framework of members joined together in the same geodesic pattern comprising hexagons and pentagons but in this case the hexagons will be initially open, rather than closed as was the case with the surfacing components of the Fig. 3 construction. The form of the members used in making the structure of Fig. 4 is shown in Figs. 9 and 10. These members may be cast of concrete with or without the use of the pre- or post-tensioning rods 19, or may be made of other suitable materials. Where the local pre- or post-tensioning rods 19 are used in the manner shown in Fig. 10, it will be observed that such rods substantially conicide with the geodesic pattern of the dome. As before, we have tension rings comprised of the tension elements 20 linked together by bolts and clevises 21 and tightened as by means of turnbuckles 22. These tension rings likewise follow the geodesic pattern of the dome. I describe them as tension elements disposed around the perimeters of areas within the geodesic pattern whose hexagonal or pentagonal borders are defined by edges of adjoining members comprised in the groups of members surrounding the respective areas. Also as before, the tension elements 20 are arranged to transmit forces from member to member throughout the geodesic pattern. Notice that the locally stressed pre- or post-tensioning elements 19 are, or may be, sustantially parallel with the tension elements 20 of the comprehensive geodesic network.

The members 2 of the embodiment of Figs. 4, 9 and 10 preferably are formed with troughs 23 in their outer surfaces to receive the tensioning structure 20, 21, 22. Following tightening of the turnbuckles 22, the troughs 23 may then be filled with a cement or other grouting mate-

FIGURE 12

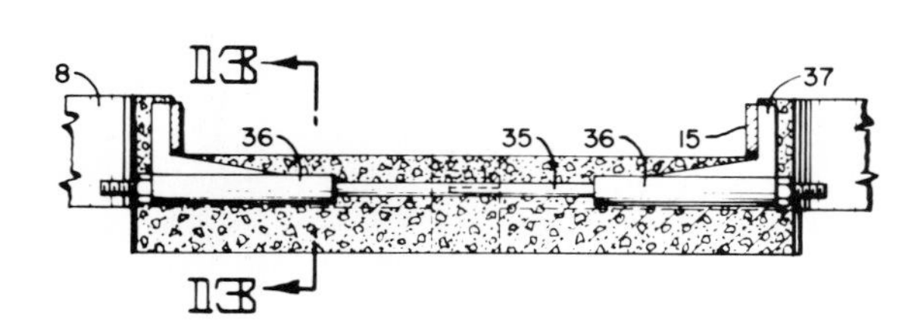

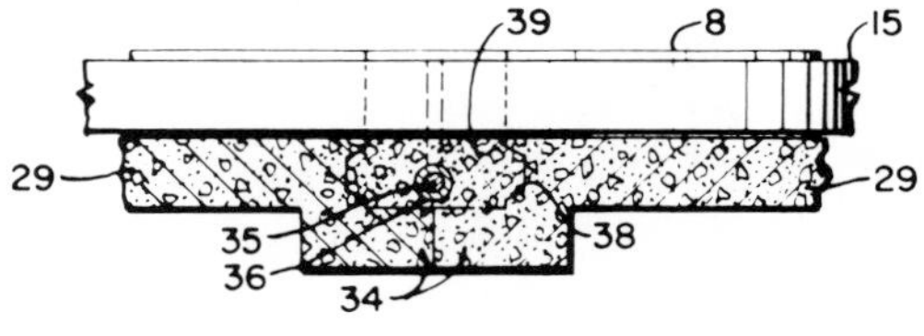

FIGURE 13

rial 24, resulting in the smooth surfaced framework shown in the central portion of Fig. 4. Later, with the use of suitable forms, the framework may be covered with a monolithic shell of concrete or the like 25, shown at the right-hand of Fig. 4. This covering shell may if desired be reinforced with a network of reinforcing rods 26. When this is done, I prefer to arrange the reinforcing rods 26 harmoniously with the 3-way grid pattern of the basic structure of the framework. Thus the rods 26 may conform to a 3-way grid pattern similar to that shown at ABCD, ABCD in Fig. 1.

The sides 4 of the members 1 (Fig. 1) and the ends 27 of the members 2 (Fig. 9) may be provided with complementary dowels and recesses 28 as indicated for example in Fig. 7, to assist in aligning the several members during the assembly thereof.

The modified construction of Figs. 11 to 13 comprises a sub-assembly in which the main component 1*e* includes three sub-components 29 all of which can be of identical size and shape. By varying the spacing between the three crete, plastic resin, light-weight insulating concrete, etc., and in such cases may include reinforcing rods projecting into meeting or overlapping relation to those of the other associated subcomponents as at 30. The width of the spaces 31, 32 and 33 is variable according to the geometry of the geodesic breakdown. Upwardly and downwardly extending flanges 8 and 9 may be provided as before. Flanges 9 are preferably extended laterally as at 34 to meet or approach corresponding extensions of the flanges of adjacent main components.

After the sub-components 29 have been brought into the desired predetermined relation one to another they are grouted or cemented together. Then they are ready for assembly in the manner of Fig. 3 with the use of tension rings 15 and/or manhole covers 18. Tension elements 15 preferably are interconnected one to another as by means of rods 35, Figs. 11–13, and suitable anchors 36 having projections 37 whch are arranged to engage tension elements 15 in the manner shown in Fig. 12. The ends of rods 35 are threaded to receive the nuts shown

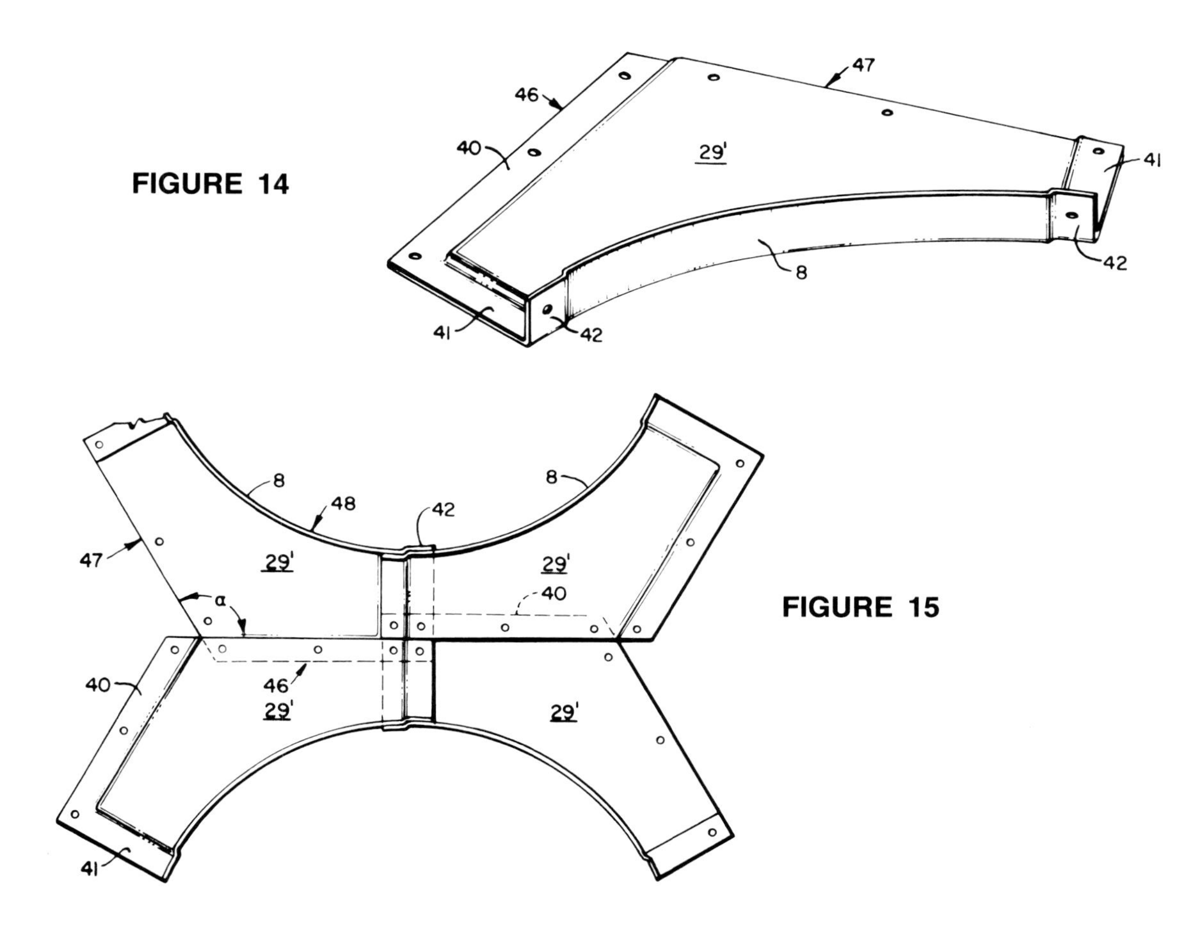

FIGURE 14

FIGURE 15

sub-components along the parts of the Y formed by the proximate edges thereof, it is possible to assemble main components of a variety of forms and sizes so as to accommodate them for use in constructing spheres or domes of even the higher frequencies of geodesic breakdown. Thus a single sub-component can be used to make a plurality of different main components wherever the geodesic construction is not adapted to construction for one size and form of main component. Where desired, the sub-components 29 may be molded or cast of con- for drawing the anchors together to produce an effective comprehensive stressing of the structure through the interaction of the tension elements 15 and rods 35.

Recesses 38 may be provided in the ends of member 1*e* to interlock with the grout or cement used for connecting and sealing the proximate ends of the adjacent main components of a structure made from this form of component. When so cemented together, the grout or cement will fill the groove between the adjacent members formed by their proximate ends and the meeting lateral extensions 34 thereof which close or partly close the bottom of the groove to make a mold or partial mold for the grout. The grout 39 surrounds, and seals in, the tension rod 35 and anchors 36.

The sub-components of the embodiment shown in Figs. 14–16 may conveniently be made of sheet metal or fiberglass, but other suitable materials may be employed as desired. According to this embodiment the principal advantages described with reference to Figs. 11–13 can be secured for component members used in either the general type of construction shown in Fig. 3 or that shown in Fig. 4, or in other constructions wherein the hexagonal members are shrunk by enlargement of the circular areas of Fig. 3 to produce diaper-form components (Fig. 16). Again we have three identical subcomponents 29′. Each subcomponent 29′ consists of an elongated member having two sides which meet at an angle *a* which is on the order of 120°, these two sides forming a generally convex longitudinal edge. Opposite this edge is a generally concave edge 48 which may be arcuate as shown, or polygonal. The opposed convex and concave longitudinal edges create a shape which resembles that of the familiar Australian boomerang. (This also describes the shape of the sub-components 29 shown in Fig. 11.) The sub-component 29′ preferably has offset flanges 40, 41, 42 for flush assembly of the several subcomponents, with overlaps between four adjoining subcomponents (Fig. 15) forming a member generally similar to the one shown in Fig. 10, or between three adjoining sub-components (Fig. 16) forming a member generally similar to those of Figs. 5–8 but of different proportioning. Notice the variable overlaps at 43, 44, 45 in Fig. 16. The sub-components are secured together at the overlaps, and sealed at the iris 46. Adhesives or mechanical fastenings, or both, may be used at the overlaps as desired. In some cases pressure sensitive adhesives will be found satisfactory, particularly when supplemented by mechanical fastenings. The components represented in Figs. 15 and 16 will as a general rule be assembled at the factory, using jigs for predetermining the several different relationships of the sub-components.

The structures I have described have been based upon what is known as a 6-frequency regular triacon, meaning that the major axes of the faces of the tricontahedra are divided into six modules according to a layout in which the sides of the tricontahedra are evenly subdivided (AB = BC = CD). If the frequency is increased to 12, using the same regular triacon breakdown, the basic components of the Fig. 3 construction will require only two primary types of which the number two's are left and right, one being a mirror image of the other. The longest diameter of the largest hexagon in the pattern will have a chord factor of 0.22; therefore using hexagons of a maximum 20 feet in diameter, a dome or sphere of 182-foot diameter will be produced. A 150-foot diameter dome in the 12-frequency breakdown would have hexagons of 16.5-foot maximum diameter. If we reduce the maximum diameter of the components to 10 feet for practicable delivery by truck, the 12-frequency layout will permit construction of a dome 88 feet in diameter at the equator. Similarly, a 24-frequency layout, using only four main types of components, will permit construction of a dome 176 feet in diameter whose components are of a size to be delivered by conventional motor transport. The boomerang-form sub-components of Figs. 11–16 serve so to simplify the sub-componentation that, regardless of size and frequency, even the largest domes can be constructed with the use of just one design of sub-component.

The terms and expressions which I have employed are used in a descriptive and not a limiting sense, and I have no intention of excluding equivalents of the invention described and claimed.

FIGURE 16

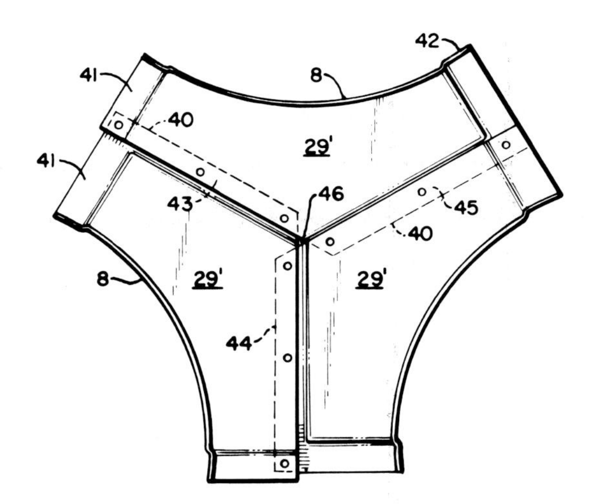

19▲LAMINAR DOME

U.S PATENT—3,203,144

APPLICATION—May 27, 1960

SERIAL NO.—32,268

PATENTED—AUGUST 31, 1965

FIGURE 1

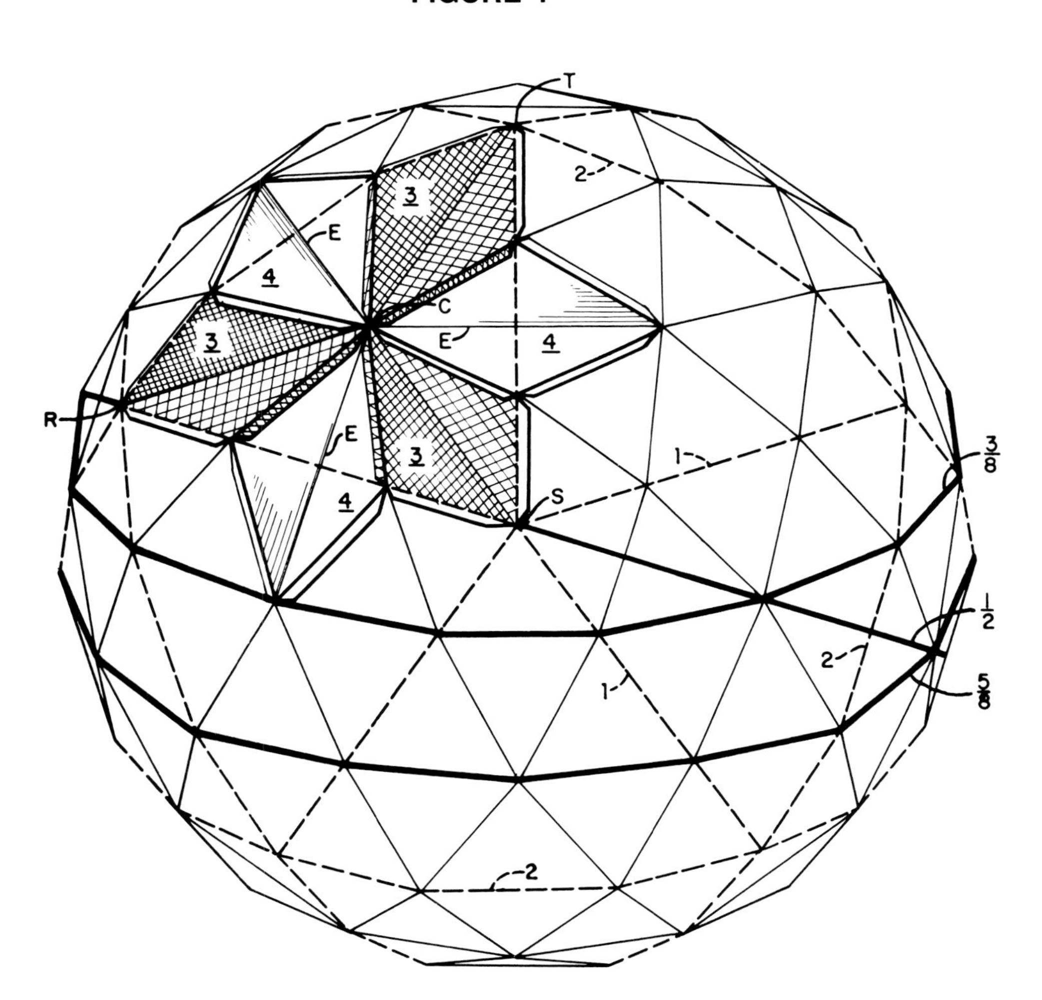

SINCE GEODESIC DOMES ARE a consistent tally of triangles, any two adjacent triangles can be combined to form a diamond shape, and it's possible to fold that diamond on its axis, then, by having the four outer edges of the diamond continued into flanges, those flanges can overlap the adjacent triangles of these diamond forms. The long diamond valley produces one edge of a tetrahedron and the four outer edges of the long valley diamond act as four other edges of a tetrahedron. When these are all joined together to form a sphere, an invisible tetra is established across the short diagonal of a folded-in-half diamond. These complete the tetrahedron, and make a very powerful geodesic dome in aluminum, plastics, fiberglass, paperboard, or steel. A number of very successful laminar domes have been constructed.

UNITED STATES PATENT OFFICE

Richard Buckminster Fuller, 407 S. Forest St., Carbondale, Ill.

LAMINAR GEODESIC DOME

The invention relates to improvements in geodesic dome construction, particularly with reference to the fabrication of extremely light-weight domes made of paperboard or plastic laminar parts comprising thin inner and outer facing sheets fastened to, and spaced by, a core of strong feather-light material such as expanded polystyrene.

When such laminar parts are arranged in overlapping fashion and held together by adhesive means, the stresses between two adjoining parts must be transmitted through those faces thereof which are in direct adherence to one another. The one part will have an overlying lap, the other an underlying lap, so that stress transfer is from the *inner* facing sheet of the one part to the *outer* facing sheet of the other. The effect of this is to place the laminae in shear. This shear loading will be transmitted to the core material of the composite laminae and through it to the opposite facing sheets, whereas the edges of these opposite facing sheets, i.e. the edges which lie opposite the adhered laps, cannot be utilized for direct transmission of the stresses. It is noted that this problem is peculiar to structures comprised of that particular type of overlapping panels which have inner and outer facing sheets with a spacing core between them. I have discovered how to weave together the inner and outer facing sheets of alternate panels of such a type in a strengthening pattern which is characterized by an improved continuity of both the inner and outer sheets extending substantially throughout the structure of the dome. A part of my discovery in this respect may be briefly summarized as follows:

FIGURE 3

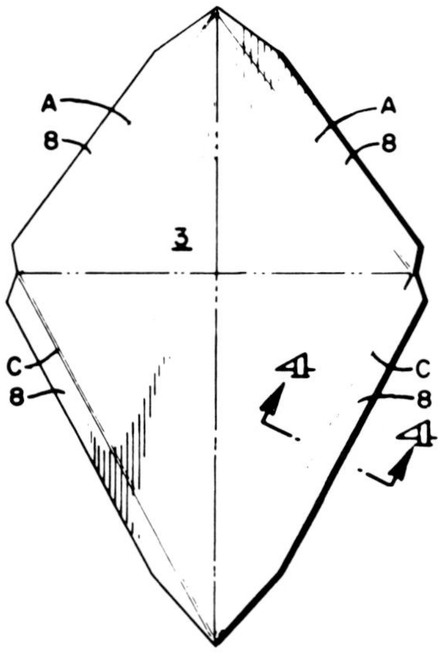

SUMMARY

In a goedesic dome comprised of diamond panels arranged in overlapping relation to one another and having inner and outer sheets and spacing means between these sheets, the diamond panels will be assembled so that their overlaps are symmetrically arranged in oppositely disposed pairs. Thus a single pair will have one pair of opposed overlying laps and one pair of opposed underlying laps. The overlapping portions of the panels are adhesively secured together with the inner sheet of one

FIGURE 2

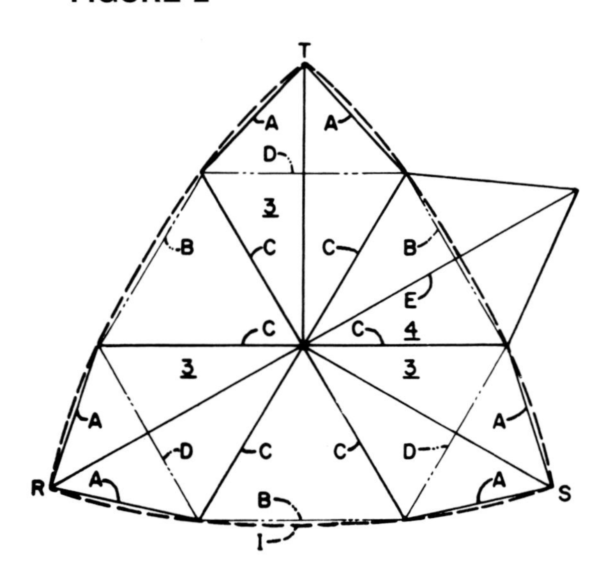

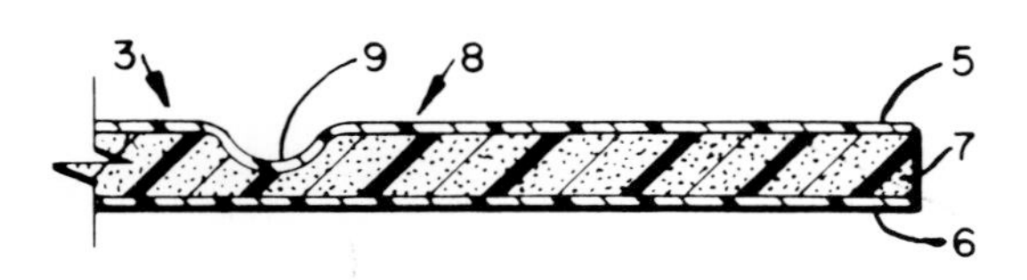

FIGURE 4

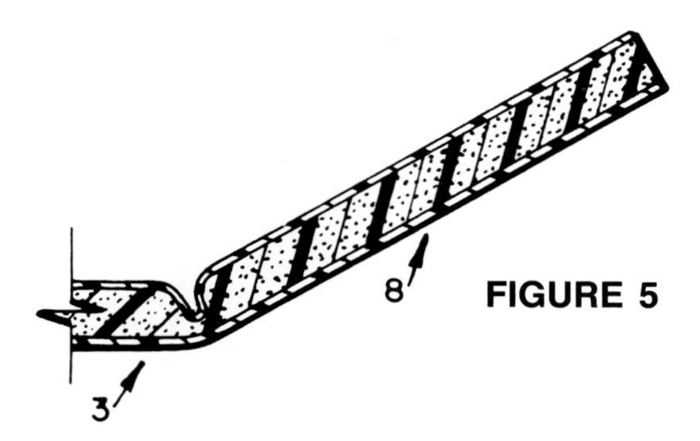

FIGURE 5

FIGURE 6

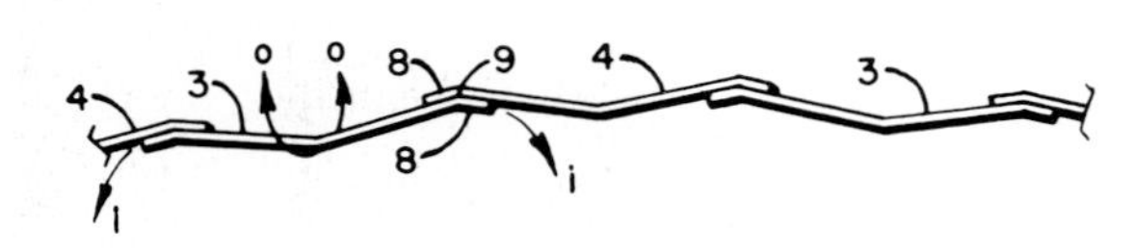

panel adhered to the outer sheets of a pair of adjacent panels and the outer sheet of the one panel adhered to the inner sheets of another pair of adjacent panels. This creates a "weave" of the inner and outer facing sheets which may be likened to the warp and woof of a woven fabric, and this woven pattern creates a continuity in the shear-connected inner and outer facing sheets. In this fashion each composite diamond panel will have an outer face carrying stresses along one continuous path, or direction, and an inner face carrying stresses along another continuous path, or direction. These stress-carrying paths are so interrelated in the overall pattern that the tensile strength of both facing sheets can be utilized to fuller advantage than has heretofore been disclosed in a laminar goedesic structure.

I have discovered how to obtain such a continuous woven stress pattern in a structure wherein opposed laps at the upper ends of the diamonds are underlying laps whereas those at the lower ends of the diamonds are overlying laps so that the construction is inherently shingled to shed water while affording the stress continuity of the woven inner and outer sheets of the panels.

Other advantages are obtained by utilizing panels of the foregoing composite type comprised of flat sheets which in their assembled relationship exhibit an overall pattern of planar triangular facets which are paired in diamond-shaped sections, the triangular facets of a plurality of pairs being arranged at an angle to one another to form diamond-shaped sections bent into outward concavity about the long axis of the diamond and presenting trough-shaped depressions in the outer surface of the dome. The triangular facets of other pairs of facets are arranged at an angle to one another to form diamond-shaped sections bent into outward convexity about the short axis of the diamond and present ridges alternating with the trough-shaped depression in the outer surface of the dome whereby said outer surface is faceted with intersecting ridges and valleys which stiffen the panel structure of the dome.

These and other advantages of my invention will appear more fully from the ensuing description of the best mode contemplated by me for carrying out my invention.

DESCRIPTION

In the drawings:

Fig. 1 is a diagrammatic view of a sphere upon which is shown the spherical icosahedron and the arrangement of the panels upon one of the faces of the spherical icosahedron. The arrangement disclosed is based upon a breakdown of the triangular faces of the icosahedron which is known to those familiar with the art of geodesic dome construction as a "three-frequency" breakdown, by reason of the fact that the sides of the icosahedron triangles are divided into three segments, or modules. "Frequency" is further described in my fundamental geodesic construction Patent No. 2,682,235.

Fig. 2 is a diagram covering approximately the area of the panels in Fig. 1.

Fig. 3 is a face view of one of the panels shown in Fig. 1 before bending.

Fig. 4 is an enlarged cross-sectional detail taken on the line 4—4 of Fig. 3 across the score line for one of the flaps of the panel.

Fig. 5 is a view similar to Fig. 4 showing a flap bent at an angle such as it assumes when the panel is fastened to an adjacent panel of like construction.

Fig. 6 is a diagrammatic cross-sectional view extending through a series of connected panels.

Fig. 7 is an exploded view of a double panel comprising an outer panel bent into outward convexity about its short axis and an inner panel bent into inward convexity about its long axis.

Figs. 8a and 8b are diagrammatic cross-sectional views extending across a series of connected double panels of the type shown in Fig. 7.

Fig. 9 is a perspective view of six inter-connected panels, each of which is bent into outward convexity about its short axis.

Fig. 10 is a similar view in which three of the panels are bent into outward concavity about their long axis.

FIGURE 7

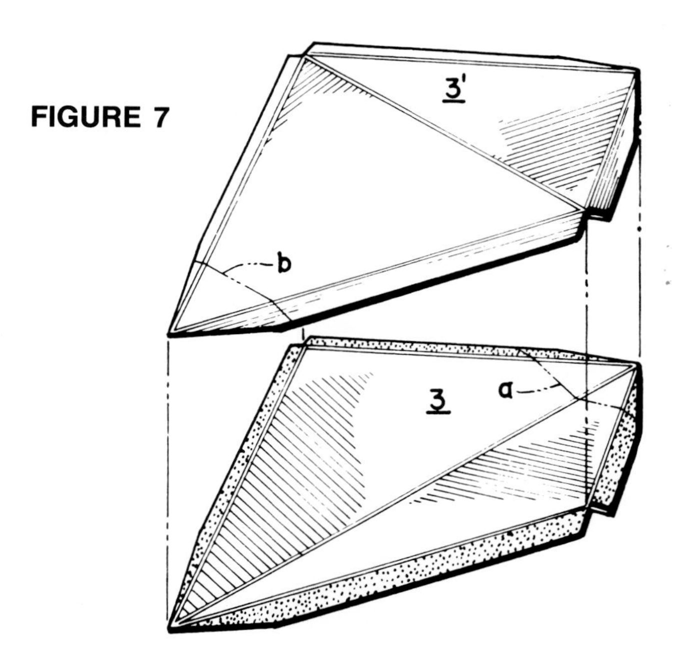

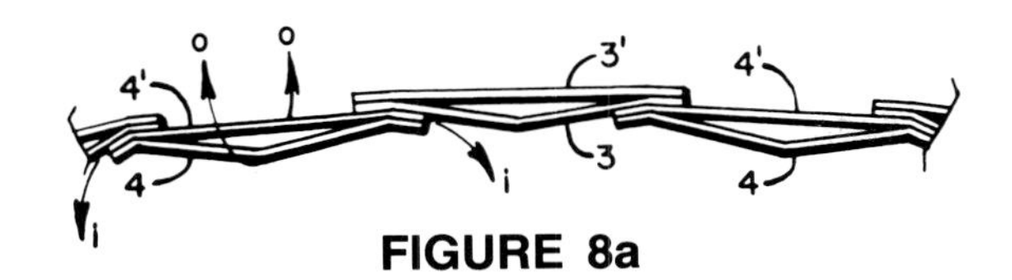

FIGURE 8a

FIGURE 8b

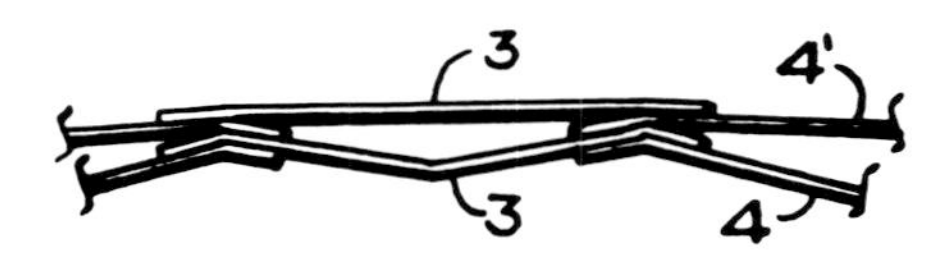

FIGURE 9

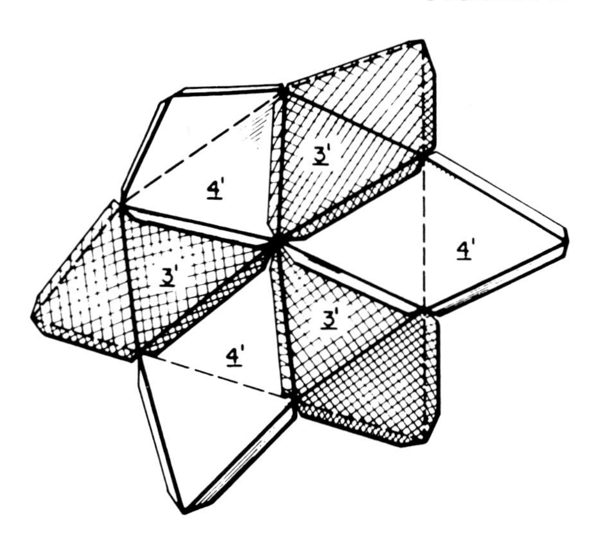

Fig. 11 is a similar view of another arrangement of three convex and three concave panels.

Fig. 12 is a side elevational view of a ⅝ sphere constructed in accordance with the panel arrangement depicted in Fig. 1 in which all of the panels are bent into outward concavity about their long axes.

Fig. 13 is a side elevational view of a ½ sphere having the identical panel construction of Fig. 12 but with the line of truncation for the base occurring as shown in Fig. 1 at the line designated "½."

Fig. 14 is a plan view of the Fig. 12 structure taken as indicated at 14—14 therein.

Fig. 15 is a plan view of the Fig. 13 structure taken as indicated at 15—15 therein.

Fig. 16 is a side elevational view of a structure comprising ⅝ spherical ends, each comprising one-half of the ⅝ sphere of Fig. 12 and a center section comprising two spherical segments duplicating a segment truncated, and extending between, the lines of truncation designated "⅜" and "⅝" in Fig. 12.

Fig. 17a is a ⅝ sphere incorporating the 3-frequency panel arrangement of Fig. 9 with modified arrangement of the overlaps between the panels, and upon which is superimposed a diagrammatic representation of the continuous paths of stresses transmitted by the inner and outer facing sheets of the interlaced panels.

Fig. 17b is a schematic exploded view, in perspective, of a particular grouping of panels within a structure similar to that shown in Fig. 17a.

Fig. 18 is a diagrammatic view of a sphere truncatable alternately at a ⅜, ½ and ⅝ sphere showing a 4-frequency arrangement of panels of the type which is bent into outward concavity about its long axis.

Fig. 19 is a diagram covering approximately the area of the panels in Fig. 18.

Fig. 20 is a perspective view of panels similar to those shown in Fig. 18 except that they are of the type which is bent into outward convexity about the short axis.

Fig. 21 is a similar view showing a combination of outwardly convex and concave panels.

Fig. 22 is a similar view showing another combination of outwardly convex and concave panels.

Fig. 23 is a perspective view of a ⅝ 4-frequency sphere showing a modified overlapping arrangement of the panels.

Fig. 24 is a plan view of the ⅝ 4-frequency sphere of Fig. 25 taken as shown at 24—24 in that view.

Fig. 25 is a side elevational view of the ⅝ sphere of Fig. 24.

Fig. 26 is a side elevational view of an elongated structure having ⅝ spherical ends, each comprised of one-half of the structure shown in Fig. 25, and a center section

FIGURE 10

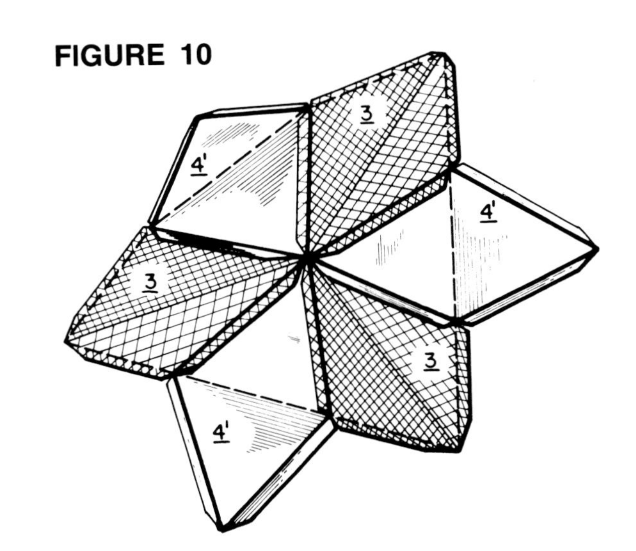

FIGURE 11

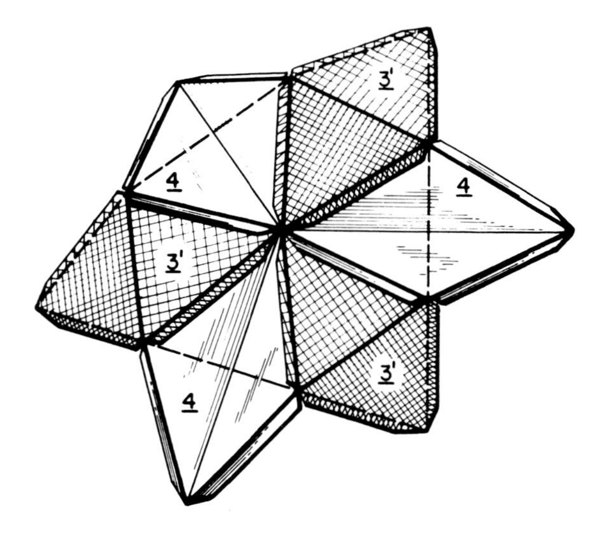

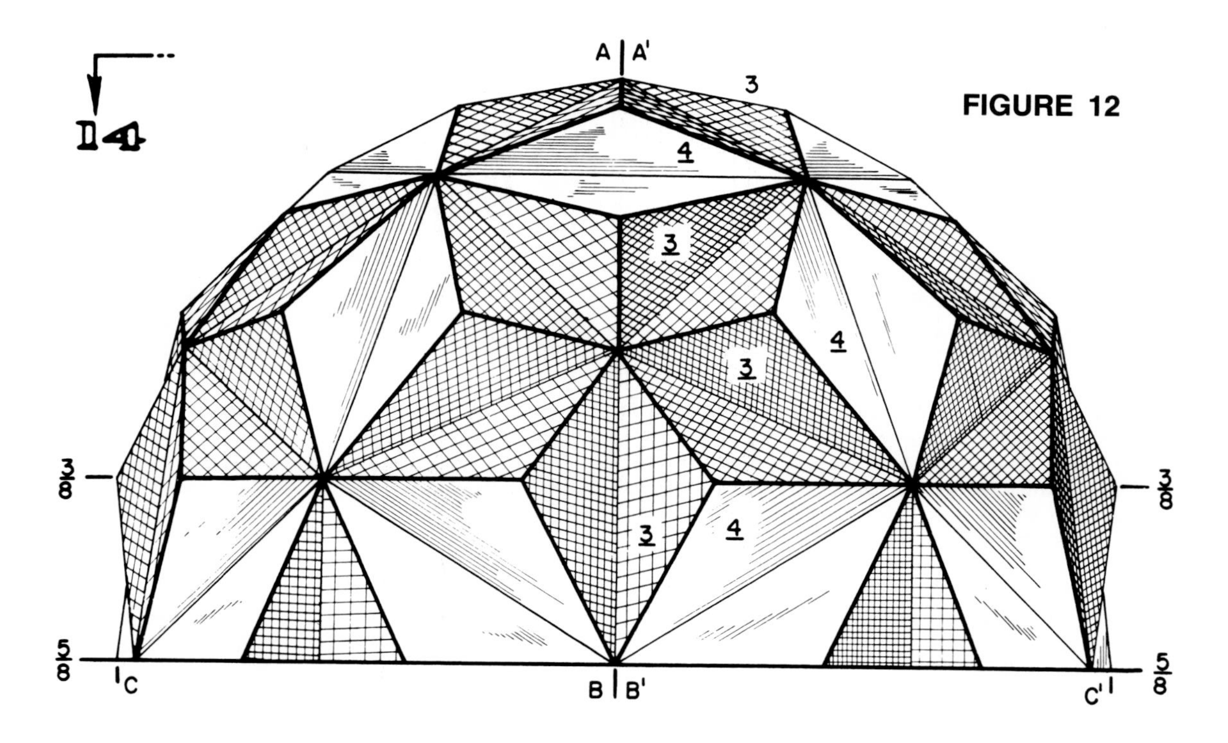

comprising two spherical segments duplicating a segment truncated, and extending between, the lines of truncation designated "3/8" and "5/8" in Fig. 25.

Geodesic construction adapted to be truncated

In Fig. 1 the dotted lines 1 represent a spherical icosahedron comprising twenty equilateral spherical triangles, the sides of which are all great circles of the sphere. If we consider point S as the zenith, the five spherical triangles whose apexes come together at the zenith are sometimes referred to by persons skilled in the art of geodesic construction as the "icosacap." The five great circle lines radiating from S may be read either as the edges of the spherical triangles, i.e. the intersections with a sphere of planes passing through the center thereof and containing the point S, or as chords or chordal modules of such great circle intersections. The dotted lines 2 forming the peripheral edges of the icosacap are shown in Fig. 1 as chordal modules of great circles. The structure of the spherical icosahedron is known in the art of geodesic structures and has been described in my Patent No. 2,682,235, granted June 29, 1954, in conjunction

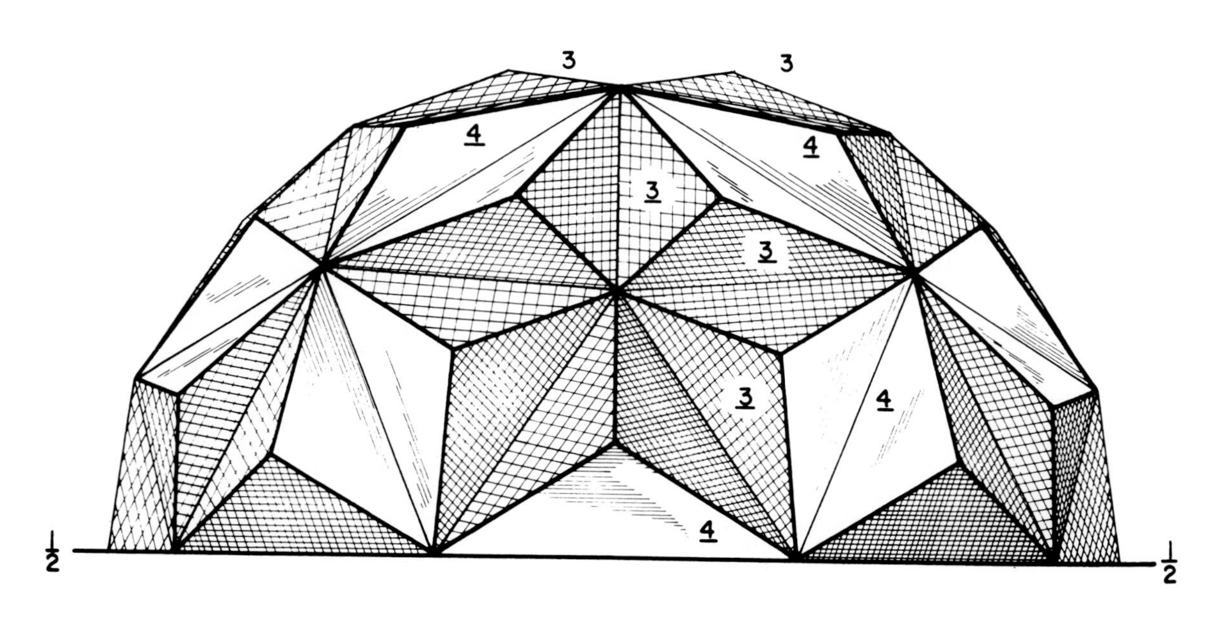

with a description of my 3-way grid, according to which each of the spherical triangles comprised in the spherical icosahedron is subdivided into substantially equilateral triangles or diamonds. According to one aspect of my present invention, the structuring of the 3-way grid is modified in a manner which provides for easy truncation of the sphere along lines at the edges of panels which will yield either a one-half sphere, a three-eighths sphere or a five-eighths sphere without in any way increasing the number of different kinds of panels used in the construction, a coplanar ground line being provided by edges of the diamond panels or half panels formed by dividing a panel in two on one or the other of its axes.

ring, however, as incidents of true great circle, i.e. geodesic, construction.

With reference to Fig. 2, I shall now describe how to obtain, in a 3-frequency breakdown, the particular 3-way grid pattern which will yield the several planes for truncation as above described and which will produce remarkable simplicity of componentation in that the entire structure may be assembled from only two kinds of parts. This view is a diagram of one of the spherical icosahedron triangles RST corresponding to triangle RST in Fig. 1. The spherical triangle is shown in dotted lines. Superimposed upon the edges of the spherical triangle are chordal modules thereof designated A and B. Chords

FIGURE 14

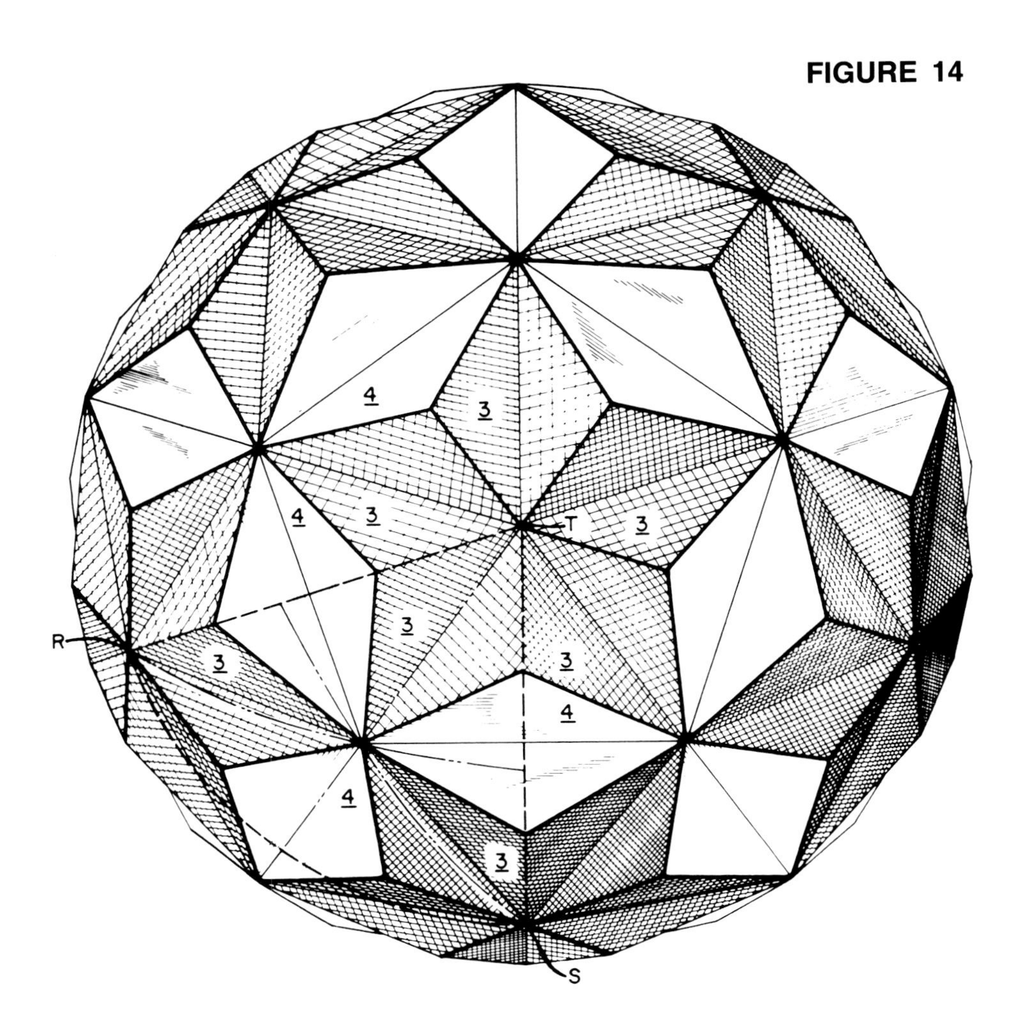

Particular attention is directed to the fact that the chordal modules of the lines of truncation designated "3/4" and "5/8" in Fig. 1, when viewed in one aspect, appear to be chordal modules of lesser circles. However, by construction upon the spherical icosahedron wherein all of the vertexes, and therefore both axes, of the diamond panels lie in great circle planes, these chordal modules in reality lie in planes passing through the center of the sphere whose intersections with the sphere describe great circle arcs. The phenomenon of alignment of panel edges for truncation may be described as the ancillary appearance of small circles, which may be likened to the parallels of latitude of a standard globe of the earth, at the three-eighths and five-eighths lines of truncation occur-

C intersect the sphere at the center of the spherical triangle and at the points where the meeting ends of chords A and B intersect the sphere. Chords D intersect the sphere at the points of intersection of chords A, B and C. Chord E forms the major axis of that one of the diamond types whose minor axis lies in the great circle plane which contains one of the sides (ST) of the spherical triangle. In Fig. 1, E = E = E. Thus the six diamond panels which are comprised within the shaded area of Fig. 1 consist of two types, the one type, 3, being represented by the darker shading, the other type, 4, by the lighter.

Given the diameter of the sphere upon which the desired geodesic dome is to be constructed, the critical dimensions of the diamond panels are ascertained as

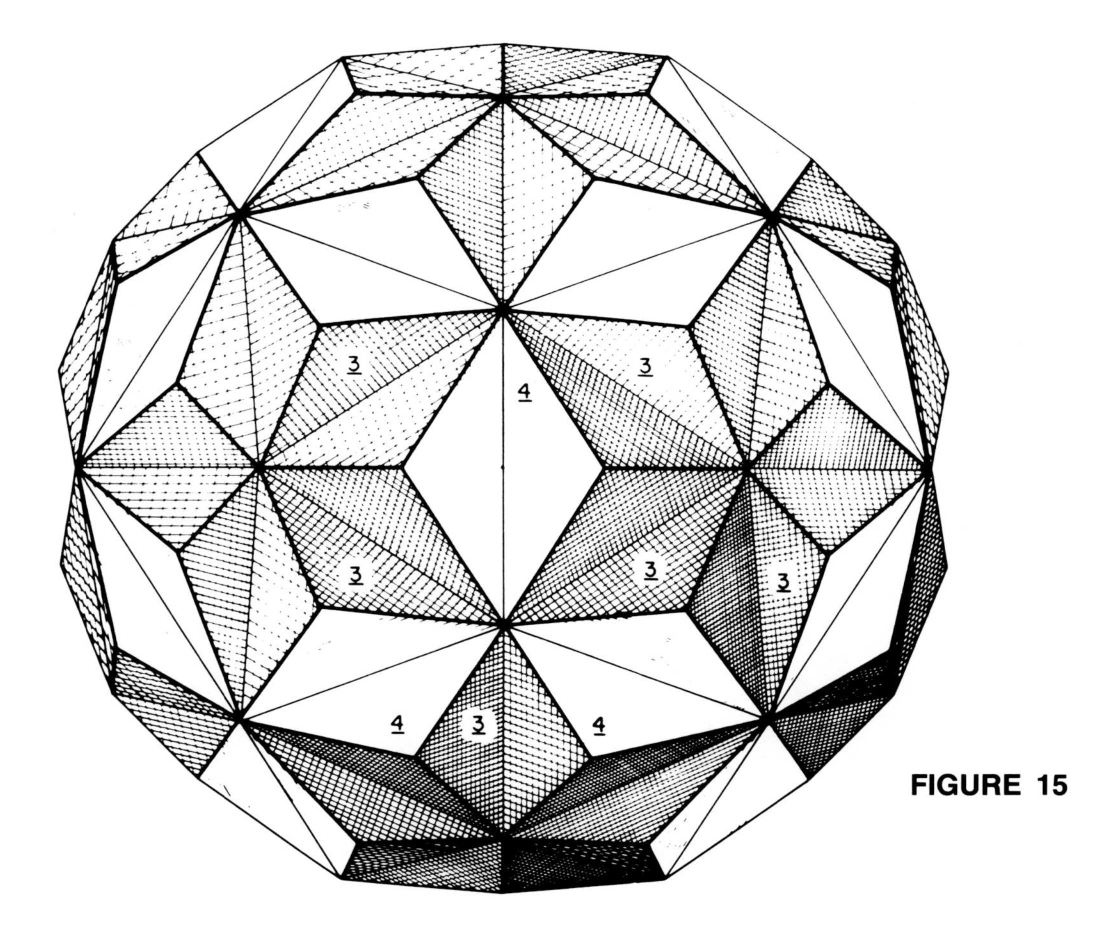

FIGURE 15

follows for the chords as designated with reference to FIG. 2:

A = 0.32968   D = 0.38216
B = 0.44112   E = 0.71364
C = 0.42154

Multiply each of these factors by the radius of the desired dome expressed in whatever units may be desired, such as feet, inches, meters or centimeters. The product will give the lengths of the chords according to the units of measurements selected, thus providing the dimensions for the lengths of the sides of each of the two diamond types. In the case of the diamond whose major axis is shown at E in Fig. 2, the dimensions of the outer pair of sides derive from the lengths of the other chords as determined above.

Panel construction

Figs. 3, 4 and 5 illustrate the construction of one of the panels of which type 3 is taken as representative. The panels comprise inner and outer sheets 5 and 6 and spacing means 7 between the sheets. The inner and outer facing sheets 5 and 6 may be made of any suitable material such as paperboard, metal, foil, plastic, etc. The spacing material of core 7 preferably is made of some foam material such as expanded polystyrene, although other types of filling material may be employed, such as corrugated sheet. A preferred composite sheet is one made of an expanded polystyrene core with liners of facing sheets of Kraft paper, a material sold under the trademark "Fomecor." The Kraft paper or paperboard may be waterproofed and strengthened by a plastic or other coating. Composites employing the expanded polystyrene cores have been found to possess exceptionally good characteristics for dome construction according to my present invention, affording laminae which are tough, stiff and almost feather-light. Each panel 3 (and 4) is of the general diamond form shown in Fig. 3 and has four extending flaps, 8 along its margins providing means for attachment to adjacent panels. The panels may be scored as at 9, Fig. 4, along the lines of the chords A, C, etc., which define the edges of the panel proper and the lines of fold of the flaps 8. Each panel may also be scored along either the major or minor axis of the diamond, as indicated by the dot-dash lines in Fig. 3. Thus, in Fig. 1, each of the panels 4 is scored and bent about its long axis which represents the chord E previously referred to, this fold being in the direction to produce an outward concavity in the surface of the dome. (In Fig. 1 all of the panels shown are bent into outward concavity about their long axes).

Fig. 6 shows a series of connected panels of this configuration. The score lines 9 of adjacent panels are arranged in juxtaposition, affording a double thickness of the panel structure at the bend of the joint between them. The panels are adhesively secured together. I have found that a contact type adhesive such as a neoprene base adhesive is well suited for this construction.

Tetrahedral panels

Some or all of the panels may be comprised of two sheets 3 and 3′, Fig. 7, sheet 3 being bent into inward convexity about its long axis and sheet 3′ being bent into outward convexity about its short axis as shown. The flaps of the two panels are adhesively secured together. This results in a hollow panel having four faces and six edges, an irregular tetrahedron which is very strong and stiff and furnishes insulation against heat and cold. The

FIGURE 16

outer panel 3′ may have its lower end cut off at the line b, and the inner panel have its upper end cut off at the line a, furnishing weatherproof ventilation for the structure. Figs. 8a and 8b show a series of connected double panels of the form shown in Fig. 7 except that some of the panels will be of the general type 3 and others of the general type 4, having reference to the preceding description of these general types. When the double panels are made by connecting the individual panels in alternating sequence, e.g., panel 3′ to panel 4′ to panel 3 to panel 4, as shown in Fig. 8b, a particularly strong interlaced construction is afforded.

"Inner" and "outer" panels

Figs. 9, 10 and 11 illustrate several different arrangements of the type 3 and type 4 panels, some of such panels being bent about the short axis and others about the long axis. Thus Fig. 9 shows a group of panels 3′ and 4′ all bent into outwardly convex form. In Fig. 10 some of the panels, namely those designated 3, are bent into outward concavity about their long axes. In the arrangement of Fig. 11, the outwardly concave panels 3 of the Fig. 10 construction are replaced by outwardly convex panels 3′, whereas the outwardly concave panels 4′ of Fig. 10 are replaced by outwardly concave panels 4, showing the complete interchangeability of the concave and convex forms. Also, it may be considered that the construction shown in each of Figs. 9, 10 and 11 may comprise double panels of the form exemplified in Fig. 7. For simplicity of description, if we call the outwardly convex panels "outers," the outwardly concave panels "inners," and the combination of these two forms as "doubles," this aspect of my invention may be summarized as including all of the following possible combinations:

(1) all "inners"
(2) all "outers"
(3) all "doubles" (inners and outers)
(4) part "inners," part "outers"
(5) part "inners," part "doubles"
(6) part "outers," part "doubles"
(7) part "inners," part "outers," part "doubles"

Elongated geodesic domes (Part I)

In Fig. 12 and 14, we see a completed five-eighth sphere constructed in the manner I have described with reference to Figs. 1–5 inclusive.

Figs. 13 and 15 represent the same construction as applied to a one-half sphere truncated as shown at the line designated "½" in Fig. 1.

If we take the structure shown in Fig. 12 and divide it in two along the line designated A–B (or A′–B′), another aspect of the invention is revealed. This produces two half sections of a five-eighth sphere, ABC and A′B′C′. In Fig. 16 we see one of these sections ABC at the right and the other, A′B′C′, at the left, the one having been rotated 90° clockwise from the position shown in Fig. 12 and the other 90° counterclockwise. Between the planes BC and B′C′ of Fig. 16, we have a center section BB′C′C comprising two spherical segments duplicating a segment truncated, and extending between, the lines of truncation designated "¾" and "⅝" in Fig. 12. In the completed Fig. 16 dome, the center section gives an appearance of being generally cylindrical in form but from the discussion which has preceded, it will be understood that it is in reality comprised of two segments of a spherical geodesic dome. Thus I have discovered a way in which sphericity can in effect be extended or "stretched" to produce elongated geodesic domes. If desired, the center section BB′C′C may be omitted from the Fig. 16 construction and the end portions thereof be brought together with a coincidence of the planes BC and B′C′, producing an elongated structure somewhat shorter than shown in this view of the drawing. Or as a further alternative, it is pos-

FIGURE 17a

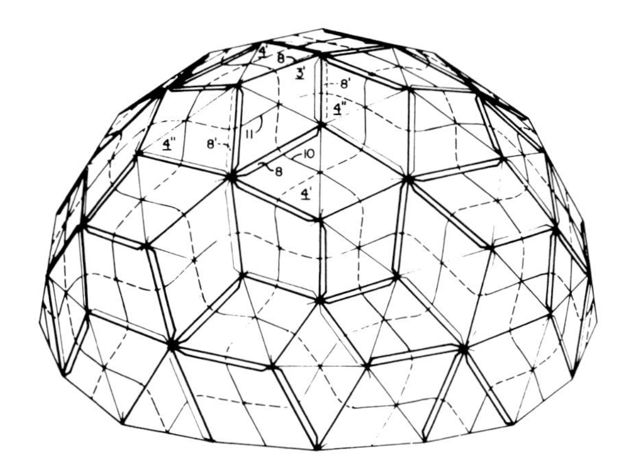

construction in respect to its unique abilities for maximum utilization of the tensile properties of materials and the provision of a more favorable weight-strength ratio than has ever been attained with the use of conventional ideas in building construction.

FIGURE 17b

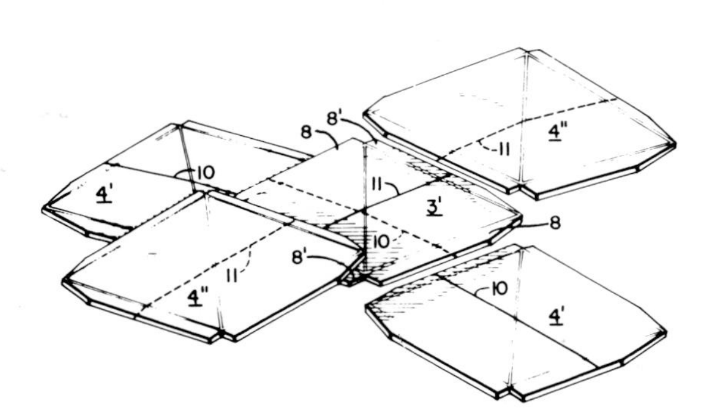

Weaving of panel overlaps

Again referring to Fig. 6, it will be observed that panels 3 have underlying laps whereas 4 has overlying laps. Hence stress transfer in the plane of the surface of the structure occurs between the inner facing sheet of panel 4 and the outer facing sheets of the adjoining panels 3. The effect of this is to place the panel laminae in shear, the shear loading being transmitted to the core material of the composite laminae and through it to the opposite facing sheets. The edges of these opposite facing sheets, i.e. the edges which lie opposite the adhered laps, cannot be utilized for direct transmission of the stresses. This problem is peculiar to structures comprised of that particular type of overlapping panels which have inner and outer facing sheets with a spacing core between them. This brings us to that aspect of my invention which concerns my discovery of how to weave together the inner and outer facing sheets of alternate panels in a strengthening pattern which is characterized by an improved continuity of both the inner and outer sheets extending substantially throughout the structure of the dome as will now be ex-

FIGURE 18

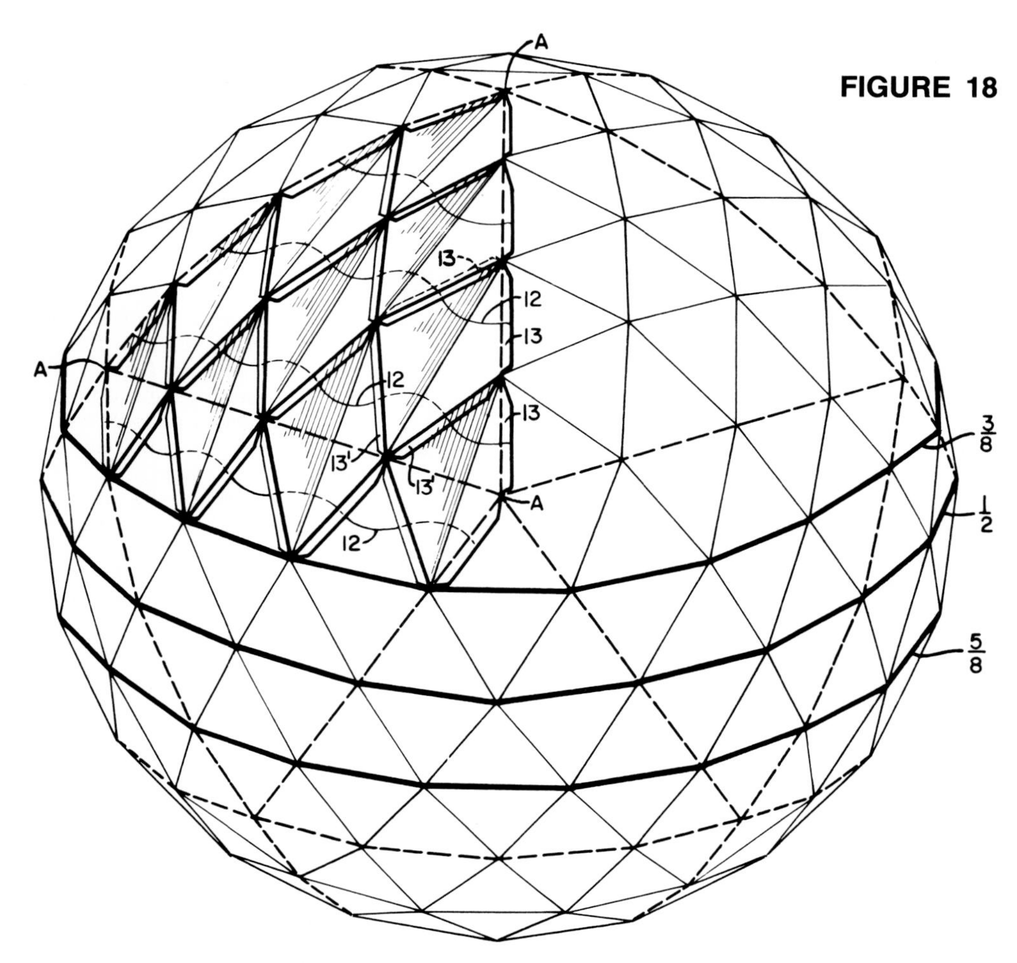

sible to use two or more of the sections BB′C′C to produce a tunneled geodesic structure of any desired length. Again, by removing the five-eighths sphere half section which lies to the left of plane B′C′ of Fig. 16, and considering the plane B′C′ as the base of the remaining structure we have one-half of a silo-like structure which will mate with another structure of identical form along plane AB′ to produce a tower having a five-eighths spherical top and generally cylindrical walls. These several constructions possess in common a generally cylindrical midsection having the primary attributes of true geodesic

plained with reference to Figs. 17a and 17b. Here the diamond panel laminae are assembled in such a way that their overlaps are symmetrically arranged in oppositely disposed pairs. Thus a single panel 3′ will have one pair of opposed overlying laps 8 and one pair of opposed underlying laps 8′. The overlapping portions of the panels are adhesively secured together with the inner sheets of panel 3′ adhered to the outer sheets of a pair of adjacent panels 4′, and with the outer sheet of panel 3′ adhered to the inner sheets of another pair of adjacent panels 4″. Thus we have a direct transfer of stress between the inner sheet of panel 3′ and the outer sheets of panels 4′ along a path represented diagrammatically by the line 10 (Fig. 17b). Similarly there is a direct transfer of stresses from the outer sheet of panel 3′ to the inner sheets of panels 4″ as represented diagrammatically by the line 11 showing a cross weave of the stress pattern. Stress paths 10 and 11 are made up of full lines wherever the stress is in the outer sheet of the pattern, by dotted lines where the stress is in the inner sheets. These paths of stress transfer create a "weave" of the stresses in the inner and outer facing sheets which may be likened to the warp and woof of a woven fabric, and this woven pattern produces a continuity in the shear-connected inner and outer facing sheets which extends substantially throughout the structure of the dome. The intersecting paths of stress themselves create a 3-way grid of stress lines, being so interrelated in the over-all pattern that the tensile strength of both facing sheets can be utilized most effectively in what appears to be a new synergy complementing the fundamental synergetic phenomenon which has come to be recognized by scientists and builders familiar with present day geodesic building construction and strategy.

FIGURE 19

Another advantage of the crisscross stress weaving pattern exemplified in Figs. 17a, 17b and 23 may be explained with reference to Fig. 6 which shows how, with reference to selected panel 3, the stresses in the inner sheets of two adjacents panels 4 create an inward pull at the overlapped edges of panel 3 as suggested by the arrows i, whereas the stresses in the outer sheets of two other adjacent panels (one above and one below the plane of the drawing and therefore not shown in Fig. 6, but compare the two sets of adjacent panels 4′ and 4″ in Fig. 17b) create an outward pull at the underlapped edges of panel 3 as suggested by the arrows o. The crisscrossed inward and outward pulls on a given panel exert compressive stresses in the core material 7 of the panel, and the combination of such inward and outward pulls has been found to have a constraining effect which tends to draw the edges of the panel toward one another so that the inner and outer facing sheets of the panel are somewhat constrained against peeling apart. Cf. also the arrows i and o in Fig. 8.

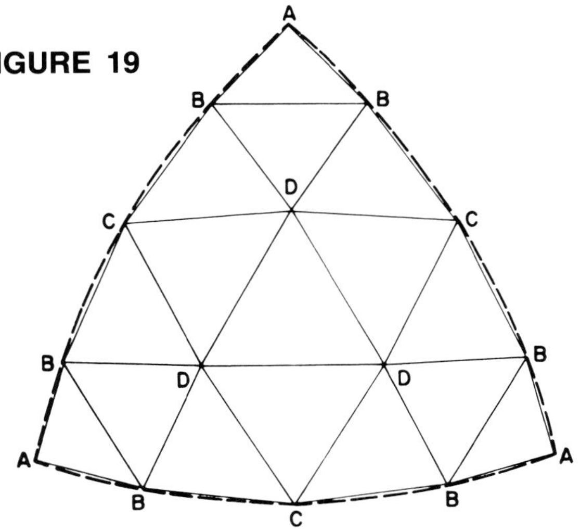

FIGURE 20

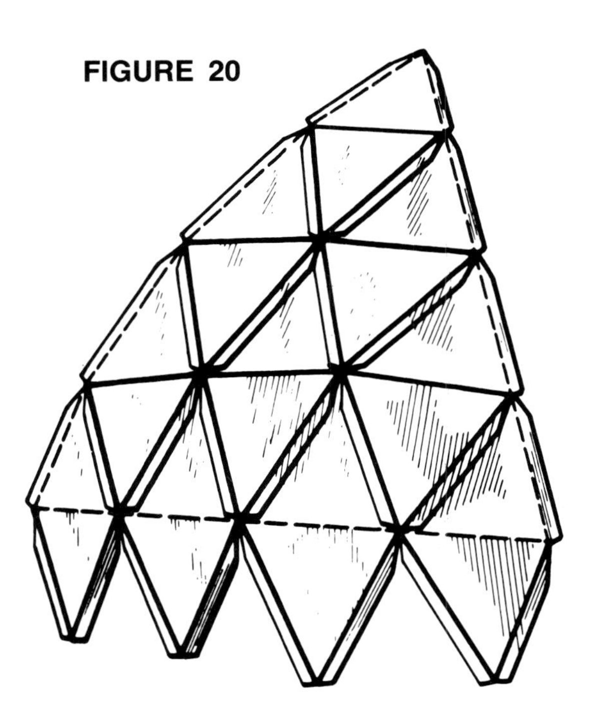

FIGURE 21

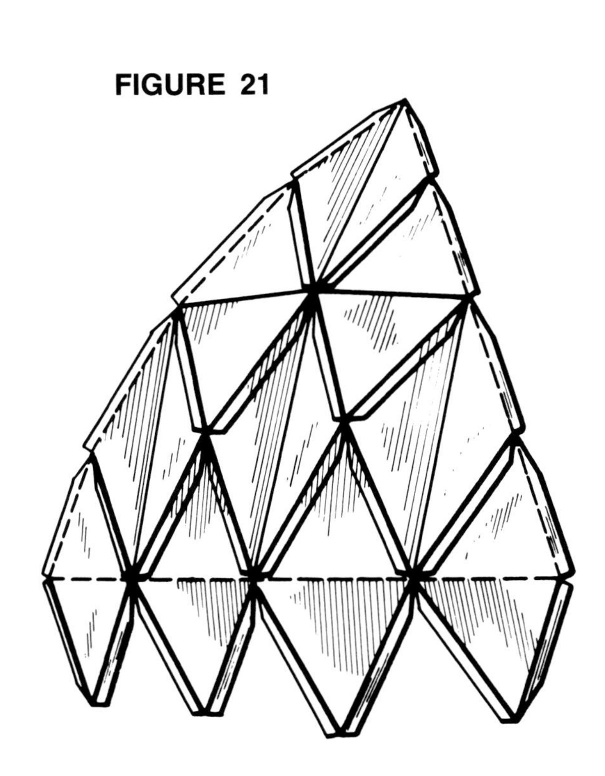

In Fig. 18 I have disclosed how to obtain a continuous woven stress pattern in a structure wherein opposed laps at the upper ends of the diamonds are underlying laps, whereas those at the lower ends of the diamonds are overlying laps so that the construction is inherently shingled to shed water while affording the stress continuity of the inner and outer woven sheets of the panels. The paths of the continuous lines of stress are indicated diagrammatically by the several lines 12. Here we have a pair of opposed underlying laps 13 and a pair of opposed overlying laps 13′.

In the Fig. 18 construction, the panels are all bent into outward concavity about their long axes. In Fig. 20 the same weaving of stress patterns in a shingled construction is obtained with the use of panels bent into outward convexity about their short axes, i.e. the panels are all "outers," to use the short terminology heretofore adopted.

Fig. 21 utilizes the same stress weaving pattern and shingling in a construction comprising a combination of outers and inners.

Fig. 22 illustrates a different combination of the outers and inners. As in the discussion which has preceded, it will be understood that wherever outers are depicted, the drawing may be read as illustrating either outers or a combination of outers and inners ("doubles"), referring to the double panel construction described with reference to Fig. 7.

FIGURE 22

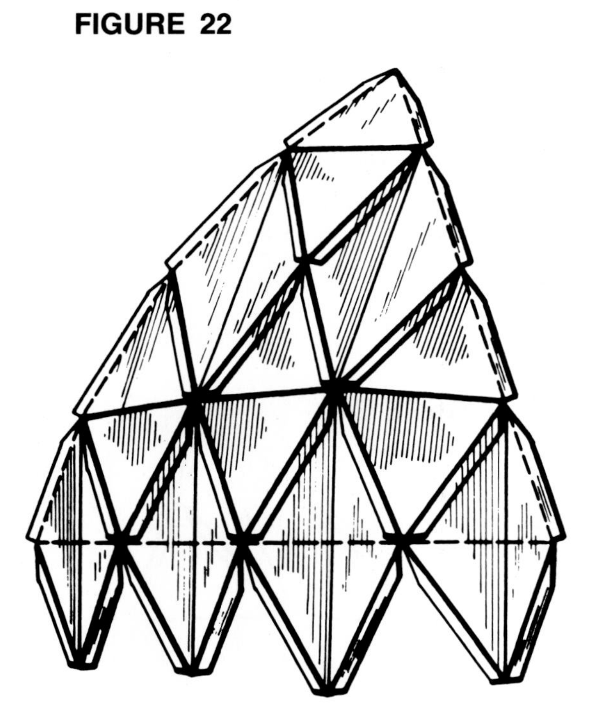

Fig. 23 illustrates a still further modification of the arrangement of the panels and their overlaps. As before, a single panel has one pair of opposed overlying laps and one pair of underlying laps. The paths of direct transfer of skin stresses are shown diagrammatically by the lines 14 and 15 interlaced in a continuous woven pattern.

Recapitulating with respect to Figs. 17 to 23 inclusive it may be stated that Fig. 17 typifies a stress weaving pattern in a 3-frequency dome whereas Figs. 18 to 23 inclusive exemplify stress weaving patterns in a 4-frequency dome. (See the explanation of "frequency" as given hereinabove.) The 4-frequency dome possesses the inherent lines of truncation which have been described with reference to the 3-frequency dome as denoted in Fig. 18 by the lines of truncation designed "3/8," "1/2" and "5/8." It will be observed that in Fig. 1 the line of truncation for a one-half sphere is at an angle to the lines of truncation for the three-eighth and five-eighth spheres whereas in Fig. 18 all three lines of truncation occur in parallel planes.

FIGURE 23

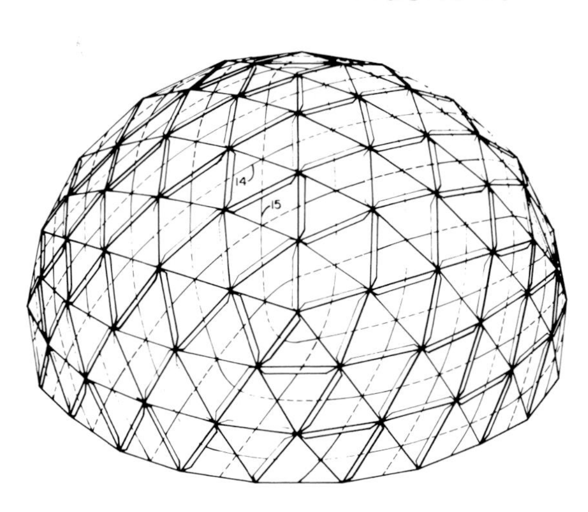

Fig. 19 shows the breakdown for the particular 3-way grid employed in the 4-frequency dome. This view is a diagram of one of the spherical icosahedron triangles AAA corresponding to triangle AAA in Fig. 18. As before, the spherical triangle is shown in dotted lines. Superimposed upon the edges of the spherical triangle are chordal modules thereof designated AB, BC, CD, etc. Given the diameter of the sphere upon which the desired geodesic dome is to be constructed, the critical dimensions of the diamond panels are ascertained as follows for the chords as designed with reference to Fig. 19:

AB = 0.22019
BB = 0.25958
BC = 0.32942
BD = 0.30907
CD = 0.31287
DD = 0.32492

Multiply each of these factors by the radius of the desired dome in whatever units may be desired, such as feet, inches, meters or centimeters. The product will give the lengths of the chords according to the units of measurement selected, thus providing the dimensions for the lengths of the sides of each of the diamond types. Trun-

FIGURE 24

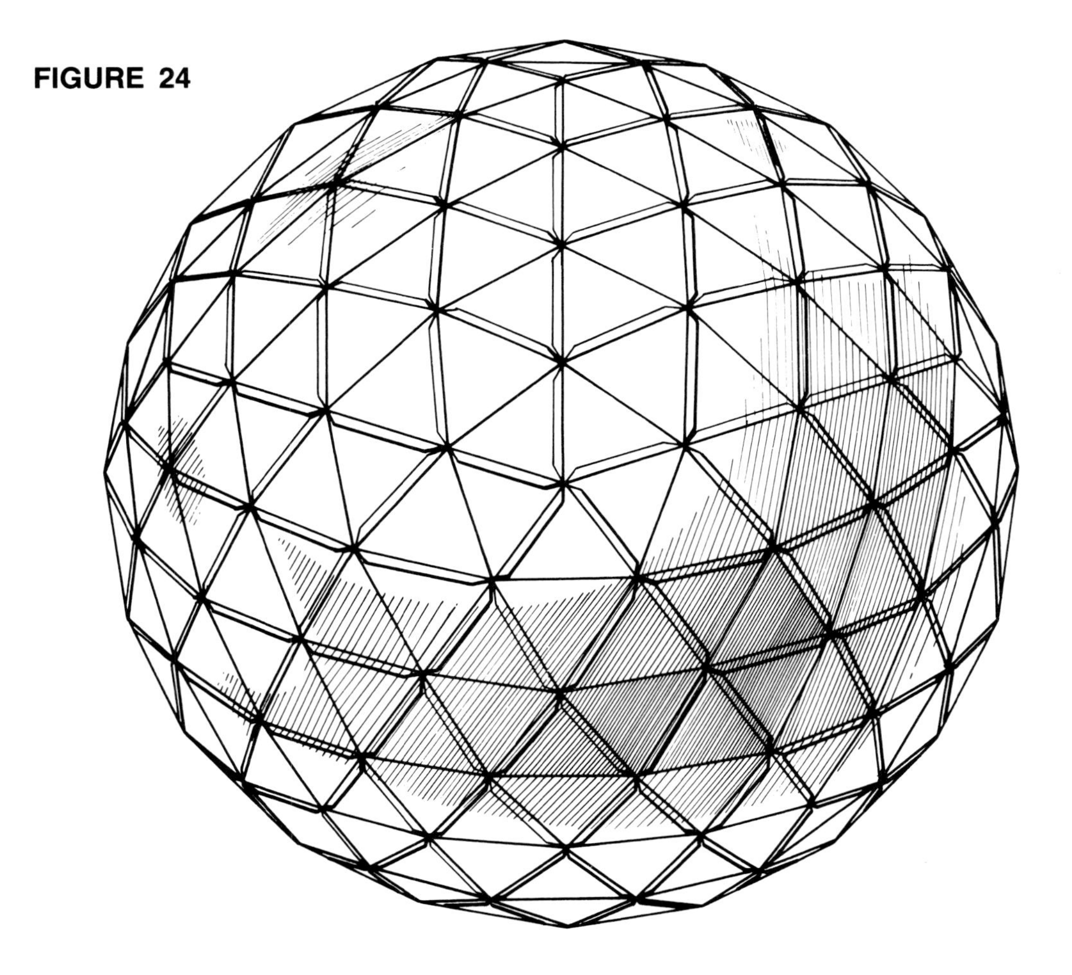

cation may be obtained in domes in which the 3-way grids of the icosahedron triangles are constructed at frequencies of 2, 3, or 4, or multiples thereof, for example frequencies of 6 and 12. It will be remembered that frequency has been defined as the number of modules into which each side of an icosa triangle is divided in laying out the 3-way grid pattern. In these patterns all of the points of the diamonds lie in great circles. It follows that all of the center lines of the diamonds, i.e. the long and short axes, are chords of great circles. Consequently all of the domes which I have described possess, in addition to the special advantages herein described, the fundamental advantages of true geodesic or great circle construction.

Elongated geodesic domes (Part II)

In a 4-frequency dome constructed in the manner which has been described, and pursuing the panel arrangement disclosed in Fig. 24 as a plan of the five-eighth 4-frequency sphere of Fig. 25 taken as shown at 24—24 in that view, and with that orientation in which the center of the icosacap is the zenith, we obtain a modified form of the truncation segment which extends between the 3/8 and 5/8 lines of truncation. This in turn produces a modified form of the elongated structure heretofore described with reference to Figs. 12 and 16. In this case, if we take the structure shown in Fig. 25 and divide it in two along the line designated DE (or D′E′), this produces two half sections of a five-eighth sphere DEF and D′E′F′.

In Fig. 26 we see one of these sections DEF at the right and the other D′E′F′ at the left, the one having been rotated 90° clockwise from the position shown in Fig. 25, and the other 90° counterclockwise. Between the planes EF and E′F′ of Fig. 26, we have a center section EE′F′F comprising two spherical segments duplicating a segment truncated, and extending between, the lines of truncation designated "3/8" and "5/8" in Fig. 25. In the completed Fig. 26 dome, the center section gives an appearance of being generally cylindrical in form but as has been explained with reference to Fig. 16, it will be understood

FIGURE 25

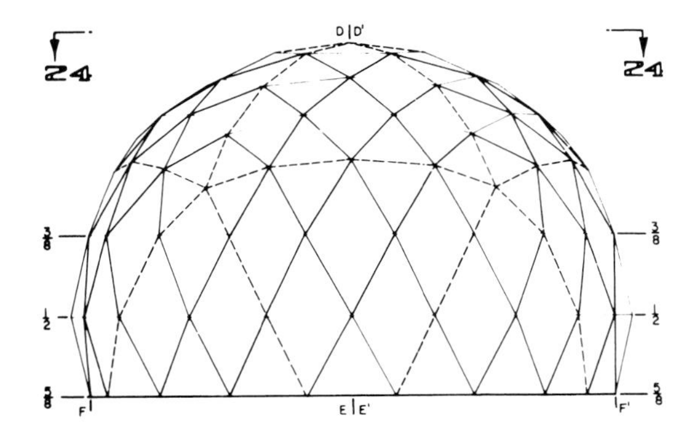

FIGURE 26

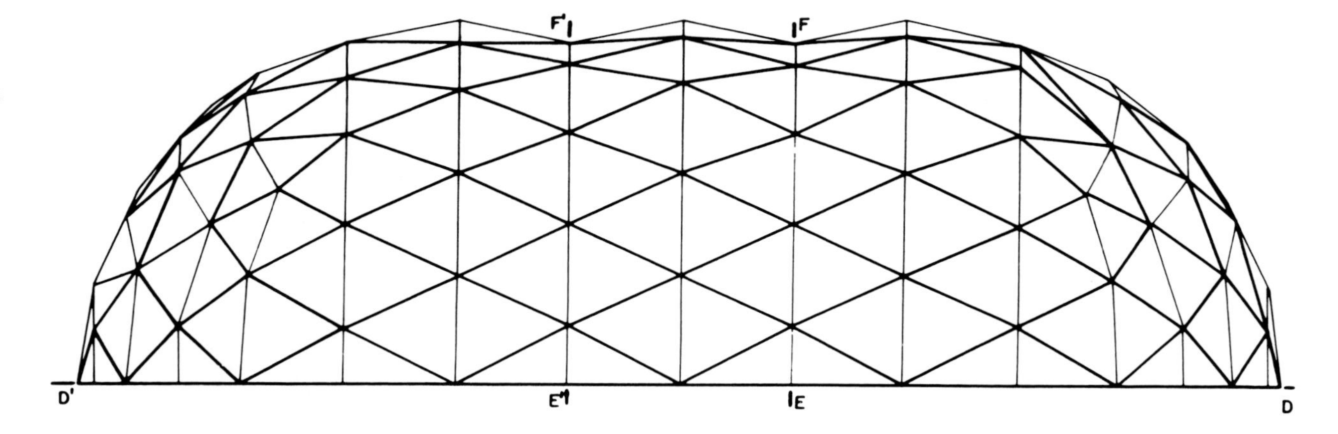

that it is in reality comprised of two segments of a spherical geodesic dome. The mid-section of the Fig. 26 construction, in contradistinction to that shown in Fig. 16, is made up entirely of "outer" diamond panels bent about their short axes and half sections of "inner" diamond panels truncated at their short axes. This results in circumferential corrugations as will be readily apparent from an examination of Fig. 26. If desired, the center section EE′F′F may be omitted from the Fig. 26 construction and the end portions thereof brought together with a coincidence of the planes EF and E′F′, producing an elongated structure somewhat shorter than shown. Or as a further alternative it is possible to use two or more of the sections EE′F′F to produce a tunneled geodesic structure of any desired length. Again, as was explained with reference to Fig. 16, these several constructions possess in common a generally cylindrical mid-section having the primary attributes of true geodesic construction.

As an example of the exceedingly favorable weight-strength ratio obtainable with my peculiar form of laminar construction, I may cite a dome made of "Fomecor" comprised of Kraft paperboard facing sheets upon expanded polystyrene core material having a composite thickness of one quarter in., utilizing single panels in a geodesic structure 19′ in diameter and having the following specifications:

Configuration	⅝ sphere.
Head room use	98%.
Covered floor area	275 square feet.
Shelter volume	2150 cubic feet.
Weight (including accessories) . .	175 pounds.
Volume of components as packaged for shipment	25 cubic feet (a package 4′ square and 1′6″ thick).
Weight per unit covered floor area	0.636 pounds per sq. foot.

When my invention is applied to larger sizes of domes, the weight of the dome per square foot of covered floor area can be decreased still further due in large measure to the form and the arrangement of the overlapping composite panels in which the inner and outer facing sheets of alternate panels are woven together in a strengthening pattern which complements the inherent strength of geodesic 3-way grid construction.

The terms and expressions which I have employed are used in a descriptive and not a limiting sense, and I have no intention of excluding such equivalents of the invention described as fall within the scope of the claims.

20▲OCTA SPINNER (1965)

USING A MACHINE, I FOUND we could produce octahedra, which have twelve edges, in such a manner that the length of any one of the twelve edges could vary, because the octahedron has six vertices, therefore three axes, and the octahedron could be spun on any one of those three axes. You can spin sheet metal to be bent into angular sections to produce a stiff octahedron.

Octahedra joined together edge to edge produce tetrahedral bases/faces between them, so octahedra can be joined together, to produce the octet truss. I did not go through with the octet spinner patent after filing because the expense of patent work is very great, and I'm not in the manufacturing world, and I felt that it would not be worth carrying any further.

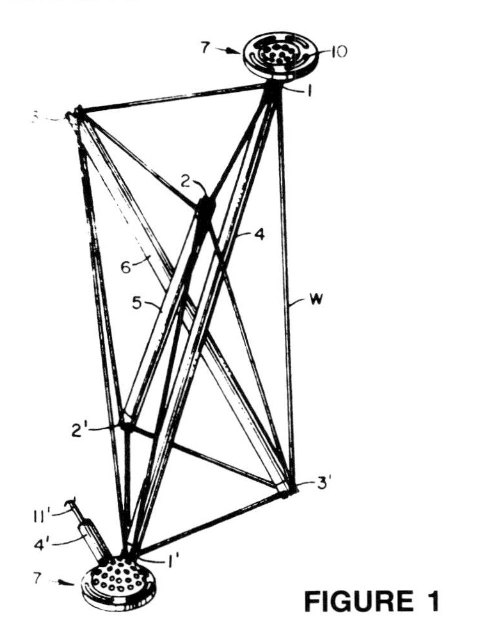

FIGURE 1

FIGURE 2

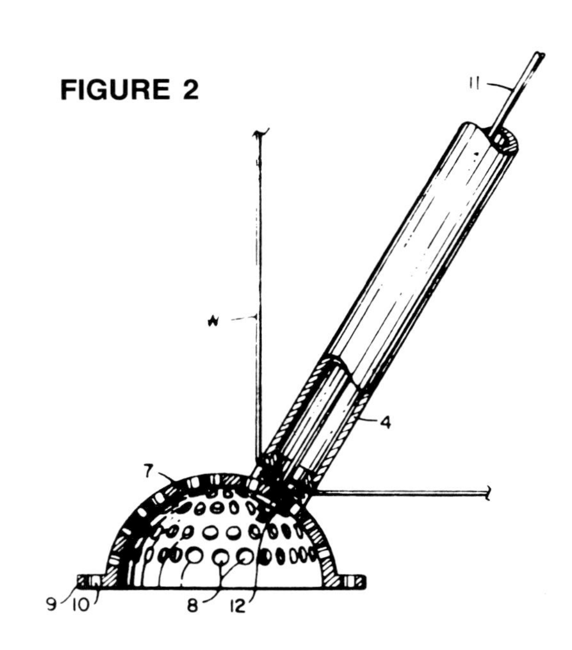

PATENT APPLICATION OF RICHARD BUCKMINSTER FULLER FOR METHOD AND APPARATUS FOR MAKING THREE-DIMENSIONAL TRUSS COMPONENTS

Case 349.021

The invention relates to the fabrication of building trusses and components thereof.

In my co-pending application for Patent, Serial No. 416,228, filed December 7, 1964, I have described a truss construction which is capable of utilizing more efficiently the tensile strengths of the materials from which the truss is constructed. In such construction, it has been found possible to use many elements loaded purely in tension, indeed, one in which such purely tensioned elements predominate so that relatively few compression members are needed. The construction is one in which a plurality of units, conveniently made of criss-crossed struts bound together by a network of tension elements, form the basic components, or "building blocks" used in putting together the truss. While such truss components, once assembled, are self-contained units that are easy to handle, their fabrication can become rather complex due to the fact that they are made up of compression struts which are virtually suspended in a network of wire. During fabrication, it is essential to maintain the proper angular relationship between the criss-crossed struts, and the relative dispositions of the ends of the struts for a given predetermined angular relation.

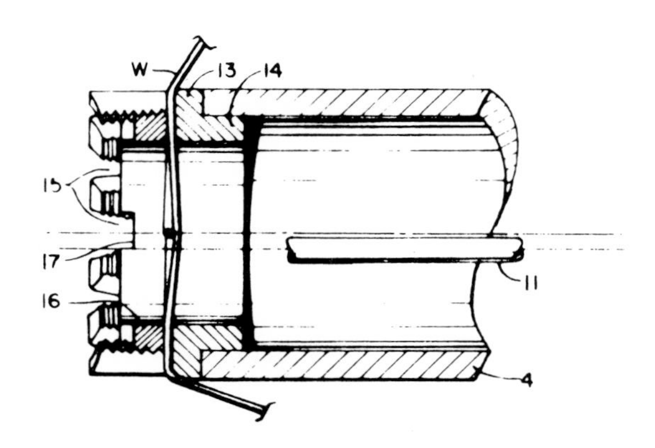

FIGURE 3

My present invention is concerned with the solution of the particular problems involved in the fabrication of these peculiar strut-and-wire components such as those exemplified in my prior application aforesaid, and to the means by which such components may be interconnected to form a building truss.

According to a preferred form of my invention, the truss structure produced comprises a plurality of interconnected three-dimensional components each of which has a plurality of struts and flexible edge portions extending between the ends of the struts to form an initially self-supporting unit. The plurality of such components is joined together by connecting the ends of the struts of one component to the ends of the struts of adjacent components through partly spherical fastening elements perforated by a plurality of apertures providing selective adjustment of the angular relationship between the struts of the interconnected components.

Fabrication of the truss components comprises the steps of arranging the struts in predetermined angular relation to one another to form a preliminary strut assembly, rotating the strut assembly, feeding a wire for attachment to the ends of the struts, and producing relative movements between the rotating strut assembly and the wire feed to bring the wire into engagement with first one strut and then another and thus form the flexible edge portions of the truss component. My apparatus includes means for performing these several steps in the desired sequence, and for programming the wire feeding device according to predetermined design patterns for components of varying form.

The invention has particular application to the fabrication of components of octahedral form comprising eight triangular faces. The eight triangular faces are defined by the wire network, and there are three compression struts which are arranged along the three axes of the octahedron.

In the drawings, wherein I have illustrated the best mode contemplated by me for carrying out my invention:

Fig. 1 is a perspective view of an octahedral component fabricated in accordance with the invention.

Fig. 2 is a detail sectional view showing the partly spherical fastening element secured to an end of one of the struts of the component of Fig. 1.

Fig. 3 is an enlarged detail view showing an end portion of one of the struts in longitudinal section, with associated wire-fastening means.

Fig. 4 is an isometric perspective view of the apparatus which "spins" the wire on the struts.

Fig. 5 is a detail view of an optional form of means for binding the struts together after removal of the completed octahedral component from the spinning apparatus.

Fig. 6 is an isometric perspective view of the apparatus of Fig. 4, inclusive of the wire feeding means and means for producing relative movements between the rotating strut assembly and the wire feeding means, in combination with a programming control means.

Fig. 7 is an isometric perspective view of a portion of the apparatus of Fig. 6 as it appears following removal of the stylus frame of the programming device and substitution of the photoelectric "playback" device.

Fig. 8 is a diagram of the control circuit for one of the two motors of the wire feed control.

Fig. 9 is a diagram of a special case in which the truss component is a regular octahedron.

Reference is first made to Fig. 1 in which I have illustrated the application of my invention to the fabrication of a building truss component of octahedral form. This component comprises three compression struts 4, 5 and 6 which are arranged along the three axes of the octahedron, 1–1′, 2–2′, 3–3′. The upper ends of the compression struts 4, 5 and 6 lie in one plane and the lower ends lie in another plane below the first. The seemingly complex but truly simple form of the octahedral unit will be understood by identifying the faces of the octahedra, the tension elements and the compression struts as follows:

Eight faces of the octahedra:

1–2–3	1–2′–3
1′–2′–3′	1–2′–3′
1–2–3′	1′–2–3′
1′–2–3	1′–2′–3

Twelve tension elements:

1–2	1–2′
2–3	1–3′
3–1	2–3′
1′–2′	2–1′
2′–3′	3–1′
3′–1′	3–2′

Three compression struts:

4	5	6

From the foregoing tabulation of the truss elements, the student of this disclosure will appreciate the preponderance in tension elements over compression elements and the significant improvement thus obtained in the direction of increased utilization of the high tensile properties of the improved materials and alloys available today.

In the preferred construction shown, we have a three-dimensional truss component having a plurality of struts 4, 5 and 6 and flexible edge portions extending between the ends of the struts, these being the twelve tension elements as listed above. The struts 4, 5 and 6 are comprised of tubular members. Partly spherical fastening elements 7 are secured to selected ends of the struts for connection to selected ends of the struts of similar components in forming a truss structure. The spherical fastening elements 7 are perforated with a plurality of apertures 8, Fig. 2, and are secured to the struts by means of the tension wire 11 extending through the tubular members and through selected ones of the apertures 8. Wires 11 are stressed in tension, as by means of nuts 12 threaded onto the ends thereof.

The fastening elements may be provided with flanges 9 in which are formed arcuate slots 10 for securement to cladding sheets such as described in my co-pending application aforesaid.

The wire W is attached to the ends of the struts in any suitable manner, preferably by the means shown in Fig. 3. Here a collar 13, provided with an attaching flange 14 is suitably secured to each end of the strut as by welding or brazing. Collar 13 is interiorly threaded as shown, and is castellated to provide tapered notches 15 to receive and position the wire W as it is spun onto the strut complex. Thereafter a locking ring 16, having suitable notches 17 for engagement by a tightening wrench is secured into the end of collar 13 for clamping the wire into engagement with the bases of the notces 15 wherever the wire W lies.

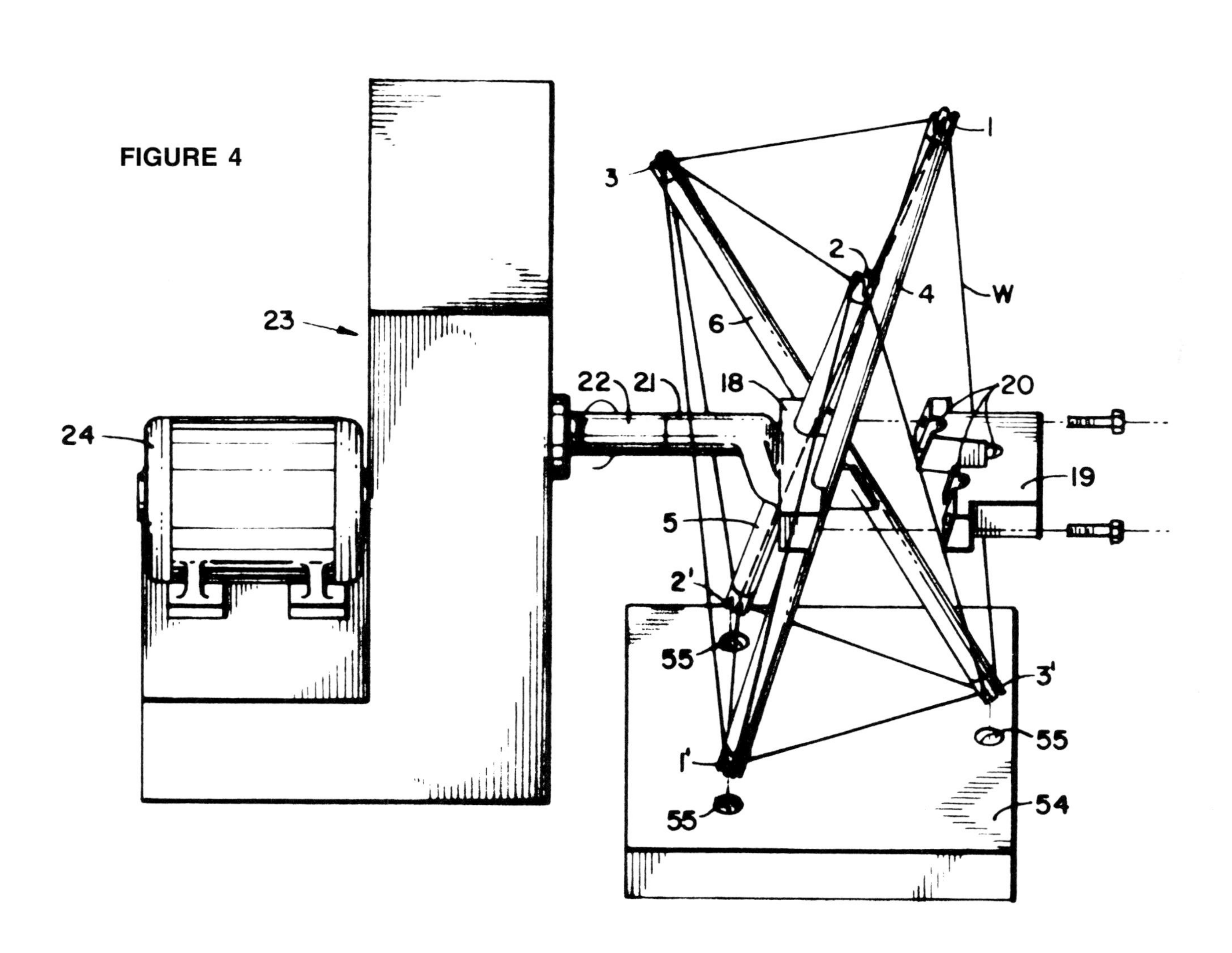

FIGURE 5

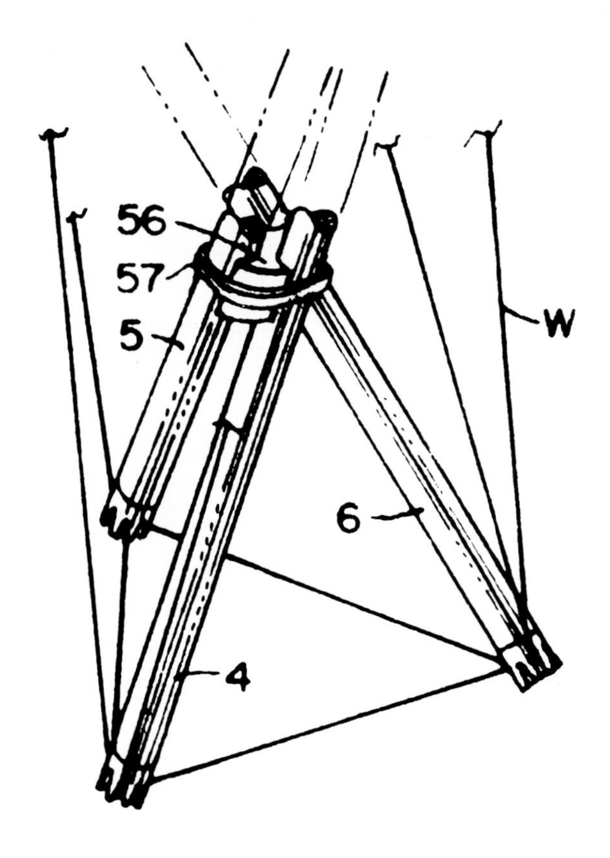

Notice that the wire crosses each end of each strut twice. The partly spherical fastening elements 7 are fastened to selected ends of the struts after the locking rings 16 have been set. Fig. 1 shows a completed truss component with two such fastening elements in place, in this instance at the ends of strut 4. Struts 5 and 6 will be secured to similar partly spherical fastening elements initially forming a part of adjacent components, and struts of adjacent components whose ends are not initially provided with such fastening elements will be secured to fastening elements 7 of the Fig. 1 component as indicated at 4′, 11′ in Fig. 1. The plurality of apertures 8 in the partly spherical fastening elements provides selective adjustment of the angular relationship between the struts (such as struts 4 and 4′) of the interconnected components.

With reference to Figs. 4–8, I shall now describe a preferred form of apparatus for spinning the tension wire over the ends of the struts to make a three-dimensional truss component. This apparatus will be described in its particular application to the fabrication of a particular truss component which is in the form of an octahedron having eight triangular faces defined by the tension wires of the complex. In the case of the octahedron there will be three struts, 4, 5 and 6, extending between the three pairs of vertexes as has been described with reference to Fig. 1. The apparatus comprises means for holding the stuts 4, 5 and 6 in predetermined angular relation to one another to form a preliminary strut assembly. This means comprises the clamping members 18 and 19 fixed to a hub 21 for attachment to shaft 22 of suitable means for rotating the strut assembly such as the rotator 23 driven by motor 24. The apparatus further comprises means for feeding a wire W for attachment to the ends of the struts 4, 5 and 6, and means 25 for producing relative movements between the rotating strut assembly and the wire feeding means to bring the wire into engagement with first one strut and then another and thus form the flexible edge portions of the truss component.

The means for producing relative movements between the rotating strut assembly and the wire feeding means includes means such as the "reader" drum 27, Fig. 6, for storing data for successive relative positions of the rotating strut assembly and wire feeding means, and means, such as the photoelectric sensor 53, Fig. 7, for translating the stored data into controlled operation of the aforesaid relative movements in timed relation to rotation of the strut assembly.

In Fig. 4 the strut clamp 18, 19 is shown exploded, i.e., with member 19 removed to permit removal of the completed truss component. Grooves 20 in member 19 are disposed in a predetermined angular relationship one to another and cooperate with complementary grooves in member 18 in determining the predetermined angular relationship between the several struts. In the particular construction shown, the rotator 23 may turn the strut assembly about the axis of shaft 22 at a constant speed, and will turn three revolutions to spin one octahedron. The reader drum 27 may be suitably driven by the rotator 23 through a 3:1 chain drive (not shown) so as to turn one revolution to each three revolutions of the rotator. Thus each revolution of the drum 27 can be made to direct the movements of the wire guide 25 throughout the spinning of one complete truss component of octahedral form.

Wire leader 25 is capable of moving the guide 28 horizontally and vertically, or both simultaneously. Wire W passes through a suitable aperture 28′ in the guide 28 after being fed from a reel. The wire being fed into the apparatus may be placed under controlled tension, as by means of any of the well known devices for tensioning feed wires. Horizontal movement of guide 28 is achieved by a reversible motor 29 which is geared to screws 30 and 31, driving lead unit 32 horizontally. Shafts 33 and 34 comprise guides for this horizontal movement. Vertical movement of guide 28 is produced by a reversible motor 35 which is geared to screw 36, 37 being a guide rod for such vertical movement. When motor 29 is operated for horizontal movement of guide 28, rod 38 fixed thereto slides through sleeve 39 so as not to produce any movement of the vertical rod 40. During this horizontal movement, wire 41, which is fixed at one end to lead unit 32, is wound or unwound on drum 42 fixed to a rotatable shaft which also carries spool 43. The wire 44 and spring 45 move a stylus unit 46 one way or the other to produce a line on the graph paper on drum 27. When vertical movement of guide 28 is produced, shaft 38 acts to raise or lower vertical rod 40 to wind or unwind wire 41′ on reel 47 which is mounted on a shaft 48 connected to spool 49 on which is wound a wire connected to stylus unit 50 biased by a spring 51. The shaft 48 of spool 49 is hollow and concentric with the shaft for spool 43. Stylus 46 records a trace for horizontal movement and stylus 50 a trace for vertical movement of guide 28.

When setting up the control pattern on reader drum 27, the rotor motor 24 and leader motors 29 and 35 can be operated by manual switching (not shown) so as to bring the aperture 28′ of guide 28 into the proper successive positions to connect the wire W to the respective ends of the struts 4, 5 and 6 as each is presented in turn to the wire leader. Errata and irregularities in the graph on reader drum 27 as produced during this manual pilot operation for a given design of truss unit can be straight-

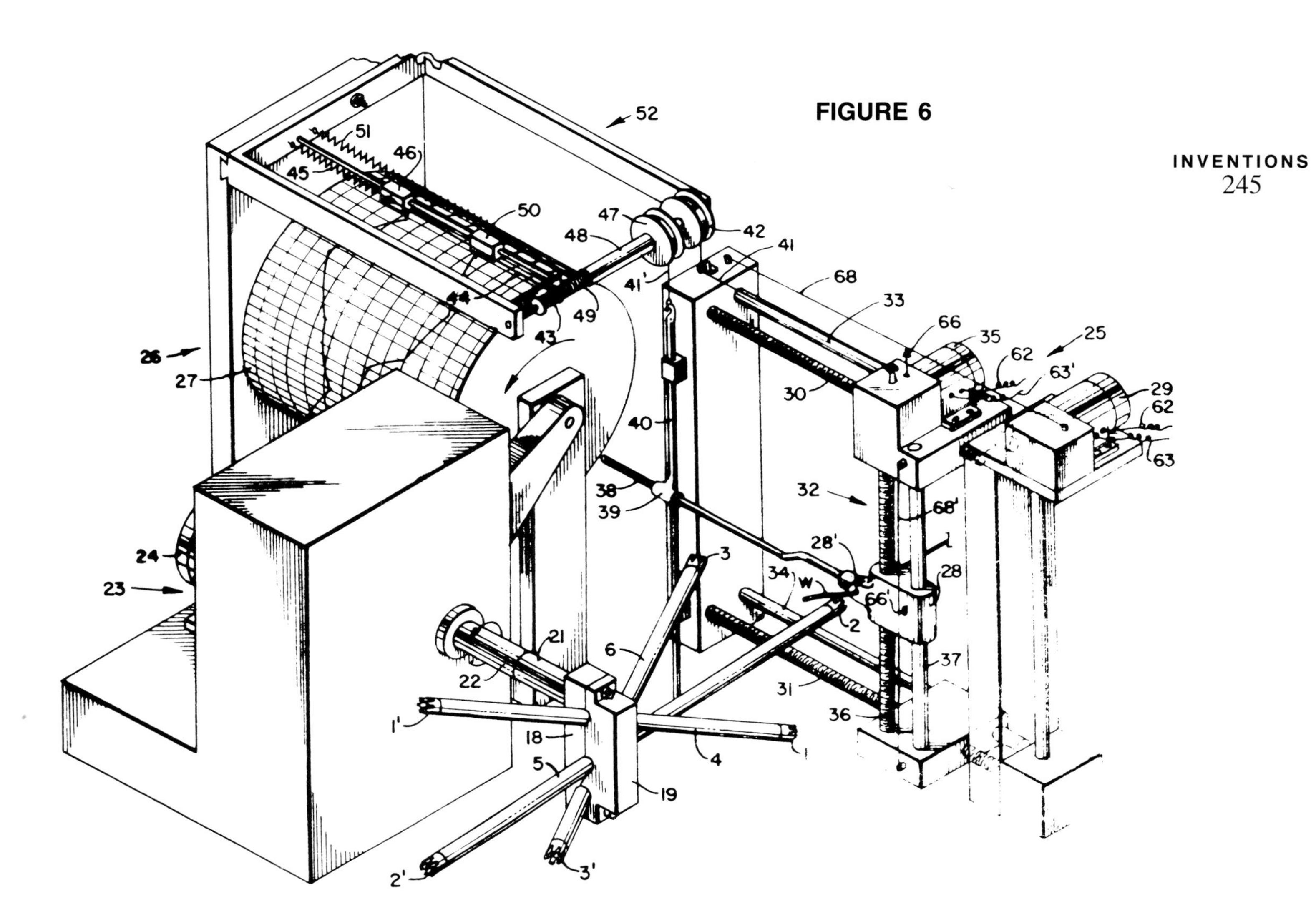

FIGURE 6

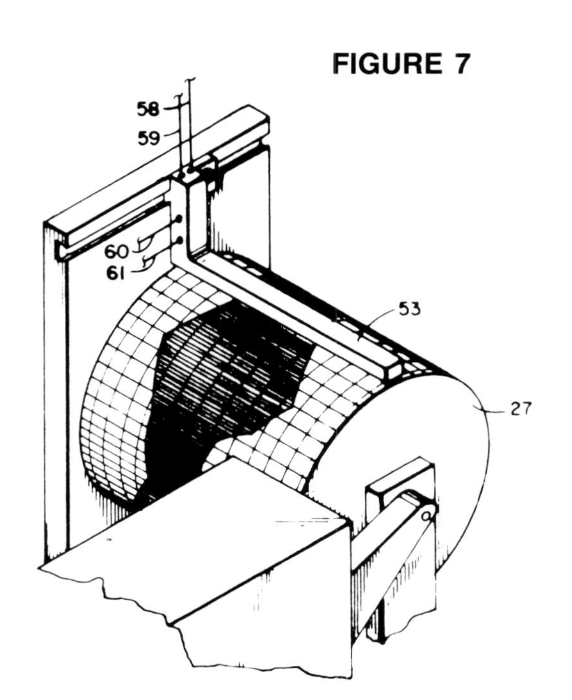

FIGURE 7

ened out on the graph by manual editing. Thereafter the area between the two traces may be blacked-in as illustrated in Fig. 7 to complete the photoelectric playback pattern.

Once the data for the desired design of truss component has been stored in the manner described, stylus frame 52 is removed and the wires 41 and 41′ disconnected. If desired, the rods 38 and 40 may also be removed. Then the photoelectric sensor 53 is installed over the drum 27 in the manner shown in Fig. 7 to give orders to the leader motors 29 and 35 according to light variation in the drum pattern. It may be observed at this point that the graph sheet upon which the data is stored is preferably removable so that the spinning program for each different design of truss component can be filed for later use as needed. Thus the program would need to be made only once for each design of truss component. Also, while I have described one preferred method of programming by means of manual settings from a prototype setup, it will be understood that the programs could be calculated through mathematical or graphic solution, and with the aid of conventional computers as desired.

The playback control of the lead motors 29 and 35 by the photoelectric sensor 53 may employ conventional circuits such as the one illustrated in Fig. 8 which is a diagram for one of the two photocell and motor hook-ups, here considered to be the one which controls the operation of motor 29 for producing horizontal movements of wire guide 28. A photocell P in the sensor 53 provides a voltage between a pair of leads 58 and 59 connected in a bridge circuit 64 containing conventional resistors, as shown, and energized by a suitable direct current source 65. The voltage output from the photocell P varies in accordance with variations in the outline of the trace defined by the edge of the darkened area on the playback drum which is adjacent the sensor 53. The voltage appearing between a contact 66 and a lead 67 of the bridge 64 varies in accordance with the horizontal position of the aperture of wire guide 28, Fig. 6. This contact 66 may be mounted on the horizontally movable wire lead unit 32, as shown, and has a sliding contact with a resistance wire 68 which may conveniently be attached to insulator supports on fixed portions of the leader 25. The resistance wire 68 is energized by a direct current source 69, Fig. 8, to form a potentiometer.

When the wire guide 28 is located to one side or the other of the desired position as directed by the trace, an unbalanced voltage occurs across the bridge 64. This unbalanced voltage is fed through a pair of leads 70 and 71 into an amplifier. The output from this amplifier is supplied by connections 62 and 63 to the motor 29 so as to move the wire guide 28 toward its desired position as directed by the trace. The direction of the direct current produced in the leads 70 and 71 by the unbalance voltage will be determined in accordance with the position of the wire guide whether to one side or the other of its correct position at any particular moment in the programmed spinning cycle. The motor will thus be operated in a direction determined by the direction of the current produced by the unbalance voltage. When the wire guide reaches its correct position, the bridge becomes balanced, and the voltage across the leads 70, 71 drops to zero, stopping the motor 29.

Similarly, for control of the operation of the motor 35 to produce vertical movements of wire guide 28, there is a contact 66′ mounted upon the wire guide 28 so as to follow the veritical movement thereof. This contact 66′ has sliding engagement with a resistance wire 68′ attached to insulator supports on horizontally movable portions of the lead unit 32. It will be understood that contact 66′ and resistance wire 68′ are included in a vertical control circuit corresponding to that shown in Fig. 8. Contact 66′, Fig. 6, corresponds to the contact 66, and resistance wire 68′ corresponds to resistance wire 68. The output from the vertical control circuit is supplied through connections 62′ and 63′ to the motor 35.

The design of the particular component to be fabricated will determine the form of the clamping members 18, 19 and the disposition of the complementary grooves 20 therein. A series of different clamp designs may be provided for this purpose or, if desired, the clamps may be made adjustable so that the angular dispositions of the grooves 20 relative to one another can be varied at will. In either case, the means for holding the struts in any one of a number of predetermined angular relations to one another is thus adjustable.

In addition to the means for holding the struts in predetermined angular relation, I have provided means for indexing the relation between the struts to predetermine the relative dispositions of the ends of the struts for a given predetermined angular relation thereof. For this purpose, my preferred form of apparatus includes an indexing fixture 54, Fig. 4, having recesses 55 to receive the ends of the struts and properly position them within the clamp members 18, 19. Either by regulating the relative depths of the recesses 55 in the fixtures 54 or by predetermining the angular disposition of the fixture while the struts are being placed in the clamping members, the extent to which each strut projects to one side or the other of the clamp is predetermined. Thus angular disposition is governed by the clamp, and lengthwise position within the clamp is determined by the fixture 54. A series of fixtures 54 of differing patterns may be used interchangeably to secure a variety of designs of truss components each being related to a given set of clamp members 18, 19, or to a given adjustment in the case of an adjustable clamp. Alternatively, fixture 54 may be provided with suitable adjusting means for altering the relative positions of the ends of the struts. The positions of the recesses 55 in the fixture may be predetermined by mathematical or graphic solution, or with the use of a computer, as desired. My present invention is not concerned with the computation of the form of truss component, the

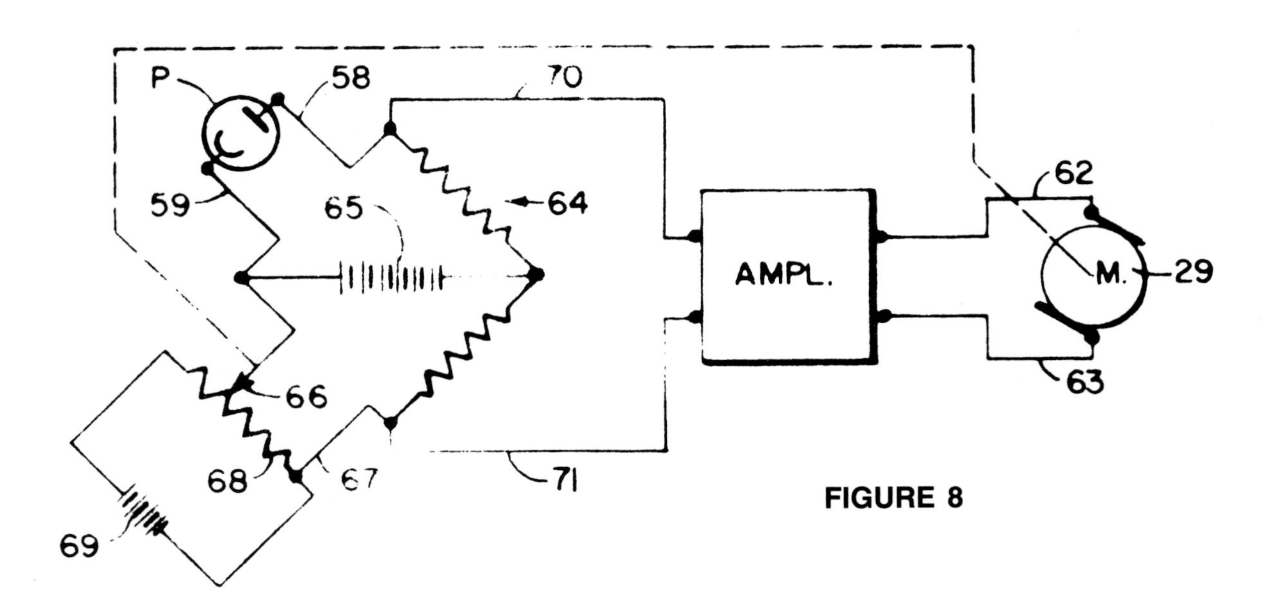

FIGURE 8

apparatus and method being designed to be used regardless of the particular design of such component. Following indexing and clamping of the struts, indexing fixture 54 is removed, and the operation of spinning the wire W onto the strut ends can begin.

In the particular octahedral form of truss component described and shown, it is possible to spin a wire around the six ends of the criss-crossed struts without reversing the direction of rotation of the strut assembly during the process. One feasible order of accomplishing this is to begin at vertex 2, locking the end of the wire to this vertex, carrying the wire from vertex 2 to vertex 1, thence to vertex 2′, to vertex 1′, etc., according to the following sequence:

Order of spinning:

2–1	3′–2
1–2′	2–3
2′–1′	3–2′
1′–3	2′–3′
3–1	3′–1′
1–3′	1′–2

This order of spinning will be found suitable when the rotator is turned in the direction of the arrows shown around the shaft 22 in Figs. 4 and 6. A reverse order might be followed and I do not wish to be limited with respect to the particular spinning sequence disclosed above. For example, as another order of spinning particularly suited to a regular octahedral unit having eight equilateral triangular faces as represented in the diagram of Fig. 9, I may proceed as follows: set up the three struts in the clamp 18, 19; then, instead of rotating about the axis of hub 21 (Fig. 4), set up the strut 1–1′ in arbors for rotation about the axis of strut 1–1′ and spin in the order:

2–3	2′–3′
3–2′	3′–2

Then set up with strut 2–2′ in arbors and, rotating about axis 2–2′, spin:

1–3	1′–3′
3–1′	3′–1

Finally place strut 3–3′ in the arbors and spin:

1–2′	1′–2
2′–1′	2–1

This method will also work for spinning irregular octahedral units so long as the wire feed is moved to and fro in the manner described hereinabove. It will be appreciated, however, that in the case of the regular, or substantially regular, octahedral unit the movements and

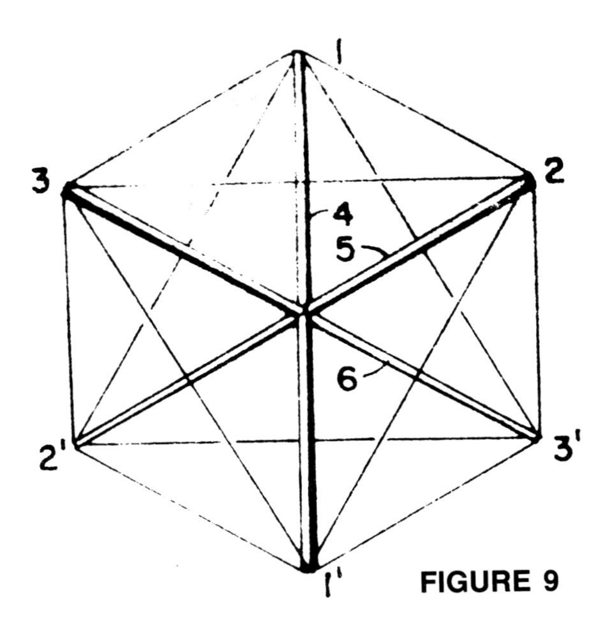

FIGURE 9

apparatus can be simplified for the reason that the step of producing relative movements between the rotating strut assembly and the wire feed can be performed by rotating the strut assembly without appreciable movement of the wire feed, if any.

Attention is directed to the fact that the wire leader 25 comprises means for moving the wire guide 28 to and fro in different directions of linear movement to produce two-dimensional movements of the guide, the control means 26 serving to control such two-dimensional movements of the wire guide in timed relation to rotation of the strut assembly to produce the three-dimensional truss components. Movements of the wire guide 28 are not necessarily restricted to the horizontal and vertical as shown in Fig. 6, as it will be appreciated that the wire feed can be disposed in any position relative to the rotating strut assembly which will serve to bring the guide 28 into proximity with each strut end successively.

After completion of spinning, and the application of the locking rings or collars 16, Fig. 3, clamp 18, 19 is opened and the truss component will be self-supporting. If desired, spacer member 56, Fig. 5, may be inserted between the criss-crossed portions of the struts and a tie 57 applied around the struts and spacer.

The terms and expressions which I have employed are used in a descriptive and not a limiting sense, and I have no intention of excluding equivalents of the invention described and claimed.

21▲STAR TENSEGRITY (OCTAHEDRAL TRUSS) (1967)

U.S. PATENT—3,354,591

APPLICATION—December 7, 1964

SERIAL NO.—416,228

PATENTED—NOVEMBER 28, 1967

TETRAHEDRA COULD BE MADE by having compression members representing the *x*, *y*, and *z* axes of the octahedron with its six terminals held in tension by cables or rods, making possible an infinite variety of octahedra. These could be assembled one with the other to produce the octahedron/tetrahedron truss with very high tensible capability. I used this octahedron pattern in the United States Pavilion in Montreal at Expo '67.

UNITED STATES PATENT OFFICE

Richard Buckminster Fuller, 407 S. Forest St., Carbondale, Ill. 62901

OCTAHEDRAL BUILDING TRUSS

The invention relates to a building truss particularly adapted to the construction of geodesic domes and having special advantages for use in other kinds of structures as well.

Advances in the technology of materials have resulted in the discovery of the means for producing remarkable increases in tensile strength properties of the materials. Noticeably this has been true in the field of metal alloys, ferrous and nonferrous. Materials of great tensile strength have been developed also in plastics. Glass fibers of enormous strength have become available and are widely used. Notwithstanding the general availability of such high tensile properties in materials, comparatively little has been done in the direction of utilizing pure tension elements in building construction. For building purposes, engineers have clung tenaciously to age-old concepts which rely primarily upon the compressive strength of the materials used so that structures have been erected stone upon stone, beam upon column, all with the utilization of a vast deadweight of materials. With the use of the somewhat lighter weight girders now employed, for example, in the construction of floors and roofs of conventional buildings, some increase in use of the tensile properties of materials has been made, but still relying to a great extent on the presence of large and heavy compression members.

SUMMARY

It has been a primary object of my invention to provide a truss construction which is capable of utilizing more efficiently the tensile strengths of the materials from which the truss is constructed. I have found a way of building a truss which allows the use of many elements loaded purely in tension, indeed, one in which such purely tensioned elements predominate, so that relatively few compression members are needed. Further, I have discovered how to do this in a way which provides a smooth surface well adapted to cladding in the construction of floors, roofs and walls, and which is remarkably well adapted to the construction of spherical form buildings inclusive of buildings known as geodesic domes.

A particular characterizing feature of my present invention resides in putting together in a new way a plurality of units of octahedral form each having eight

triangular faces, such units being connected face to face so that each pair of adjacent units has one face thereof in congruity. In the preferred form of my truss, each of the interconnected units comprises twelve flexible members capable of being stressed only in pure tension and three compression elements consisting of criss-crossed stiff members capable of being loaded as compression columns. Regarding the fundamental purpose of the invention, it is, of course, of the the utmost significance that we have here a ratio between pure tension and pure compression of four to one. (If when six units are interconnected in the manner shown in Fig. 3, only one set of tension elements is used where the congruent faces are found, some of the tension elements will be eliminated and the ratio between tension and compression elements

cladding elements removed to reveal all of the elements of the truss.

Fig. 4 is a fragmentary cross-sectional view taken generally as indicated at 4—4 in Fig. 1, the scale being somewhat enlarged.

Fig. 5 is a detail view of one of the six octahedral units which together make up the truss section of Fig. 3.

Fig. 6 is a view similar to Fig. 5 illustrating a modified form of the octahedral unit.

Fig. 7 is an exploded view of one pair of octahedral units used in the Fig. 3 truss.

Fig. 8 is a top perspective view of a six-sided capping member which forms a part of the cladding shown in Fig. 2.

Fig. 9 is a top perspective view of a triangular capping

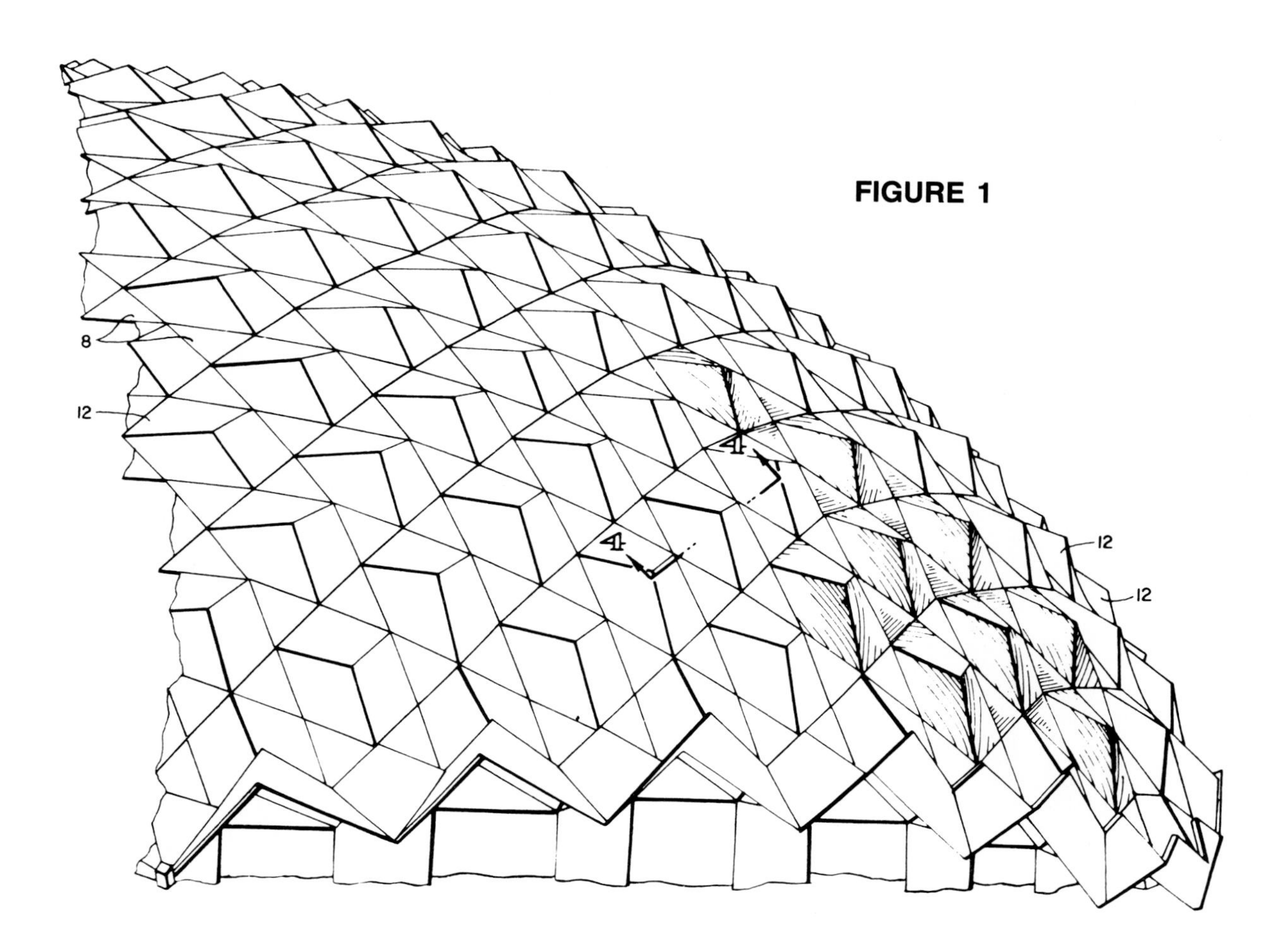

FIGURE 1

will become three to one.) Yet, as will be explained hereinbelow, this predominance of tension elements—which can even be slender wires or cables—is accomplished in a way to provide a smooth planar or spherical surface for cladding in a most convenient manner.

Description

The accompanying drawings show the best mode contemplated by me for carrying out my invention.

Fig. 1 is a side elevational view of a portion of a geodesic dome constructed in accordance with my invention.

Fig. 2 is a face view of a portion of the surface of the structure of Fig. 1 with some of the cladding panels removed to reveal a portion of the underlying truss.

Fig. 3 is a view similar to Fig. 2 but with all of the

member which forms another part of the cladding shown in Fig. 2.

Fig. 10 is a diagram showing the relationship between several types of elements which can form a part of a geodesic dome constructed in accordance with the invention.

Fig. 11 is a view similar to Fig. 4, showing a modified construction.

Fig. 12 is a side elevational view of a mast structure embodying my invention in another form.

Fig. 13 is a view similar to Fig. 12, showing a modified form of mast structure.

Fig. 14 is a view illustrating an application of the invention to a structure of asymmetrical form.

Fig. 15 is a perspective view, partially exploded, showing elements of two associated octahedral units con-

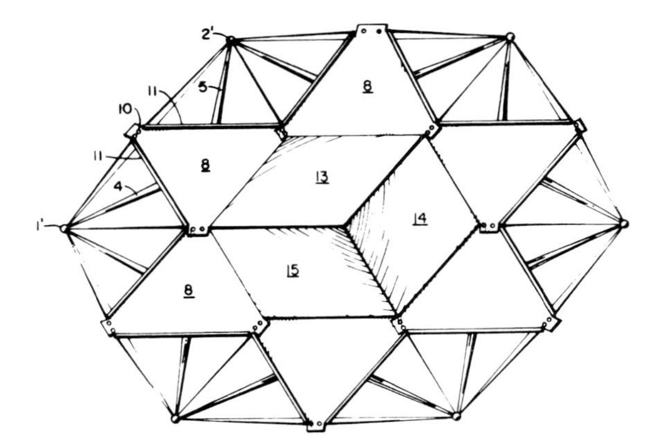

FIGURE 2

FIGURE 3

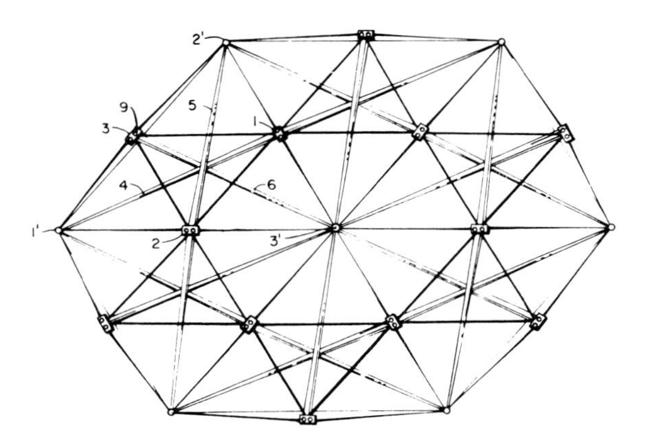

structed in accordance with another embodiment of the invention.

With reference to Fig. 5, I shall first describe one of the octahedral "building blocks" which form the basic unit of the construction. It has eight triangular faces and is, therefore, an octahedron. The eight triangular faces are defined by the tension elements which normally will be made of flexible wires or cables. There are three compression elements, 4, 5 and 6, which are arranged along the three axes of the octahedron, 1–1′, 2–2′ and 3–3′. In reading Figs. 3, 5, 6 and 7, it will be helpful to note the following: The reader is looking top down on the truss and truss units. The upper ends of the compression elements 4, 5 and 6 lie in one plane and the lower ends lie in another plane below the first. Unprimed numbers are used for those points which lie in the upper surface, and primed numbers for those which lie in the lower surface. Further, the points lying in the upper surface are marked by rectangles (rectangular elements) and those in the lower surface by circles or spheres. As an additional aid to the reader, the perspective has been heightened by an exaggerated fore-shortening of the compression members 4, 5 and 6 so that they will appear larger at the end near to the reader.

Interconnection of adjacent units in face to face relationship will be understood from Fig. 7 in which two adjacent units have been drawn apart slightly. When brought together, the faces 1′–2–3′ of the two units shown will be congruent. In the unit which appears in the lower part of Fig. 7, the tensile lines 1′–2, 2–3′, and 1′–3′ have been shown as imaginary lines, for when the two units are brought together the tension elements 1′–2, 2–3′ and 1′–3′ of the upper unit can serve as common tension elements of the two associated units as may in some cases be desired.

In Fig. 3 we see six associated octahedral units of Fig. 5, each adjacent pair of such units being interconnected in the manner I have described with reference to Fig. 7. The interconnected units form a truss structure comprising a network of tension elements arranged in a pattern defining a plurality of the Fig. 5 octahedra in face to face relationship as depicted in Fig. 7. A compression member 4, 5 or 6 extends between each of the three opposed pairs of vertexes of each octahedron. This description of my construction may now be summarized more simply as a truss structure comprising tension elements defining octahedra arranged face to face, and compression elements arranged along the three axes of each octahedron.

The seemingly complex but truly simple forms of the octahedral units will be further understood by identifying the faces of the octahedra, the tension elements and the compression elements as follows:

Eight faces of the octahedra:

1–2–3	1–2′–3
1′–2′–3′	1–2′–3′
1–2–3′	1′–2–3′
1′–2–3	1′–2′–3

Twelve tension elements:

1–2	1–2′
2–3	1–3′
3–1	2–3′
1′–2′	2–1′
2′–3′	3–1′
3′–1′	3–2′

Three compression elements:

4	5	6

FIGURE 4

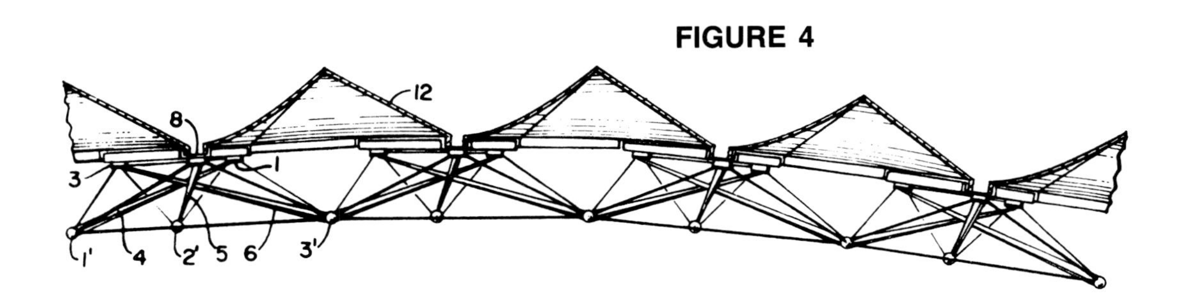

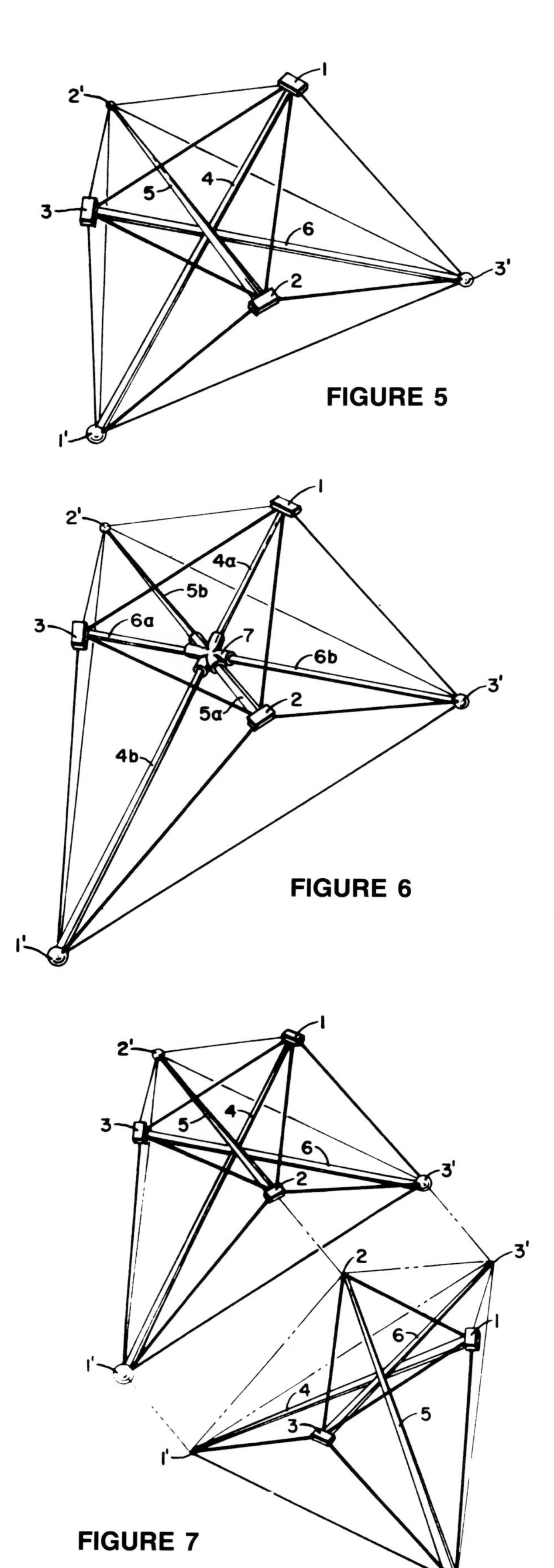

FIGURE 8

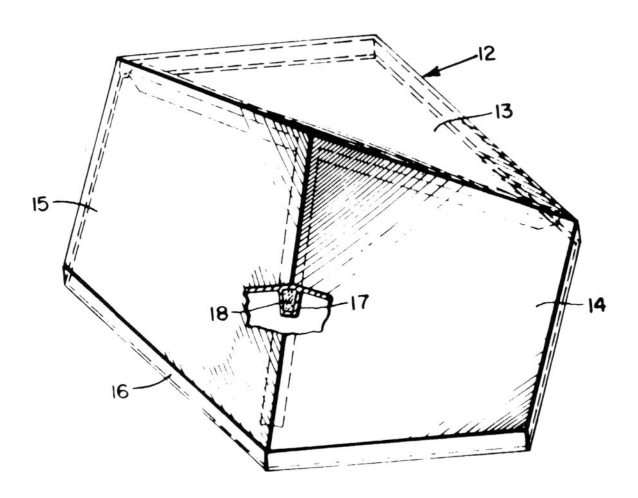

From the foregoing tabulation of the truss elements, the student of this disclosure will appreciate the preponderance in tension elements over compression elements and the significant improvement thus obtained in the direction of increased utilization of the high tensile properties of the improved materials and alloys available today. The problem heretofore has not been the availability of materials of good tensile properties, but rather the question of how such properties could be utilized more fully in the realm of building architecture; and how this could be done in a practical manner which would provide surfaces which could conveniently be cladded—even though such surfaces seem to be made up of a maze of wires and to be inherently "spikey." How this can be done is now further disclosed with more particular reference to Figs. 2, 4, 8 and 9 of the drawings.

If we take a truss section of the extent comprised of six associated octahedra, it will be noticed that one face (1–2–3) of each of the six octahedra lies in a common surface. These faces of the six octahedral units are joined together vertex to vertex, and together encircle a hexagonal area. The six triangular faces 1–2–3 are enclosed by triangular capping members 8 whch may be secured in any convenient manner as by bolting them to fittings 1, 2 and 3, Fig. 5, which may be apertured as at 9, Fig. 3, to receive bolts passing through aligned apertures 10 in overlapping corner portions of the triangular capping members. By this means, the capping members are secured to each other and to the underlying truss. In the preferred construction shown, the triangular capping members 8 are formed as panels having upwardly extending marginal flanges 11, Fig. 9. The hexagonal area within the circle of triangular panels 8, Fig. 2, may then be enclosed in any suitable manner. This may be done most advantageously with the use of a hexagonal capping member 12, Fig. 8, having downwardly extending marginal flanges 16 which fit over the upwardly extending flanges 11 at the edges of the hexagonal area. The hexagonal capping member 12 is advantageously formed with three adjoining diamond-shaped surfaces 13, 14 and 15 of hyperbolic-paraboloid form. Various materials may be employed in the cladding of the structure including formed or molded sheets of steel, aluminum, plastics, fiberglass or other suitable materials. In the representative construction shown, the triangular capping members 8 are made of steel and the six-sided capping members 12 are made of fiberglass-reinforced polyester resin. In the construction shown in Fig. 8, this capping member is further reinforced along the ridges formed by the adjoining edges of the hyperbolic-paraboloids by hollow molded ribs 17 filled with a urethane foam 18. (The cladding construction which I have described forms an effective watershed by reason of the interlocking flanges of the several cladding members and the arrangement by which the water can flow from one triangular panel to another. These panels together form a maze of gutters which will serve to clear a heavy flow of rainwater.)

In Fig. 6, I have illustrated a modified form of the fundamental octahedral unit in which the compression members extending between each of the three opposed pairs of vertexes 1–1′, 2–2′ and 3–3′ of the octahedron are each made in two parts joined together centrally of the unit by a hub member 7 provided with three pairs of fittings aligned with the respective axes of the compression members. It will be observed that in the embodiment shown in Fig. 5, the compression members 4, 5 and 6 are arranged to bypass one another in the manner of the poles of a tepee. At the point of bypass, the members may be spaced from one another or be in contact. They may or may not be secured to one another at the point of bypass as desired. In the embodiment of Fig. 6, the compression members can be arranged so that their axes intersect, the single members 4, 5 and 6 here being replaced by the compound compression members *4a–4b. 5a–5b, 6a–6b,* respectively.

When my invention is utilized in geodesic dome construction, some of the octahedral units will be arranged so that their triangular capping members encircle a pentagonal area as distinguished from the hexagonal area described with reference to Figs. 2 and 3. This occurs at each vertex of the icosahedron, dodecahedron or tricontahedron on which the design of the geodesic dome is based. This aspect of geodesic construction is well known to architects familiar with such construction, and will be understood from the disclosure of my fundamental geodesic building construction Patent No. 2,682,235 where the pentagonal areas can be seen clearly in Figs. 1 and 2. Thus in a geodesic dome constructed in accordance with my present invention, some of the octahedral units are arranged so that one face of each of six associated units lies in a common spherical surface, such faces of the six associated units being joined together vertex to vertex and enclosed by triangular capping members which encircle a hexagonal area in the common spherical surface,

FIGURE 9

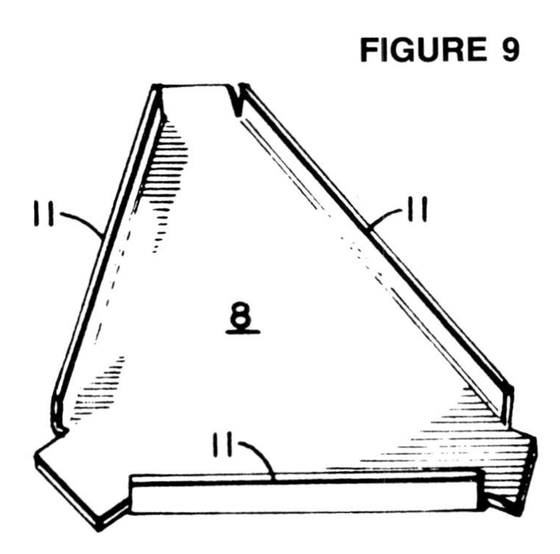

FIGURE 10

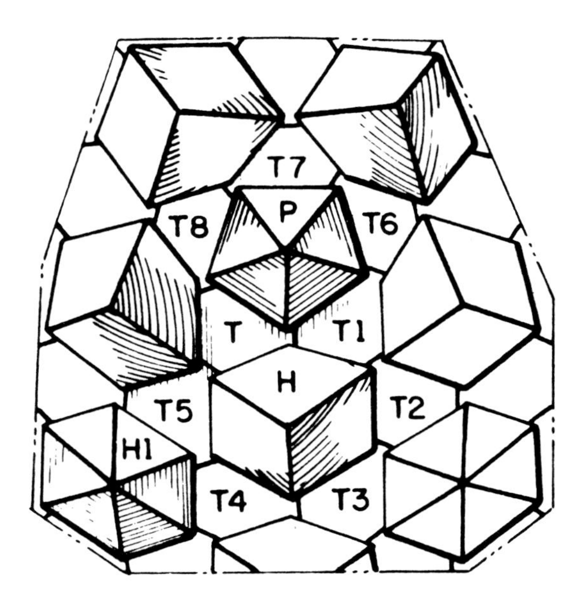

such hexagonal area being enclosed by a six-sided capping member adjoining each of the six triangular capping members; whereas, others of the octahedral units are arranged so that their triangular capping members encircle a pentagonal area in the common spherical surface, such pentagonal area being enclosed by a five-sided capping member adjoining each of the five triangular capping members. This will be further understood from the diagram of Fig. 10. The area covered by this diagram can be compared with the area at the zenith Z of my geodesic patent aforesaid. Triangular capping members T, T1, T2, T3, H4, T5 encircle a hexagonal area which is enclosed by a six-sided capping member H which might, for example, be of the construction which has been described with reference to Fig. 8. Triangular capping members T, T1, T6, T7, T8 encircle a pentagonal area enclosed by a five-sided capping member P. Capping member P may be of simple pyramidal form as distinguished from the hyperbolic-paraboloid form of Fig. 8, or H in the diagram. Further, the form of the six-sided capping members may be modified according to the wishes of the architect; for example, they may be of the simple pyramidal form shown at H1 in the diagram.

Fig. 11 illustrates a further development of the construction of Fig. 4, in which the proportions of the octahedral units are altered to create a flat floor or roof truss. This truss is then built up into two or more thicknesses (two as shown) in which every other truss is inverted. Here the smaller faces 1–2–3 of the two trusses are joined congruently. Similarly, two trusses may be joined in a manner which brings the larger faces 1′–2′–3′ into congruent relation.

Fig. 12 illustrates the application of my invention to a mast or tower structure. In this case the opposed faces 1–2–3 and 1′–2′–3′ of the octahedrons are of the same size, and parallel to one another.

Fig. 13 illustrates a modification of the Fig. 12 structure in which opposed faces 1–2–3 and 1′–2′–3′ are of different size, and are not parallel to one another.

Fig. 14 illustrates an application of my invention to an asymmetrical structure which reveals the comprehensive

FIGURE 11

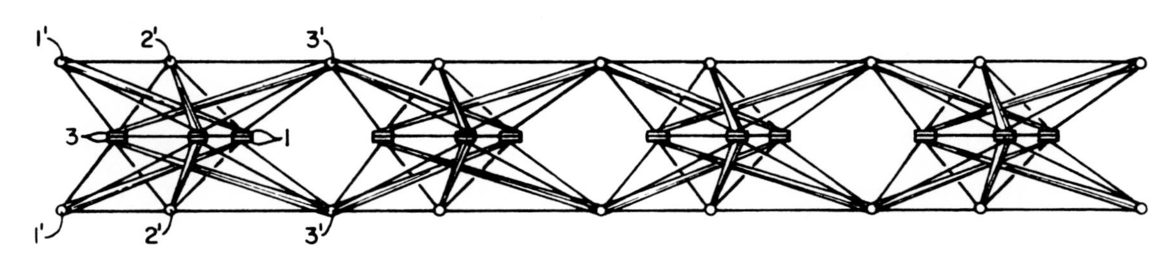

FIGURE 12

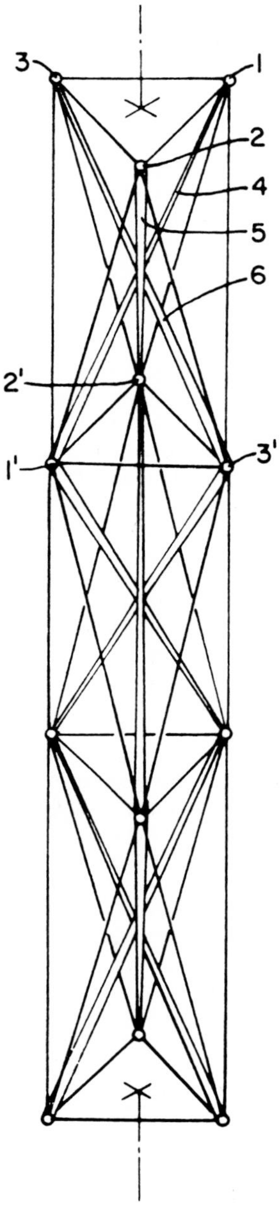

FIGURE 13

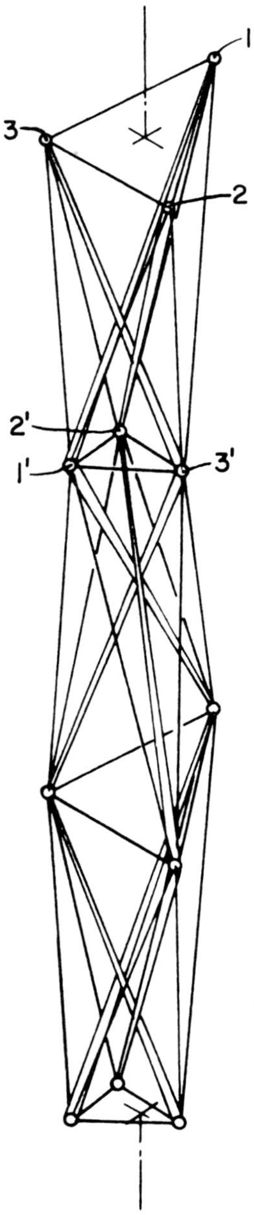

FIGURE 14

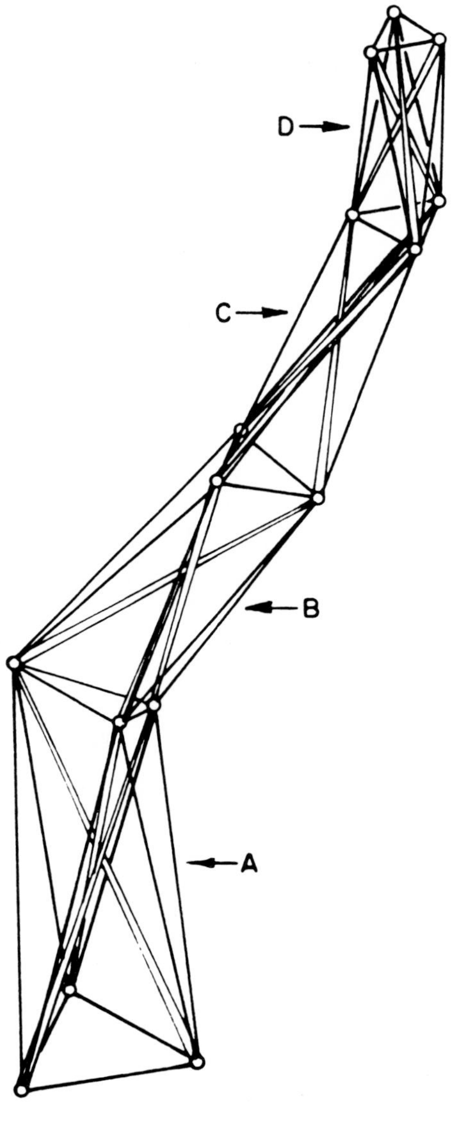

FIGURE 15

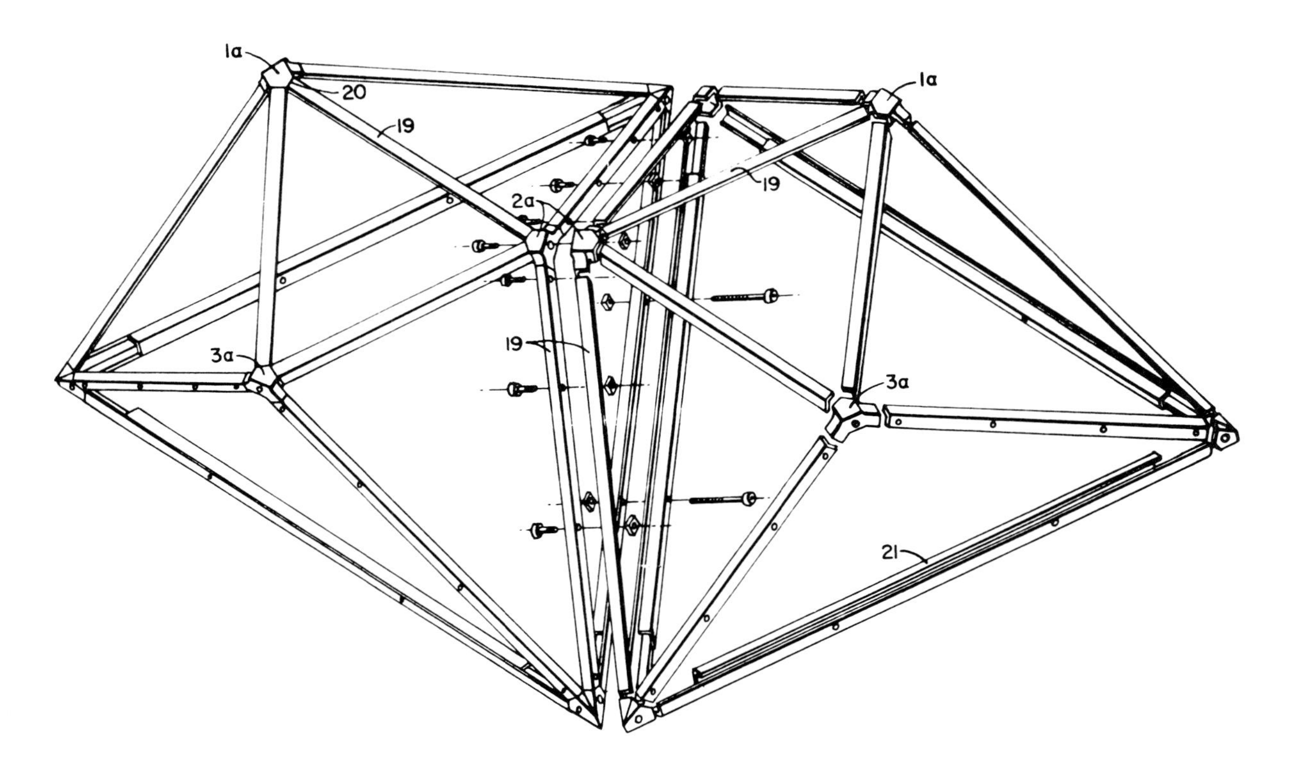

adaptability of my invention to structures of irregular form. The octahedrons A, B, C, D, etc., can be varied in their relative proportions and shapes, grouped and extended as desired, while availing of the favorable weight-strength ratio obtained by reason of the preponderance of tension elements over compression elements.

Fig. 15 illustrates a further modification of the octahedral units in which the twelve edges of each octahedron are in the form of structural shapes such as the "angle irons" of steel or aluminum shown at 19. Vertex members 1*a*, 2*a* and 3*a* correspond to members 1, 2 and 3 of Figs. 5, 6 and 7. These are advantageously made as metal castings to which the structural shapes 19 are welded as 20. The construction thus comprises a plurality of interconnected units each of which comprises twelve members interconnected to form the edges of a self-supporting unit having eight triangular faces, the units being connected face to face so that each pair of adjacent units has one face and three members thereof in congruity. The faces which are connected in congruity are those which are shown spaced slightly apart, these faces being designed to be bolted together in the manner indicated by the exploded bolting elements, which pass through aligned apertures in the congruent edge members.

In each of the embodiments described, my structure comprises a plurality of interconnected units each of which includes twelve members interconnected to form the edges of a figure having eight triangular faces, an outer face of the several units lying in a common surface and an inner face of the several units lying in a second common surface spaced from and parallel to the first. For example, in the embodiment of Figs. 1–5 and 7, and with particular reference to Fig. 5, we have an outer face 1–2–3 and an inner face 1′–2′–3′, these two faces lying in surfaces which are spaced from and parallel to one another. Notice that the outer face 1–2–3 is approximately one-half the size of the inner face 1′–2′–3′ and that the two faces are so oriented relative to one another that the vertices of the outer face lie substantially above the midpoints of the sides of the inner face. "Above" is used here in the sense of "beyond" or "outside of"; and it will be understood that if the truss extends in a generally horizontal direction, vertex 2, for example, of the outer face 1–2–3 will lie above the side 1′–2′–3′ of the inner face 1′–2′–3′, whereas, if the truss is located so as to extend in a generally vertical direction, vertex 2 will be beyond or outside of side 1′–3′. If the inner and outer surfaces of the truss are parallel flat planes arranged horizontally, vertex 2 will be directly above side 1′–3′. Again, if the common surfaces are concentric spherical surfaces, vertex 2 will be disposed radially outside of the midpoint of side 1′–3′. So in all cases it can be considered that the orientation of the outer and inner faces of each unit is such that the vertices 1, 2, and 3 of the outer face lie substantially "above" the midpoints of the sides of the inner face 1′–2′–3′.

The relative size and orientation of the inner and outer faces of each unit as described in the preceding paragraph makes it possible to interconnect the several units in a manner which joins the outer faces together vertex to vertex and which joins the inner faces together edge to edge. If the truss is inverted or turned inside out, the inner faces will be joined together vertex to vertex and the outer faces edge to edge.

The terms and expressions which I have employed are used in a descriptive and not a limiting sense, and I have no intention of excluding equivalents of the invention described and claimed.

22▲ROWING NEEDLES (WATERCRAFT) (1970)

U.S. PATENT—3,524,422

PATENTED—AUGUST 18, 1970

FIGURE 1

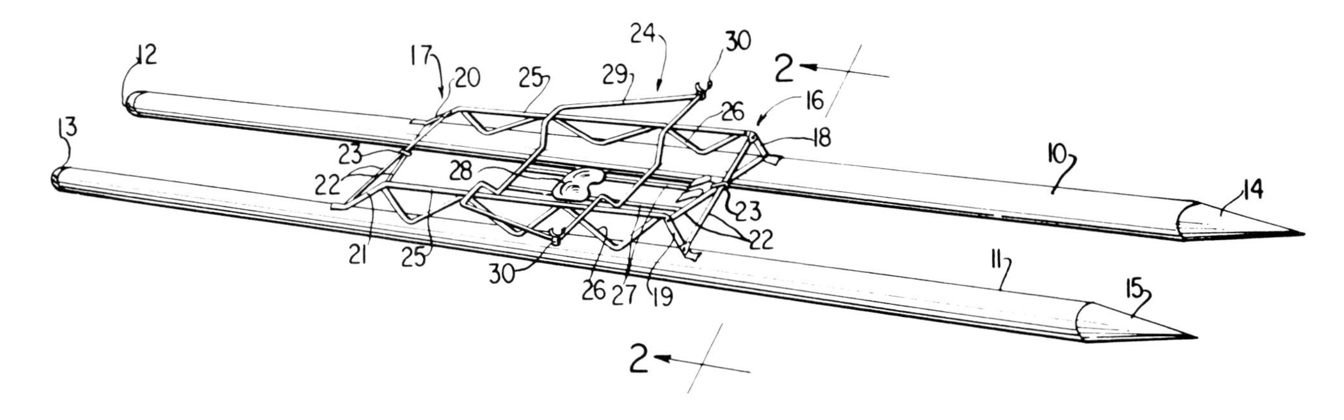

I AM AN OARSMAN; I've found in all my life nothing that employed my total muscles more effectively and put me into such a healthy condition as rowing in a sliding seat outrigger, what is called a "single" shell. These single shells average around twenty-two feet in length, and have a beam of only eighteen inches. They are kept balanced entirely by the oarsman, and his oars are passed rigidly through oarlocks and outriggers. I've done much such rowing.

On one occasion, I was rowing in Wiscasset, Maine, using a "single" that was loaned to me by my friend Charles Chase. I rowed what is called "upriver," which is from the float in Wiscasset. This is about two miles to the windward of the Wiscasset anchorage and dock, when my right steel oarlock broke (it had apparently been rusting), and immediately, instinctively to keep myself balanced, my left-hand oar became more and more vertical, which tended to push more and more of the bulk of the wooden oar to greater and greater depth. I suddenly found myself sitting using just the buoyancy of that one oar to self-balance against completely rotating. My boat was pouring in water. Then I realized that I was holding on to the oar whose oarlock had broken. I realized that I could gradually rotate that broken oarlock oar out along the outrigger. The more I brought out along the outrigger, the more my left-hand oar was able to emerge. Finally, I was able to get the broken oarlock with its woodenness impinged on the sharp vertical breakage point of the oarlock itself. Now my two oars were out horizontally again, but I obviously could not row with a broken-oarlocked oar. I could row with the left-hand oar. I figured out with the wind blowing me back toward the float at the Wiscasset anchorage, that I might be able to steer myself downwind by from time to time rowing myself in a circle, and then coming off on the right part of that circle directly to windward of the float. I did succeed in getting home, in getting my "single" out of the float and stored away. The water that day was about 58° F. If I had rolled over, that coldness would probably have done me in before I would have been able to swim home.

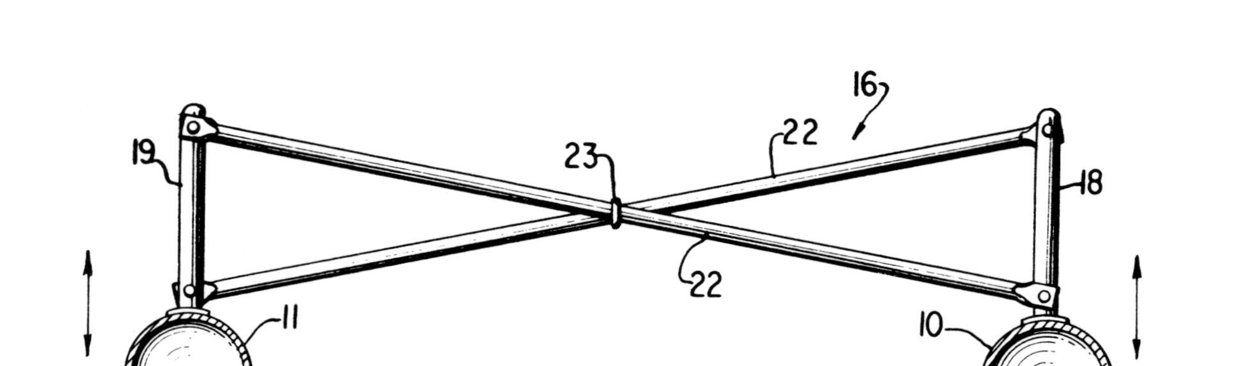

FIGURE 2

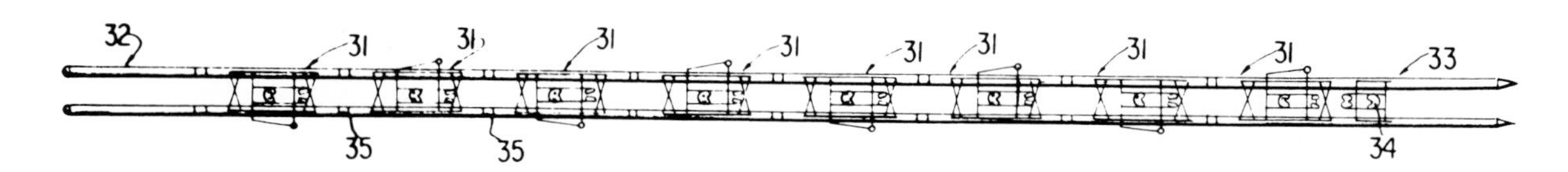

FIGURE 3

On another occasion, I was rowing in Detroit, on the Detroit boat club on Belle Isle. The waters from Lake Sinclair to the north of Belle Isle come down by Belle Isle on both sides, and by the city of Detroit, and gradually out to Lake Erie. There is quite a current, then, that runs by Belle Isle. Rowing in a "single" out of the Detroit Boat Club on Belle Isle, I would go northward to Lake Sinclair. In doing such rowing there, one has to avoid several things. There are very large boats and oar ships coming down from Lake Superior to bring steel to the various mills of Ohio and Michigan. They usually keep on the eastern shores of Belle Isle, and I would row on the western side of Belle Isle between the city of Detroit and the island. Once we got into Lake Sinclair it was easy to stay clear of other boats. I was coming back to Belle Isle from Lake Sinclair when suddenly what was called a "gold cup power boat," a racing boat, went very thoughtlessly so close to me that its enormous wave rolled my shell over.

I was now in the water, and the oars were locked into the oarlocks on the outriggers, and

FIGURE 4

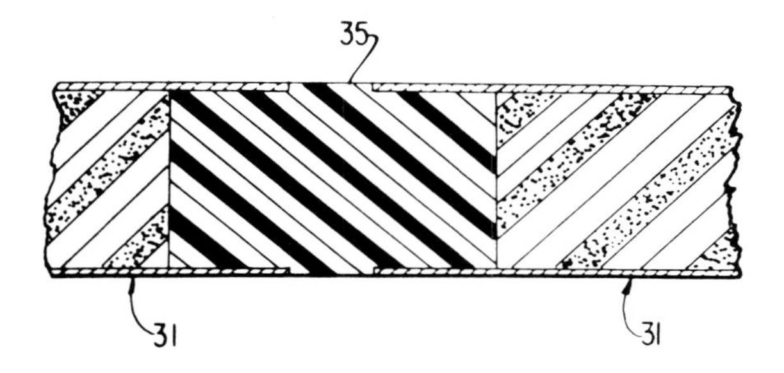

I could take hold of the boat and swim with my legs pushing it. I felt it was very touch and go whether I could get in close enough to Belle Isle to get up on the beach before I was swept down through the city of Detroit and out toward Lake Erie. Luckily, I did get it over to the shore of Belle Isle, and I was then able, standing in the shoal water, to remove the oars from the outriggers, to pick the shell up and dump the water out of it, to re-rig the oars, to mount back into my sliding seat, and to bring the shell back to the Detroit Boat Club, to whom the "single" belonged.

Because, then, I had found on two occasions how dangerous a "single" really can be, because I cared so much about rowing, I saw I could make a catamaran by putting two "singles" together, which would prevent the "singles" from being rolled over, and if my two shells were absolutely watertight, what I call needles, then it could not be foundered in any way.

It will probably be very important in years to come for people to be able to have a rowing machine like this to put on top of their cartops and be able to take to appropriate areas to row. If you have what is called a rowing ma-

FIGURE 5

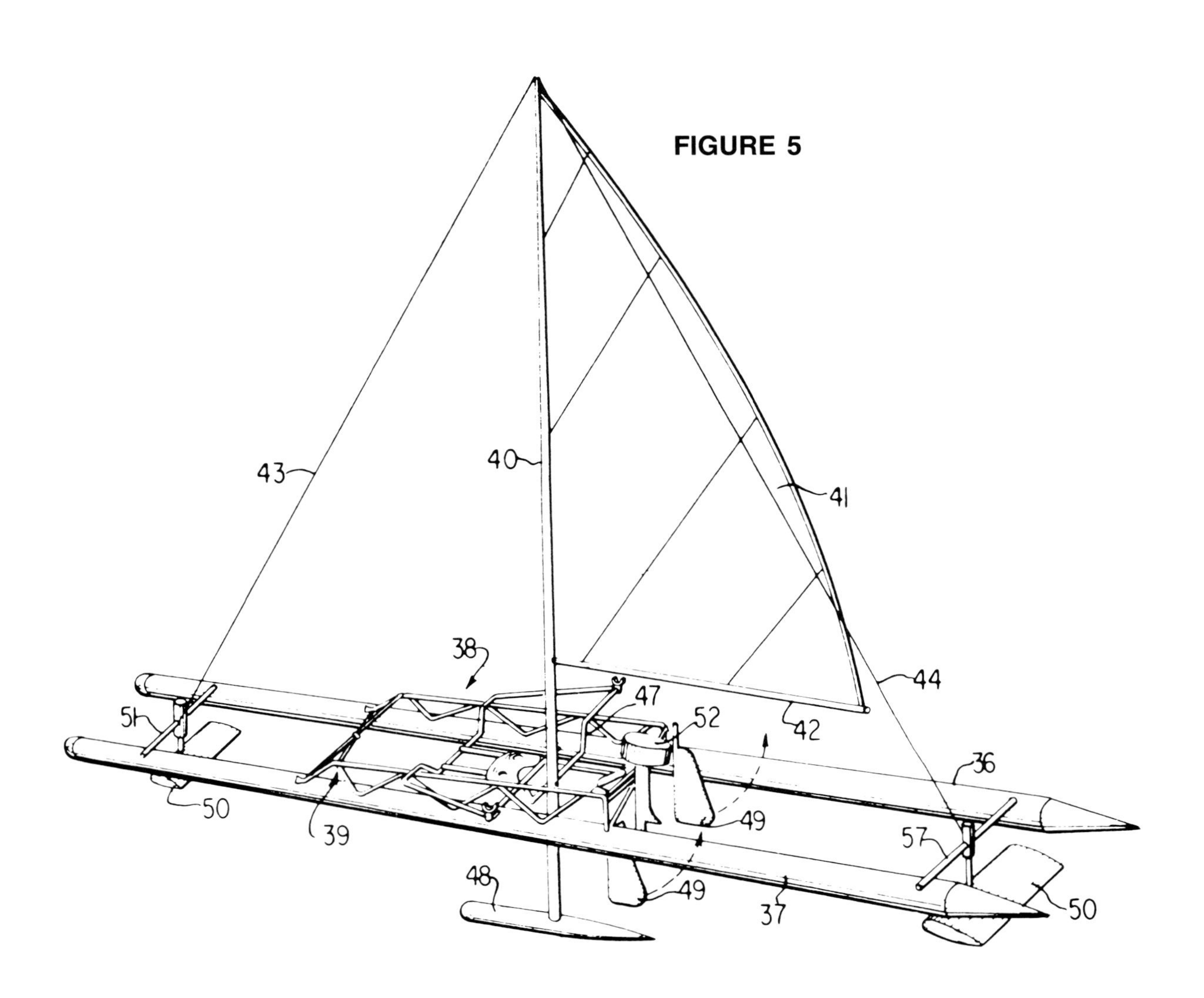

chine inside your house, and you start to really work it, you'll find you'll breathe more powerfully, and if there's lots of dust in the house, you'll find you'll really feel and taste that dust badly. There is nothing quite so pure as the air we get when rowing a "single" on a remote lake or river, and I want to be able to get out where the waves are large, so I developed what was then called rowing needles. They are now quite successful, and they have a beautiful hull. They may become popular.

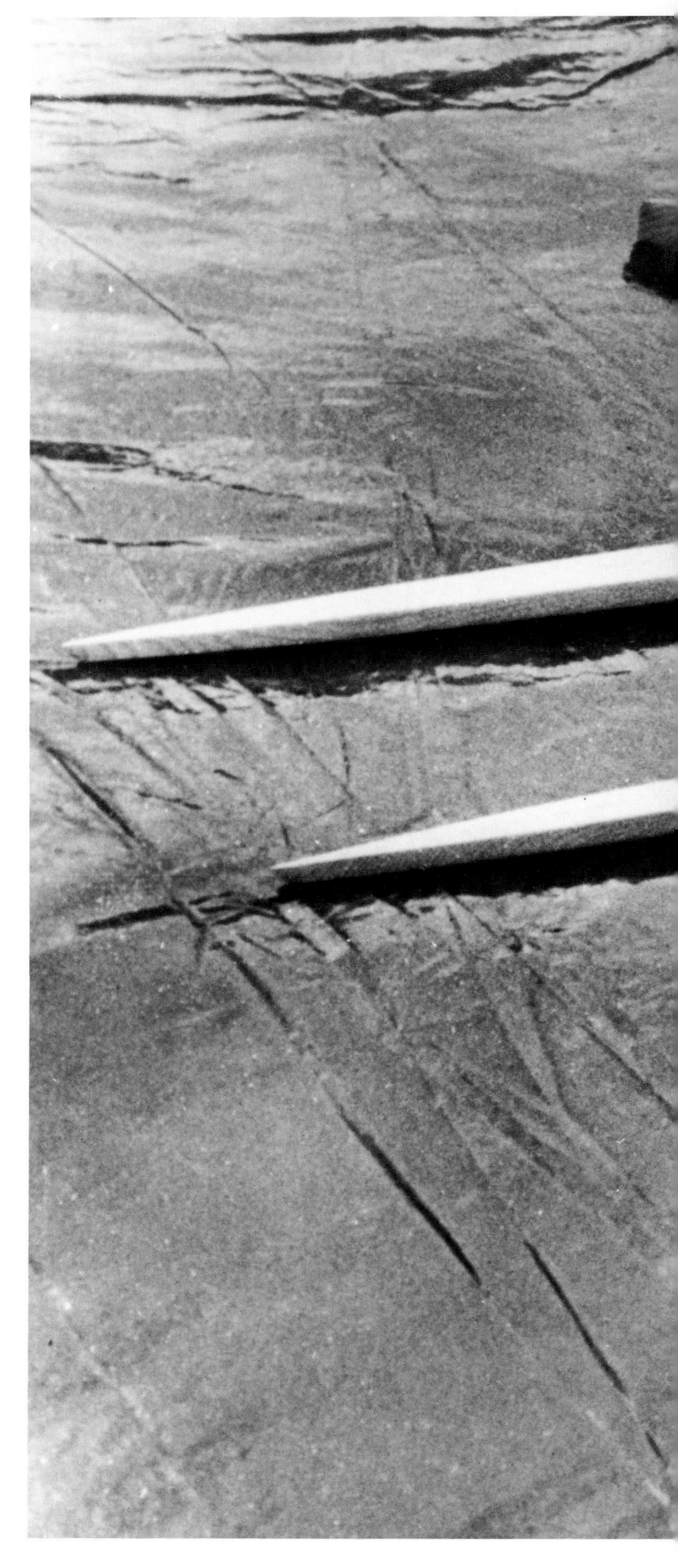

UNITED STATES PATENT OFFICE

WATERCRAFT

This invention relates to watercraft having pairs of floats and especially to an arrangement for maintaining the floats in desired space relation.

In the prior art, catamarans and similar type craft have been known, but they are unwieldy and do not have the necessary characteristics for use for various purposes. It is also desirable to be able to control directionally the floating units of such a vehicle so as to either cause them to stay in parallelism for streamlining purposes or to be able to change the angle for blocking or braking. Also, there has not been a satisfactory arrangement for connecting a plurality of units wherein there is sufficient flexibility so that the structure is in effect articulated.

One of the objects of the invention is to provide an improved multi-hull watercraft which will maintain the desired relationship of the hulls or float members.

Another of the objects of the invention is to provide a watercraft which can be adapted for various uses.

Another of the objects of the invention is to provide an efficacious manner of making an articulated watercraft.

In one aspect of the invention, at least a pair of floating members or hulls are used which are kept in a substantially parallel arrangement by transverse structural members longitudinally spaced from each other, one toward the bow and one toward the stern. The transverse structural members are arranged so that forces involved in keeping the floats aligned are directed from a zone or point on a post vertically over each of the floating elements toward the top of the transversely located vertical post on the other float. The structural members can be formed in several manners. In one arrangement, there can be a cross bar connecting the top of the posts. In another form, a pair of crossed elements can be used extending from the top of one of the posts to the float portion. Also, various shaped members can be used to connect the two posts. The critical consideration is that these elements may have both torsional effect and column effect in maintaining the floats in desired relationship. The positioning of the floats is accomplished by the described arrangement of the vertical posts or vertical portions so that, as the floats twist or take angular relationships other than horizontal in adjusting to the water, the floats will be maintained in their parallel longitudinal axis relative to each other.

The arrangement can be used for various purposes. For

FIGURE 6

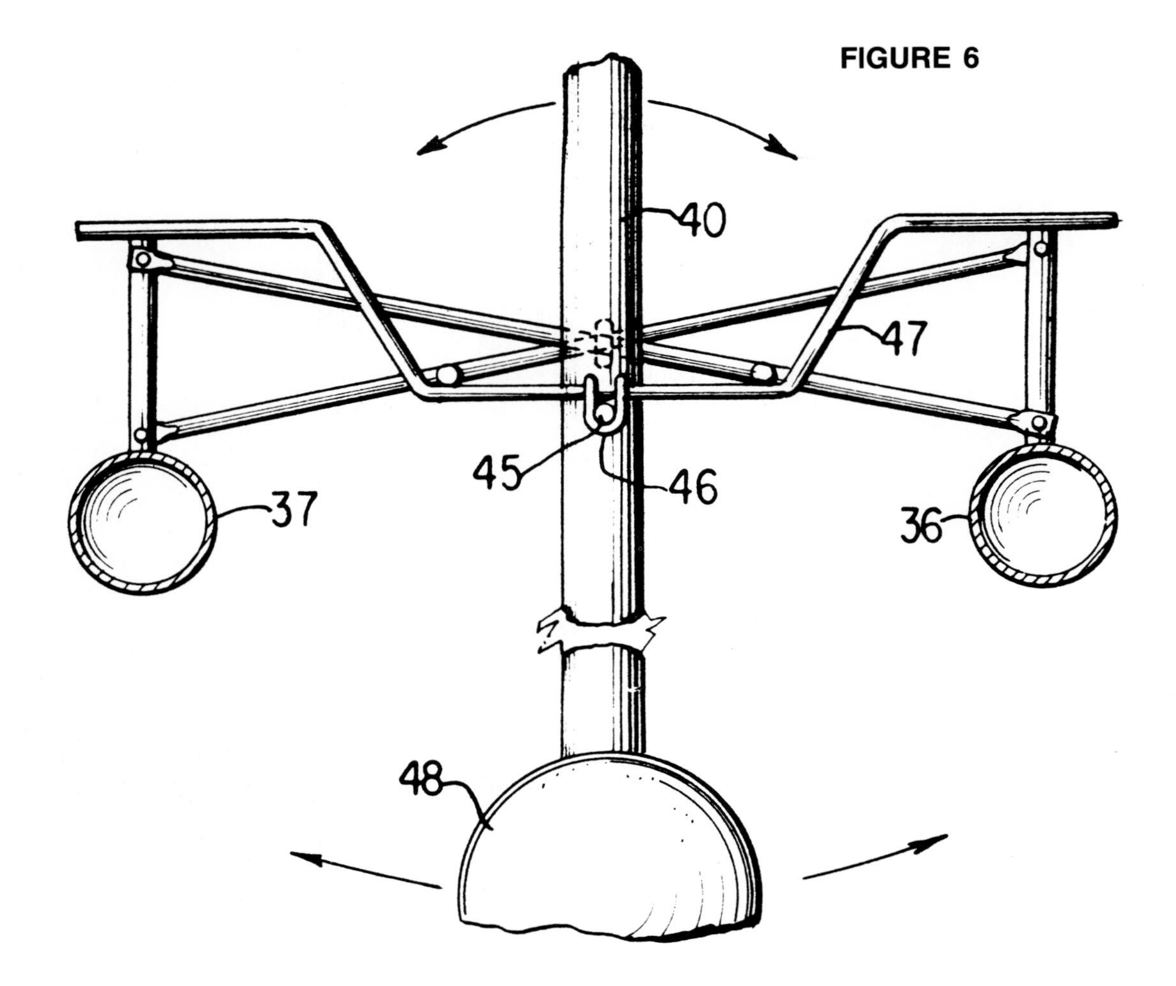

example, track and sliding seat means, similar to those used in a rowing shell, can be placed thereon with the oars and outriggers fastened to a platform arrangement connected between the vertical posts or the floats.

Also, sailboat rigging and gear can be placed on the float member combination with the mast arranged to be pivoted and having a keel at the lower end thereof so that it will always remain vertical. Further, hydrofoil arrangements can be placed at either end of the craft between the two float members.

In a still further aspect, a plurality of units can be connected longitudinally with a flexible connection means between float units so that the float members will in effect be articulated.

These and other objects, features and advantages of the invention will become apparent from the following description and drawings which are merely exemplary.

In the drawings:

Fig. 1 is a perspective view of one form of use of the invention;

Fig. 2 is an enlarged view taken along the line 2—2 of Fig. 1;

Fig. 3 is a view of the invention wherein a plurality of units are joined;

Fig. 4 is an enlarged fragmentary sectional view showing one form of connection for use in Fig. 3;

Fig. 5 is a perspective view of the invention when used as a sailing craft;

Fig. 6 is a fragmentary sectional view of Fig. 5 showing one form of mast arrangement;

Fig. 7 is a reduced size fragmentary side view of Fig. 5; and

Fig. 8 is a sectional view of another form of structural cross member.

The invention will first be described in conjunction with a watercraft for rowing use as seen in Figs. 1 and 2. Where appropriate, similar numerals will be given like parts in the various figures. Float members 10, 11 may be of a suitable material such as aluminum tubing. They may

FIGURE 7

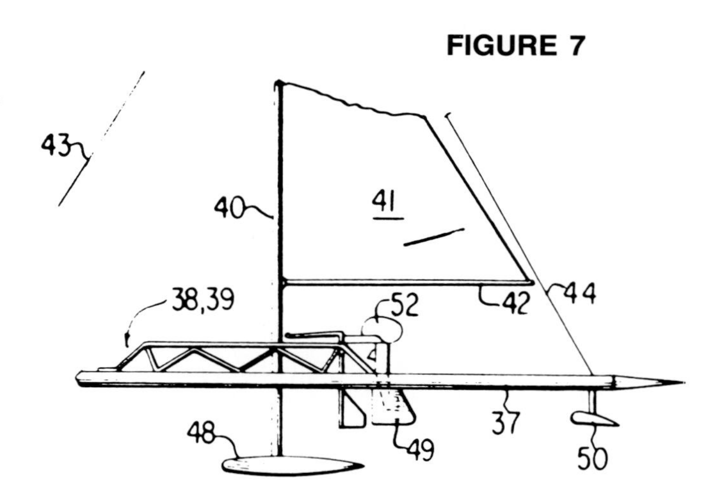

FIGURE 8

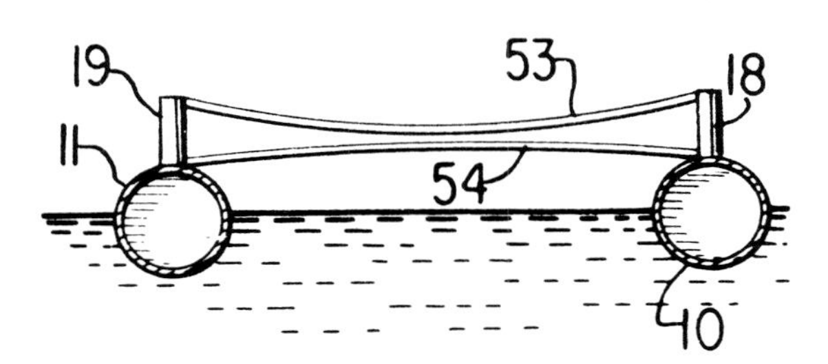

be hollow or may have a foam plastic therein. The leading or bow end of each has rounded heads or plugs 12, 13 and conically shaped trailing plugs 14, 15. The heads and plugs can be given contours to provide the best hydrodynamic effects.

In the form shown in Figs. 1 and 2, the floats 10, 11 are maintained in their desirable substantially parallel relationship by transverse longitudinally spaced structures 16, 17. Vertical posts or members 18, 19, 20, 21 are in spaced pairs and project upwardly from respective floats upon which they are mounted.

From the tops of each post there are cross members 22 extending to the lower part of the opposite post. Preferably a loose link 23 encircles the cross members 22 adjacent the crossing point thereof.

The rowing or shell assembly 24 is carried by side bars 25 which can have truss arrangement 26 therealong. Tracks 27 support sliding seat 28. Outriggers 29 are also carried by the shell assembly, the outriggers having the usual rowlocks 30 mounted thereon.

As the floats tend to follow wave motion, the twisting and movement thereof will be inhibited by compressional and torsional forces exerted on cross members 22 through vertical bars or posts 18.

In another form, a plurality of units 31 may be joined with bow unit 32 and a stern unit 33. The stern unit 33 may have a coxswain seat assembly 34 thereon. The bow and stern units have heads and stern plugs similar to that described for Figs. 1 and 2. The ends of the floats of units 31 can be joined by flexible connectors 35 which can be made of any suitable flexible material. The connector also could be made of a tube.

As can be seen in Figs. 3 and 4, a form similar to a rowing shell arrangement is the result. The flexible connectors give an articulated effect so the watercraft can follow movement of the water. As is known, the stiffness of a long narrow wooden boat, such as a rowing shell, makes it difficult to operate in rough water. The connector can be made of neoprene, nylon, polypropylene and the like.

In a still further form, the float members of the hull can be used to carry a sailing rig such as shown in Figs. 5, 6 and 7. The floats 36, 37 thereof have spaced transverse structural members 38, 39 similar to those previously described. Mast 40 can carry sail 41 and boom 42 in the usual manner. Fore and aft stays 43, 44 can support the mast in a fore and aft direction. The mast is pivotally mounted by pin 45 and yoke 46 to a crossbar 47. Keel or weight 48 at the bottom of the mast 40 will tend to urge the mast to a vertical direction and yet will allow movement thereof in response to wind pressure on the sails. Rudders 49 can be mounted on the assembly. The transverse structural members 38, 39 will permit movement of the parallel floats in a vertical plane because of water movement.

As a further arrangement, hydrofoils 50, 50 can be mounted on cross bars 51 as seen in Figs. 5 and 7. The hydrofoil shaft can be raisable and lowerable as needed. Also, an outboard motor 52 could be attached to suitable brackets on the assembly.

The transverse structural member can take various forms. For example, as shown in Fig. 8, floats 10, 11 could have the vertical posts or members 18, 19 joined by curved tubing means or struts 53, 54. Also, a single bent member (not shown) rising vertically from each float and then extending across could be used. It also is conceivable that one of the transverse members could be adjustable in water so as to apply a braking force by changing the parallelism of the floats.

It should be apparent that details in construction and arrangement of parts could be made without departing from the spirit of the invention except as defined in the appended claims.

23▲HEXA-PENT (1974)

U.S. PATENT—3,810,336

PATENTED—May 14, 1970

POPULAR SCIENCE MAGAZINE ASKED for a design of a geodesic dome that would be tremendously simple for Eskimos to produce and deploy. My partner, Shoji Sadao, developed what we then called the Hexapent Dome, made out of plywood. *Popular Science* published the design and sold drawings for $5 to those who wished to produce the dome. They were soon adopted by Eskimos above the Arctic Circle, and have proved tremendously satisfactory.

FIGURE 1

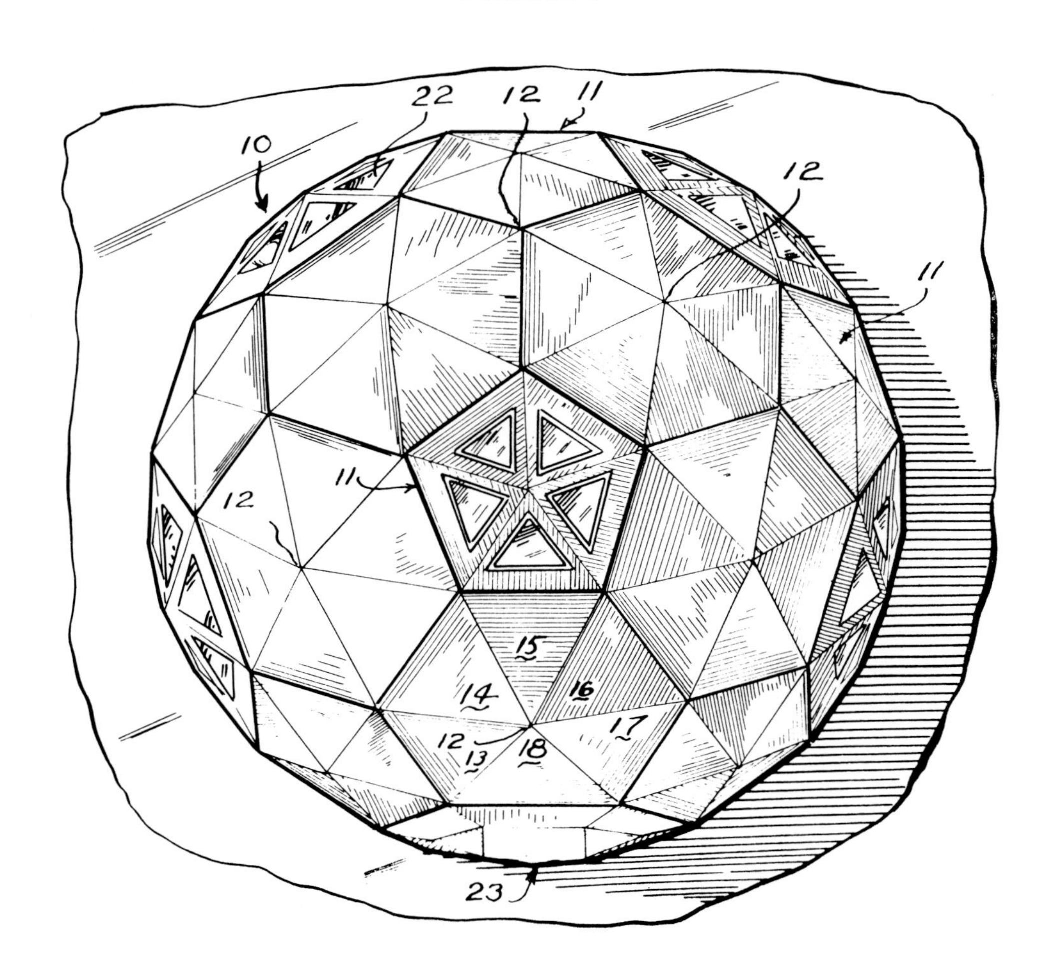

FIGURE 2

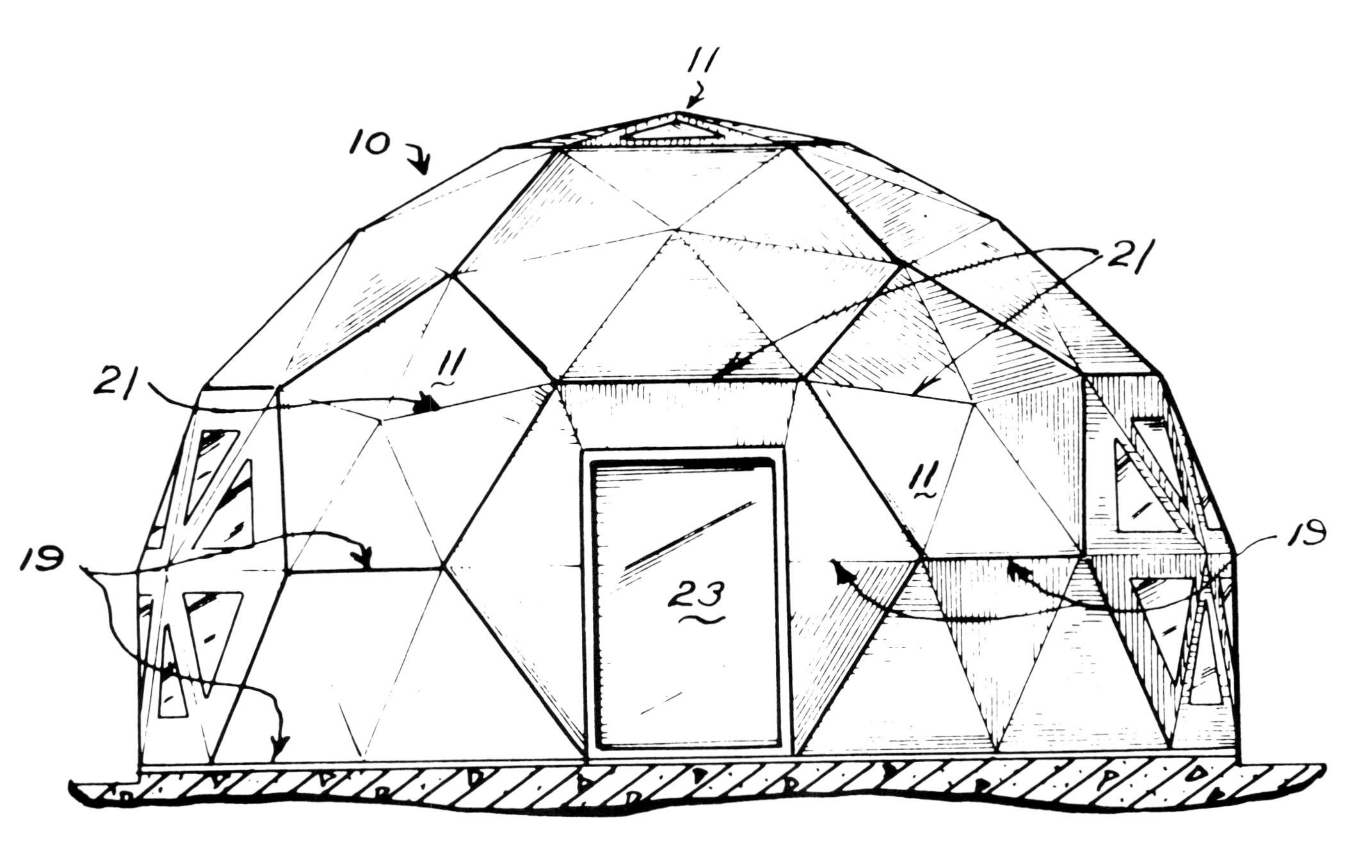

GEODESIC HEXA-PENT*

GEODESIC PENTAGON AND HEXAGON STRUCTURE

This invention relates to building structures of the geodesic-type dome.

It is desirable in building structures and frames to obtain maximum strength and space with a minimum of materials. An example of such is described in U.S. Patent No. 2,682,235 to R. Buckminster Fuller. Geodesic dome structures have been used extensively for various purposes, such as radomes, houses and shelters. Another example of a larger dome is the geodesic dome at the Montreal "Expo '67" exposition.

It is desirable to simplify the structures as much as possible so that individuals can build the same or so as to reduce cost of the materials involved.

One of the objects of the invention is to provide an improved structural arrangement of the modules.

Another of the objects of the invention is to provide a structural arrangement that will not require special modules at the lower or supporting edge.

Still another of the objects is to provide an improved joint arrangement.

In one aspect of the invention, the building framework of generally spherical form has main structural elements connected in a pattern of great circle arcs and lesser circle arcs of at least three frequency. The arcs intersect to form a three-way grid of isosceles triangles, and the modules include hexagons and pentagons. The joints can be secured by using gusset plates that are perpendicular to the radial from the center of the sphere formed by the framework.

These and other objects, advantages and features of the invention should become apparent from the following description and drawings which are merely exemplary.

---

*Inventor: Shoji Sadao, patent assigned to Fuller & Sadao, Inc.

FIGURE 3

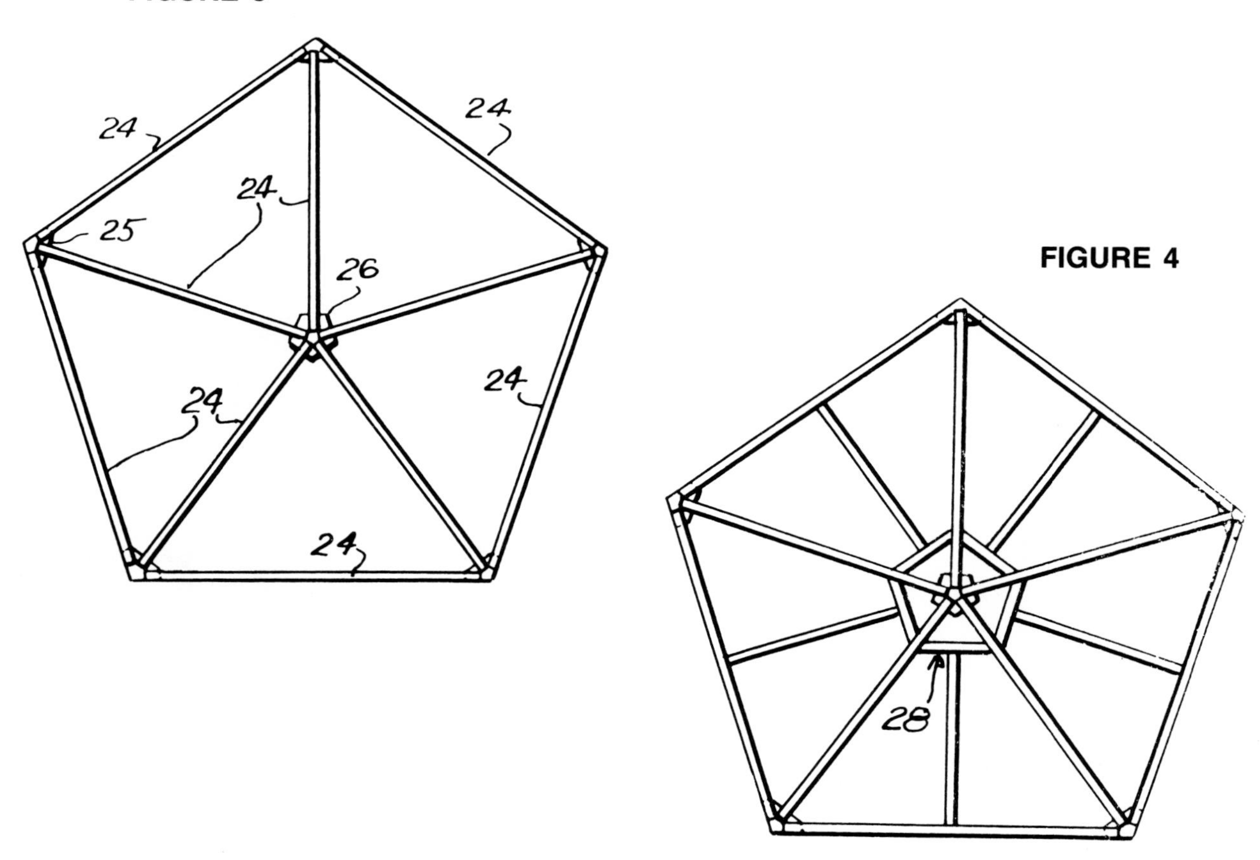

FIGURE 4

FIGURE 5

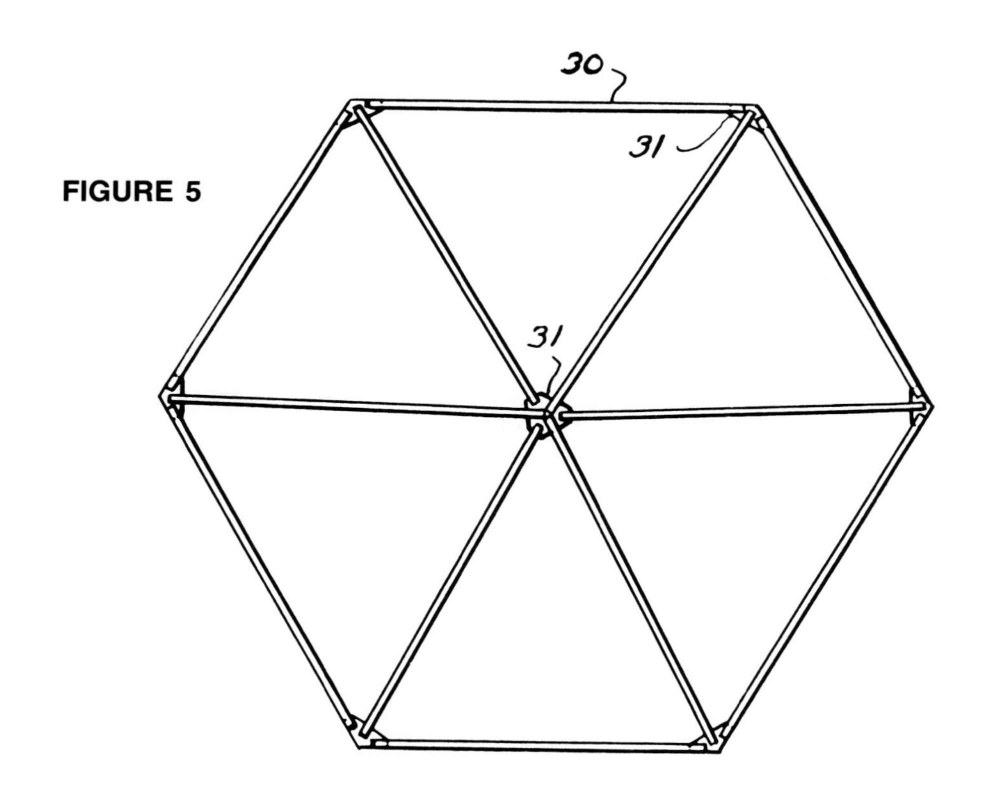

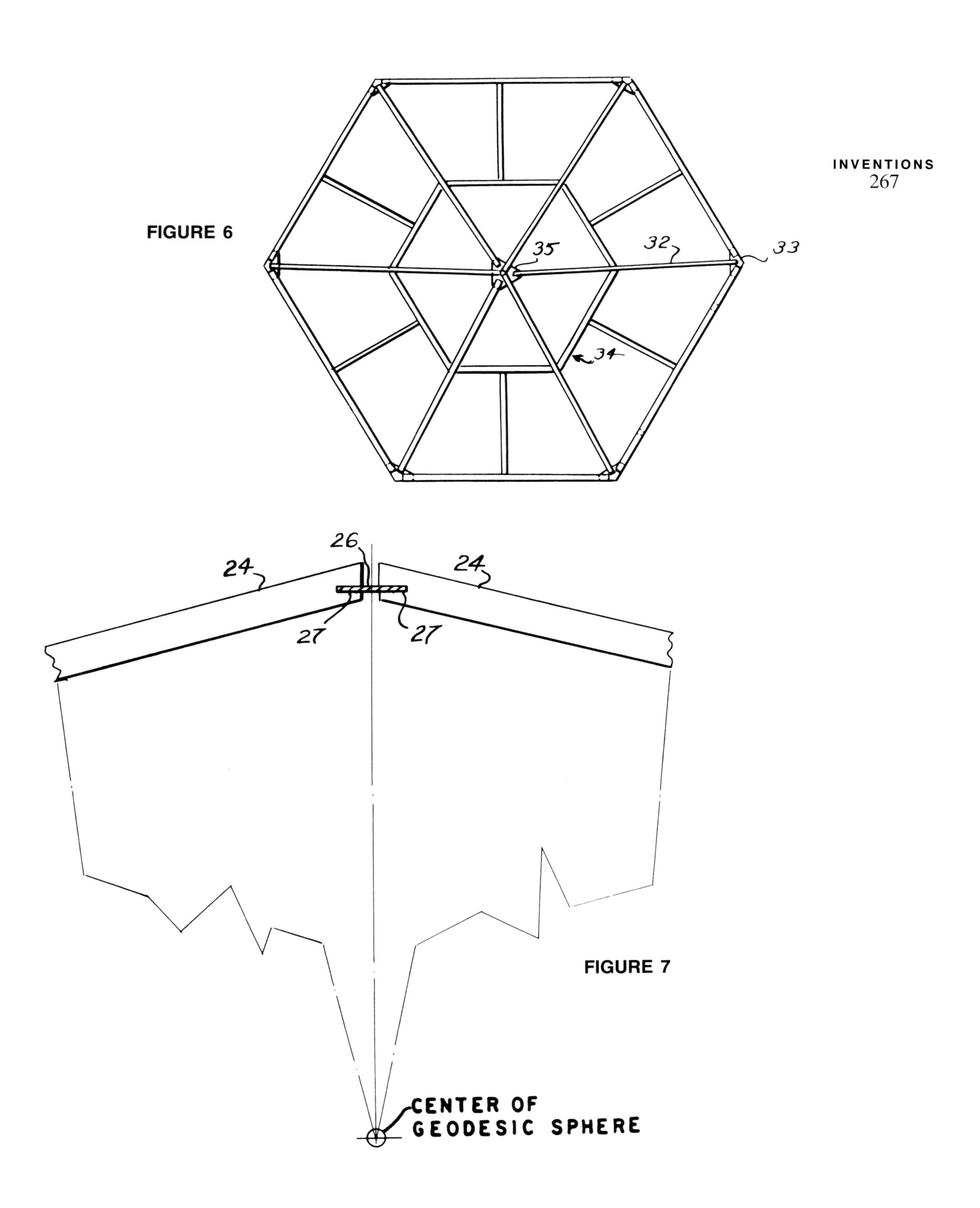
FIGURE 6
35
32
33
34
26
24
24
27
27
FIGURE 7
CENTER OF
GEODESIC SPHERE

In the drawings:

Fig. 1 is a top plan view of a structure utilizing the present invention;

Fig. 2 is an elevational view looking in the direction of line 2—2 of Fig. 1;

Fig. 3 shows the joined struts for a pentagon module;

Fig. 4 is similar to Fig. 3 except there is an additional pentagon adjacent the center;

Fig. 5 shows the joined struts of a hexagon module;

Fig. 6 is similar to Fig. 5 except that there is an additional hexagon adjacent the center; and

Fig. 7 is a schematic view showing an arrangement of the gusset plates relative to the struts and center of the dome.

Where appropriate, like parts will be given the same reference numerals in the various figures.

Referring to Fig. 1, a plurality of modules forms the surface of the structure. The elements or edges of the modules form a substantially spherical icosahedron form structure 10 or portion thereof. The pentagon-shaped modules 11 are at the vertices of the icosahedron-like structure. The hexagon-shaped modules 12 fill in the space between the pentagons.

The hexagon modules 12 are composed of a plurality of isosceles triangles 13, 14, 15, 16, 17 and 18.

As mentioned, the centerlines of some of the struts lie generally in a lesser circle plane such as 19, and others lie generally in a great circle plane 21.

If desired, windows 22 can be placed where needed in various of the modules. A door 23 also can be inserted in one of the modules.

By using great and lesser circle arcs, it is possible to have the bottom edge of the structure terminate in one of the planes as desired without requiring special edge or bottom module construction. As an example, the structure in Fig. 2 shows the cutoff being below the equator of the structure.

A strut framework for the pentagon modules is seen in Fig. 3 having struts 24 of suitable material, such as wood. The ends of the struts are joined by gusset plates 25, 26. As shown schematically in Fig. 7, the flat face or plane of gusset plate 26 is fitted and fastened into slots 27 cut into the ends of the struts 24 and at the angle needed, so that the struts are positioned properly relative to each other. The plane of the various gussets is arranged so as to be substantially perpendicular to the radial to the center of the sphere.

Fig. 4 illustrates a structure wherein there is an internal pentagon 28 to support the edges of the plywood or other covering material when the distance involved is such that one sheet of standard material is not wide enough and two sheets or pieces must be used.

Fig. 5 is similar to Fig. 3 except it shows the arrangement of struts 30 and gusset plates 31 for a hexagon module.

Fig. 6 is similar to Fig. 4 except it shows an internal hexagon 34 with struts 32 and gusset plates 33 for a hexagon module for use where more than one piece of covering material must be used for a facet.

The exterior or edge struts of the various modules can be assembled to appropriate hexagon or pentagon modules by bolts (not shown) or any desired means. The covering facet material may be plywood or a suitable plastic sheet. By employing the hexagon and pentagon as basic modules rather than triangles, fewer bolts and less strut material are needed.

It should be apparent that variations may be made in constructional details and configurations without departing from the spirit of the invention except as defined in the claims.

24▲FLOATABLE BREAKWATER (1975)

U.S. PATENT—3,863,455

PATENTED—FEBRUARY 4, 1975

THERE ARE MANY, MANY ISLANDS around the world that would be happily inhabited had they harbors. It is a question of whether the direction of the wind means that boats left at moorings on what would be the lee side of the island may, a few hours later, be on the windward side. Breakwaters consisting of rock or masonry cost vast amounts of money. I began to realize that the inertia of water itself, if trapped, could serve as a breakwater. In the last few decades, large rubber tubes ten or twenty feet in diameter have been filled with water, and have served very effectively as dams in rivers. I saw that there was a possibility that the mass inertia of trapped-water dams would lift with the approaching waves, and that the elevated water could be made to flow 90° to serve as a propellant of the water's motion to generate power. So I undertook to produce such floating breakwaters and found that they work successfully.

UNITED STATES PATENT OFFICE

FLOATABLE BREAKWATER

The present invention relates to breakwaters and, more particularly, to a floatable breakwater.

The use of breakwaters to shelter selected bodies of water, such as harbors, has been long known. Such breakwaters are generally permanent in construction and are disposed in the water at such a depth so as to have an attenuating action upon the waves. Various forms of floating breakwaters have also been proposed in an attempt to provide a less expensive and relatively temporary breakwater construction. Such floating breakwaters are generally intended to be constructed at a central location and then towed into position at which point they are suitably anchored. However, it has been found that such floating breakwaters are generally unsatisfactory since they are relatively complicated in structure so as to be expensive to manufacture and do not truly fulfill their function of effectively diminishing wave action so as to protect a harbor or the like.

It is known that the vertical motion of a wave is caused by a substantially elliptical movement of particles of water. An effective breakwater, floating or anchored to the bottom, must have a structure which effectively breaks up this elliptical movement of the water particles.

One of the objects of this invention is to provide an improved floatable breakwater.

Another one of the objects of this invention is to provide a floating breakwater which is substantially filled with water which effectively breaks up the wave motion.

FIGURE 1

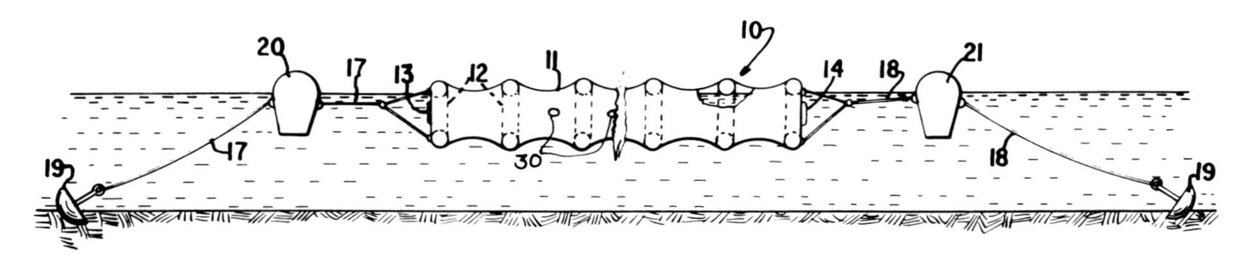

According to one of the aspects of this invention, a floatable breakwater may comprise a flexible open-ended tubular envelope enclosing a plurality of axially spaced buoyant annular members or rings therein. The rings are movable axially with respect to each other while supporting the envelope so that the movement of the envelope creates an accordion effect. Means are connected to the open ends of the envelope for anchoring each end of the envelope. The breakwater is positioned in the water with its longitudinal axis perpendicular to the direction of the wave movement and is floating at such a depth that the breakwater will be substantially filled with water. The filled envelope is expandable and contractable axially because of the relative movement of the rings therein when the breakwater is acted upon by the waves.

The ends of the envelope may be provided with a suitable drawstring-like attachment which can restrict the opening in the ends of the tubular member when it is filled with water and control flow of water therethrough.

Other objects and advantages of the present invention

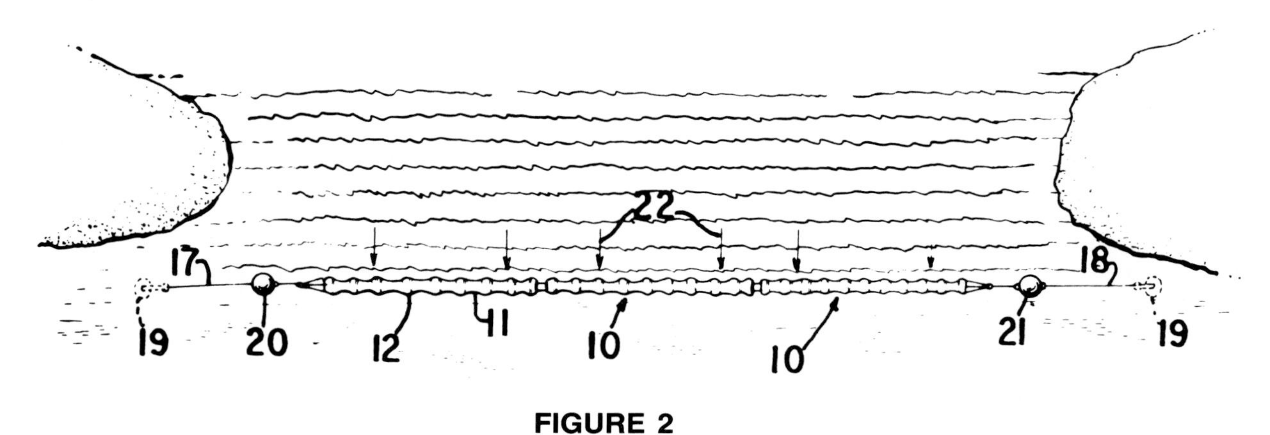

FIGURE 2

FIGURE 3

will become apparent upon reference to the accompanying description and drawings which are merely exemplary.

In the drawings:

Fig. 1 is a longitudinal vertical view of one form of the breakwater made in accordance with the present invention and anchored in position in the water;

Fig. 2 is a top plan view of the breakwater shown in Fig. 1;

Fig. 3 is a fragmentary top view of the breakwater of Figs. 1 and 2;

Fig. 4 is a schematic sectional view taken along the line 4—4 of Fig. 3;

Figs. 5A and 5B are schematic side views of the invention showing the manners of operation;

Fig. 6 is a top plan view of the breakwater showing waves acting on the breakwater in a direction angularly disposed to the central axis;

Fig. 7 is a vertical view of Fig. 6 showing the action of the waves thereon;

Fig. 8 is a vertical view of a breakwater with a portion broken away to show an alternate form of securing means;

Fig. 9 is a sectional view taken along the line 9—9 of Fig. 8; and

Fig. 10 is a perspective view of the form shown in Figs. 8 and 9.

Proceeding next to the drawings wherein like reference symbols indicate the same parts throughout the various views, a specific embodiment of the present invention will be described in detail.

According to the present invention, the floatable breakwater is indicated generally at 10 which comprises a flexible open-ended tubular envelope 11 supported on a plurality of buoyant rings or annular members 12. The envelope 11 is formed of a relatively strong fabric or synthetic plastic material and is mounted upon the rings 12 so that the rings are capable of a relative axial movement between each other. The movement of rings 12 toward and away from each other creates an accordion-like effect in the surface of the envelope.

Rings 12 may be formed of a suitable water-buoyant material or construction and may comprise tubes from the tires of motor vehicles, such as automobiles, trucks or tractors. The material for the envelope 12 may be a transparent plastic material so that the motion of the water within the breakwater can be observed.

The ends 13 and 14 of the tubular envelope are both open, and each end has attached thereto drawstring-like arrangements 15, 16, comprising a valve-like closure with drawstrings 15A, 16A, such as resilient lines adjacent the inner peripheral edge of the closure. As water runs in or out, it will cause the drawstring in the inner periphery of the opening to tend to open, the extent of the opening depending upon the velocity or rate of volume flow. It is also possible to arrange the drawstrings so as to operate in the opposite direction, i.e., to attach the anchor cables

FIGURE 4

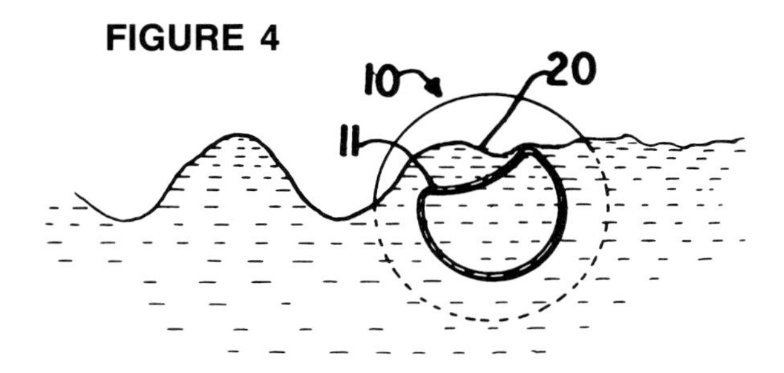

FIGURE 5A

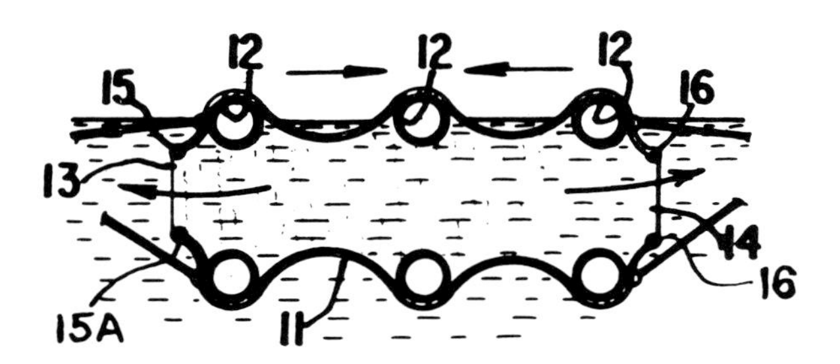

FIGURE 5B

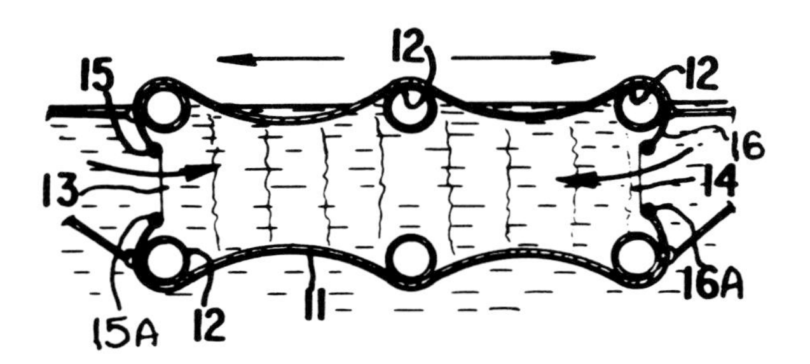

FIGURE 6

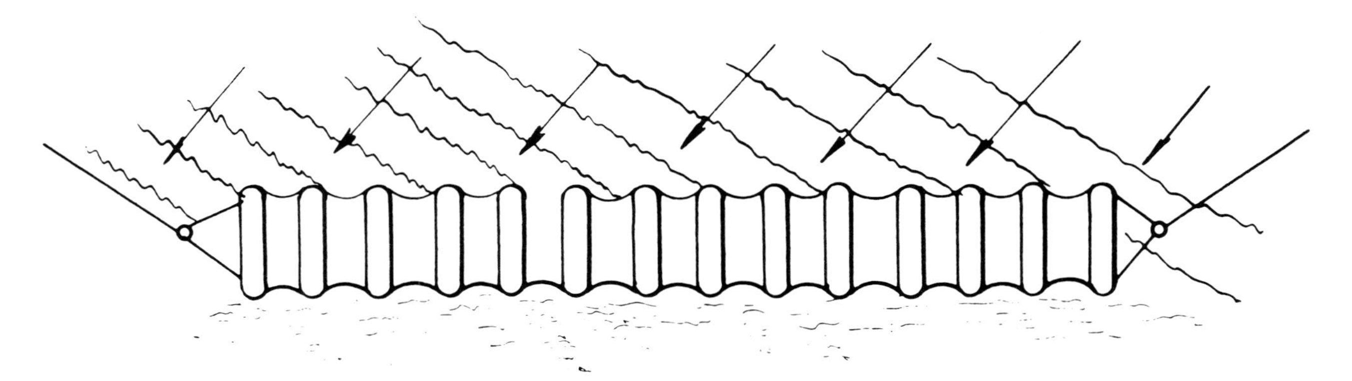

FIGURE 7

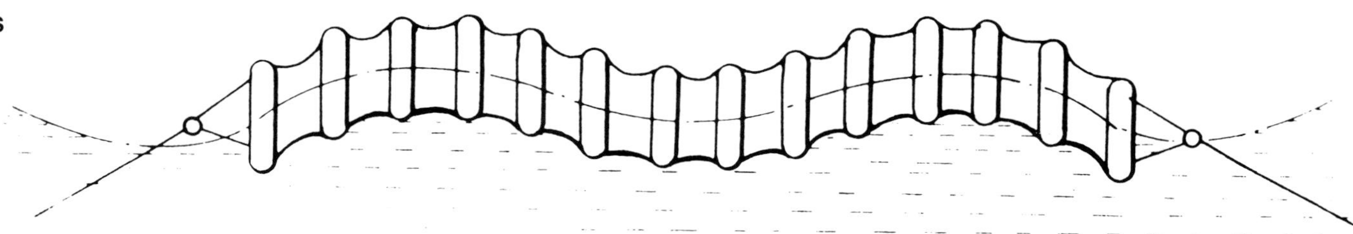

FIGURE 8

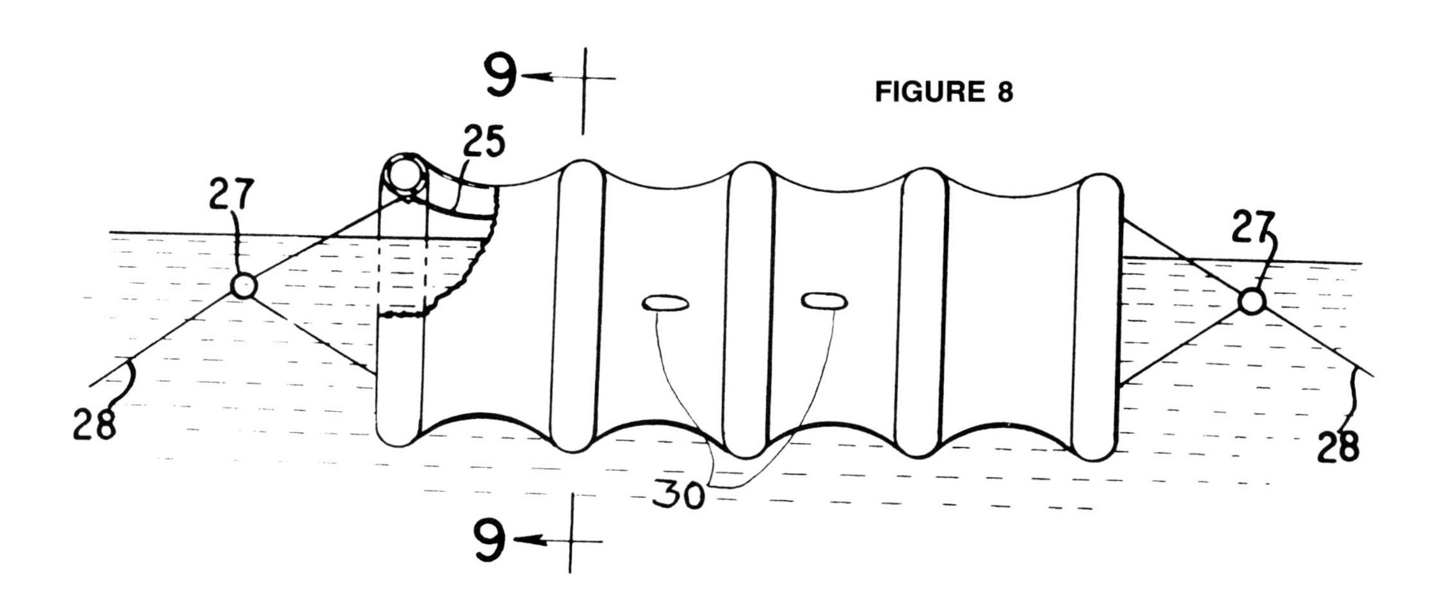

17 or 18 to the drawstring (not shown) so that it will pull closed. Anchoring cable or cables 17 and 18 are connected to the ends of the envelope and to suitable means such as anchors 19 as shown in Fig. 1. It should be evident that by use of an anchor of the type illustrated, that the breakwater can be readily moved to another location.

Suitable water buoys of buoyant material may be attached at 20 and 21. The buoy material may comprise cork or foamed plastic as known in the art.

In operation, the breakwater can be constructed elsewhere and transported to its point of use. The breakwater is anchored within the water so that its longitudinal

FIGURE 9

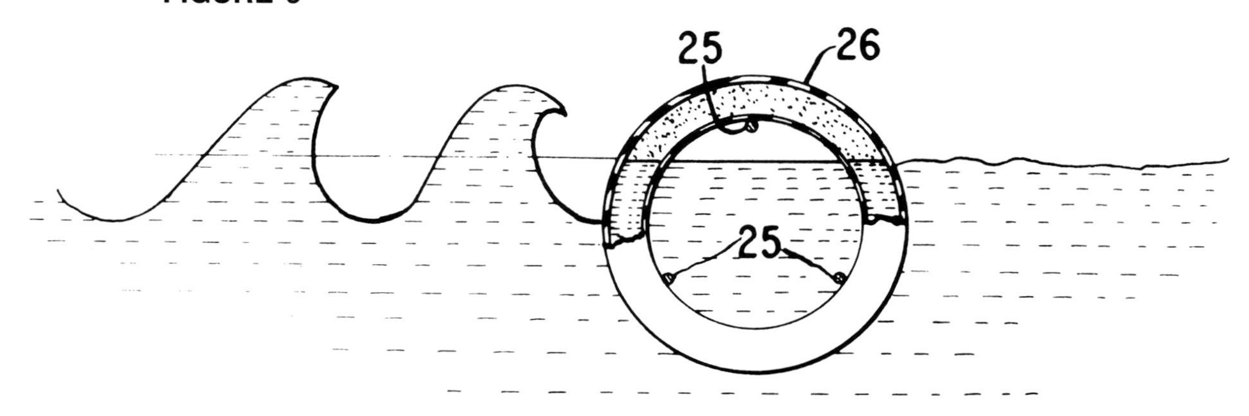

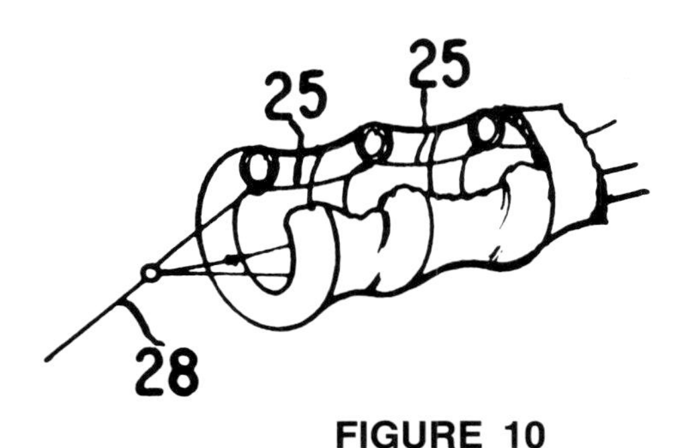

FIGURE 10

axis is perpendicular to the direction of the wave movement as indicated by the arrows 22 in Fig. 2. There is sufficient flotation in the breakwater so that when it becomes substantially filled with water, as can be seen in Fig. 1, it will float with preferably a portion of the breakwater protruding above the surface of the water.

When waves hit the breakwater, they will cause the breakwater to move in the direction of the waves. By positioning the breakwater as shown in Fig. 1, the waves will hit upon the top of the breakwater and the force thereof will be distributed along the entire length of the breakwater.

It has been found that for effective operation, the rings should be positioned a distance apart which is substantially equivalent to the diameter of the respective rings. The breakwater can be made in 50-foot and longer sections, depending upon the intended use.

Fig. 3 illustrates movement of the breakwater in the direction of the waves as it is hit by the waves. Figs. 5A and 5B show the accordion-like movement of the envelope 11.

Fig. 4 illustrates the manner in which a wave at 20 may change the shape of the envelope 11 and cause movement of the breakwater elements in an accordion-like fashion.

It has been found that it may be desirable to fill the tubes about two-thirds full of water so as to provide inertia. Waves of about one-half the overall height of the tube diameter will be broken.

The effect of breaking waves moving at an angle to the axis of the breakwater is shown in Figs. 6 and 7. The inert water mass captured by the sausage-like structure will convert water force into an attempt to lift and move at 180° in the case of Fig. 2 or at an angle (Fig. 6) to the impingement of the waves on the breakwater. The local caterpillarlike lifting of the elements will cause water to surge toward the ends whose flexible openings tend to valve the in and out flow so as to further dissipate energy.

Figs. 8, 9 and 10 show another means of anchoring the breakwater. The lines or cables 25 are fastened at spaced points to the rings 26 and then are led to a common connector 27. Liner or cables 28 then are connected to the anchoring means.

The force of the waves in lifting the large body of water enclosed in the envelope would result in water pouring out of the ends. Thus, the vertical impingement of force would be dissipated at 90° to the impinging waves.

Flap or check valves 30 (Fig. 1) could be used, if desired, to permit water to flow in at the center but prevent water going out at the center. Similarly, flap valves (not shown) would be provided on the outside of the ends which would let water go out easily but would not let water come in except at the center.

It is therefore apparent that the present invention discloses a floatable breakwater which effectively distributes the force of waves along its length and which is resilient so as to be somewhat compressible in response to the force of the waves. A resilient and compressible characteristic of the breakwater effectively dissipates the wave force while at the same time permits construction of the breakwater to be relatively light.

It will be understood that various details of construction and arrangement of parts may be made without departing from the spirit of the invention except as defined in the claims.

25▲NON-SYMMETRICAL TENSEGRITY (1975)

U.S. PATENT—3,866,366

PATENTED—FEBRUARY 18, 1975

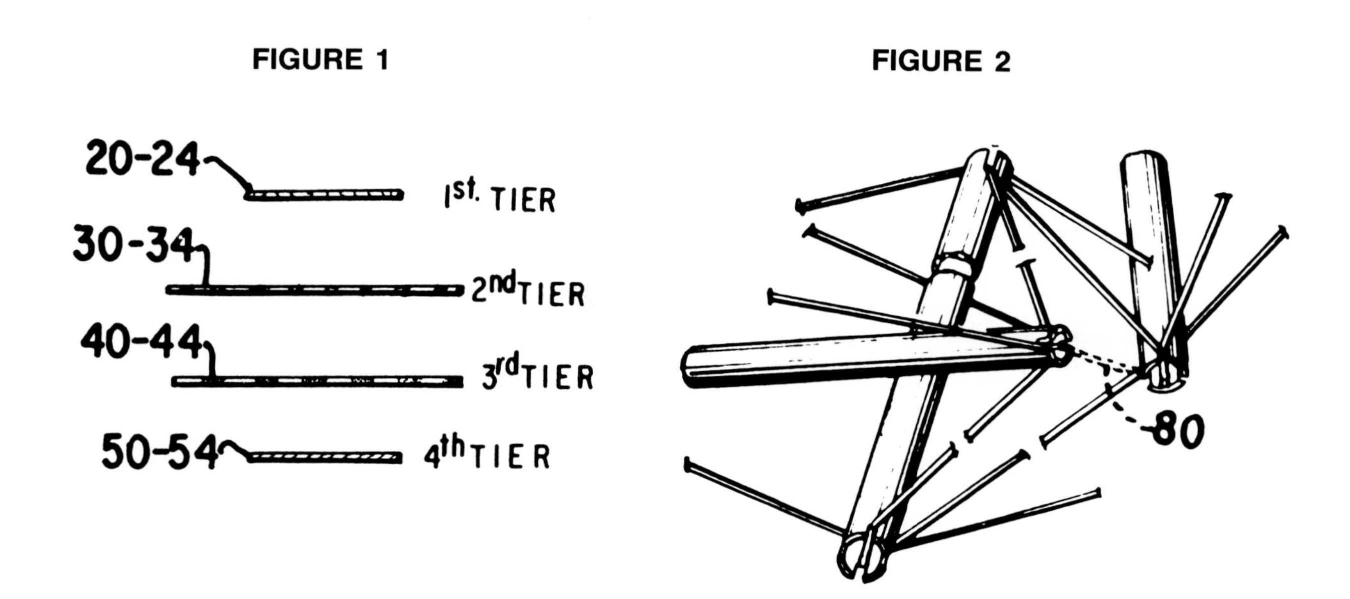

I RECEIVED A COMMISSION in 1969 from St. Peter's College in Oxford, England, to design a theater. They wished to have a portable theater, and I found that such a portable theater called for a more or less cylindrical form with proper doming and proper bowing of the bottom for the audience and stage. Because the portable theater also needed to be very light, I wished to use tensegrity structuring, which meant that the tensegrity structuring would not be symmetrically formed as I had been using it in the geodesic dome. It would be what I called asymmetrical. I found it quite possible to make the asymmetrical tensegrity.

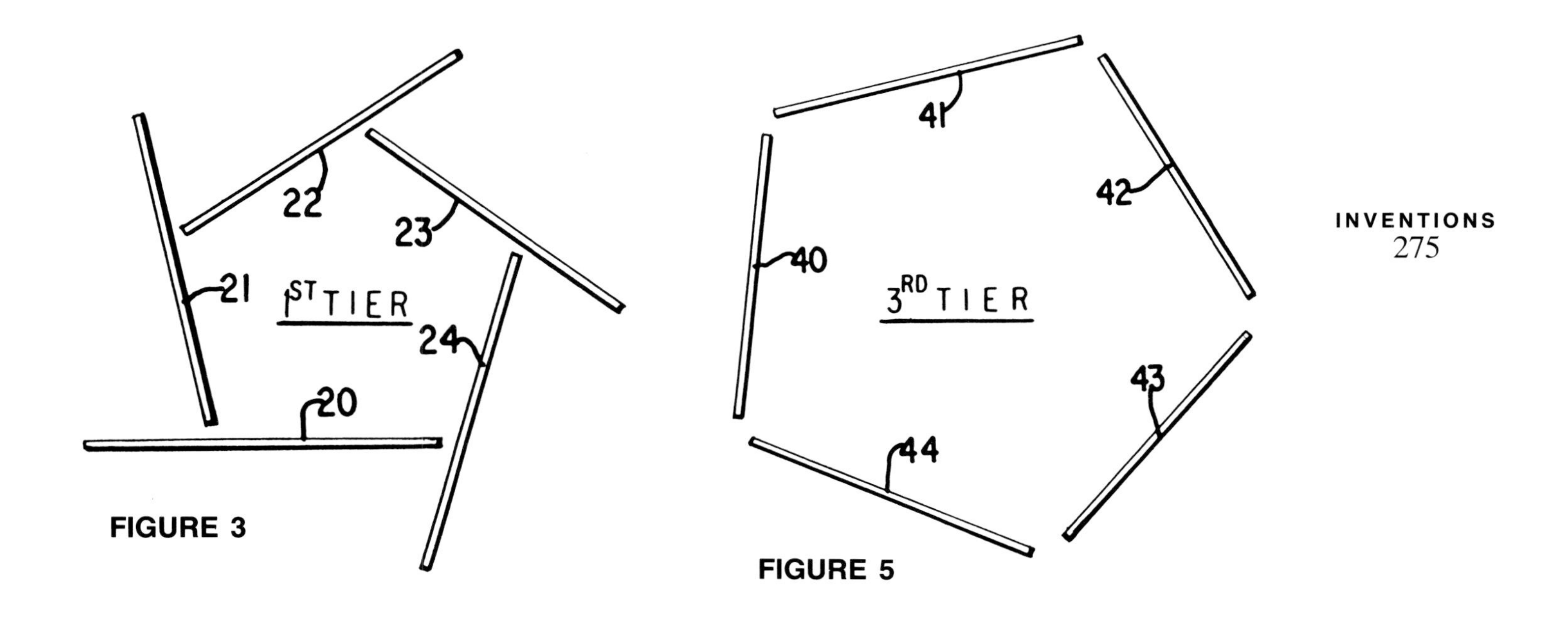

FIGURE 3

FIGURE 5

FIGURE 4

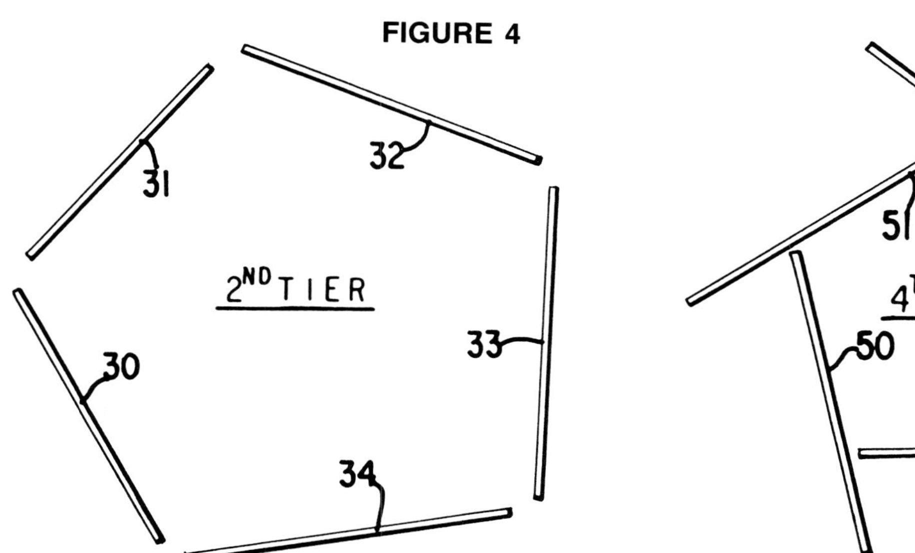

FIGURE 6

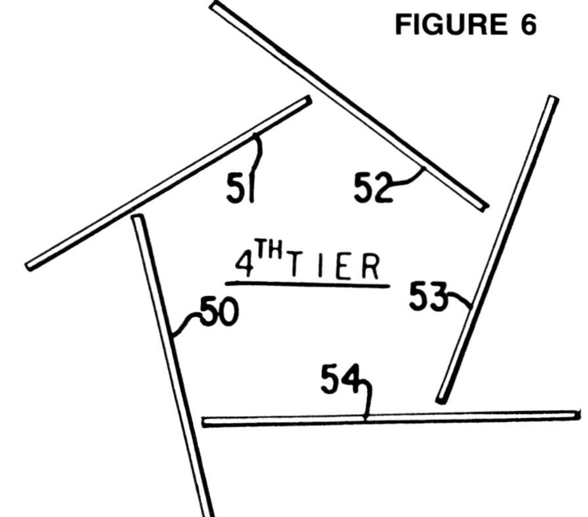

UNITED STATES PATENT OFFICE

NON-SYMMETRICAL TENSION-INTEGRITY STRUCTURES

This invention relates to building structures and particularly to one of the tension-integrity types of structure.

The present invention is an improvement on and a variation of tension-integrity structures, such as shown in prior U.S. Patent No. 3,063,521, Nov. 13, 1962. The structures involved are known generally as geodesic-type dome structures. The tension-integrity structure is one that is of generally spherical form having discontinuous compression columns joined with tension elements in a manner to provide the aspect of discontinuous compression and continuous tension sometimes referred to as "Tensegrity" structures. In some instances, it has been found desirable in a structure of the type involved herein to have rectangular-like areas or zones for windows, walls, doors, or the like.

One of the objects of the present invention is to provide an improved tension-integrity type structure having zones of rectangular-like configuration.

Another of the objects of the invention is to provide a tension-integrity type of structure wherein a portion can be prefabricated.

In one aspect of the invention and in the form shown, the generally spherical-like building structure can have a substructure at the top and a substructure at the bottom,

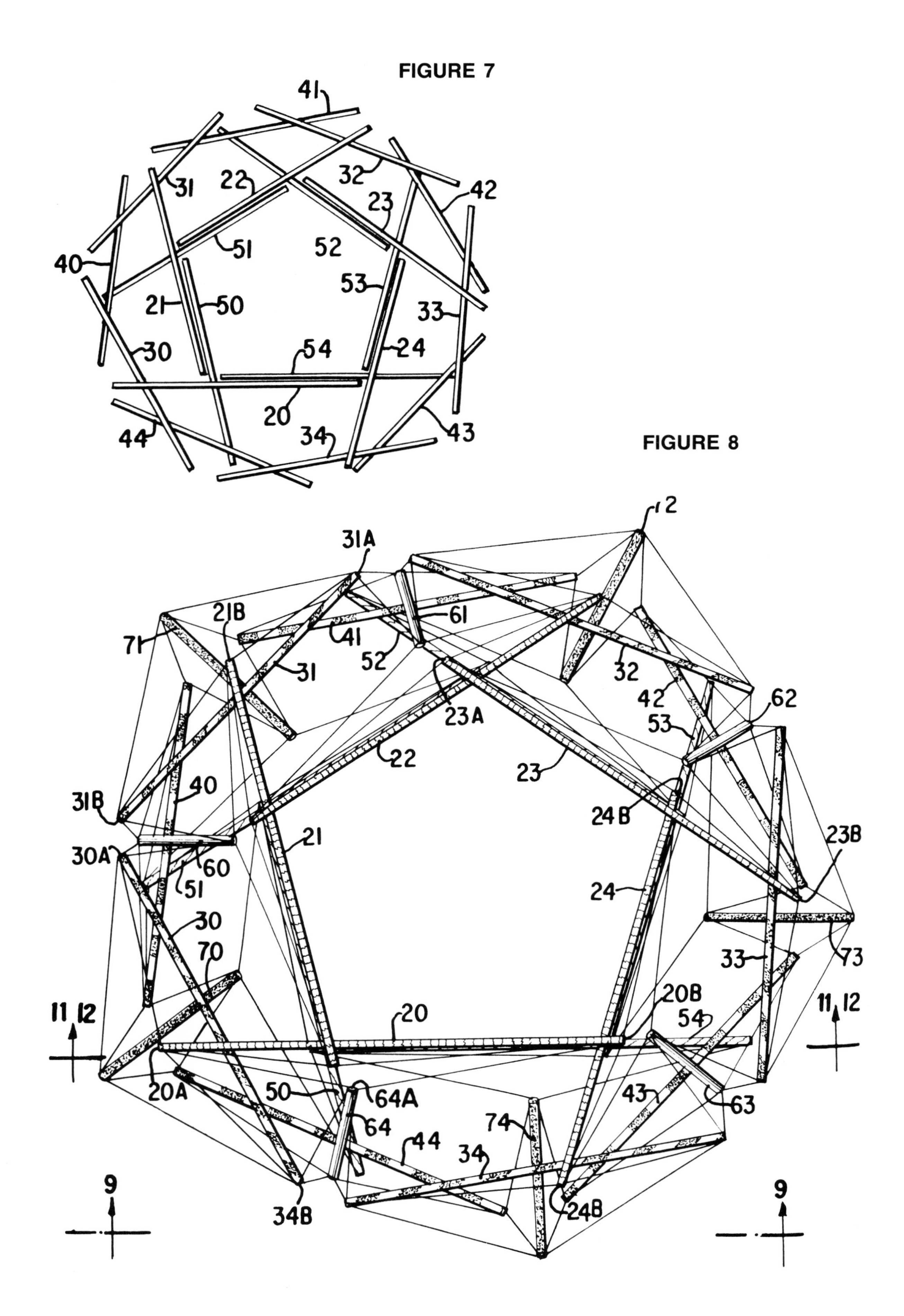
FIGURE 7
41
31
22
32
42
23
40
51
52
53
33
21
50
30
54
24
20
44
34
43
FIGURE 8
12
31A
21B
71
41
61
52
31
32
42
62
23A
53
22
23
31B
40
24B
21
30A
60
51
23B
24
30
70
73
33
20
20B
54
11 12
11 12
20A
50
64A
43
63
64
74
44
34
9
9
34B
24B

the bases of the substructures being spaced from each other, the bases being of pentagonal configuration which are twisted relative to each other in a manner to provide rectangular-like facets therebetween. The substructures and spacing of the pentagonal configuration, which are lesser circles as compared with great circles, are formed by a plurality of column-like compression members joined by a plurality of tension elements, such as wire or rope, the column-like members being axially spaced relative to each other by the tension elements which are attached near the ends of spaced column-like members. The pentagon of the top substructure above the aforementioned lesser circle has five column-like members forming a generally pentagonal configuration. The bottom substructure has five column-like members in a pentagonal configuration and below its lesser circle which is the reverse of the top pentagon. Column-like members extend between the two substructures and are joined thereto with tension elements, the reversal of the top and bottom pentagons resulting in the twisting of the lesser circle pentagons relative to each other. It is to be understood that various frequencies can be used and various types of enclosures or panels employed.

Other objects, advantages and features of the present invention will become apparent from the accompanying description and drawings, which are merely exemplary.

In the drawings:

Fig. 1 represents four tiers, the top two representing the top substructure and the lower two representing the lower substructure;

Fig. 2 is a fragmentary perspective of a column-like member with four tension elements attached adjacent the ends thereof for attachment to other column-like members;

Fig. 3 is a schematic representation of the relationship of the column-like members of the top pentagonal configuration;

Fig. 4 is a schematic representation of the upper substructure pentagonal lesser circle or base;

Fig. 5 is a schematic representation of the lower substructure pentagonal lesser circle or base showing the same in its rotated position relative to the lesser circle pentagon of Fig. 4;

Fig. 6 is a schematic representation of the column-like members of the bottom pentagonal configuration which is similar to Fig. 3 but is in a reversed relationship; Fig. 7 is a top view showing generally the manner in which the column-like members appear;

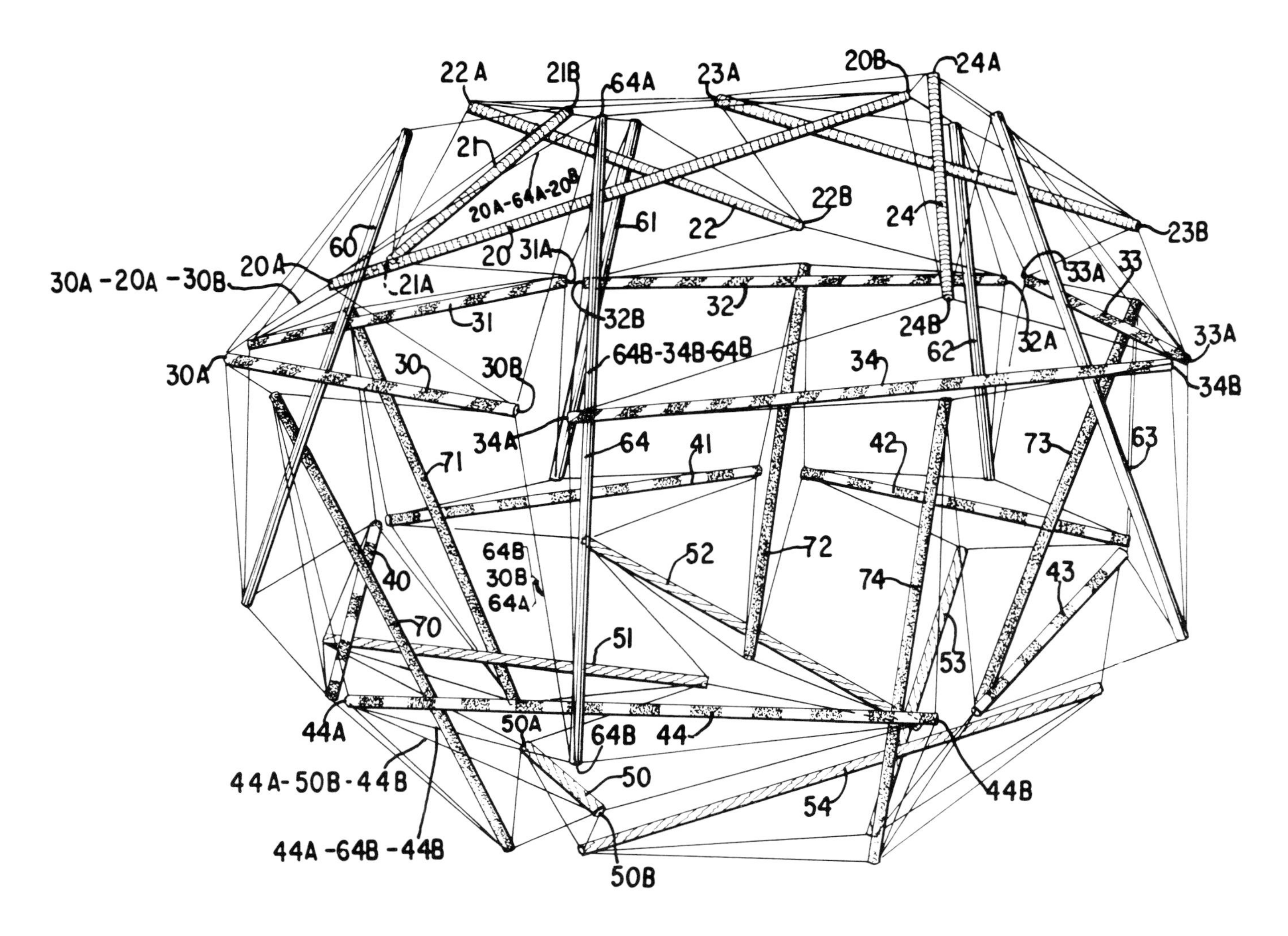

FIGURE 9

FIGURE 10

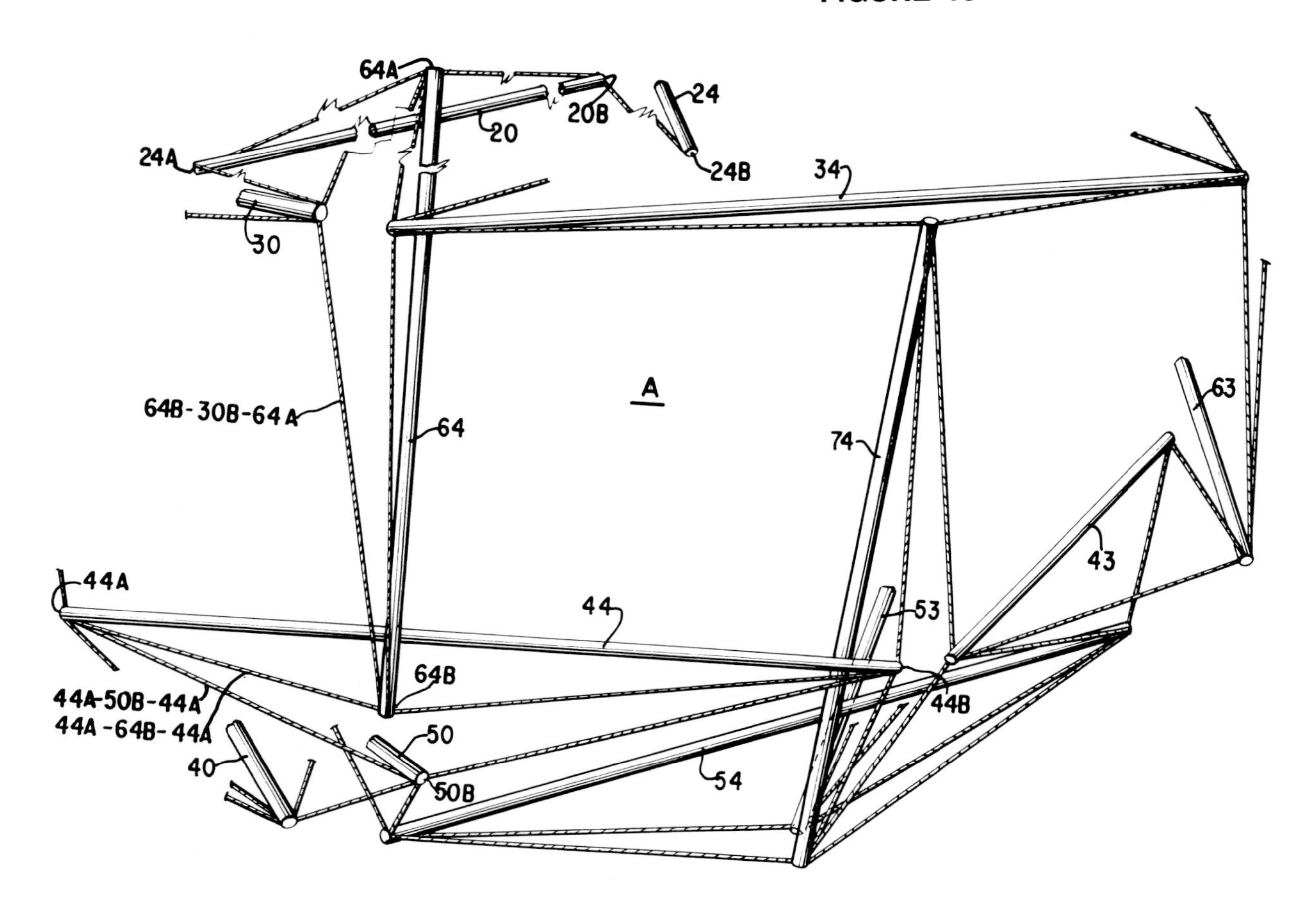

Fig. 8 is a top view generally similar to Fig. 7 but showing the tension elements between the column-like members;

Fig. 9 is a perspective side view of Fig. 8 taken in the general direction 9—9 of Fig. 8;

Fig. 10 is an enlarged fragmentary view of the upper and lower vertical columns together with related columns and tension elements, there being five of each around the periphery;

Fig. 11 is a fragmentary view of the top substructure looking generally in the direction 11—11 of Fig. 8; and

Fig. 12 is a fragmentary view of the lower substructure looking generally in the direction 12—12 of Fig. 8.

Proceeding next to the drawings wherein like reference symbols indicate the same parts throughout the various views, a specific embodiment of the present invention will be described in detail. In the interest of simplicity, reference symbols have not been included in all of the figures.

The column-like members can be made of wood, aluminum, plastic or any suitable material. The ends of the column-like members can be notched or otherwise fitted to hold the tension elements or wires at the ends of said column-like members. Each column-like member will have four tension elements emanating from each end thereof as shown in Figs. 2 and 10.

In Figs. 3, 4, 5 and 6, for purposes of description, the general relation of the column-like members before they are joined together with the tension elements is shown. As can be seen, the top pentagonal configuration, illustrated in Fig. 3, is oriented in a clockwise configuration as compared with the counterclockwise configuration of the bottom pentagonal configuration shown in Fig. 6. As will be explained hereafter, the upper substructure base lesser circle pentagon configuration of Fig. 4 is twisted in the final assembled structure relative to the lower base lesser circle pentagon configuration of Fig. 5.

Fig. 7 illustrates the relation of the members in the final assembled structure, the elements of Figs. 3 to 6 being combined therein.

Referring now particularly to Figs. 8 and 9, the top pentagonal configuration will be described. Column-like members 20, 21, 22, 23 and 24 are attached or supported by tension elements to other of the column-like members so as to be axially spaced therefrom. For example, column-like member 20 has a tension element 20A–64A–20B leading from end 20A to an end 64A of vertical column-like member 64 (Fig. 9) and thence to end 20B. Additionally, a tension element 30A–20A–30B from end 20A to end 30A and end 30B of the column-like member 30 of the lesser circle pentagon of the top substructure.

Vertically extending column-like members 60, 61, 62, 63, 64 extend between tension elements fastened to the column-like members 40, 41, 42, 43 and 44 of the lower substructure lesser circle pentagon and tension members between their opposite ends and column-like members of the top pentagon configuration column-like members 20, 21, 22, 23, 24.

Similarly, vertically extending column-like members 70, 71, 72, 73, 74 extend from tension elements attached

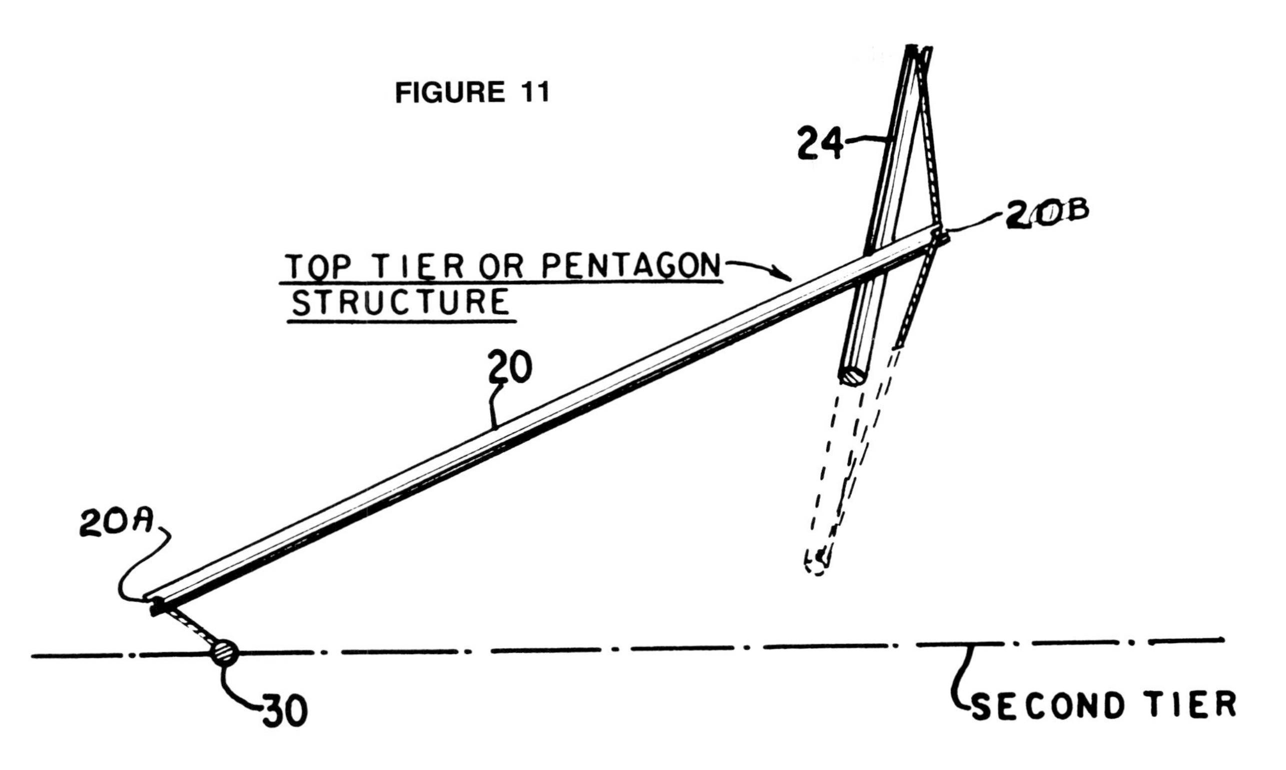

to the ends of the upper lesser circle pentagon members 30, 31, 32, 33, 34 and tension members attached to column-like members 50, 51, 52, 53, and 54 of the lower pentagonal configuration.

The lower pentagonal configuration is similar to the top configuration except that it is reversed, as can be seen in Figs. 3 and 6.

In Fig. 10, the column-like members are numbered so as to agree with Figs. 8 and 9. One set of members will be described, the others being similar thereto. Upwardly extending member 64 has end 64B connected to member 44 of the lower lesser circle pentagon by tension element 44A–64B–44B. End 64B of member 64 is also connected to end 30B of upper lesser circle pentagon member 30 and

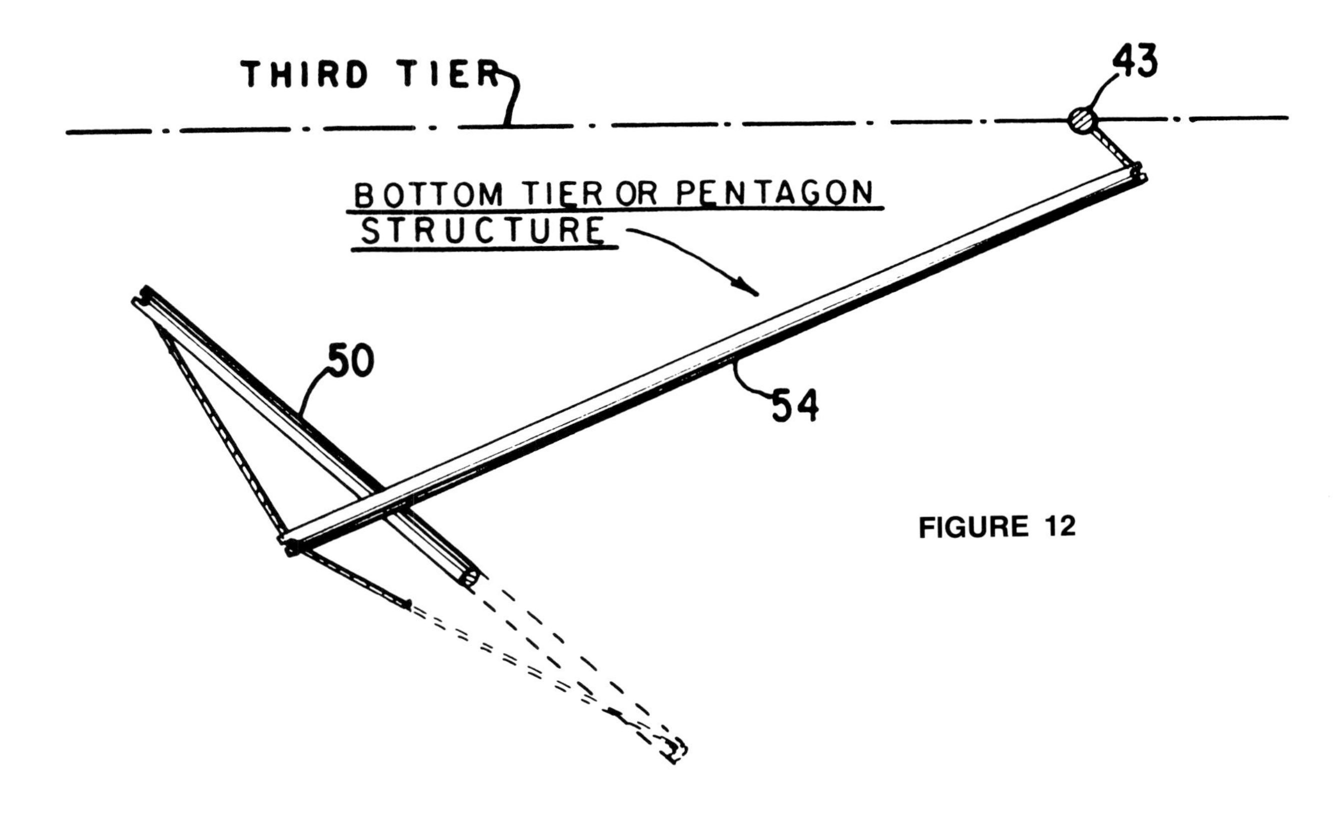

end 64A by tension element 64B–30B–64A as well as to lesser circle member 34 by tension element 64B–30B–64A. The upper end 64A of 64 also is connected to the ends of the upper pentagonal configuration member 20 by tension element 20A–64A–20B. The end 50B of member 50 connects to member 44 by tension element 44A–50B–44B. There are five lower and five upper vertical column-like members repeated in a similar manner. The term "vertical" means generally vertical when the structure is oriented as shown.

It can be seen that the column-like compression members are held in axially spaced relation to each other by the tension elements.

As a result of the combination of column-like elements and tension elements shown and described, there will be rectangular-like facets such as depicted at A (Fig. 10) between the upper lesser circle including elements 30, 31, 32, 33, 34 and the lower lesser circle formed by elements 40, 41, 42, 43, 44.

Figs. 11 and 12 show the relation of 3 of the elements of the top tier and of the bottom tier.

As seen in Fig. 2, a loose connecting wire or elements 80 can be used so as to limit outward movement.

Merely by way of example, the structure of the present invention can be formed of thirty aluminum struts or column-like members, each 36¾ inches long and 1 inch outside diameter with one-fourth inch slots formed in each end thereof for receiving the tension elements. The tension elements can be made of synthetic fiber line having a length of about 37½ inches between ends of the column-like member, there being a distance of about 5 inches from the middle point of the tension element to the column-like member when the line is stretched tight and engages other elements of the combination. It is understood that these dimensions will vary according to material of the line or rope and other dimensions of the structure. Also, the tension elements combination can be prefabricated and the column-like members inserted therein when and where the structure is to be erected.

It will be understood that various details of construction and arrangement of parts may be changed without departing from the spirit of the invention except as defined in the claims.

26▲FLOATING BREAKWATER (1979)

U.S. PATENT—4,136,994

PATENTED—JANUARY 30, 1979

MY FIRST FLOATING BREAKWATERS were very severe on the materials used, especially if they were pressured, so my second patent was designed to reduce those pressures without, in any way, losing the principle of waterpower generation, as well as the interference with the onshore wind and waves protection. This second system employs the inherent strength of inter-hinged boom triangles and also employs the distributed lifting strains of the enormous masses of water which are progressively elevated. The elevated water is then directed at right angles to the waves to pass through turbines, thus generating power.

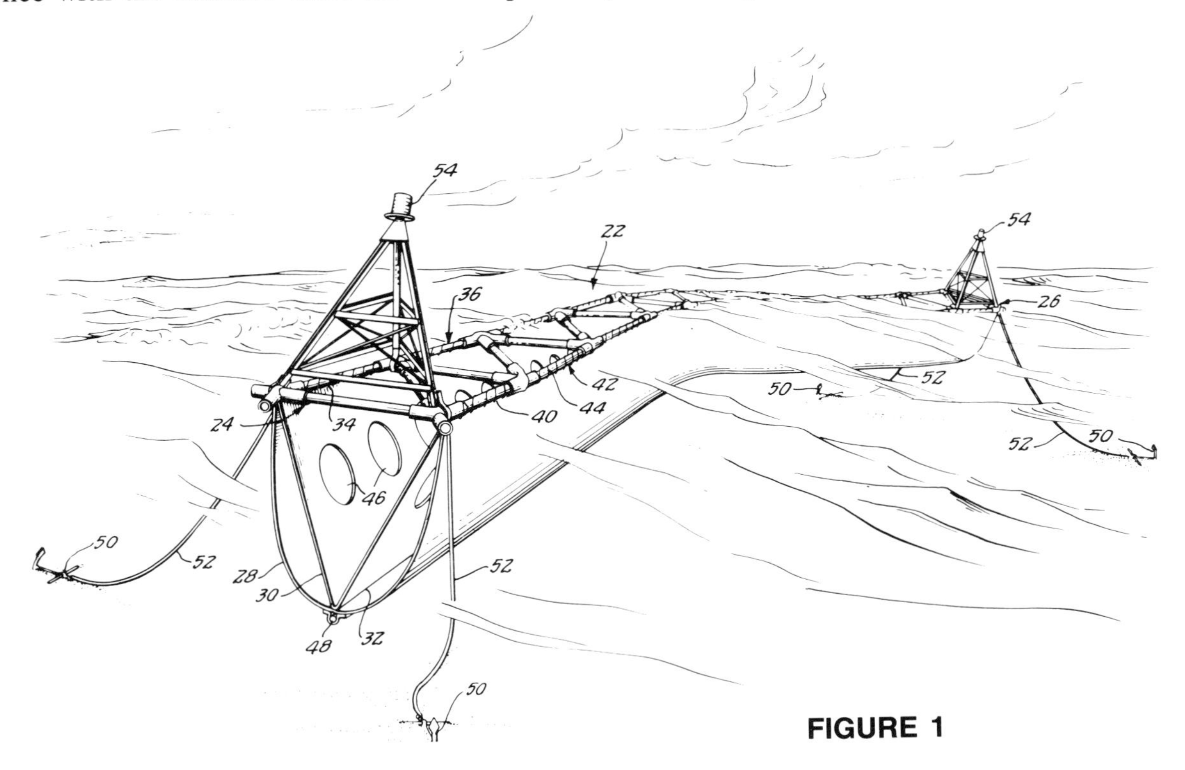

FIGURE 1

UNITED STATES PATENT OFFICE

FLOATING BREAKWATER DESCRIPTION

This invention relates to breakwaters and more particularly to breakwaters of the floatable type.

The use of breakwaters to dissipate wave energy and to thus protect shorelines is well known. Although past breakwaters are generally solid structures, portable breakwaters are also known. Of the portable breakwaters, the most efficient, in terms of effective use of materials and ease of transport, are those which are floatable and use the action of the wave to dissipate itself. An example of such a breakwater is found in my prior U.S. Patent No. 3,863,455. In that patent, I disclosed a flexible tubular breakwater which floats at the water surface. A wave lifts the tubular structure as well as water therein and causes a flow of water within the tube. The energy required to lift and to move the water within the tube is supplied by the wave, and thus the energy of the wave is dissipated and reduced.

An object of this invention is to provide a portable breakwater which makes efficient use of materials.

A further object is to provide a breakwater which is easily transported, assembled and installed.

Yet another object of this invention is to provide a portable breakwater which effectively dissipates wave energy by churning water contained therein through internal ports.

A still further object of this invention is to provide a breakwater which is adaptable for use with a turbine to generate usable power from the wave energy.

According to the present invention in one of its aspects, a floatable breakwater comprises a floatable open grid with a troughlike membrane suspended below the grid. Inner membrane components which are also suspended from the floatable grid serve to define various ports or apertures through which the water is moved and churned as the breakwater is lifted by a passing wave and thereby dissipate energy from the wave.

According to other aspects of the invention, these inner membrane components may have a truncated tetrahedral configuration converging downwardly in the manner of multiple funnels such that water contained within the respective tetrahedron is ejected downwardly into the outer troughlike membrane when each passing wave momentarily elevates the breakwater.

In accordance with the present invention in certain of its aspects, a turbine may be positioned at one end of the

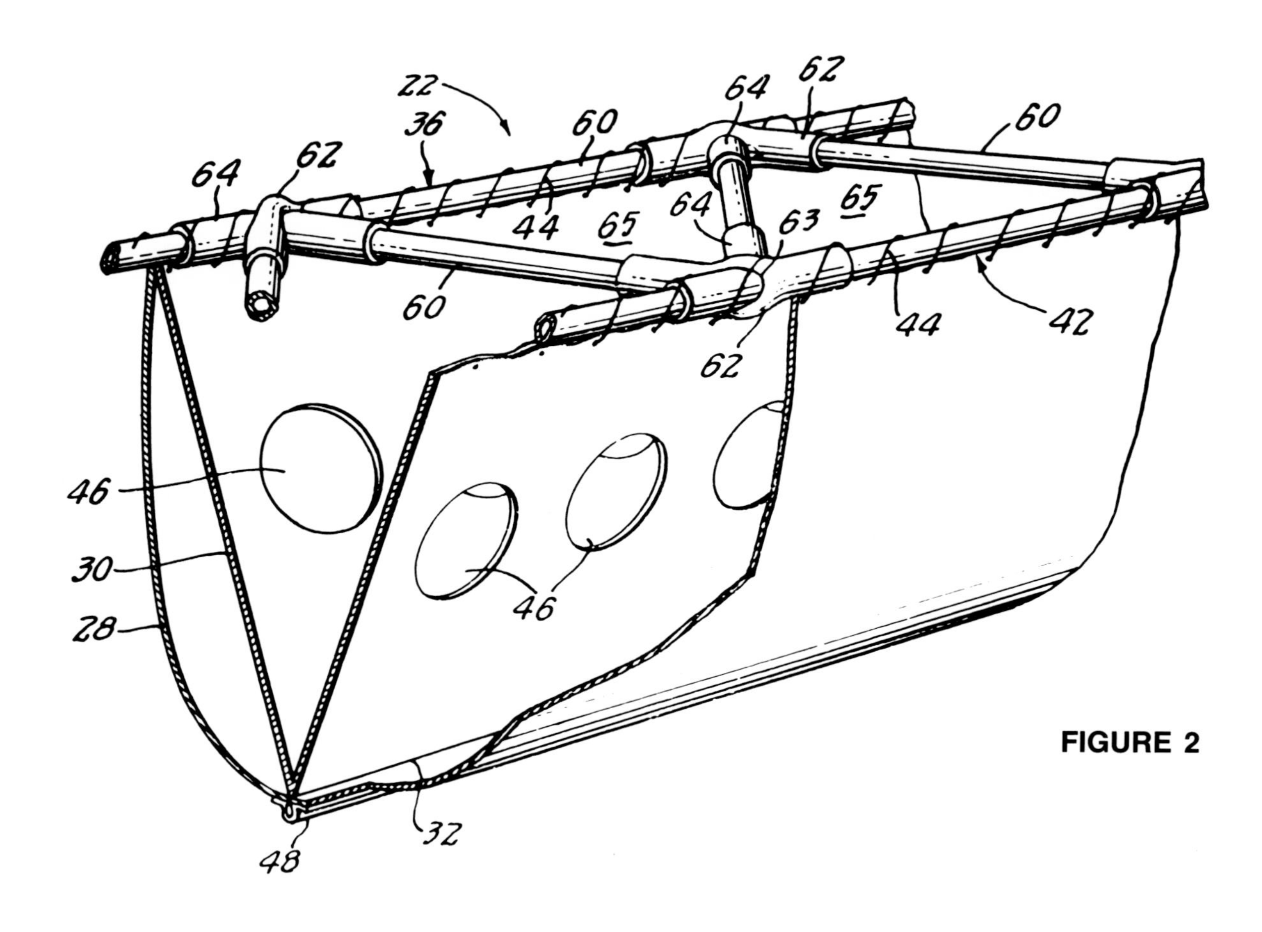

FIGURE 2

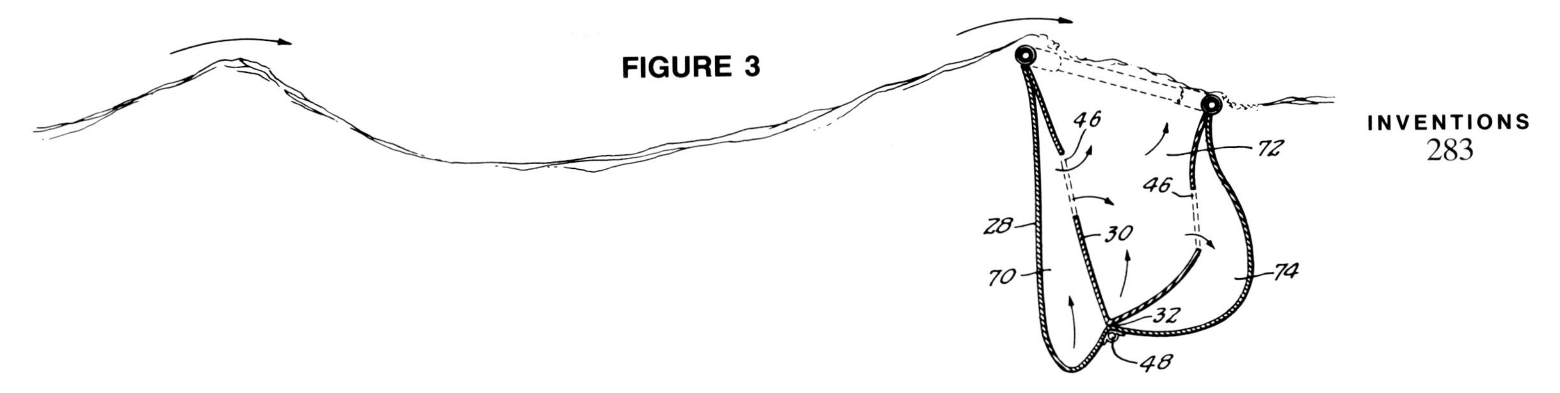

FIGURE 3

breakwater for converting the energy of the longitudinal flow of water out of the end of the troughlike membrane into electric power.

BRIEF DESCRIPTION OF THE DRAWINGS

The foregoing and other objects, features, and advantages of the invention will become more fully understood from the following description of preferred embodiments of the invention, as illustrated in the accompanying drawings in which like reference characters refer to the same parts throughout the different views. The drawings are not necessarily to scale, emphasis instead being placed upon illustrating the principles of the invention.

Fig. 1 is a perspective view of a preferred embodiment of the floating breakwater of this invention;

Fig. 2 is a perspective view of a section of the breakwater of Fig. 1, partially broken away;

Fig. 3 is a sectional view of the breakwater of Fig. 1 demonstrating the effect of wave action on the breakwater;

Fig. 4 is a plan view of an alternative embodiment of the invention in which the inner membrane components define truncated tetrahedral compartments;

Fig. 5 is a perspective view, similar to Fig. 2, of the alternative embodiment of Fig. 4;

Fig. 6 is an isometric view, partially broken away, of yet another embodiment of the invention;

Fig. 7 is a plan view of the embodiment of Fig. 6;

Fig. 8 is a sectional view of the embodiment of Figs. 6 and 7 taken along lines 8—8 in Fig. 7;

Fig. 9 is a side view, partially in section, of the embodiment of Fig. 6 having a turbine positioned at one end for utilizing the longitudinal flow of water out of this end of the breakwater.

DESCRIPTION OF PREFERRED EMBODIMENTS OF THE INVENTION

Referring to Fig. 1, the breakwater comprises a floating grid structure 22 having ends 24 and 26. Additional grid structures may be added on to either or both ends 24 and 26 to extend the breakwater.

A first flexible generally U-shaped membrane 28 is suspended below the floatable grid structure 22, and a second inner flexible membrane 30 is suspended below the grid structure within the first membrane. These two membranes 28 and 30 are joined along a line 32 which extends along the center bottom of the breakwater. They may be joined continuously along this line 32 or at selected points spaced therealong. The upper edges 34 of the membranes 28 and 30 are secured to a membrane support portion 36 of the grid structure which extends longitudinally along a first side of the breakwater between its ends 24 and 26. The other upper edges 40 of these membranes are secured to a second support portion 42 of the grid structure extending longitudinally along the second side of the breakwater. The membranes are secured to these support portions of the grid 22 by binding lines 44 which are shown helically wrapped around the respective support portions.

The second membrane 30 is perforated with holes 46 shown as ports spaced therealong to permit flow of water therethrough. A line of weights 48 extends the length of the membranes. This line of weights 48 is located along the center bottom of the breakwater in the manner of a keel weight and, by weighting down both membranes 28 and 30, retains them in the breakwater configuration

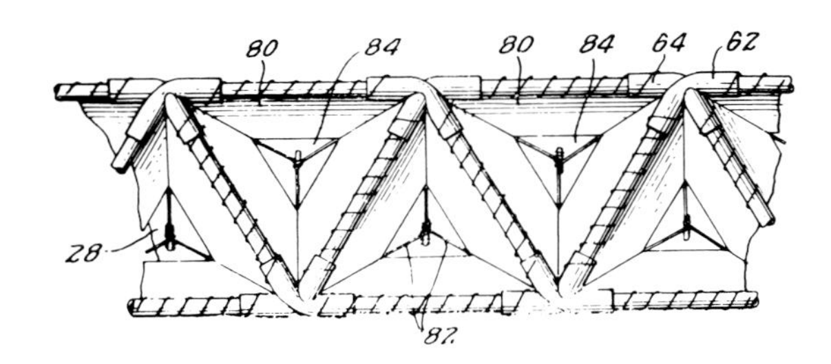

FIGURE 4

shown. Because the surface width of membrane 30 is substantially less than the width of membrane 28, membrane 30 is pulled taut by weights 48 into a V-shape while membrane 28 assumes a more relaxed position defining a generally U-shaped trough.

The breakwater is moored in position by anchors 50 fixed to the breakwater by anchor lines 52 which are long enough to permit some movement of the breakwater. Lightweight towers with warning lights 54 ward off any boats in the area.

The grid structure 22 is best shown in Fig. 2. The grid comprises a plurality of rigid tubular flotation segments 60 which are hingedly connected at their ends to form a line of alternately inverted equilateral triangles. These rigid segments 60 are shown as being each of the same length and serve as floatable booms. At each nodal point, four of these segments 60 are joined by two flexible sleeve-like connectors 62 and 64. These connectors 62 and 64 are generally tubular, and each tubular connector 62 includes a transverse opening 63 therethrough for retaining its companion connector 64.

FIGURE 5

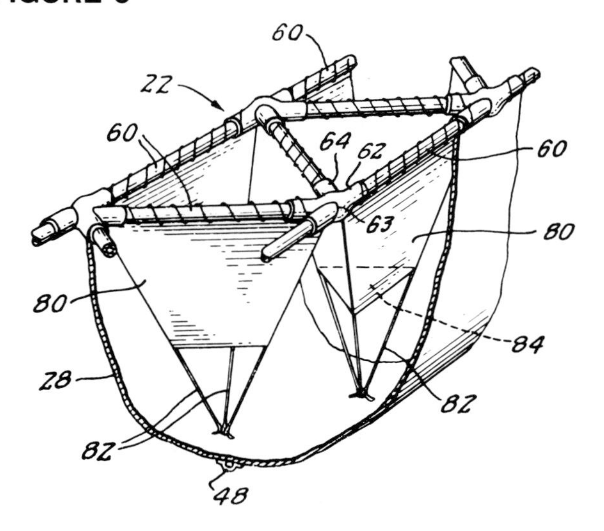

As seen most clearly in Fig. 2 the tubular connector 64 extends through the transverse opening 63 in the other tubular connector 62. These tubular connectors are formed of flexible material such as rubber, tough plastic, and the like. They in effect serve as hinge connectors between the respective boom segments 60. The triangular arrangement of these boom segments 60 provides an overall stiffness to the grid structure 22 in the plane of each resultant equilateral triangle 65. In other words, this grid structure 22 acts as a truss for resisting lateral deflection or bending in the plane of the triangles 65 except to the extent that the flexible tubular connectors 62 and 64 may permit some flexibility. However, these hinged interconnections permit the grid structure 22 to deflect or bend readily in directions perpendicular to the planes of the successive equilateral triangles 65, thereby providing an articulated grid which can be heaved up and twisted in response to wave movements, as illustrated in Figs. 1 and 3.

Boom segments 60 may be sealed containers having a gas therein or may contain floatable foam or other buoyant filler. Membranes 28 and 30 are preferably formed of tough, strong fabric, such as canvas but may be formed of strong plastic or the like. The membranes 28 and 30 extend down such that the line of keel weights 48 are positioned at depth in the range from approximately one and a half to approximately twice the horizontal distance between the grid support portions 36 and 42.

The operation of the breakwater in dissipating wave energy can best be understood by reference to Fig. 3. With the floatable grid riding the waves, wave energy is dissipated as the breakwater, and thus the water therein, is lifted and churned by each wave. Due to the law of conservation of energy, as the potential energy of the breakwater and the water contained therein is increased by its elevation to a higher position, the kinetic energy at the source, that is of each wave, becomes decreased. The water lifted by the outer trough-like membrane 28 and by the inner V-shaped membrane 30 is surged and churned through the respective ports 46. Additional water may be poured from the crest of the dissipated wave down into the trough-like breakwater itself. When the breakwater returns again to its lower potential, the energy will not, for the most part, be returned to the original wave action but will be dissipated as thermal energy and as wave action across and against the wave motion.

As the grid flexes with wave action and one portion of the breakwater rises in potential with respect to another, fluid flow along the length of the breakwater is initiated. Again due to the conservation of energy, the wave energy is decreased by the kinetic energy of the fluid flow along the breakwater as well as by the thermal losses resulting from friction between the fluid and the membranes.

The present invention further utilizes the phenomenon that, due to thermal losses, a substantial amount of energy is extracted from a liquid as it passes through a restriction such as an orifice. When one support portion of the grid is raised above the other by the wave action as in Fig. 3, the volume of the region or space 70 between the outer and inner membrane components 28 and 30 is reduced, forcing water through port holes 46 in membrane 30 into the region or space 72 and then through other port holes 46 into the region or space 74. Hence, additional energy is absorbed as the water is churned back and forth through the apertures or port holes 46.

By virtue of the fact that the grid 22 is open, the breaking wave does not break completely over the breakwater but may break into it, as illustrated in Fig. 3. The water entering through the grid then forces water already in space 72 out through holes 46 into the spaces 70 and 74. Thus, water in all three spaces is forced toward the open ends of the breakwater. Because the energy required to move the water longitudinally along the length of the breakwater and to churn water through the holes 46 is taken from the wave, the wave energy is substantially dissipated.

The dimensions of the breakwater are of some importance. Because the breakwater is most efficient when it

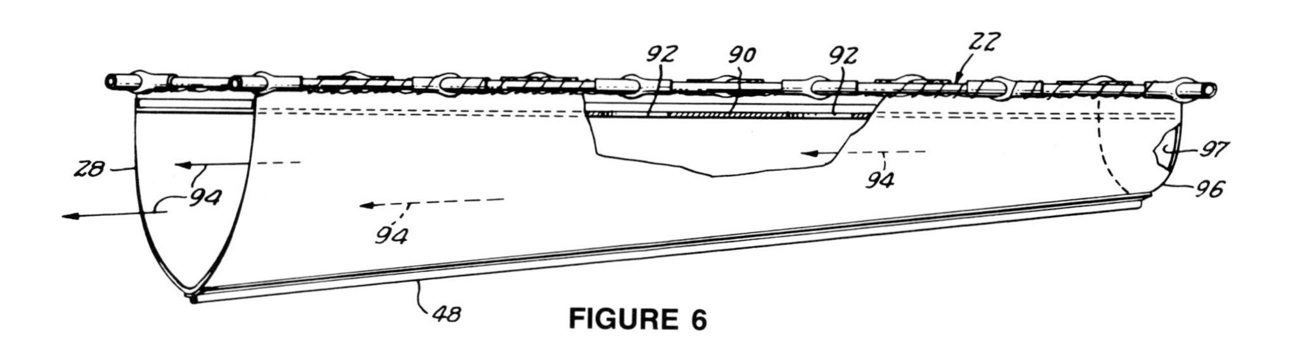

FIGURE 6

reaches below the elliptical motion of the water molecules comprising the wave into stable water, its depth should be at least approximately twice the height of the largest wave intended to be effectively intercepted. It is also important that the membrane support portions 36 and 42 extending along opposite sides of the grid structure be close enough to each other that they will both rise, i.e. become elevated, on the largest wave to be effectively encountered. If the grid were wider, water would be jostled from one space to another but, because one portion would be moving upward while another portion moves downward, there would be little overall lifting of the water enclosed by the breakwater; thus, energy dissipation due to the lifting phenomenon would be reduced. As larger waves are the most troublesome and damaging to shorelines, the dimensions of the breakwater should be set to make it most efficient for those waves.

An alternative embodiment of the breakwater is shown in Figs. 4 and 5. In that embodiment, the second membrane components 80 are suspended below each equilateral triangle section of the grid. Each membrane component 80 defines a truncated tetrahedron converging

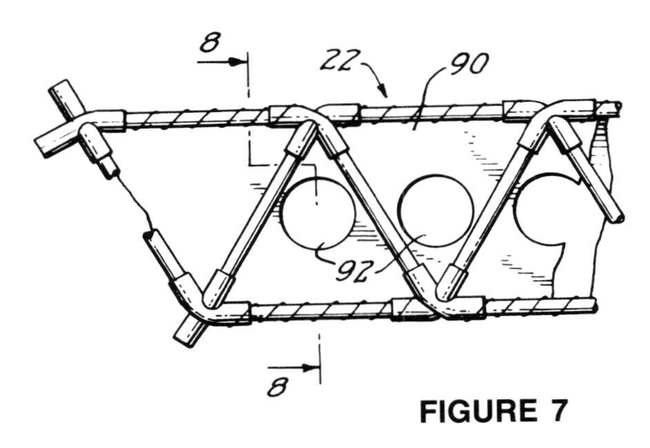

FIGURE 7

downwardly when maintained taut by three lines 82 extending down as the apex of each tetrahedron. The port or aperture 84 formed at the bottom of each truncated tetrahedron serves the purpose of churning the water within the breakwater for dissipating energy. Water forced through ports 84 as the breakwater rises and falls will take energy from the wave as will water which breaks into the tetrahedron and flows through these ports. In other words, the water from the crests of waves breaking into the open grid structure 22 enters the open-top tetrahedral components 80 and funnels out of the ports 84.

This second embodiment as shown in Figs. 4 and 5 also has the advantage that multiple flow paths are set up in and around and in between the respective tetrahedra resulting in multiple eddy currents which extract energy from the waves.

Another embodiment of the invention is shown in Figs. 6–9. In this embodiment, the second membrane 30 is removed and a membrane or flexible sheet 90 is extended horizontally below the grid. Ports 92 are formed in this sheet 90 to permit waves to break into the breakwater. Because the second membrane 30 is not present in this embodiment to shape the first membrane 28, some alternative must be used if it is desired to hold the membrane 28 in a generally U-shaped configuration. For example, the membrane 28 may be shaped by stiffly flexible U-shaped ribs, such as wire ribs, sewn into pockets in the fabric.

FIGURE 8

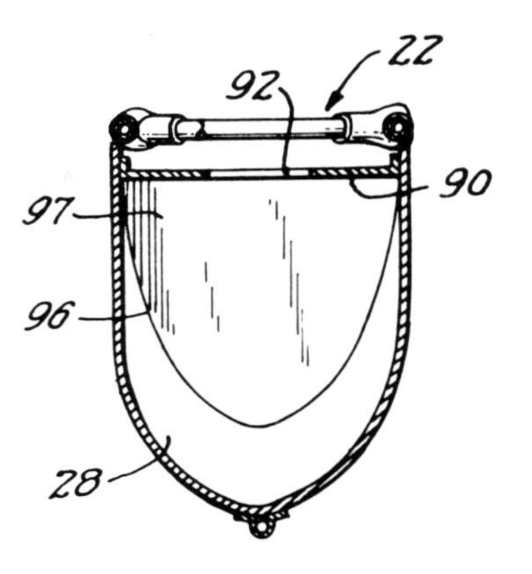

Alternatively, the line of keel weights 48 may be permitted to pull the membrane 28 into a more generally V-shaped configuration than that shown in Figs. 6 and 8.

In this final embodiment, the surface width of membrane 28 is less at one end 96 than at the other; thus, at one end the weights 48 will be held closer to grid 22 than at the other end. The incline in the trough formed by membrane 28 causes fluid flow in the direction indicated by arrows 94 as the breakwater rises and falls. As an alternative to the inclined trough or as an addition thereto, the end 96 of the trough is completely closed off by the

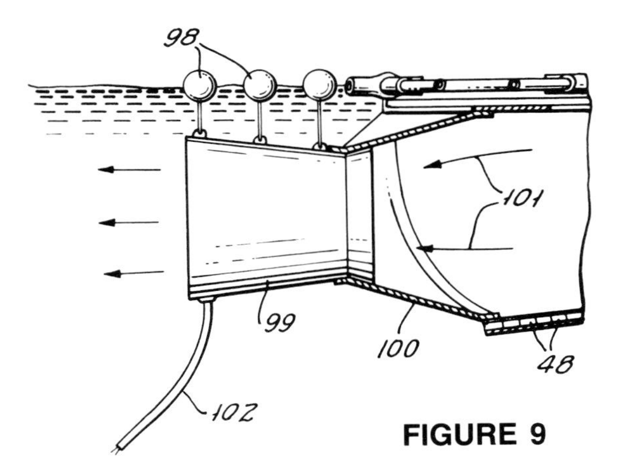

FIGURE 9

end 97 or at least is substantially restricted by such an end closure.

As shown in Fig. 9, a turbine 99 is suspended in the water from buoys 98. The turbine is coupled in sealed fluid flow relationship to the end of the trough formed by membrane 28 by means of a funnel-shaped duct 100. Water flowing longitudinally through the breakwater trough due to wave action passes as shown by arrows 101 through the duct 100 and through the turbine 96 generating electrical power to be used elsewhere. The electrical power may be fed to shore through a flexible power cable 102. Thus this embodiment serves the dual purpose of breaking waves and generating power.

While the invention has been particularly shown and described with reference to preferred embodiments thereof, it will be understood by those skilled in the art that various changes in form and details may be made therein without departing from the spirit and scope of the invention as defined by the claims.

27▲TENSEGRITY TRUSS(1980)

U.S. PATENT—4,207,715

PATENTED—JUNE 17, 1980

MENTIONED EARLIER WAS THE octahedron-tetrahedron truss. The octahedra could be edge-fastened (edge to edge) to produce the octahedron-tetrahedron truss. What I found to be very desirable in relation to the octahedra that we mentioned earlier would be an octahedron supported from the interior by three compression axes between its vertexes. Chris Kitrick then tried fastening the vertexes of the octahedra compression members to the mid-edge tension members of adjacent octahedra. This produced an extraordinarily rigid spheric, a structure that will cut down weights very greatly. I authorized Chris to take out a patent on his invention, which by agreement I paid for and on the basis it be assigned to me.

UNITED STATES PATENT OFFICE

TENSEGRITY MODULE STRUCTURE AND METHOD OF INTERCONNECTING THE MODULES

Background of the Invention

This invention relates to a tensegrity structure and more particularly to such a structure which is constructed from a number of tensegrity modules.

In his U.S. Patent No. 3,063,521, Richard Buckminster Fuller introduced the tensile integrity, or tensegrity, contruction technique. Tensegrity construction is based on the realization that most building materials are much more efficiently utilized, smaller cross-sectional areas can be employed, and the materials can often withstand higher forces, when in tension than when in compression. In tensegrity construction there is a high ratio of tension to compression elements. The tension elements provide continuous lines of tension throughout a structure; whereas there is separation of the compression forces such that the compression members are discontinuous. The compression members in effect float within a sea of tension.

In the above-mentioned Fuller patent, the basic tensegrity element is an octahedron formed of three column-like compression members arranged in a tepee fashion with the ends thereof interconnected by tension wires or cables. The end of each compression member is positioned near the vertex of the octahedron where tension elements intersect, and each tension element runs along an edge of the octahedron. The octahedrons are joined vertex-to-vertex, the compression members thus being in "apparent" continuity. The term apparent continuity is used, for although the continuity in a structural sense is real, because of the resultant tension forces at each intersection of the vertices, the result is discontinuity in re-

FIGURE 1A

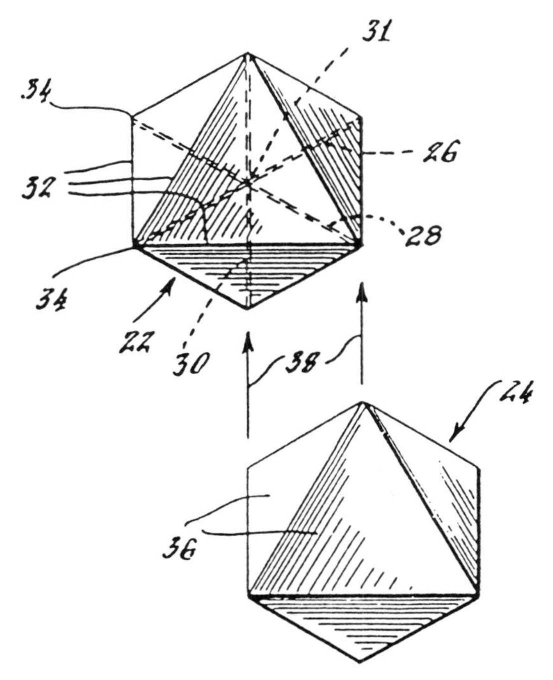

spect of functions in compression at that point. The compression members are joined end-to-end but no compressive force is transferred through the intersection.

In a more recent Fuller U.S. Patent No. 3,354,591, octahedral tensegrity modules are joined face-to-face with compression members of each two interconnecting modules connected at the three vertices of the abutting faces. Each tensegrity module advantageously has a pure tension to pure compression element ratio of four to one. However, with the modules joined face-to-face, the tension elements along the edges of each abutting face become redundant; that is, two tension wires are provided along a single edge where only one is required. If one of those redundant tension elements were eliminated for each edge, the ratio between tension and compression elements is reduced to three to one.

An object of the present invention is to provide tensegrity structure which makes optimum use of each tension element within the structure while keeping a high ratio of tension to compression elements.

FIGURE 1B

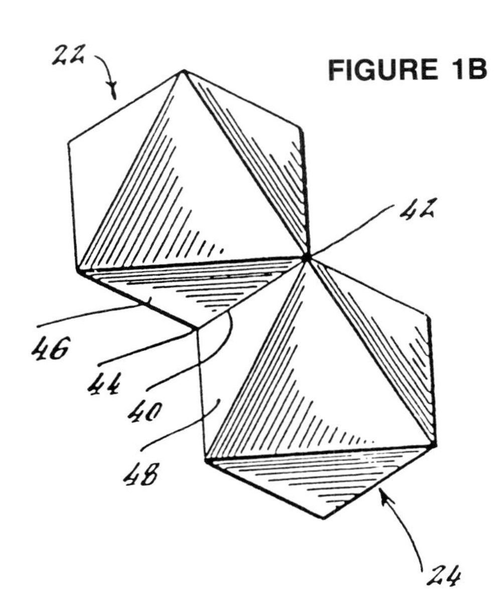

SUMMARY

In accordance with the invention in one of its aspects, a tensegrity structure is formed by interconnecting a plurality of tensegrity polyhedral modules, each two interconnected modules having a vertex of a first module joined to an edge of a second module and a vertex of the second module joined to an edge of the first module.

In accordance with the invention in another of its aspects, each tensegrity module includes column-like compression members and tension elements extending between ends thereof. The tension elements define the edges of a three-dimensional geometric figure having at least four triangular faces and they intersect at the vertices of the geometric figure. Each module is joined to another module with faces abutting but with edges of the abutting faces being nonaligned. The faces of the abutting triangular surfaces of the respective modules are rotated 180° out of superposition. A vertex of each abutting triangular face is joined to an edge of the abutted face, being joined at a point located one-half or one-third of the way along the length of such edge.

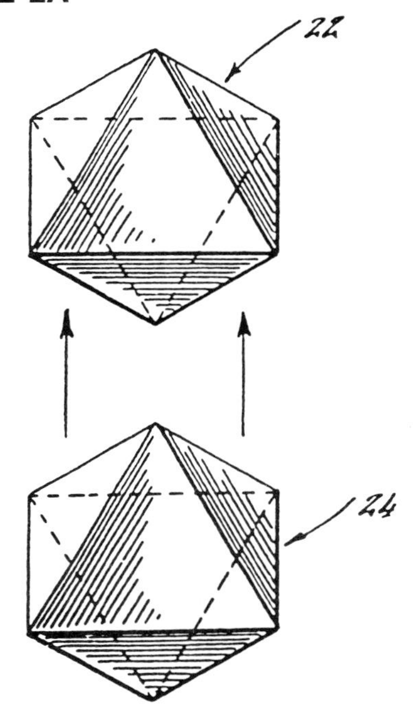

In one embodiment of the invention, each tensegrity module is an octahedron formed of three compression members and all of the interconnected modules are arranged with their faces co-planar such that the tensegrity structure has a face characterized by an array of triangles interlaced with three-sided dimples.

In accordance with another embodiment of the invention, each tensegrity module is a tetrahedron including four compression members joined at a center point, and the modules are interconnected such that the face of the tensegrity structure is characterized by an array of pyramidal dimples. By offsetting the vertices from the midpoint of the joined edges, intersection of vertices is avoided and the tensegrity structure is formed of an array of pyramidal dimples of two sizes.

In yet another embodiment of the invention, each tensegrity module is an icosahedron.

BRIEF DESCRIPTION OF THE DRAWINGS

The foregoing and other objects, features, and advantages of the invention will be apparent from the following more particular description of preferred embodiments of the invention, as illustrated in the accompanying drawings in which like reference characters refer to the same parts throughout the different views. The drawings are not necessarily to scale, emphasis instead being placed upon illustrating the principles of the invention.

Fig. 1A is a plan view of two octahedral tensegrity modules of the prior art;

Fig. 1B is a plan view of the two tensegrity modules of Fig. 1A interconnected in one arrangement of the prior art;

FIGURE 2B

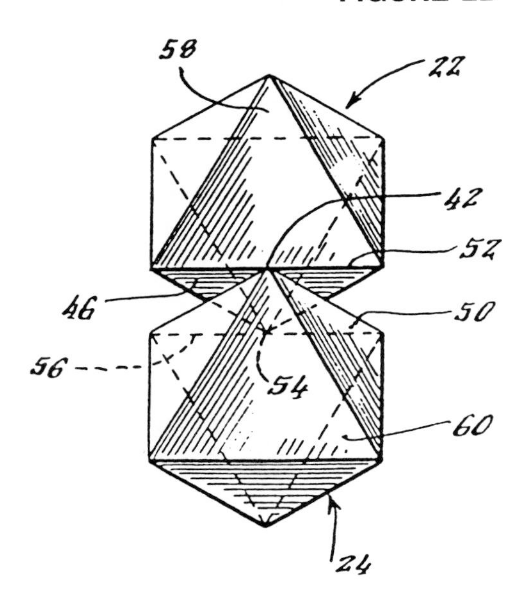

Fig. 2A is a plan view similar to Fig. 1A but of two octahedral tensegrity modules positioned for interconnection in accordance with the present invention;

Fig. 2B is a plan view similar to Fig. 1B but with the tensegrity modules interconnected with overlapping faces and with vertices joined to edges in accordance with the present invention;

Fig. 3 is a plan view of seven octahedral tensegrity modules with the compression members and tension elements shown, six of the elements are interconnected and the seventh is positioned in readiness for interconnection;

Fig. 4 is an angled view of three of the octahedral tensegrity modules of Fig. 3;

Fig. 5 is a plan view of the seven octahedral tensegrity modules of Fig. 3 but with each module shown as a solid figure;

Fig. 5A is a diagram on enlarged scale showing the relative orientation of two abutting triangular faces for purposes of explanation;

Fig. 6 is a plan view similar to Fig. 5 but showing a large number of modules shown on a reduced scale;

Fig. 7 is a view of a tetrahedral tensegrity module having four compression members joined at a center point;

Fig. 8 is a plan view of six interconnected tetrahedral tensegrity modules shown as solid figures;

Fig. 9 is a plan view similar to Fig. 8 of six tetrahedral tensegrity modules but with vertices joined at other than the midpoints of edges;

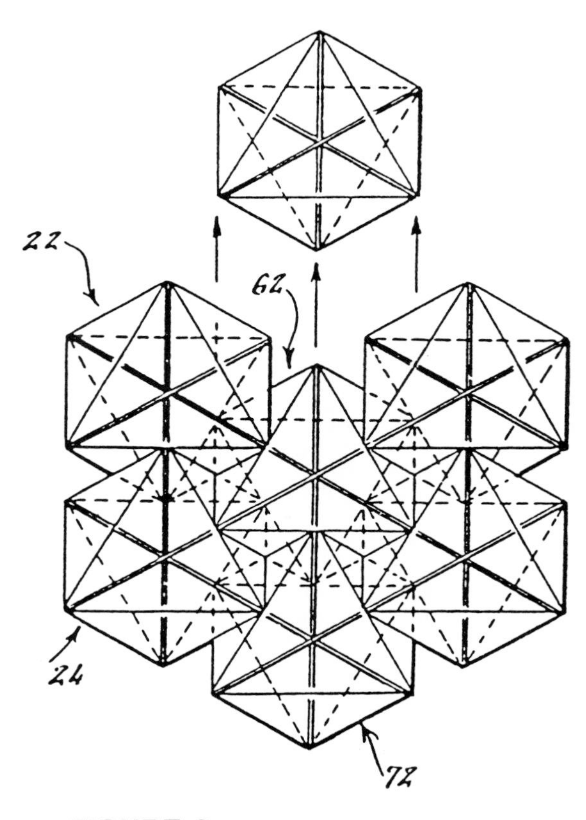

FIGURE 3

Fig. 9A is a diagram, shown on enlarged scale, illustrating the orientation of two abutting triangular faces for purposes of explanation;

Fig. 10 is a plan view similar to Fig. 9 but of a large number of tetrahedral tensegrity modules shown on a reduced scale;

Fig. 11 is a view of an icosahedral tensegrity element viewed as a solid figure;

Fig. 12 is a view similar to Fig. 11 but with compression members and tension elements being illustrated;

Fig. 13 is a plan view of seven icosahedral tensegrity modules interconnected in accordance with the present invention.

DESCRIPTION OF PREFERRED EMBODIMENTS OF THE INVENTION

Two octahedral tensegrity modules 22 and 24 of the prior art are shown in Fig. 1A. As shown in broken lines on the module 22, each module comprises three column-like compression members or struts 26, 28 and 30 which are positioned to form three orthogonal axes. Tension wires 32 extend between the ends of the compression members. Each wire can be seen as an edge of an octahedron having triangular faces and the tension wires 32 intersect at the vertices of the octahedron. If each triangular face is covered with a membrane or skin, then each of the octahedral modules 22 and 24 can be perceived as two solid pyramidal figures joined base-to-base.

When the modules 22 and 24 are moved together as indicated by the arrows 38 in Fig. 1A they may be interconnected as shown in Fig. 1B. The two modules are joined along a common edge 40 and at vertices 42 and 44. In this arrangement one of the tension wires along the edge 40 is redundant and can be eliminated. But by eliminating the tension wire, the ratio of tension elements to compression member is reduced.

In another prior art arrangement shown in the Fuller U.S. Pat. No. 3,354,591, the modules 22 and 24 are interconnected face-to-face with faces 46 and 48 abutting, and the edges of those faces including edge 40 joined. In that interconnection of the modules, three tension wires are rendered redundant with each interconnection.

Fig. 2A shows the two octahedral tensegrity modules 22 and 24 positioned for interconnection in accordance with the present invention. As shown in Fig. 2B, the module 24 overlaps module 22 with portions of the faces

FIGURE 4

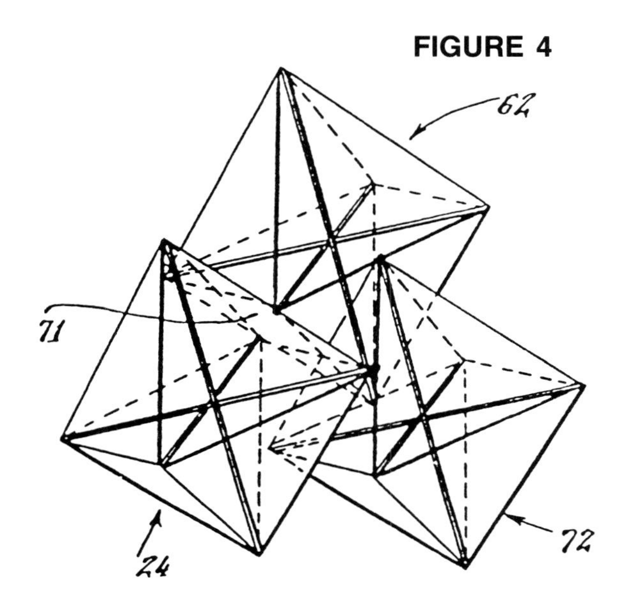

46 and 50 abutting each other. The edges of the faces 46 and 50 are nonaligned with the vertex 42 of module 24 joined to the edge 52 of module 22 at the midpoint thereof. Similarly, a vertex 54 of the module 22 is joined to the edge 56 of module 24 at the midpoint of that edge. With the modules similarly oriented as they are shown in Fig. 2B, the triangular faces 58 and 60, which may form a segment of a structural surface, lie in a common plane.

Multiple tensegrity modules may be similarly oriented and interconnected as shown in Figs. 3 through 6. Each two interconnecting modules, such as 22 and 24, 22 and 62, and 24 and 62, are interconnected with portions of the triangular faces abutting each other. A vertex of each abutting module face is joined with an edge of the abutted face. Thus, as best shown in Fig. 5, with module 22 abutting module 62, vertex 64 of module 22 is joined to edge 66 of the module 62. Conversely, with module 62 abutting module 22, a vertex 68 of module 62 is joined to edge 70 of the module 22.

When the vertex of a first octahedral module is joined to the midpoint of the edge of a second abutting octahedral module and a vertex of the second octahedral module is joined to the midpoint of an edge of the first module with a portion of a triangular face of the first module abutting against a portion of the face of the second (as seen in Figs. 2 through 5), then the portions of the two respective triangular faces which are abutting are

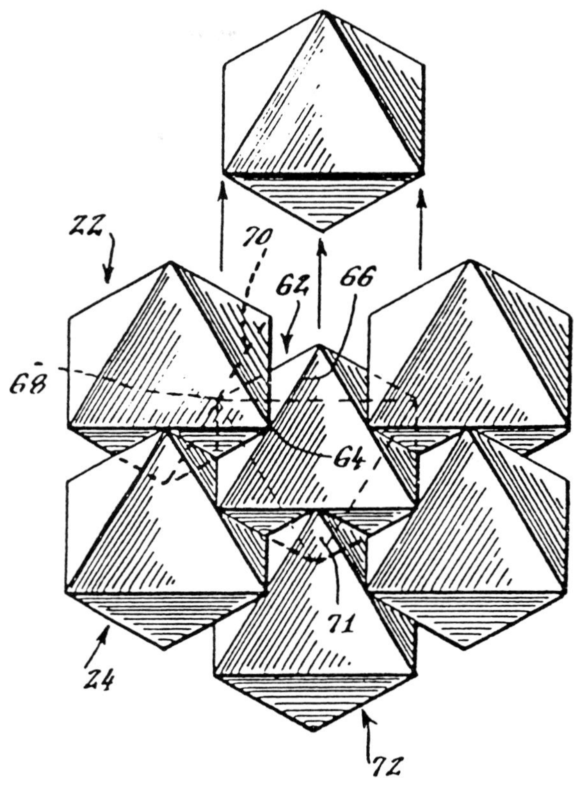

FIGURE 5

each four-sided, namely each being a rhombus. Such an abutting rhombus-shaped area portion is seen clearly in Fig. 4 in dotted outline at 71. For purposes of illustration, such a rhombus-shaped abutting area 71 is shown in darkened outline in Fig. 5.

The diagram of Fig. 5A shows a triangular face 46, for example such as the face of the octahedral module 22 shown in Fig. 2B, abutting against a triangular face 50 of the octahedral module 24. The rhombus-shaped area 71 of the abutting portions of the triangular faces is seen clearly in Fig. 5A. Also, it is to be noted that each triangular face 46 or 50 is rotated 180° out of superposition with respect to the other triangular face.

As best seen in Fig. 4, in which the modules are viewed from a different angle than in Fig. 3, the compression members of the interconnected modules do not intersect.

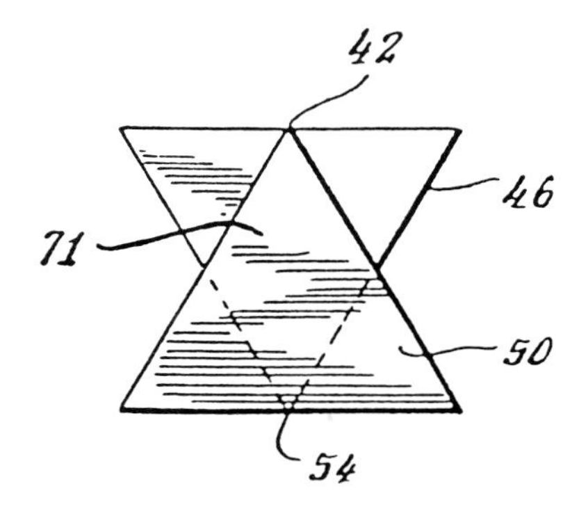

FIGURE 5A

Thus, the respective compression members are completely in compression discontinuity, even apparent continuity being avoided. Although tension elements cross, none of those elements are aligned. Thus, none of the tension elements are redundant, and a full four to one ratio of tension to compression elements is maintained in the structures shown in Figs. 2 through 5.

Fig. 6 shows a floor or wall or layer of tensegrity modules arranged as in Figs. 2 through 5. It can be noted that each module has a triangular face 74 lying in the same plane as the corresponding triangular face of each of the other modules. This common plane in which lie all of the corresponding triangular faces 74 of the modules can be considered to be the face surface of the tensegrity structure. The regions between these corresponding co-planar triangular faces 74 contain three-sided dimples 76, i.e. tetrahedral configured dimples, which give the tensegrity structure an unusual and pleasing appearance. Each dimple is formed from the exposed surfaces of the abutting faces of three modules.

Thus, advantageously the octahedral modules can be so arranged and interconnected as to form a wall or floor truss. As shown in Fig. 6 each module, or the outer surface of a completed structure, has a membrane, or skin, across the octahedron faces to provide a floor surface. If a completely uniform and flat face is desired, then the membrane, which may in fact be formed of concrete, metal, plastic, wood or the like, can be extended across the dimples as well as the co-planar triangular faces for forming a continuous flat surface.

It is not necessary that the outer surface of the modular structure be planar. By suitably dimensioning the compression members and tension elements within each module, the interconnected modules may form a dome or an arch or the like.

Further, the tensegrity modules need not be octahedrons. For example, as shown in Figs. 7 through 10, tetrahedral modules may be interconnected with nonaligned abutting faces such that a vertex of each abutting triangular face is joined to an edge of the abutted face.

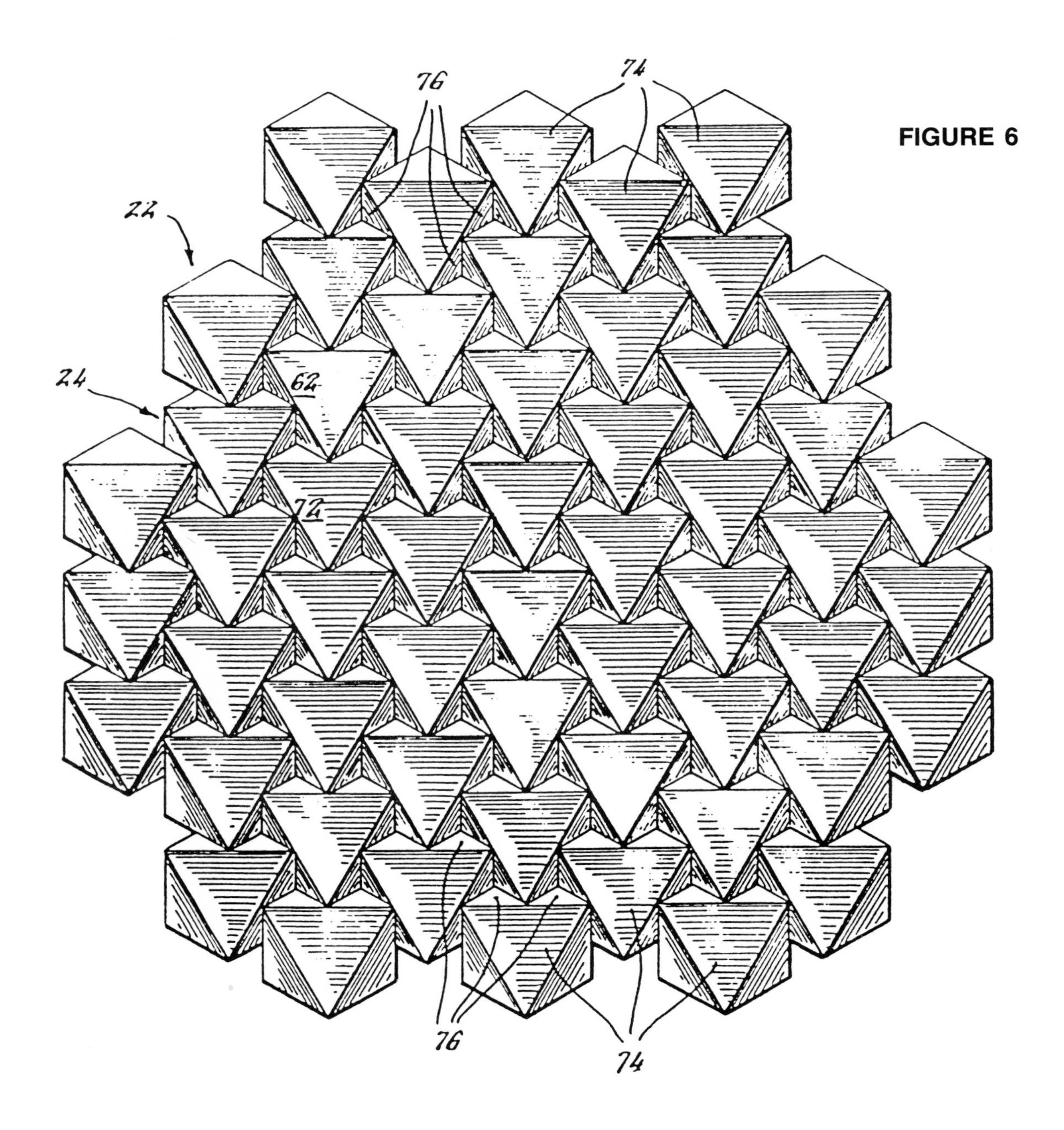

FIGURE 6

FIGURE 7

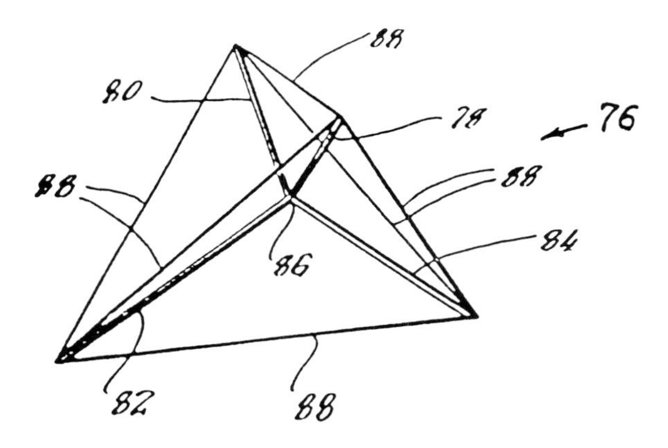

Fig. 7 shows the basic tetrahedral tensegrity module 76. The module 76 comprises four compression members 78, 80, 82 and 84 joined at a center point 86. The outer ends of the compression members remote from the center 86 are interconnected by six tension elements 88 which form the edges of four equilateral triangular faces. Three tension elements 88 intersect at each of the four vertices of the tetrahedron.

As with the octahedrons, each tetrahedron may at least be partially covered with a skin so that at least the outer surface of a structure composed of the interconnected modules is closed.

Six tetrahedral modules shown in the form of solid figures, that is, with a skin over each face, are interconnected in accordance with the present invention as shown in Fig. 8. A first tetrahedral module 90 has two upper triangular faces 92 and 94. It is to be understood that these two faces 92 and 94 slope downwardly and diverge from an upper edge 96, said upper edge 96 appearing much like the ridge of a roof. Two lower triangular faces (which are not seen in Fig. 8) slope downwardly and converge to an edge orthogonal to and spaced below the edge 96. The module 90 abuts and is interconnected with another module 98 having upper faces 100 and 102 sloping downwardly and diverging from an edge 104. A portion of the face 100 of module 98 is abutted by a portion of the face of the module 90 not shown, the latter face having edges 106 and 108 as well as the edge orthogonal to and spaced below the edge 96.

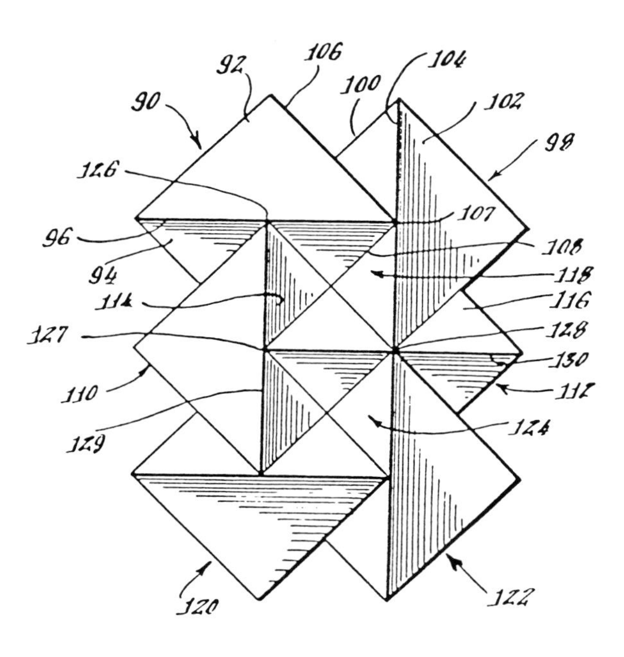

FIGURE 8

In the structure as shown in Fig. 8 in which a vertex 107 of the tetrahedron module 90 is joined to a midpoint of the edge 104 of the tetrahedron module 98, then the abutting portions of the equilateral triangular faces are rhombus-shaped, similar to the relationship shown in Fig. 5A. Also, each abutting triangular face is rotated 180° out of superposition with respect to the other triangular face, similar to the relationship shown in Fig. 5A.

Similarly, a third module 110 overlaps and abuts the face 94 of module 90 and a fourth module 112 overlaps and abuts the face 114 of module 110. The face 116 of module 112 is in turn overlapped by the module 98 and abuts a face thereof. Thus, the four modules 90, 98, 110 and 112 can be seen to form a continuous overlapping arrangement defining and encircling a pyramidal dimple 118, with a vertex of one being joined to a midpoint of an edge of another.

The structure can be continued by interconnecting additional modules in a similar fashion, each set of four interconnecting modules encircling a pyramidal dimple. The face of the completed structure is an array of such dimples 118, 124, and so forth. Each two interconnected modules have nonaligned abutting faces, and a vertex of each abutting triangular face, such as vertex 107, 126, 127 or 128 is joined to a midpoint of an edge of the abutted face, such as the respective edge 104, 96, 129 or 130, which encircle the pyramidal dimple 118.

As with the octahedral tensegrity arrangement, there is no redundancy of edges in the tetrahedral tensegrity arrangement shown in Fig. 8; that is, there are no two edges of adjoining modules which are aligned to render one of the edges unnecessary.

It is interesting to note that in the structure shown in Fig. 6 formed of octahedral modules the dimples 76 are tetrahedral in shape; while in the structure shown in Fig. 8 formed of tetrahedral modules, the dimples are pyramidal in shape, thus each being one-half of an octahedron.

Unlike the octahedral embodiment shown in Fig. 6, the tetrahedral embodiment, as shown in Fig. 8, does include vertex-to-vertex interconnections between the modules. For example, module 98 intersects module 122 at a common vertex 128. The result is an apparent continuity between the two modules joined to the edge 130. This apparent continuity requires a flexible hub or the like for joining two compression members to a tension element at a single point. This requirement of a flexible hub can be avoided by the use of the modular arrangement shown in Fig. 9.

In the embodiment of Fig. 9, each vertex is joined to an edge of another module at a one-third-point of the edge.

For example, the vertex 132 of the module 134 is joined to the edge 136 of module 138 at a point which is located one-third of the length of the edge 136 from vertex 140. The two modules 134 and 138 are thus interconnected with nonaligned abutting triangular faces but with the respective vertices intersecting edges of adjoining modules at a one-third-point. Similarly, the vertex 140 of module 138 intersects the edge 142 of module 144 at a one-third-point. The modules have a continuous overlapping arrangement as in the embodiment of Fig. 8 around a pyramidal dimple 146. However, the sides of that dimple 146 are only two-thirds the length of the sides of the dimples 118 and 124.

FIGURE 9

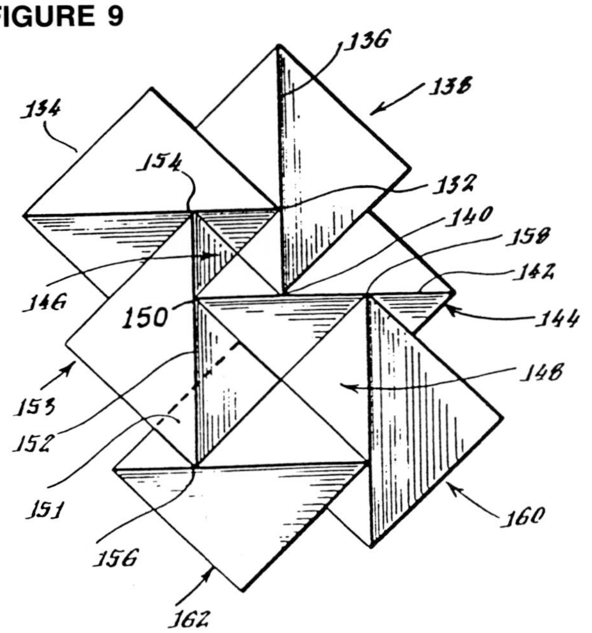

FIGURE 9A

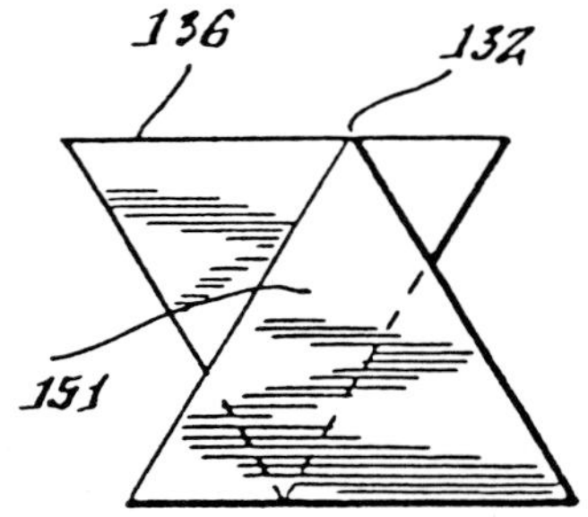

The next set of overlapping modules then form a larger dimple 148 which has sides 1⅓ times the length of the sides of dimples 118 and 124, that is, being twice the length of the sides of the smaller dimple 146. The vertex 150, being located at a one-third of edge 152 from the vertex 154, is correspondingly located at a two-thirds-point of edge 152 from the opposite vertex 156. This two-thirds line segment of the edge 152 forms one boundary of the dimple 148. Continuing in the same fashion, vertex 158 is joined to the edge 142 at a point which is two-thirds of the length of that edge from vertex 150 and so on. The abutting portions of the equilateral triangular faces of adjoining modules are parallelogram shaped as illustrated in dark lines at 151 in Fig. 9.

In the diagram shown in Fig. 9A the vertex 132 of one triangular face is joined at the one-third point of the edge 136 of the abutting triangular face. The parallelogram shape 151 of the abutting portions of the two triangular faces of the respective modules 138 and 134 is seen clearly in Fig. 9A. The interrelationship between the two abutting triangular faces as seen in Fig. 9A is that each triangular face is rotated 180° away from superposition with respect to the other and then is shifted laterally by one-sixth of its width so that its vertex joins the edge of the other at a point one-third of the way along the length of such edge.

By similarly interconnecting a large number of tetrahedral tensegrity modules with vertices joined to one-third-points of respective edges of the modules, the completed structure assumes the attractive configuration shown in Fig. 10. The entire face of the structure is formed of large and small pyramidal dimples 146 and 148, respectively. As before, these dimples may be covered with a continuous membrane to form a flat plane. The dimensions of the compression members and tension elements may be set such that the structure assumes other than a planar configuration, such as a dome or arch, or the like.

An icosahedron, that is a solid figure having twenty planar triangular faces, is shown in Fig. 11. Each of the triangular faces is defined by three tension elements. An icosahedral tensegrity element may be constructed as

FIGURE 10

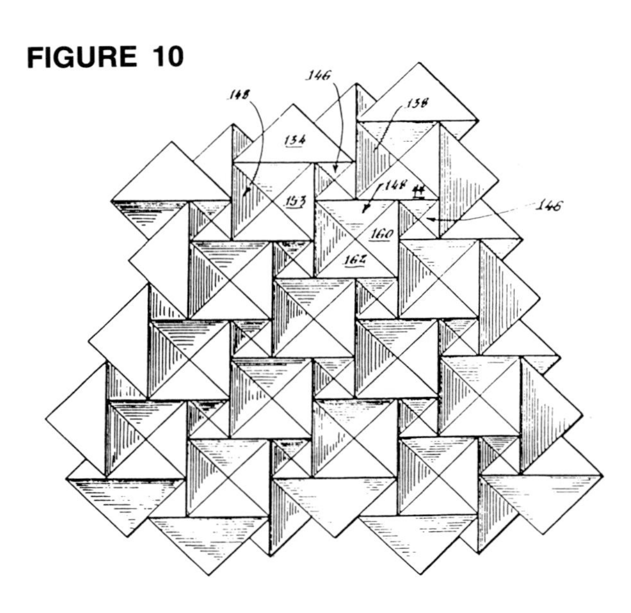

FIGURE 11

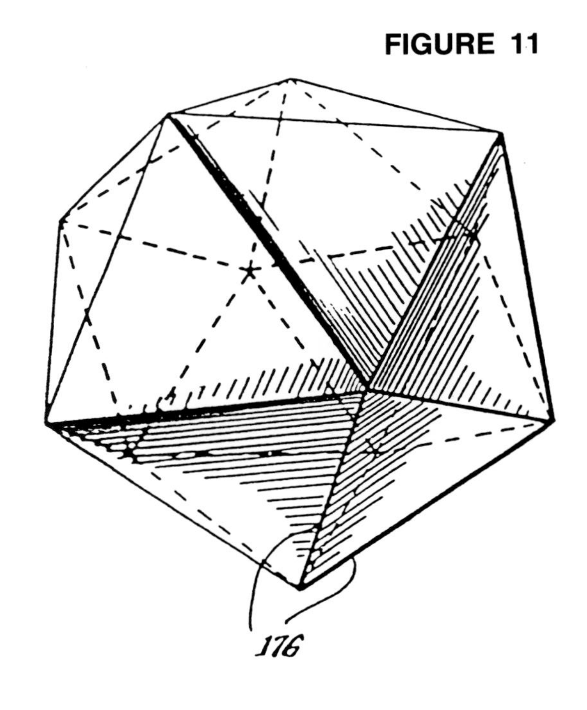

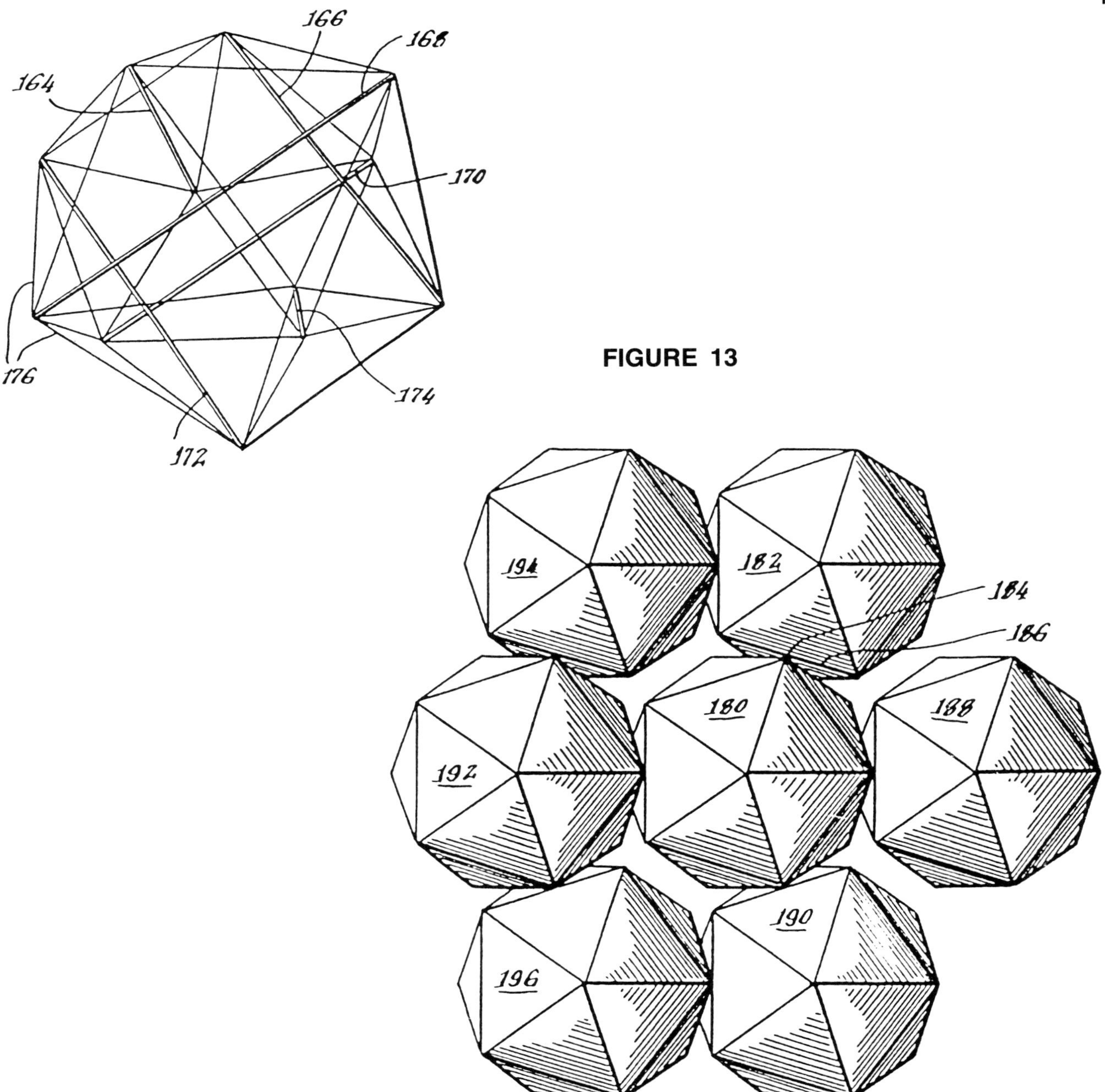

shown in Fig. 12. Six nonintersecting compression members 164 through 174 are interconnected by tension elements 176. As before, tension elements 176 form the edges of the twenty triangular faces of the icosahedron and intersect at the vertices.

In Fig. 13 seven icosahedral tensegrity modules are shown interconnected in a face overlapping arrangement in accordance with the present invention.

With all of the icosahedrons similarly aligned, each module interconnects with four other modules at overlapping faces. For example, module 180 in Fig. 13 is positioned face-to-face with module 182, the vertex 184 of module 180 being joined to the midpoint of the edge 186 of module 182. Similarly, module 180 is interconnected to modules 188, 190 and 192 with faces abutting and vertices joined to the midpoints of respective edges. A tensegrity structure can be built by interconnecting additional modules such as modules 194 and 196.

While the invention has been particularly shown and described with reference to preferred embodiments thereof, it will be understood by those skilled in the art that various changes in form and details may be made therein without departing from the spirit and scope of the invention as defined by the claims.

28▴HANGING STORAGE SHELF UNIT (1983)

U.S. PATENT—4,377,114

PATENTED—MARCH 22, 1983

I FOUND IT WOULD OFTEN BE DESIRABLE to have a storage shelf unit that does not stand on the floor and would not be up against the wall. A hexagonal-patterned bookshelf or storage shelf, which should be capable of carrying great weights, became desirable. I found that if I hung such a storage unit from three points, with the tension going tangentially from the hanging point to either side of the storage unit, the unit develops very great stability—the heavier the loading, the more stable. I made such a unit and found that people seeing this storage unit hanging assumed that it could be readily pushed or moved. A lot of people would walk up to it in a bookstore, touch it, and, finding that it did not yield, lean or push up against it with more force and find that it still did not yield. The storage unit has a fascinating quality about it.

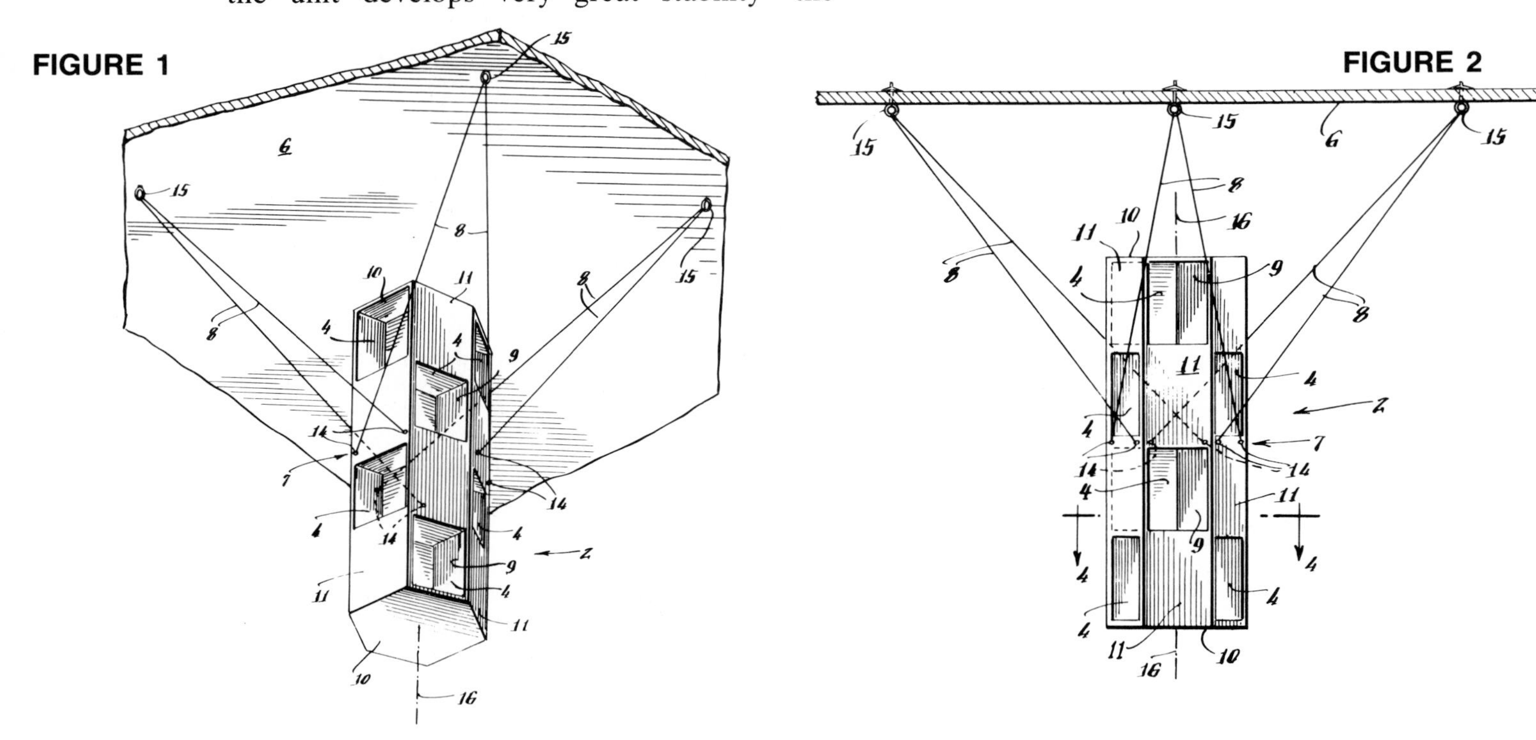

FIGURE 1

FIGURE 2

UNITED STATES PATENT OFFICE

HANGING STORAGE SHELF UNIT

BACKGROUND OF THE INVENTION

The present invention relates to storage shelf apparatus adapted to be suspended from an overhead structure such as a ceiling. The shelf is suspended by a plurality of tension lines and the weight of the storage shelf and its contents keep these tension lines taut to provide the stability for the overall suspended structure. The storage shelf embodying the present invention is intended for use in homes and offices for the same purposes as conventional storage shelf units are now used.

Conventional storage units or shelves commonly used in homes or offices fall into two general categories. The first includes storage units or shelves supported directly on a lower supporting surface, such as a floor. A bookcase is a good example of this first type. The second includes storage units or shelves which are mounted above floor level by mounting to the side of a supporting structure, such as to a wall by brackets.

One disadvantage of such conventional storage units is that access to the interior of the unit is restricted to one side only. That is, storage units supported directly on a floor usually are designed so that entry therein is possible only from the front. The back of the unit is usually closed or abuts against a wall, rendering access through the back surface impossible. Likewise, access to a storage unit mounted to a wall by brackets can only be made through the front of the unit.

In addition to the restriction on access into conventional storage units, there is a problem of stability. With respect to the storage units supported directly on a floor, the taller the unit, the less stable it becomes. Often the top of a tall bookcase is fastened to a wall to prevent its leaning or falling away from the wall. Regarding the type of storage units mounted to a wall by brackets, the higher the unit is, the less stable it will be. A larger number of increasingly stronger brackets and wall anchors are required to support taller wall-mounted storage units of the latter type.

Moreover, both types of known storage units are generally positioned in a remote area of a room. Those mounted on a wall must, by necessity, be adjacent to the wall. Those supported on the floor are usually placed at the side or in the corner of the room to avoid occupying the limited floor space in the center of the room. Furthermore, many storage units are unattractive, thereby providing a further incentive for placing them in a remote location.

The present invention provides a suspended storage shelf unit of attractive appearance in which access to the various storage areas defined therein is available from all sides of the unit. The storage unit embodying one aspect of the present invention is rotatable relative to a suspended stationary member for the convenience of the user in selecting the specific portion of the unit through which access is desired. Because the unit is suspended by tension lines in an advantageous pattern, the forces resultant from gravitation exerted on the unit keep the unit stable. Additionally, the elevation of the storage unit can be selected by the user as desired by setting the length of the supporting tension lines.

In a further embodiment of the invention, a nonrotatable storage unit is suspended from an overhead structure such as a ceiling. This unit may be positioned such that it is centrally located in a room to provide access to its storage compartments from all sides. Because the unit is elevated, it does not occupy floor space in a room, and the weight of the unit keeps its supporting lines taut to enhance overall stability.

Accordingly, the suspended storage unit embodying the invention to be described herein overcomes the problems of access and stability inherent in the conventional type storage units discussed above and provides an attractive appearance.

SUMMARY OF THE INVENTION

The present invention is embodied in a storage unit defining storage compartments or shelves and adapted to be suspended from an overhead structure such as a ceiling, by tension lines. In one embodiment, the storage unit includes a suspended stationary member, such as a platform or flat ring, and at least one storage compartment is rotatably mounted to this suspended member. The suspended member includes means for securing the ends of the tension lines, such as cables or wires, for hanging the storage unit from a ceiling. Means are provided for allowing rotation of the storage compartment relative to the suspended stationary member, as for example, a ball-bearing ring, interposed between this member and the storage compartment and located in alignment with the axis of rotation.

In a further embodiment of the invention, an upper storage compartment is rotatably mounted above the suspended stationary member, while a lower storage compartment is rotatably mounted below this stationary member. A rotatable shaft passing through the suspended stationary member concentric with the rotation bearing means couples the upper storage compartment to the lower storage compartment. Accordingly, rotation of one of the compartments relative to the suspended stationary member simultaneously causes a corresponding degree of rotation of the other storage compartment.

In another embodiment of the invention, a hexagonal non-rotatable storage unit is suspended from an overhead structure, such as a ceiling. This unit may be suspended in a central location of the room, but by virtue of its elevated position, it does not occupy any floor space.

In all embodiments of the invention, the storage unit is suspended from an overhead structure by tension lines. By the advantageous interaction of three pairs of these tension lines in an equilateral triangular arrangement, the storage unit advantageously exhibits excellent stability. Moreover, access to the storage unit is available from all sides.

As used herein, the term "tension line" is intended to include any suitable wire, cable, strand or rope of strong, essentially non-stretchable material capable of supporting a heavy load under tension for an indefinitely long period of time without changing length, i.e., without "creeping" and without breaking.

Further features, aspects and advantages of the suspended storage unit will become more fully understood when the following description is considered in conjunction with the accompanying drawings.

BRIEF DESCRIPTION OF THE DRAWINGS

Fig. 1 illustrates in perspective the hanging storage apparatus in accordance with one aspect of the present invention;

Fig. 2 is a front elevational view of the hanging storage compartment apparatus of Fig. 1, showing the storage apparatus suspending by three pairs of tension lines from a ceiling;

Fig. 3 is a top plan view as seen looking down on Fig. 2, showing the arrangement of the three pairs of tension lines in an equilateral triangular pattern;

Fig. 4 is a plan sectional view taken along the line 4—4 in Fig. 2, and shown enlarged;

Fig. 5 is a perspective view of the hanging, rotatable storage compartment apparatus illustrated in Fig. 6;

Fig. 6 is a front elevational view illustrating hanging, rotatable storage compartment apparatus in accordance with another aspect of the present invention.

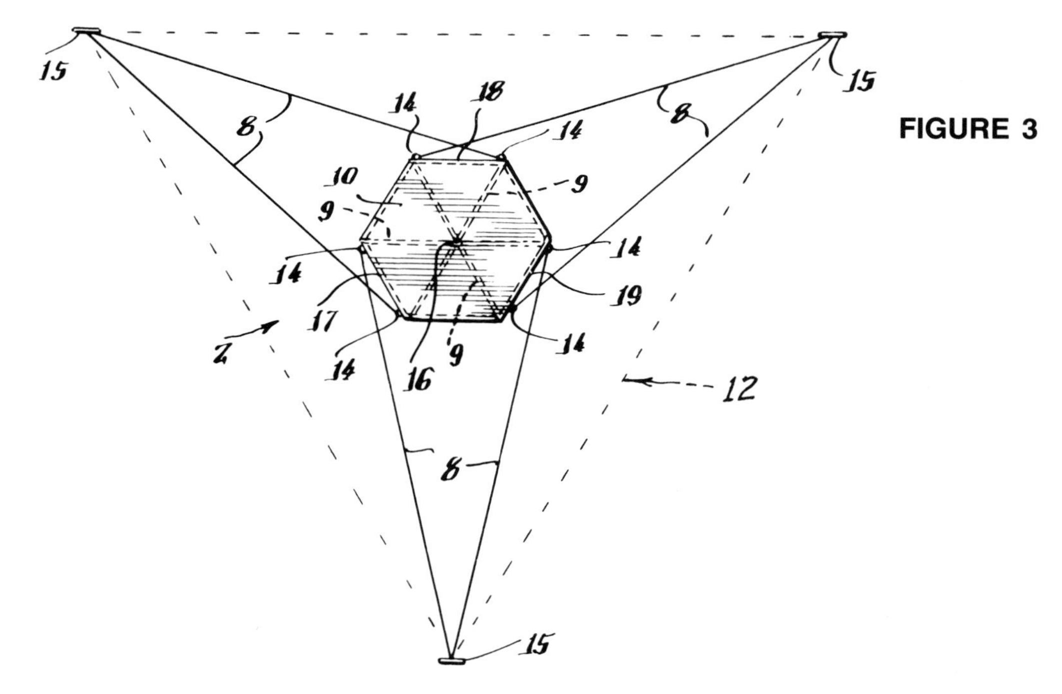

FIGURE 3

DESCRIPTION OF THE PREFERRED EMBODIMENTS

Figs. 1, 2, 3 and 4 illustrate a first embodiment of hanging storage compartment apparatus in accordance with the present invention. The storage compartment apparatus, generally indicated by the reference numeral 2, includes a plurality of storage shelf areas 4. In the disclosed embodiment, the storage compartment apparatus is formed as an attractive integral unit, having a hexagonal configuration in plan view, made from suitable structural material, such as wood, rigid plastic or metal, and includes four separate levels or tiers defining the respective storage shelf areas 4. Each storage area 4 has an opening in the outer surface of the storage compartment 2. Doors (not shown) may be mounted proximate to each opening to selectively open and close the entrance to each storage area, if desired by the user.

However, the regular hexagonal configuration with three open shelf areas 4 in each level is attractive as shown, and therefore, it is believed that most users will prefer to have the shelf openings uncovered to provide an eye-pleasing contrast with the exterior panels 11 located on each level alternating with the openings into the respective shelf areas 4. The alternating sequence of the three open shelf areas 4 and the three exterior panels 11 on each level is most clearly seen in plan cross section in Fig. 4. Thus, each of the three identical shelf areas 4 has

FIGURE 4

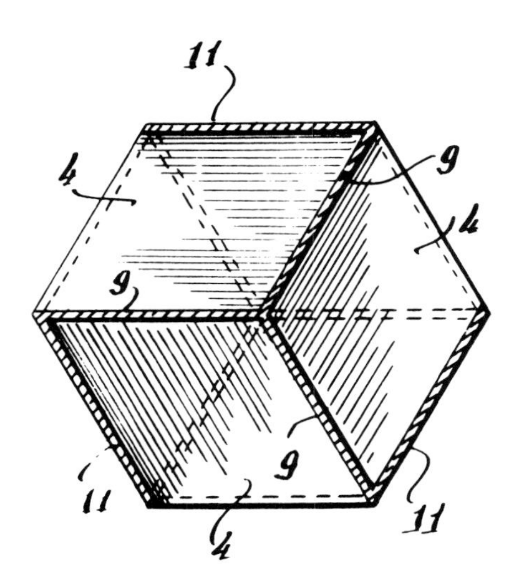

a rhombus or diamond shape with 60° and 120° interior angles at the respective vertices as seen in Fig. 4. None of the space in each level is wasted. The three rhombic-shaped shelf areas 4 nest neatly together to form the hexagonal plan configuration of the overall unit 2. These three shelf areas are separated by the three internal partitions 9. On the next adjacent levels, the location of the shelf areas 4 and of the exterior panels 11 are shifted one-sixth of a circle, i.e., by 60°. Thus, on the next level each shelf opening is located above an exterior panel on the level below and each exterior panel is located above a shelf opening, yielding a pleasing checkerboard contrast between panels 11 and openings into the shelf areas 4. At the bottom of each level, there is a horizontal hexagonal deck 10 which forms the three respective shelves and which forms the bottom of the overall unit. There is a similar deck 10 forming the top of the unit.

The storage compartment apparatus 2 is mounted to an overhead supporting structure, for example, a ceiling 6 by six tension lines 8. One end of each such tension line is secured directly to the approximate mid-level region 7 of the storage compartment 2 through suitable fastening means 14 thereon, such as screw eyes or cable clamps mounted on the apparatus 2. As illustrated in Figs. 1 and 2, the tension lines 8 are secured to the storage compartment at the mid-level region 7, namely at the longitudinal mid-section of the apparatus 2. This longitudinal mid-section is located between the second and third tiers of storage areas 4. The other ends of each such tension line are mounted to the ceiling 6 by suitable overhead cable attachment means 15, such as screw eyes, screw hooks or eyebolts.

Fig. 3 illustrates one arrangement in which the tension lines 8 advantageously can be used to suspend the storage compartment apparatus 2 in a very stable manner. As illustrated in Fig. 3, the storage compartment in the present embodiment of the invention has a regular hexagonal configuration as seen in plan view. The tension lines 8 are secured to the storage compartment in a three-attachment-point, six-line arrangement. Specifically, three overhead attachments 15 are mounted to the ceiling 6 and spaced apart from each other so as to define an equilateral triangle 12 therebetween. Thus, these three overhead attachments 15 are located at the respective vertices of the equilateral triangular pattern 12. Preferably, each leg of the equilateral triangle 12 is approximately 7 feet long when the tension lines 8 are used to suspend a hexagonal storage compartment apparatus 2 having a lateral expanse of approximately twenty inches. That is, the hexagonal configuration fits within a circumscribed circle having a diameter of twenty inches.

As illustrated in Fig. 3, there are a pair of the tension lines 8 secured to the ceiling by each of the attachment means 15. The pair of tension lines extend diagonally down from each of the attachments 15 to a respective pair of fasteners 14 located near diametrically opposite corners of the hexagonal midsection 7. Thus, as seen in plan view the pair of tension lines 8 extending from each attachment 15 define the two legs of a long, narrow isosceles triangle. By virtue of this isosceles triangular configuration with the pair of tension lines 8 connected to diametrically opposite fastening points 14 on the midsection level 7, the overall unit 2 is stabilized against turning about its vertical central axis 16. Also, by virtue of the fact that there are the three overhead attachment points 15 located at equilateral vertex points, namely at points spaced 120° apart around the vertical central axis 16 of the overall unit 2, it is stabilized against lateral motion or swaying motion in any direction.

The tension lines 8 extending from the opposite connections 14 on the three sides 17, 18 and 19 of the hexagon crisscross each other and are mounted to opposed attachments 15. It is to be noted that the sides 17, 18 and 19 of the hexagon to which the tension lines 8 are fastened, are parallel to the adjacent legs of the equilateral triangle 12 defined between their respective attachment points 15. The weight of the storage compartment apparatus 2 maintains the tension lines 8 taut, and the described three-point, six-line equilateral suspension arrangement provides excellent stability for the suspended storage compartment unit 2.

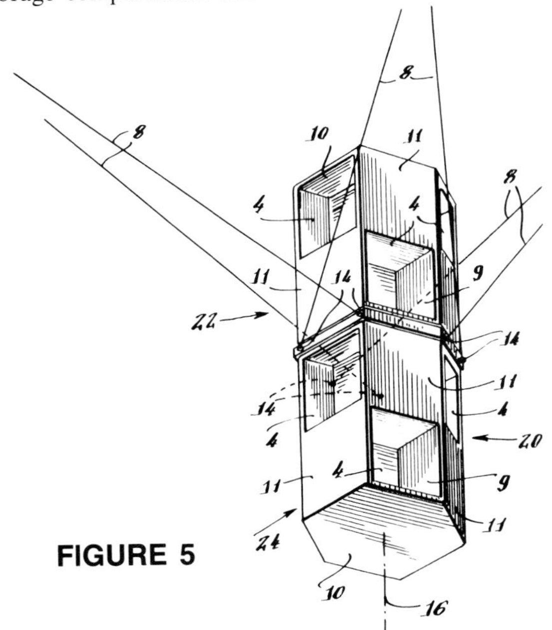

FIGURE 5

The stationary storage compartment unit 2 described herein can be suspended at any convenient position in a room to provide access to all sides thereof, and as previously discussed, does not occupy valuable floor space.

Figs. 5 and 6 illustrate a second embodiment of hanging storage compartment apparatus in accordance with the present invention. In this embodiment, the whole storage compartment apparatus, indicated generally by the reference numeral 20, includes two different units 22 and 24. These different units may be called an upper rotatable unit 22 and a lower rotatable unit 24. As in the first embodiment discussed above, the present storage compartment has a hexagonal plan shape. A suspended, stationary hexagonal ring member or platform 26 is disposed between the top deck 10 of the lower storage unit 24 and the bottom deck 10 of the upper storage unit 22. Rotation bearing means 28, for example, such as a ring of ball bearings concentric about the central vertical axis 16, are positioned between the top of the suspended platform member 26 and the bottom deck 10 of the upper storage unit 22. The peripheral shape of the suspended stationary member 26 desirably matches the plan configuration of the storage compartment, which in the disclosed embodiment, is hexagonal. If desired, the suspended stationary platform member 26 may be made circular in peripheral configuration, for providing an eye-attracting plane of demarcation between the upper and lower hexagonal rotatable units 22 and 24, respectively. Such a circular platform member 26 has a diameter at least as large as the lateral expanse or width of the apparatus 20 as measured diametrically from corner to opposite corner in order to provide clearance between the upper rotatable unit 22 and the diagonally extending tension elements 8.

This suspended stationary member 26 may be a hexagonal ring having sufficient radial area for mounting the ring bearing means 28 concentric around the vertical central axis of rotation 16.

One end of a vertical shaft 30 located in the axis 16 is rigidly affixed to the upper storage unit 22. The shaft 30 extends upwardly through the platform 26 and through the rotatable bearing means 28, and the lower end of this shaft is affixed to the lower storage unit 24. If a solid platform 26 is used instead of a ring-shaped platform, then a suitable axial opening is provided to permit free rotation of the shaft 30.

The ends of six tension lines 8 are fastened to the stationary member 26 through suitable fasteners 14 thereon. The other ends of these tension lines are mounted to an overhead supporting structure, as for example, a ceiling 6 by the attachment means 15. Preferably, a three-point, six tension line equilateral suspension arrangement is provided identical to that as described above for the unit 2 shown in Figs. 1–4. This advantageous suspension arrangement is illustrated in Figs. 5 and 6.

In operation, the upper and lower storage units 22 and 24 sections are rotatable relative to the stationary member 26 as a result of the rotation bearing means 28. Because the upper and lower storage compartment sections are joined together by the shaft 30, rotation of one unit relative to the stationary support simultaneously rotates the other unit a corresponding amount. Accordingly, access to all of the respective storage shelf areas 4 within both storage units 22 and 24 is facilitated by merely rotating either the upper or lower storage compartment section.

It is to be understood that each of the storage units 22 and 24 may be constructed similarly to two levels of the unit 2 shown in Figs. 1–4, with decks 10, interior partitions 9 and exterior panels 11.

It is also within the scope of the present invention to provide rotatable storage compartment apparatus similar to that described in connection with Figs. 5 and 6, except that both the upper and lower storage units 22 and 24 are independently rotatable. To accomplish this independent rotation, the shaft 30 is omitted and a second rotation bearing connection is provided between the upper deck 10 of the lower section 24 and the lower surface of platform member 26.

It is further within the scope of the present invention to modify the rotatable storage compartment apparatus of Figs. 5 and 6 by eliminating either the upper unit 22 or the lower unit 24. This modification is, however, less advantageous than the apparatus 20 as shown in Figs. 5 and 6, because less storage capacity is provided, and the symmetrical balance of the two units (upper and lower) is not achieved.

Three pairs of tension lines 8 employing three overhead attachment points 15 located at the respective vertices of an equilateral triangular pattern 12 is the preferred suspension arrangement. It is the optimum arrangement, because it provides a very stable suspension action involving the least number of tension lines 8 and

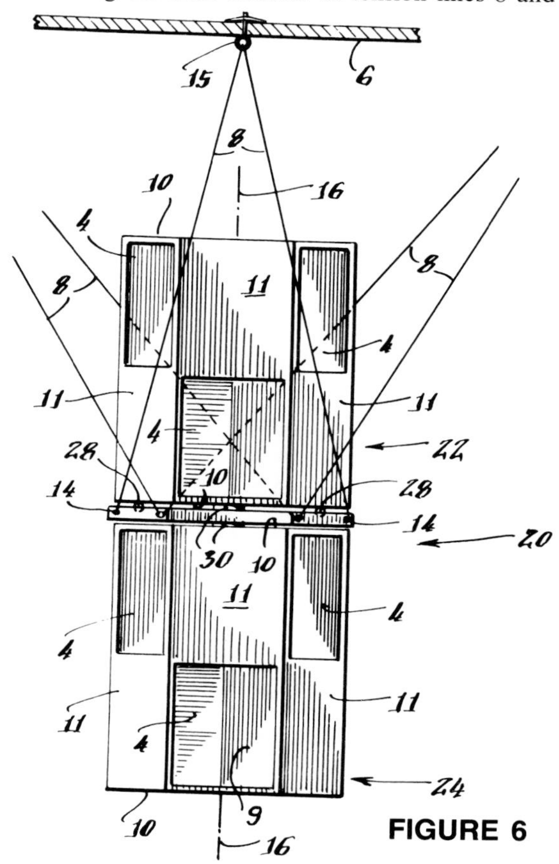

FIGURE 6

the least number of attachment points 15. It is to be understood that more than six tension lines and more than three attachment points 15 can be employed, if desired. However, in my view, using more than the minimum is wasteful and not so attractive as the optimum which is described above. Furthermore, when more than the optimum number are employed, the suspension may not be so stable. For example, using four attachment points 15 and eight tension lines is not likely to be so stable as using three attachment points and six tension lines, because the precise length and tension in each line then becomes much more critical, just like the leg length of a four-legged table is more critical than the leg length of a three-legged table in achieving stability of the table.

It is understood that other modifications of the present invention may become apparent to those skilled in the art. The description of the preferred embodiments illustrated in the drawings and discussed herein are intended to be illustrative only and not restrictive of the scope of the invention.

APPENDIX

HONORARY DOCTORATE CITATIONS

1954–1983

IN AN ERA OF INCREASING SPECIALIZATION, when the validity of intellectual achievement is judged in the first instance by an ever narrowing group of one's scientific or scholarly peers, it is particularly interesting to consider the academic recognition granted to a man who is a relentless and thorough-going comprehensivist. Buckminster Fuller's well-known determination to think for himself, to disregard the conditioned thinking of our inherited scientific disciplines in favor of direct contact with an intuitioned reality, unavoidably exposed him to the danger of being labelled a crack-pot by the scientific and educational establishment. Such social pressures in favor of intellectual conformism undoubtedly cause many young people to falter in their determination to think for themselves.

Thus it was considered relevant to include in this book the record of Fuller's honorary doctorate citations. Written by members of various university or college faculties, these informed assessments of Fuller's work serve as critical appraisals of the historical relevance, practicality, and relative effectiveness of Fuller's only-apparently-foolish decision to disregard conventional conceptionings and think things through for himself. Like the record established by his patents, these public recognitions are an integral part—akin to experimental data—of what Bucky calls "Guinea Pig B," his half-century experimental commitment to see what a single individual can accomplish without any organizational backing—including the organizational backing of our knowledge establishment.

It is Fuller's hope that the public results of "Guinea Pig B" documented in this book will offer encouragement to the young world to persevere in its commitment to absolute intellectual integrity.

Michael Denneny, Editor

DOCTOR OF DESIGN

June 6, 1954

North Carolina State College of Agriculture and Engineering

Raleigh, North Carolina

Mr. Chancellor, I have the honor to present RICHARD BUCKMINSTER FULLER of Forest Hills, Long Island.

A native of Massachusetts, Mr. Fuller attended school in Milton and has been a student at Harvard and the United States Naval Academy. As a distinguished engineer, mathematician, inventor, designer, mechanic, writer, and philosopher, he has become one of the most influential and controversial personalities of the machine age. He has invented a new system of geometry; he holds the only United States patent for a new kind of map projection; he has written several books; he has published a magazine; he has developed the most radical cars and houses in the world; he has made a significant contribution to the study of housing as a crucial world engineering problem; and he has developed revolutionary new ideas of lightness and economy in construction never before conceived of in engineering and architectural thinking. As Visiting Professor and Lecturer in the leading schools of architecture and design throughout the country, he has exerted a dynamic influence on the creative youth of our time.

Because of these most distinguished accomplishments, I commend Richard Buckminster Fuller for the Honorary Degree of Doctor of Design.

DOCTOR OF ARTS

June 11, 1955

University of Michigan

Ann Arbor, Michigan

R. Buckminster Fuller, author, inventor, and architect. Educated at Harvard and at the United States Naval Academy, whence he was commissioned in the regular Navy, he has since risen to a place of remarkable eminence in the art and science of architecture. In two world wars he gave himself unstintingly in the service of his country. As an author, he has published books in a variety of fields and edited a magazine. As a creative artist, he has become one of the most controversial and original intellects of our time. A new system of geometry was born of his thinking; he holds the only United States patent for a new kind of map projection. Applying his Dymaxion Theory, he has constructed automobiles and houses of a most radical design. He has developed revolutionary ideas of lightness and economy wholly new to the engineering and architectural world. Withal, he has found time to give unselfishly of his energy to creative youth in schools of architecture from coast to coast. Honor and affection have come his way in profusion. In 1952 he received the Award of Merit from the New York Chapter of the American Institute of Architects; in 1954 the Citation of Merit from the United States Marine Corps, at Princeton, New Jersey. The University of Michigan is happy to show him its gratitude and affection, for as much as any it has benefited from an exciting personal contact with his genius.

DOCTOR OF SCIENCE

June 12, 1957

Washington University

St. Louis, Missouri

Mr. Chancellor, I have the honor to present R. Buckminster Fuller, author, industrialist, architect, prophet, and humanist.

Developer of advanced designs for motor cars, building equipment, long span structures, his geodesic structures house our far-flung arctic radar networks, and have become standard equipment with the Marine Corps. The State Department regularly uses his geodesic domes in foreign lands to demonstrate the highest achievement of American building technique.

I consider it a distinct privilege to present R. Buckminster Fuller for the degree of Doctor of Science, *honoris causa*.

DOCTOR OF ARTS

June 17, 1959

Southern Illinois University

Carbondale, Illinois

Author, inventor, and architect; creative artist, controversial and original intellect; inventor of the dymaxion house and the geodesic principle of structural design; originator of a new system of geometry for architecture; holder of United States Patent for a new kind of map projection; developer of a revolutionary idea of lightness and economy new to the engineering and architectural world; indefatigable in the service of his country in war and peace; vigorous teacher and lecturer, frequently honored for his genius and admired for his devotion to teaching youth in schools of architecture from coast to coast.

DOCTOR OF HUMANE LETTERS

February 22, 1960

Rollins College

Winter Park, Florida

Richard Buckminster Fuller was born in Massachusetts in 1895. He was educated at Milton Academy, Harvard, and the United States Naval Academy. As engineer, researcher, and writer, he has held a wide variety of posts, such as technological editor of *Fortune* magazine, chairman of the Board of Trustees of the Fuller Research Foundation, President of Geodesics, Inc., President and Chairman of the Board of Synergetics, Inc., professor or lecturer at eight of the nation's leading universities (Cornell, Massachusetts Institute of Technology, University of California, Yale, and others). He is an inventor of distinction; among his latest is the astonishing and revolutionary geodesic dome, which was one of the most exciting contributions to the American national exhibition at Moscow. His public recognitions include Doctor of Design, University of North Carolina; A.F.D., University of Michigan; Award of Merit, A.I.A.; citation, USMC; Centennial Award, Michigan State University; Gran Premo Triennale of Milano, Italy; Fellow A.A.A.S.; member, Harvard Engineering Society, and many others. He is author of *Nine Chains to the Moon.*

Richard Buckminster Fuller: engineer extraordinary; creative mind merging theory and thing into new and dynamic forms; fabricator of structures impressive for their useful immensity; weaver of mathematical systems beyond the range of the accepted and the known; determined practitioner of a technology that brought envisaged structures into being in spite of the inertia of skepticism and disregard; by the spoken word, and by written symbol you have gained widening authority as designer, inventor, innovator, teacher, administrator and leader.

It is with the profoundest sense of honor that by virtue of the authority vested in me by the Trustees of Rollins College I confer upon you the degree of Doctor of Humanities, invest you with the hood which from medieval days has been symbolic of the degree and bestow upon you all the honors, privileges, and obligations appertaining thereto.

DOCTOR OF LETTERS

May 30, 1964

Clemson College

Clemson, South Carolina

Richard Buckminster Fuller—creative genius, author and world-renowned lecturer—because, in quest of nature's true design, you have explored extensively and discovered prolifically new forms and patterns, including, among many others, the geodesic dome; because you have left and are continuing to leave a heritage of simple and complex verities for immediate or future reference to the betterment of mankind; and because you have most effectively communicated and dramatized your innovations in the thinking process as an inspiration to thousands of students on hundreds of college campuses, I, by authority of the Board of Trustees, confer upon you the honorary degree of Doctor of Letters.

DOCTOR OF FINE ARTS

June 5, 1964

The University of New Mexico

Albuquerque, New Mexico

Engineer, comprehensive designer, scientist, inventor, and teacher, his prodigious intellect and abounding energy have added impressively to the progress of mankind. From his scientific and philosophical explorations have evolved such imposing and revolutionary concepts as synergetic geometry, dymaxion projects, and geodesic structures. Today, more than two-thousand geodesic domes—the world's largest form of clear span enclosure—stand in forty-one countries as lasting monuments to his genius.

DOCTOR OF SCIENCE

August 22, 1964

University of Colorado

Boulder, Colorado

R. Buckminster Fuller—

To the world, a creative scientist who presents new interpretations of architecture and engineering and expands the vision of human responsibility, resources and potential; an educator who inspires through personal contact, example and published work; a globe-circling representative of the American people through invited consultation in scientific endeavor.

To the University of Colorado, a personal friend, consultant and extender of horizons in the search for expanded knowledge and the striving for greater effectiveness in the promotion of man's betterment.

Mr. President, in recognition of his outstanding contribution to the expansion of human well-being through understanding, creativity and education, I have the honor to present R. Buckminster Fuller for the degree of Doctor of Science, *honoris causa.*

DOCTOR OF HUMANE LETTERS

April 24, 1965

Monmouth College

Monmouth, Illinois

Richard Buckminster Fuller, poet laureate of technology, author, dynamic dreamer, modern-day Thoreau, tireless traveler and lecturer, seeding your ideas in the fertile soil of the future; for much of your life, your advanced theories and inventions were dismissed as impractical by colleagues and critics alike. Now, dramatically brought to life through your geodesic dome, the Dymaxion principles you have shaped are bringing a new dawn to the world of design.

You are an anticipator of tomorrow, a technological liberator of mankind, inspired and exhilarated by man's potential to free himself through the use of anticipatory design. You are—as you, yourself, have said—not a thing, a noun, "but . . . a verb, an evolutionary process; an integral function of the universe." We at Monmouth College salute you.

In testimony of the admiration and acclaim which this College accords you, I confer upon you the degree, honoris causa, Doctor of Humane Letters.

DOCTOR OF HUMANE LETTERS

June 7, 1966

Long Island University, Zeckendorf Campus

Brooklyn, New York

In Nature, before any new species emerges and takes root in the realm, a prototype of that species appears.

Twenty-first Century's new earthman is already here in prototype in the person of Richard Buckminster Fuller.

Here indeed, in the chunky figure with sur-human cerebral prowess, and a dauntless imagination which has rocked our technology and sparked a scientific renaissance affecting the remotest reaches of man's endeavors, we confront a human data processor, a mathematician, an inventor, artist, poet, sociologist, educator, architect, engineer and an inspired designer bent on making the whole earth enjoyable by all men.

For Buckminster Fuller's dedicating himself as a one-man task force charging the outer ramparts and inner defenses of the natural enemies and scourges of man . . .

For his unveiling basic patterns in Nature and bringing new understanding of structure, of motion, of spherical geometrics, tensions and synergistic forces in our universe . . .

For providing man with new techniques for the achieving of maximum quality of living standards through the use of minimal output of energy . . .

For his broadening and extending the scope and vocabulary of modern scientific and technological language . . .

For his compilation of a world resources inventory and for firing and inspiring the scientifically inclined youth of all countries with a non-political comprehensive design initiative in hopes of making the world's total energy resources serve all humanity . . .

For pointing the way not merely to a material utopia for man but for teaching and promoting the study, investigation and the improvement of the inner man by man himself . . .

For generally lighting the way to a better life in the Twentieth and Twenty-first Centuries for all men on an order higher than ever before envisioned . . .

Long Island University, in fullest admiration for his contributions to mankind's general welfare, invests Richard Buckminster Fuller with this high symbol of its esteem, the degree of Doctor of Humane Letters, honoris causa.

DOCTOR OF HUMANE LETTERS

June 10, 1966

San Jose State College

San Jose, California

Mr. Chancellor, I present for an honorary degree R. Buckminster Fuller.

Courageously defying tradition and patiently surmounting man-made obstacles, he has successfully pursued his independent course toward basic truths while outlining a new and brighter promise for the human race.

His life-long pursuit of knowledge, spurred by a deep love of humanity, has exemplified the finest in the scholar, teacher, inventor, scientist, engineer, mathematician, architect, cartographer and designer.

His demonstrated triumphs in these varied endeavors emphasize the kinship of all knowledge. They are also a testimonial to his abiding faith in ultimate peace and in the rich destiny of mankind—a faith which he not only has enthusiastically stimulated in his students but also personally transmitted throughout the world.

From practicality, he has created beauty; from despair he has brought hope.

For these notable achievements—and fittingly also for his contributions as a visiting lecturer at San Jose State College—we salute him and proudly present him for the honorary degree of Doctor of Humane Letters to be conferred by the California State Colleges.

DOCTOR OF FINE ARTS

June 11, 1966

California College of Arts and Crafts

Oakland, California

In recognition of his contribution to our world of a vision of a finer way of life; for his efforts to bring it about; for the courage and generosity of spirit that guide his work as architect, engineer, mathematician, inventor, philosopher and teacher; and for his inspiration to those who in future years may also become dedicated and valued "world citizens";

The Board of Trustees of the California College of Arts and Crafts is proud to confer upon Mr. R. Buckminster Fuller the honorary degree of Doctor of Fine Arts.

DOCTOR OF ENGINEERING

June 4, 1967

Clarkson College of Technology

Potsdam, New York

R. Buckminster Fuller—inspired thinker, creator, master of design. A native of Massachusetts, Buckminster Fuller is a descendant of strong-minded New England individualists dedicated for generations to the law and to the ministry. He has combined these interests of his forbears in the search for structure and order in Nature. His genius for seeing new and different relationships has led to applications which are changing technology and its social implications and, indeed, may drastically alter the techniques, role, and scope of education.

He has achieved honor and fame for his geodesic dome, which has been termed "an elegant modern miracle of spatial splendor and structural economy." His design symbolizes this nation's creative genius at The Universal and International Exhibition of 1967 at Montreal, Canada.

We of Clarkson College honor R. Buckminster Fuller for this achievement, but also because he has long been an advanced thinker in his field—a poet of technology who has sought to bridge the gap between the humanities and the sciences. We honor him because the edifice of his own being is a tetrahedron of intelligence, energy, and design based upon social responsibility and faith in mankind. This has enabled him to stand steadfast through adversity and neglect, pondering fundamental questions and seeking the pattern of nature amidst the detail to bring about for mankind new and useful structures. We honor him because, at 71 years of age, he is an optimist who finds in youth a promise for man's future. We honor him, finally, because he is a citizen of the world dedicated to the service of mankind through the application of technology, the purpose for which this College stands.

DOCTOR OF FINE ARTS

May 19, 1968

Ripon College

Ripon, Wisconsin

Mr. President, I have the honor to present Buckminster Fuller, candidate for the degree Doctor of Fine Arts.

Mr. Fuller was born in Milton, Massachusetts, in 1895 and there attended Milton Academy from 1904 to 1913. Following two years of study at Harvard University, he served an apprenticeship with Armour and Co. in New York City until 1917, when he spent a year at the United States Naval Academy and was commissioned an ensign in the United States Navy.

The year 1927 marked a turning point in Mr. Fuller's career. From this time to the present, he was to direct his energies to the development and synthesizing of his theories of comprehensive design. This year saw the invention of the now famed Dymaxion House, which defined his ultimate goal of finding ways of doing more with less to the end that all people everywhere can have more and more of everything.

Mr. Fuller organized the Dymaxion Corporation in 1932 and, as its director and chief designer, was to invent and produce such forms as the Dymaxion Car and the geodesic dome. From 1942 to 1944, Mr. Fuller served with the Board of Economic Warfare and as special assistant to the Director of the Foreign Economic Administration.

Following World War II, Mr. Fuller's comprehensive concepts concerning environmental controls led to the wide acceptance of his mathematical and architectural principles through the use of his geodesic domes, from the Ford Museum in Dearborn, Michigan, to the radomes which dot the arctic Distant Early Warning line. In 1954 the United States Marine Corps cited Mr. Fuller for "the first basic improvement in mobile environmental controls in 2600 years." His design for the United States pavilion at Expo '67 in Montreal, Canada, drew worldwide attention and praise.

Mr. Fuller is a prolific writer. Over the years he has authored more than ten books dealing with a range of subject matter from architecture to zoology. He has written hundreds of articles which have appeared in both lay and professional journals. His journeys have carried him throughout the world, where he has lectured on a wide variety of subjects. In 1959 he was appointed Research Professor of Design at Southern Illinois University, a position which he holds today. During the academic year 1961–1962 he was the Charles Eliot Norton Professor at Harvard University.

The honors which have come to Mr. Fuller during a life of great activity are of such number that to recount them today is almost impossible. He is the holder of thirteen honorary degrees from major academic institutions. He has been the recipient of the Gran Premio Award of the Trienniale de Milano twice, the Gold Medal Scarab of the National Architectural Society, the Gold Medal of the American Institute of Architects, the Benjamin Franklin Life Fellowship in the Royal Society of Arts, and the Plomado de Oro Award from the Society of Mexican Architecture. He is a fellow and honorary trustee of the Institute of General Semantics and a life member of the National Institute of Arts and Letters.

Mr. President, I am honored to present Buckminster Fuller, the one man fully deserving of the title of the Leonardo of the Twentieth Century, candidate for the degree Doctor of Fine Arts.

DOCTOR OF HUMANE LETTERS

June 16, 1968

Dartmouth College

Hanover, New Hampshire

R. Buckminster Fuller, you have gone your great aunt Margaret one better: she merely accepted the universe, you preach total immersion in it. And if any doubter should ask whether you really believe in salvation through such immersion you might well respond: "believe in it? Shucks, I've done it!" True, you call your kind of salvation "comprehensive design," but let's not quibble about words, you are a thinker who literally as well as figuratively rejects all walls—except, of course, those of geodesic dome design—and who believes that only as man "goes for broke" with his mind can he hope from here on out to win any hand in the human game. Few among us (indeed is there one respected other?) would dare say with you "I made a bargain with myself that I'd discover the principles operative in universe and turn them over to my fellow man." How far you've gone in keeping that bargain no man today can know for sure, and that of itself is saying an awful lot. A friend of the human future, you as builder of that future are the first man of the twenty-first, or perhaps even of the twenty-third century to whom Dartmouth presumes to award her doctorate of human letters, *honoris causa.*

DOCTOR OF HUMANE LETTERS

October 10, 1968

New England College

Henniker, New Hampshire

Inventor, engineer, mathematician, architect, philosopher, poet, and pedagogue—you who call yourself "cosmogonist" remind us of the power of the human mind to leap across its own stone walls and roam the pastures of the universe.

The great volume of your achievement has narrow academic foundations and sanctions. Your explanation of your expulsion as an undergraduate from Harvard "officially for cutting classes but actually for irresponsibility" and your subsequently becoming Charles Eliot Norton Professor at Harvard reminds us that the turnpike to truth admits of many alternative routes, with perhaps him who travels cross-lots arriving first of all. It is no wonder that you consider "today's educational process . . . totally obsolete." In the New England tradition of honoring its critics, we hope that you will probe some of our educational obsolescence with us.

As many are more impressed with things than with ideas, you are probably best known as the dome-man; your geodesics threaten to cover us all. We are grateful for these and for your many other comprehensive designs which are, as you say, "finding ways of doing more with less."

But it is especially for your provocative, ubiquitous ideas that our Faculty has called you here. You continue to show us the fertility of wedding ancient principles and new technology when the match is made by fresh eyes and a clear brain. You obfuscate us with your analysis of the present and overwhelm us with your vision of the future. You have tried, not altogether successfully, to jolt us from our parochial ways of thinking and to show us ourselves within the context of what comes next.

We return you now to the twenty-first century—but not before New England College enjoys the privilege of awarding you her Doctorate of Humane Letters, honoris causa.

DOCTOR OF HUMANE LETTERS

October 14, 1968

University of Rhode Island

Kingston, Rhode Island

Inventor-discoverer of a new mathematics, energetic-synergetic geometry; architect of the widely used geodesic dome that grew of that geometry; geographer, developer of an original, more accurate method of mapping the globe; engineer; and creator of distinctive language—you are the 20th century's highly mobile model of the universal man. Viewing the earth as "this little sun-orbiting space ship," you have been able to see the world and all that is in it as uninhibited by pre-conceived ideas as is humanly possible. Indeed, you have carried to its ultimate the academic imperative to question and to seek, and in the process have become a non-graduate, non-conforming philosopher-poet—a dreamer of great dreams whose avowed objective is "humanity's comprehensive success in universe" to be accomplished through the "do-more-with-less invention revolution" now in process.

We honor you today for your durable accomplishments in many fields, and we take pride in conferring the honorary degree of Doctor of Humane Letters upon a New England individualist in whom the infinite capabilities of the free-ranging human mind are manifest.

DOCTOR OF ARCHITECTURAL ENGINEERING

January 26, 1969

The University of Wisconsin

Milwaukee, Wisconsin

The University of Wisconsin-Milwaukee cites
Dr. Richard Buckminster Fuller

For his vision and foresight in redefining man and man's needs.

For his unique combining of engineering, architecture and the humanities to evolve a new means to problem-solution.

For his creating structures and enclosures of space that are giving physical and cultural form to the twentieth century.

DOCTOR OF FINE ARTS

June 2, 1969

Boston College

Boston, Massachusetts

DOCTOR OF SCIENCE

July 4, 1969

Bates College

Lewiston, Maine

Mr. President, it is my honor to present Richard Buckminster Fuller, Distinguished University Professor at Southern Illinois University.

There is no attribute of man more rare, more necessary, or more difficult to comprehend than the gift of original thought. Ingenuity, a very valuable but still far commoner quality, is often mistaken for it; but there is a difference between true creative originality and ingenuity that is as great as the difference between discovery and adaptation. There is also the difference that ingenuity, which derives from known principles, is usually recognized and rewarded; but the really original idea is slowly understood and often grudgingly appreciated, because it is strange to our custom and transcends our expectation. Thus creativity must find its full realization only in posterity, and one can now only guess what the future will conclude. It nonetheless seems likely that when the future compiles its small and awesomely selective list of truly original thinkers, it may include the name of Buckminster Fuller.

Born into a suburban Boston family of great tradition and propriety, Buckminster Fuller remained within the expected tracks to the extent of graduation from Milton Academy and matriculation at Harvard, but shortly thereafter his energetic and questing mind took him far outside the predictable pattern—into business and engineering ventures of great variety and daring, into invention and patent development, into architecture, and into mathematics, poetry, cartography, criticism, ecology, and philosophy. He is now probably best known as the architect of the United States pavilion at the Montreal World's Fair, which brilliantly demonstrated both the scientific and aesthetic truth of his famed geodesic principle; but in the surging galaxy of his accomplishment this may eventually be seen as a relatively minor effort of a man who the distinguished astronomer, Harlow Shapely, once described as "probably the brightest man alive."

For a stimulating and distinguished career that includes contributions in many fields of knowledge, but especially for his discovery and articulation of general principles of structure and form, which promise to revolutionize architecture and engineering, and which may prove even to be revealing of some of the secrets of nature itself, I am honored to present Richard Buckminster Fuller, for the honorary degree, Doctor of Science.

DOCTOR OF FINE ARTS

May 8, 1970

The Minneapolis College of Art and Design

Minneapolis, Minnesota

To

R. Buckminster Fuller

The Artist—

whose studio is the planet earth
whose materials are the mysteries of the universe
whose goal is the discovery of Truth
whose motivation is Love
whose Legacy is a vision of hope for the future of mankind

The Minneapolis College of Art and Design is greatly honored to confer the Degree of

Honorary Doctor of Fine Arts.

DOCTOR OF LAWS

May 24, 1970

Park College

Parkville, Missouri

Buckminster Fuller's name has become a household word because of his forthright demonstration and perseverance in his belief that all men deserve to be recognized for their own dignity as human beings, and in his untiring efforts to ameliorate the lot of the human race through constructive thinking and implementation of his ideas towards that end.

Buckminster Fuller is a native-grown product of the New England Yankee heritage. His roots go deep and are alive, and what they support is straight and tall. As a youngster living on Bear Island off the Penobscot coast of Maine, he faced nature and man in both their benign and often quarrelsome aspects. From these experiences there was inculcated in the boy's mind—and this has been with him ever since—that we are part of the whole, that man is part of nature, and that the whole is good, if man but understand it and put himself in the proper frame of reference and act. Symbolic of this attitude and action was his youthful marvelling one day at the beauty and functioning of the jellyfish near his island home. This creature had a lesson which he grasped and put to use by adapting its principle of propulsion to his then daily task of ferrying a small boat across the open waters of the bay, thus freeing him to do something more appropriate to a human being.

There is ample record for anyone to examine of his activities and of the multitudinous honors conferred upon him. He is a man whose honors have come to him; they have not been sought. Laughed at, rebuffed, he many times rose to make the laughers take notice and his rebuffers come for counsel. Buckminster Fuller was a Harvard College drop-out who later in life returned there as the Charles Eliot Norton Professor. This is but one of the extraordinary academic encounters of the gentleman we seek to honor today. Buckminster Fuller is a balanced man who knows his worth—"Who shall love thee if thou lovest not thyself?"—He knows his worth: "I have learned much," he writes, "but I don't know very much; but I have learned, I have learned by trial and error. And I have great confidence in the meager store of wisdom I have secured."

Buckminster Fuller has been, and is, engaged in "Comprehensive anticipatory design science," to employ his own definition of his *métier*. He is a hard-headed New England Universal visionary. The Geodesic Dome is but one example of this man's vision and practicality. True to his great aunt, Margaret Fuller, of the Emerson Circle fame, he holds with her that

"the public must learn how to cherish the noble and rarer plants, and to plant the aloe, able to wait a hundred years for its bloom, or its gardens will contain presently, nothing but potatoes and potherbs.

Mr. President, on behalf of the Faculty of Park College, I present R. Buckminster Fuller, for whom we want to express our esteem by recommending that you grant him the honorary degree of Doctor of Laws.

DOCTOR OF HUMANE LETTERS

June 6, 1970

Brandeis University

Waltham, Massachusetts

RICHARD BUCKMINSTER FULLER—poet, prophet, professor, inventor and scientist; world citizen by dint of convictions and creations; cosmic yankee who discerns in a regenerating universe the ideas and designs to restructure our future. His geodesic domes challenge architectural tradition from Okinawa to New Delhi, from Moscow's World's Fair to Montreal's Expo '67, exciting representations from an incandescent imagination that has transformed fantasy into reality and produced cars and maps and original ways of living bearing the enigmatic designation Dymaxion. With a vision that may yet redeem modern society, he has harnessed energies and resources so that all men living on "this space ship called the earth" may enjoy a fuller life.

DOCTOR OF HUMANE LETTERS

June 12, 1970

Columbia College

Chicago, Illinois

We honor you for a lifetime of extraordinary consequence, dedication to the liberation of the human spirit, and the design of a rational environment.

DOCTOR OF SCIENCE AND HUMANE LETTERS

September 28, 1970

Wilberforce University

Wilberforce, Ohio

DOCTOR OF FINE AND APPLIED ARTS

June 13, 1971

Southeastern Massachusetts University

North Dartmouth, Massachusetts

Richard Buckminster Fuller, because with dedalian wit and compelling finesse you have demonstrated the capacity to draw space from a very tyrant's grasp and turn it to the elegant service of all mankind, Southeastern Massachusetts University acknowledges the debt owed you by people everywhere and confers upon you the degree of Doctor of Fine and Applied Arts, honoris causa.

DOCTOR OF LAWS

May 24, 1972

Grinnell College

Grinnell, Iowa

DOCTOR OF LAWS

June 4, 1972

Emerson College

Boston, Massachusetts

In an age when so many of our fellowmen have lost all sense of the past and comprehend little meaning in the present, it is heartening to find one who has not forgotten the heritage of history nor ignored the challenge of the contemporary scene, who has himself declared that, "we must save and incorporate into the social machinery of the new all that is incontestably of continuing value in the old and must make the equally incontestable validity of the new clear to the old so that we may join those old and new forces towards gaining the speediest and smoothest readjustment of society to acquisition of the new existence."

The sixth generation of your family in unbroken succession since 1740 to attend Harvard College, you have been honored by more than a score of sister universities and recognized by countless scientific and learned societies. As architect, builder and designer you have materialized your ideals and made real your creative visions. An artist who can double as an engineer, you have given us geodesic domes and geospheres. But above all, you have spoken the universal language of reconciliation and understanding, and not without reason are you intelligible to the young and the old, the simple and the wise, the tutored and the unlettered.

It was said of your chaplain grandfather that he "was a man whom we honor the more because in public and private, he was consistent with his principles," and it was he himself who said, "I must do something for my country." You likewise have these many years lived a life consistent with your highest principles, and it has been your constant concern to do something not only for your country but for all mankind.

DOCTOR OF SCIENCE

June 5, 1972

University of Maine

Orono, Maine

RICHARD BUCKMINSTER FULLER, architect, dreamer, wizard. Since the 1950's you have carried out the highest ideals of the wizard's trade—you have been sorcerer and magician, a skilled and clever builder, a wise man, poet and sage, and, above all a caster of spells.

The University of Maine honors you for your contributions to technology—those light and sturdy structures which are the trademark of a people who send their image to the planets.

But more than that we honor you for your prodigal ideas and energy and for the generosity with which you have filled and stretched the minds of a generation of designers, writers, and thinkers. You have charmed us; you have transformed us; you have worked your magic on us all.

R. BUCKMINSTER FULLER, in recognition of your contributions, the Trustees of the University of Maine have voted to confer upon you the honorary degree of DOCTOR OF SCIENCE with all the rights and honors pertaining thereto. Your name will be forever borne upon the rolls of the University of Maine.

DOCTOR OF LAWS

May 20, 1973

Nasson College

Springvale, Maine

Richard Buckminster Fuller is a true proponent of the concept of individual creative thought. Having divested himself of all forms of patterned thinking early in his life, he has devoted his energies to a remarkable series of inventions, discoveries, and designs. As mathematician, teacher, engineer, author, architect and philosopher, he is constantly seeking answers to the challenge of critical world problems. His realization of the urgent need for wise ecological practices and his untiring efforts in their behalf, far in advance of current trends, demonstrate his commitment to the future of our civilization.

DOCTOR OF FINE ARTS

May 25, 1973

Rensselaer Polytechnic Institute

Troy, New York

RICHARD BUCKMINSTER FULLER In deeds, by example, as inspiration to your fellow man, you have fashioned a life serving the noblest aims of education. As teacher and lecturer you have shared your wisdom with young and old alike. As inventor and leader of industry you have contributed singularly to the comfort and progress of mankind. As an architect of international reknown you have fired men's minds with a new understanding of our environment. For your creative vision, for your dedication to the enrichment of our society and our civilization, Rensselaer Polytechnic Institute hereby awards you the honorary degree of DOCTOR OF FINE ARTS and causes you to be vested with the hood signifying to this degree.

DOCTOR OF LITERATURE

May 27, 1973

Beaver College

Glenside, Pennsylvania

Mr. President:

It is my privilege to present for the honorary degree of Doctor of Literature, Richard Buckminster Fuller—inventor, engineer, architect, scientist, philosopher, educator, futurist.

Born in 1895 to a distinguished New England family, Buckminster Fuller might well have become a proper Bostonian, a stereotype associated in many minds with the American tradition. Instead, he is the apothesis of another American tradition forged by such strong individualists—iconoclastics, if you will—as Emerson, Thoreau, and Margaret Fuller who saw the world with new eyes and fresh delight. Like them, Buckminster Fuller is an explorer and discoverer in the realm of ideas; like them he combines the visionary and the practical. Yet the comparison is limited, for not since the Renaissance has the world seen such many-faceted achievements on the part of one person. But even that analogy will not entirely serve, for in the space age, Buckminster Fuller's wide-ranging imagination is not only global, but cosmic.

Beaver College is, of course, honoring Dr. Fuller for all his many and varied accomplishments, but difficult as it is to be selective, we are especially emphasizing today his impact on thousands of college students through his lectures and books. To the youth of America he has brought his optimism that what man wills he can do, and his vision of a future in which technology is harnessed in the service of humanism.

Mr. President, I present Buckminster Fuller, a man of the twenty-first century, who in accepting this degree is conferring an enduring honor on Beaver College.

DOCTOR OF ENGINEERING

April 27, 1974

University of Notre Dame

Notre Dame, Indiana

At an Academic Convocation Observing 100 Years of Engineering Education The University of Notre Dame confers the degree of Doctor of Engineering, *honoris causa* on a creative genius.

He began as an apprentice machine fitter, but was to have many more interesting titles in an intriguing succession of corporate neologisms—Dymaxion, Geodesics, Synergetics, and Tetrahelix, to name a few of his companies.

Inventor of the geodesic dome, author of fifteen books, designer of United States pavilions and foreign airports, holder of more than one hundred world patents, international lecturer—he is all these things, but more than their sum.

At the heart of his astounding career is a restless original intellect, probing always, asking the simple questions—"Why?" "How?"

DOCTOR OF HUMANE LETTERS

May 13, 1974

Saint Joseph's College

Philadelphia, Pennsylvania

A modern-day version of the universal Renaissance man, R. Buckminster Fuller has given his creative attention to so many aspects of human understanding that his activities defy categorization. He is a mathematical genius, an engineer of rare and extraordinary foresight and a philosopher concerned with the most colossal problems of life and living.

A prolific writer, Mr. Fuller is the author of at least fifteen books and hundreds of articles on various subjects. He has been the recipient of no less than thirty-three honorary degrees and hundreds of other honors and lectureships, including a year's tenure as Charles Eliot Norton Professor of Poetry at Harvard and his present position, World-Fellow-in-Residence for a consortium of colleges and the University City Science Center in Philadelphia. His works were recently featured in an exhibition at the Franklin Institute Science Museum entitled, "Bucky, The Fuller Man."

The holder of numerous patents on a variety of practical inventions, Mr. Fuller is perhaps best known for his invention of the geodesic dome. By now a familiar feature of the architectural landscape, geodesic domes have been used for the United States Pavilion at the Montreal Expo '67 and for the United States Research Dome at the South Pole, among many others. The geodesic dome represents the successful application of a fundamental approach in all of Mr. Fuller's efforts, to discover ways to do more things with less material and less effort so that the good things of life might be more readily available to all people.

Saint Joseph's College is proud to recognize the universal scope of his wisdom and the practical effectiveness of his concern for the human condition.

DOCTOR OF HUMANE LETTERS

May 20, 1974

University of Pennsylvania

Philadelphia, Pennsylvania

A tireless seeker in the universe of ideas, Richard Buckminster Fuller has identified for us our spaceship, earth, renewing it through the imaginative form of fresh discovery.

Inquisitive and optomistic, he has demonstrated in his astonishing inventions a compelling vision of a rational mind-centered planet. His marvelous geodesic dome using linked triangles which through design construct a shelter of tremendous strength is an image of his belief in the power of human relatedness and comprehensive understanding.

This engineer, mathematician, philosopher, this architect and inventor, this poet of man's hopes, defies all categories as he restores us to the future and stirs anew our energies. Grateful to their Leonardo, the Trustees of the University where he is a professor welcome a World Fellow to their sphere with the honorary degree, Doctor of Humane Letters.

DOCTOR OF SCIENCE

May 31, 1974

Pratt Institute

Brooklyn, New York

R. Buckminster Fuller, internationally-known architect and environmentalist, has come to Pratt Institute before as a lecturer and has literally packed the house. Pratt students have wanted to hear whatever he had to say on whatever topic he chose to discuss. A creative, future-oriented thinker, Mr. Fuller has equally full audiences wherever he's been and he has contributed his ideas prodigiously in a variety of guises—visiting professor, lecturer, essayist, critic seminar director and principle speaker at institutions of learning throughout the world. Mr. Fuller also is the author of dozens of books on architecture and the environment.

The famous creator of the geodesic dome, Mr. Fuller today adds Pratt Institute's honorary degree to some thirty-four others he has received from other colleges. He is also the recipient of innumerable awards from architectural organizations, honor societies, and architectural and professional societies all over the globe.

Mr. Fuller is distinguished university professor at Southern Illinois University and world fellow in residence for the consortium of the University of Pennsylvania, Haverford College, Swarthmore College, Bryn Mawr College and the University City Science Center.

Mr. President, I have the honor to present R. Buckminster Fuller for the degree of Doctor of Science, *honoris causa.*

DOCTOR OF SCIENCE

June 5, 1974

McGill University

Montreal, Quebec, Canada

Cum perspicuum sit constetque inter omnes eo consilio institutas esse dignitates academicas ut viri ingenio et doctrina praestantes insignibus honoris ornarentur praeter ceteros auctoritate Senatus Buckminster Fuller virum clarum et illustrem ne dignitate careret debita meritaque Doctorem Scientiaé honoris causa renuntiavimus atque constituimus.

DOCTOR OF HUMANE LETTERS

June 1, 1975

Hobart and William Smith Colleges, The Colleges of the Seneca

Geneva, New York

Future historians may well label 1927 the *annus mirabilis* of the twentieth century. A wondrous year it was—the year that brought us talking pictures, television, and transatlantic flight.

It was also the year that science and poetry were joyously wed in the mind of a remarkable man who decided on the shore of Lake Michigan that his life belonged to everyone.

You resolved that life should be neither a subject nor an object, but quite literally a *project,* a continuous throwing forth into reality of poetic intimations and scientific anticipations.

The truth of your impossible dreams has proved far more thrilling than any fiction. Above all, for the "young world," your dreams are refreshing and efficacious antidotes to cynicism, nihilism, and other tempting philosophies of gloom.

Neither "cockeyed optimist" nor melancholy pessimist, you are in the best sense an enlightened and enlightening *meliorist,* totally persuaded that human destiny can and will be better.

There are no frontiers to your genius, no apparent limits to all the things that are drempt of in your philosophy.

By authority delegated to me, I confer upon you the degree of Doctor of Humane Letters, *honoris causa,* and admit you to all the liberties, privileges and responsibilities thereto appertaining.

DOCTOR OF SCIENCE

September 26, 1978

Hahnemann Medical College & Hospital of Pennsylvania

Philadelphia, Pennsylvania

Mr. President, I have the honor to present to you one of our world's rare geniuses, unique in creativeness, unique in philosophy, unique in teaching and inspiration, and, above all, the insightful poet of the Universe.

My candidate was born in Milton, Massachusetts in 1895, graduated from Milton Academy and attended Harvard University. In World War I he enlisted in the U.S. Navy; as a chief boatswain he showed such extraordinary qualities he was sent to the Naval Academy and was commissioned ensign, later lieutenant.

In a career spanning more than 60 years, he has risen from apprentice machine fitter to his current status as Emeritus University Professor at both Southern Illinois University and the University of Pennsylvania, and as World Fellow in Residence of the Consortium of the University of Pennsylvania, Haverford College, Swarthmore College, Bryn Mawr College and the University Science Center.

During his many years he has circled the earth tens of times, has been invited to speak in many places on all continents, most often at schools and colleges, to instruct how to participate in Nature's continuous reshaping of both the physical and the metaphysical. He has spoken personally, rather than by television and radio, face to face with approximately a quarter of a million persons every year and has fascinated and inspired many of them, especially the young.

His disciplines, penetrating insights, synthesized by

what he has named "comprehensive anticipatory design science" have generated his unique, world famed creations—the Dymaxion House, the Dymaxion Automobile, the Dymaxion World Map, the Geodesic Dome and its multitude of species now spread over our planet, also a series of basic structural elements awaiting multiplication and synthesis, and a number of vast projects of astounding courage and easily comprehensible validity.

His eleven books consummate command of widest knowledge and principles, and often the principles are his own discoveries. The earliest title is *Nine Chairs to the Moon;* two of the subsequent ones are *No More Secondhand God* and *Operating Manual for Spaceship Earth;* the latest is *And It Came to Pass—Not to Stay,* a summing-up in blank verse.

My candidate's comprehensive range and attainments are attested by: 39 honorary degrees, including one from St. Peter's College, Oxford University; innumerable awards and distinctions from innumerable prestigious institutions and societies, including the Gold Medals of the American Institute of Architects and the Royal Institute of British Architects; and awards by such diverse organizations as the American Association of Humanists, the U.S. Marine Corps, and the United Nations. He has been Harvey Cushing Orator to the American Association of Neurological Surgeons, International President of Mensa International, Charles Eliot Norton Professor of Poetry at Harvard University, and was given many other rare distinctions.

His works have been exhibited, and have been built, in numerous places throughout the world. There have been more than 60,000 references to him in journals, newspapers, magazines, books, radio and television.

Dr. Likoff, by reason of his comprehensive, universal self and achievements, I ask that you confer upon Richard Buckminster Fuller, known affectionately everywhere as "Bucky," the degree of Doctor of Science, Honoris Causa.

DOCTOR OF HUMANE LETTERS

June 8, 1979

Southern Illinois University

Edwardsville, Illinois

Born July 12, 1895, in Milton, Massachusetts, Richard Buckminster Fuller pursued his education at Milton Academy, Harvard University, and the United States Naval Academy at Annapolis. After service with the Navy during World War I, Fuller engaged upon a career in business, while commencing his own investigations into the vast scope of natural interests and humane concerns: endeavors that led to the insights which have captured the attention of the entire world and led to steps of profound importance for the future of humanity and the planet Earth.

During a long and energetic career Fuller has travelled to every corner of the world, exploring ways to keep the environment habitable and amenable to the creative evolution of all the species that inhabit the universe, with a special focus upon transforming the arts of technology for human creative evolution. During his career he has written books of both prose and poetry, examining with vision and imagination problems and proposals that impact on every aspect of human activity. As a speaker and lecturer, he has acted as an unflagging voice of optimism and hope during a century of turmoil and anxiety, always providing special sparks of inspiration to young people, many of whom are now engaged in following up on his ideas and translating them into action. His work has been the source of widespread comment and study, to the extent that he has become nothing less than a pivotal figure in the intellectual, technological, and spiritual life of humanity in the twentieth century.

At the behest of former president Delyte W. Morris, R. Buckminster Fuller directed his interest upon Southern Illinois University in many significant ways. From 1959 to 1968, he was Research Professor, Design Science Exploration; Director of Inventory of World Resources, Human Trends and Needs; and Founder and Director of the World Game. In 1968, he was named a University Professor at SIU. From his home in Carbondale, designed according to his world famous concept of the Geodesic Dome, he was a dynamic presence to many of our students active in the planning phase of Southern Illinois University at Edwardsville. His influence is revealed most dramatically in the religious center on the Edwardsville campus. He has also offered a challenging overall plan for the future of the city of East St. Louis.

In consideration of a truly unparalleled record of achievement during a long, vigorous career, and his special impact on Southern Illinois University, it would be impossible to imagine a figure more worthy of honor by this institution.

DOCTOR OF HUMANE LETTERS

December 7, 1979

Alaska Pacific University

Anchorage, Alaska

Richard Buckminster Fuller—planetary citizen, consummate catalyst and provocative teacher, wise recorder of the world's intricate and irreducible design, spokesman of the solitary individual and the universally human, intuitive interpreter of cosmic design and analyst of integral operative principles, poet, philosopher, architect, inventor, author, inspirer of youth—beloved of children, respected by statesmen, sought by humanities' leaders, and friend of a hopeful future, we celebrate your incomparable life, your unconquerable optimism about our human prospects, and your courageous invitation to make the world work for one hundred percent of humanity, in the shortest possible time through spontaneous cooperation without ecological offense or the disadvantage of anyone.

Your life and thought orchestrates that global connectedness, that internal ordering and linkage our university aspires to prompt in its students; that human helpfulness and unfailing goodwill we covet for this learning community and the wider world it serves; that mellowed maturity and ever spontaneous newness of life we seek to renew in our common life daily; that reverence for all life and all forms of creation that mirrors a spiritual design and Divine artistry we would emulate in all forms of our faith and religious devotion, and that unassuming human joy and friendship that lightens the day and load of everyone privileged to walk with you.

For these, and so many other ways, you honor and embody what is best in this university, and every university established to understand and ennoble the distinctly human. Therefore, by the authority of the Board of Trustees, and on recommendation of our faculty, we delight to commend you for the Degree, Doctor of Humane Letters, honoris causa.

DOCTOR OF HUMANE LETTERS

January 3, 1980

Roosevelt University

Chicago, Illinois

To Richard Buckminster Fuller: Geometer, Visionary, Seer; Inventor of Ideas, Tools, Technologies and the Terms to Assert Their Utility; Who Transcends the Disciplines of Design, Architecture, Engineering, Economics, and Ecology with an Integrated and Architectonic World View; Whose Work Gives Us Hope that Man's Genius Can Surmount His Most Intractable Dilemmas; Who Teaches Us to Appreciate the Order and Rhythm of Space, to Seek Harmony in the Environment We Create, and to Plan Our Future on a Scale Commensurate with What Man Is and What He Can Become.

The Trustees of the University by virtue of the authority invested in them and upon recommendation of the Faculty, have conferred on Richard Buckminster Fuller the honorary degree of Doctor of Humane Letters and have granted this certificate as evidence thereof.

DOCTOR OF HUMANITIES

May 17, 1980

Georgian Court College

Lakewood, New Jersey

Citation
for
BUCKMINSTER FULLER

Doctor of Humanities, honoris causa

Buckminster Fuller, a man born in the nineteenth century, has spent a lifetime helping to create the twenty-first century. The entire world is in his debt, not merely for major inventions such as the geodesic dome, but also for major philosophic impetus, long before the concept of ecology had become fashionable, toward conservation of space and of energy on this "Spaceship Earth" (the famous name he gave to our planet).

Author, architect, philosopher, inventor, futurist, he is a major thinker and seminal force of our time. He has been accorded such titles as "the universal genius of our age," "the modern da Vinci," "our greatest living genius and the anticipator of the world to come."

It gives me great pleasure, therefore, to bestow upon Dr. Fuller in the name of Georgian Court College the honorary degree of Doctor of Humanities. A man born before his time and at home in every place, he has taught us to conserve our space and to preserve our time.